Handbook of Research on Smart Technology Models for Business and Industry

J. Joshua Thomas
UOW Malaysia KDU Penang University College, Malaysia

Ugo Fiore
University of Naples Parthenope, Italy

Gilberto Perez Lechuga
Autonomous University of Hidalgo State, Mexico

Valeriy Kharchenko
Federal Agroengineering Centre VIM , Russia

Pandian Vasant
Universiti Teknologi PETRONAS, Malaysia

IGI Global
PUBLISHER of TIMELY KNOWLEDGE

A volume in the Advances in Computational Intelligence and Robotics (ACIR) Book Series

Published in the United States of America by
IGI Global
Engineering Science Reference (an imprint of IGI Global)
701 E. Chocolate Avenue
Hershey PA, USA 17033
Tel: 717-533-8845
Fax: 717-533-8661
E-mail: cust@igi-global.com
Web site: http://www.igi-global.com

Library of Congress Cataloging-in-Publication Data

Names: Thomas, J. Joshua, 1973- editor.
Title: Handbook of research on smart technology models for business and industry / J. Joshua Thomas [and 4 others].
Description: Hershey, PA : Engineering Science Reference, an imprint of IGI Global, [2020] | Includes bibliographical references and index. | Summary: "This book discusses relevant abstract frameworks and the latest experimental research findings in theory, mathematical models, software applications, and prototypes in the area of smart technologies"-- Provided by publisher.
Identifiers: LCCN 2019057486 (print) | LCCN 2019057487 (ebook) | ISBN 9781799836452 (hardcover) | ISBN 9781799836469 (ebook)
Subjects: LCSH: Automation. | Artificial intelligence--Industrial applications. | Big data. | Business--Data processing.
Classification: LCC T59.5 .H2853 2020 (print) | LCC T59.5 (ebook) | DDC 658/.05--dc23
LC record available at https://lccn.loc.gov/2019057486
LC ebook record available at https://lccn.loc.gov/2019057487

This book is published in the IGI Global book series Advances in Computational Intelligence and Robotics (ACIR) (ISSN: 2327-0411; eISSN: 2327-042X)

British Cataloguing in Publication Data
A Cataloguing in Publication record for this book is available from the British Library.

For electronic access to this publication, please contact: eresources@igi-global.com.

Advances in Computational Intelligence and Robotics (ACIR) Book Series

Ivan Giannoccaro
University of Salento, Italy

ISSN:2327-0411
EISSN:2327-042X

Mission

While intelligence is traditionally a term applied to humans and human cognition, technology has progressed in such a way to allow for the development of intelligent systems able to simulate many human traits. With this new era of simulated and artificial intelligence, much research is needed in order to continue to advance the field and also to evaluate the ethical and societal concerns of the existence of artificial life and machine learning.

The **Advances in Computational Intelligence and Robotics (ACIR) Book Series** encourages scholarly discourse on all topics pertaining to evolutionary computing, artificial life, computational intelligence, machine learning, and robotics. ACIR presents the latest research being conducted on diverse topics in intelligence technologies with the goal of advancing knowledge and applications in this rapidly evolving field.

Coverage

- Intelligent control
- Brain Simulation
- Robotics
- Computational Intelligence
- Evolutionary Computing
- Pattern Recognition
- Neural Networks
- Natural Language Processing
- Heuristics
- Synthetic Emotions

IGI Global is currently accepting manuscripts for publication within this series. To submit a proposal for a volume in this series, please contact our Acquisition Editors at Acquisitions@igi-global.com or visit: http://www.igi-global.com/publish/.

The Advances in Computational Intelligence and Robotics (ACIR) Book Series (ISSN 2327-0411) is published by IGI Global, 701 E. Chocolate Avenue, Hershey, PA 17033-1240, USA, www.igi-global.com. This series is composed of titles available for purchase individually; each title is edited to be contextually exclusive from any other title within the series. For pricing and ordering information please visit http://www.igi-global.com/book-series/advances-computational-intelligence-robotics/73674. Postmaster: Send all address changes to above address.

Titles in this Series

For a list of additional titles in this series, please visit:
https://www.igi-global.com/book-series/advances-computational-intelligence-robotics/73674

Applications of Artificial Intelligence for Smart Technology
P. Swarnalatha (VIT University, India) and S. Prabu (VIT University, India)
Engineering Science Reference • © 2020 • 300pp • H/C (ISBN: 9781799833352) • US $215.00

Practical Applications and Use Cases of Computer Vision and Recognition Systems
Chiranji Lal Chowdhary (Vellore Institute of Technology, India) and B.D. Parameshachari (GSSS Institute of Engineering and Technology for Women, India)
Engineering Science Reference • © 2020 • 300pp • H/C (ISBN: 9781799849247) • US $195.00

Machine Learning Applications in Non-Conventional Machining Processes
Goutam Kumar Bose (Haldia Institute of Technology, Haldia, India) and Pritam Pain (Haldia Institute of Technology, India)
Engineering Science Reference • © 2020 • 300pp • H/C (ISBN: 9781799836247) • US $195.00

Applications of Artificial Neural Networks for Nonlinear Data
Hiral Ashil Patel (Ganpat University, India) and A.V. Senthil Kumar (Hindusthan College of Arts and Science, India)
Engineering Science Reference • © 2020 • 300pp • H/C (ISBN: 9781799840428) • US $245.00

Handbook of Research on Natural Language Processing and Smart Service Systems
Rodolfo Abraham Pazos-Rangel (Tecnológico Nacional de México, Mexico & Instituto Tecnológico de Cd. Madero, Mexico) Rogelio Florencia-Juarez (Autonomous University of Juarez City, Mexico) Mario Andrés Paredes-Valverde (Tecnológico Nacional de México, Mexico & Instituto Tecnológico de Orizaba, Mexico) and Gilberto Rivera (Autonomous University of Juarez City, Mexico)
Engineering Science Reference • © 2020 • 350pp • H/C (ISBN: 9781799847304) • US $295.00

Advancements in Computer Vision Applications in Intelligent Systems and Multimedia Technologies
Muhammad Sarfraz (Kuwait University, Kuwait)
Engineering Science Reference • © 2020 • 300pp • H/C (ISBN: 9781799844440) • US $225.00

Advancements in Models and Applications of Agents and Multi-Agent Systems
Rinkaj Goyal (GGS IP University, India) and Sanjay Misra (Covenant University, Nigeria)
Engineering Science Reference • © 2020 • 300pp • H/C (ISBN: 9781799850595) • US $225.00

IGI Global
PUBLISHER of TIMELY KNOWLEDGE

701 East Chocolate Avenue, Hershey, PA 17033, USA
Tel: 717-533-8845 x100 • Fax: 717-533-8661
E-Mail: cust@igi-global.com • www.igi-global.com

Editorial Advisory Board

List of Contributors

Table of Contents

Detailed Table of Contents

one experiment. An option for solving such problems is the compilation of the experimental databases, the introduction of the smart control systems, and the management of the technological processes of plant electric stimulation.

Several current computer science applications, implemented within specific paradigms, work at different levels to solve various challenges facing particular sectors. The potential of the internet of things (IoT) in the context of fifth-generation networks (5G) is envisioned as suggesting several beneficial opportunities for companies, industries, and users to exploit this technology's applications. This chapter establishes how the IoT works, considering its 5G architecture. The emphasis is on the infrastructural characteristic in terms of transmission power, frequency, speed, security, localization, device lifetime, and others. Additionally, the chapter illustrates what the IoT entails, discussing its workability and efficiency. Furthermore, it highlights a range of newly distinguishing features that would give it much-touted success in comparison with other technologies. It also presents research issues and challenges.

The main objective of the work is to develop principles for the formation of the optimal composition of the energy complex from all the given power plants based on renewable energy sources for an autonomous consumer, taking into account the variable energy loads of the consumer, changing climatic conditions and the possibility of using local fuel and energy resources. As a result of solving this optimization problem, in addition to the optimal configuration of the power complex, it is also necessary to solve the problem of optimizing the joint operation of different types of power plants from the selected optimal configuration, that is, it is necessary to determine the optimal modes of operation of power plants and the optimal share of their participation in providing consumers at every moment in time. A numerical method for analyzing and optimizing the parameters and operating mode of the energy complex with the most accurate consideration of the schedule of changes in consumer load and software that automates the solution of this optimization problem are also presented.

For reduction of gateway bottleneck effect, data access cost, and response delay, many cooperative caching schemes are available. But in all the existing schemes, whenever any data is newly cached or any data is evicted, the information needs to be broadcast to all the mesh access points (MAPs). As a result, communication cost increases. In this chapter, a cluster and sub-cluster-based cooperative caching scheme is designed to minimize the communication cost. Clusters and sub-clusters are formed by taking advantage of the static position of Mesh routers. Cluster-heads (CHs) are selected using minimum spanning tree and minimum weighted vertex cover algorithm. Then the clusters and sub-clusters are formed. CHs and other Maps keep different cache information tables, and based on those, cache decision is taken. Client mobility is also considered. The performance of the proposed scheme and some other existing schemes have been compared. It has been observed that in the proposed scheme, communication cost in terms of number of messages is reduced.

This chapter develops a web-based automated text scoring (ATS) system that can grade essays and check for spelling errors. The main reason behind this work is to alleviate the labour-intensive marking of essays and ensures equality in scoring for high-stakes exams like TOEFL. The researcher had performed a detailed investigation on deep learning techniques used in the field of ATS and developed a recurrent neural network model that can score essays in an end-to-end approach. Using the developed deep learning model, a web application was also developed to showcase the process of ATS by letting the web application to communicate with the trained model. The model was trained using Keras framework and TensorFlow library and the web application was done using the Flask framework. This work is the LSTM network that can capture sequential dependencies. The evaluation metrics chosen to evaluate the model are the quadratic weighted kappa (QWK) score, and the trained model can achieve 0.6 in QWK score.

The prediction of stock prices has always been a very challenging problem for investors. Using machine learning techniques to predict stock prices is also one of the favourite topics for academics working in this domain. This chapter discusses five supervised learning techniques and two unsupervised learning techniques to solve the problem of stock price prediction and has compared the performances of all the algorithms. Among the supervised learning techniques, Long Short-Term Memory (LSTM) algorithm performed better than the others whereas, among the unsupervised learning techniques, Restricted Boltzmann Machine (RBM) performed better. RBM is found to be performing even better than LSTM.

One of Myanmar's problems is energy shortage. Partially, energy shortage can possibly be decreased by the construction of sizeable grid-connected offshore wind farms. Eight prospective construction sites were selected and wind turbine models chosen. This chapter describes the method for determining the optimal composition of the wind farms complex, consisting of several offshore wind farms located at a considerable distance from each other in areas with significant wind regime asynchrony. To illustrate this method, the optimal composition with an installed capacity of 47.6 MW and located off Myanmar's west coast is defined.

Nations are facing today the transition to cleaner, more reliable and affordable energy systems where power grids are becoming less centralized, more flexible and digitalized, and where not only power utilities, but also consumers are playing a significant role. Distributed Energy Systems (DES) constitute a key element in such transition, with decentralized renewable energy generation near the consumption points, energy storage, electric vehicles, and energy management systems, with the potential to ensure continuous supply and achieve higher efficiency, while reducing costs and adverse environmental impacts. This chapter presents a review of the recent advances in the design and development of DES, focusing on the effect of taking into consideration the consumption profile and behaviour of the end-users. The chapter also revises the limitations of DES and summarizes the future directions of DES development.

Detailed data on incoming solar radiation are needed in the design of solar energy systems of any scale: from large PV plants to small off-grid systems. However, in most cases, obtaining data on measurements of solar radiation is connected with difficulties due to financial or technical restrictions. Often, ground-based measurements of solar radiation are either not carried out at all or only the value of the global horizontal intensity of solar radiation is measured. The aim of the present study is to review and to verify some existing empirical models of the global solar radiation and its components for the climatic conditions of Kyrgyzstan as well as to estimate the applicability of Meteonorm database model for the available solar radiation in the territory of Kyrgyzstan. The necessity to select the most suitable models of the solar radiation is called by the lack of similar studies on this direction for the conditions of the country.

Tourism destination is the place where tourism demand and supply meet. Destination is often the main reason why people travel. In the first part of the chapter, the research is focused on the fast growth of tourism in the past decade, which does not come without problems. As one of the most discussed problems, the focus is on overtourism, which started as growth potential and moved into a sustainability issue. The scope of this chapter was to continue previous conducted research in the Croatian coastal town Šibenik and build additional scenarios using technological advancements and available data to build a cornerstone for a data-driven Destination Management System. The results of two experiments suggest that usage of simple techniques can be widely adopted in the search for sustainable management of the destination. In the second part of the research, the authors were able to combine data before outbreak of Coronavirus COVID-19 and during its early growth phase in Croatia to show what devastating impact it has on tourist arrivals in a short period of time and demonstrate in real time how fragile tourism demand is.

Application of machine learning tools in drilling hydrocarbon well is still exploratory in its stage. This chapter presents a brief review of various applied research in drilling operations using machine learning (ML) tools and develop a deep neural network (DNN) model for predicting the downhole pressure surges while tripping. Tripping in or out drill-string/casing with a certain speed from the wellbore will result in downhole pressure surges. These surges could result in well integrity or well control problems, which can be avoided if pressure imbalances are predicted before this operation is engaged. Existing analytical models focus on forecasting the pressure imbalance but requires cumbersome numerical analysis. This could be solved by integrating DNN tool with the best existing analytical model predicted dataset. Consequently, the aim of this chapter is to provide an overview of various applications of machine learning tools in drilling and presenting a step-by-step process of developing a DNN model for the prediction of downhole pressure surges during tripping operation.

The era of the Information Technology revolution has endeavored mankind with enormous technological comforts. Apart from human health care, cloud computing integrated with other technologies can be exploited for the welfare of endangered wildlife animals that are currently on the verge of facing extinction in the near future. International Union for Conservation of Nature (IUCN) has declared a huge number of wildlife species into red list falling into 'critically endangered', 'endangered' and 'vulnerable' species because of pathogenic diseases, poaching, habitat degradation, and climate change. The aim of the chapter is to propose a cloud computing based wildlife health care model to preserve, sustain and protect the global wildlife heritage and wildlife conservation by identifying and providing treatment to diseased animals, monitoring their health conditions periodically and protecting against poaching. The proposed health care model makes use of various wearable sensors implanted on and within the body of an animal. The sensors perform data gathering about different body-related parameters to be transmitted to a cloud system for analysis and reporting a required timely action. When a cloud system senses emergency with respect to the health-related or threat-related incident, the information is immediately signaled to wildlife physician and wildlife ranger respectively to act accordingly.

This chapter analyses efficiency of support vector regression (SVR), artificial neural networks (ANNs), and structural vector autoregressive (SVAR) models in terms of in-sample forecasting of portfolio inflows (PIs). Time series daily data sourced from Rand Merchant Bank (RMB) covering the period of 1st March 2004 to 1st February 2016 were used. Mean squared error, root mean squared error, mean absolute error, mean absolute squared error, and root mean scaled log error were used to evaluate model performance. The results showed that SVR has the best modelling performance when compared to others. In determining factors that affect allocation of PIs into South Africa based on SVAR, 69% of the variation was explained by pull factors while 9% was explained by push factor. Hence, SVR model is more accurate than ANNs. This chapter therefore recommends that banking sector particularly RMB should use machine learning technique in modelling PIs for a better financial solution.

The concept of a multireservoir systems in hydropower introduces the function of multiple units simultaneously to reach the peak requirements. The reservoir optimized operation is a complex, extremely nonlinear, high dimensional, and multimodal task. The options that can be evaluated manually are generally limited in numbers, which made it difficult to identify the most appropriate option and should be taken into account while making decisions. Presently, for solving the optimization problems in multireservoir system, many modern heuristic stochastic search algorithms were established. That is possible because of the aspects of artificial and computational intelligence technologies. By connecting metaheuristic algorithms, the decision options can be identified to make the most suitable utilization of the scarce resources, best natural results for a given allotment can be attained, and the best trade-offs between contending goals can be established. In this chapter, the authors review the latest meta-heuristic optimization technique and their applications to maximize the economic factors.

Over two million kilometers of power grids in Russia exist. They have to be replaced in the coming 15 years. However, nowadays we witness and participate in the development of advanced technologies. New wireless resonant electric-power systems for different power consumers are considered including stationary single-conductor transmission lines and single-trolley and noncontact high-frequency electric transport, using non-metal conductive media. The trends of the future development and application of wireless resonant systems for electric power transmission are described. In the future, electrified mobile robots with external wireless electric power supplies will allow the organization of agricultural production with full automation of technological processes. The chapter describes the method and means for electric power transmission without metal wires. In this case, several components of conventional transmission lines, such as metal wires, insulators, and cables, are excluded, which results in considerable reduction in the equipment cost.

The chapter focuses on the implementation of Box Behnken Design (BBD) to increase permeate flux of rectangular sheet membrane. Box Behnken Design (BBD) was used to optimize the membrane operation by predicting the optimum conditions. The factors such as operating pressure, feed velocity, and pore size were selected as the input of the model. The study illustrates the optimum conditions of operating pressure, feed velocity, and pore size, which was found to be 14.5Pa, 0.179 m/s, and 0.59μm respectively. Analysis of variance was used to identify the significant terms in the model equation. The effect of input parameters on the model output evaluated using Pareto analysis. It shows that operating pressure is the most significant parameter in the developed model. The BBD predicted results follow the actual results with high accuracy.

Data hiding has emerged as a major research area due to the phenomenal growth in internet and multimedia technologies. Securing data transmitted over the internet becomes a challenging issue caused in digitization and networking over the past decade. Data hiding schemes have been adopted to protect digital media content which involves confidential data such as text, video, audio, images, and compression coding. A good reversible data hiding scheme is characterized by the possession of attributes like reversible, imperceptible, high payload capacity, and robustness. By reversible, it's meant that the extraction of the payload as well as the restoration of the host image perfectly from the stego image. Secondly, the imperceptible stego image resemblance against the cover/host image. Finally, robustness counts for the ability to sustain the secret payload against both intentional and unintentional attacks; it has been observed that all the proposed algorithms are more robust and reversible against various attacks in lower bit error rate and higher normalization coefficient.

Metaheuristics has become a top research area. Numerous optimization problems have been solved by metaheuristics as they showed comprehensive improvements to solve these intractable optimization problems. Complex problems like supply chain design problems need strategic decisions, and metaheuristics can intensify the decisions while designing supply chain network. In this chapter, the authors have introduced how nature memetic algorithms (e.g., genetic algorithm and particle swarm algorithms) are implemented to solve supply chain network design problem. A discussion about the recent research in this field shows an important direction to the future research.

Foreword

The purpose of this *Handbook of Research on Smart Technology Models for Business and Industry* is to carefully inquire, holistically reflect, comprehensively explain and illustrate the current and emerging cutting-edge methodologies of Smart Technology Models in important areas of investigation and facilities of original, innovative and novel real-life application of smart technology models, in our modern times related to new and emerging areas of business and industry. During previous decades, the analytical toolbox and the heuristic tools of applied mathematics and statistics, of computer science and novel analytics and information technologies have gained the attention of many researchers and practitioners worldwide. This exercised a strong impact also on science, engineering and economics, on the IT sector and, recently, the branches of Hospitality, Tourism, Travel and the Arts. Here, Optimization, Optimal Control, Arithmetic, Neuro- and Quantum Physics have become key technologies of support from an integrated viewpoint, and they are closely related with further areas of modern Operational Research, such as Development and Developing Countries, Societal Complexity and Ethics, to also appropriately address aspects of living conditions, education, freedom and peace in relation with those new and fast growing industries, with the environment, with and towards future generations.

Advances in machine learning, improvements on sensors, and by ever-greater computing power have helped in the creation of new generations of hard- and software automation, sensors and robots with practical applications in every industrial sector. These contributions have, in reverse, excited a great interest on the sides of venture investors, technology corporations, and a growing number of clients who implement intelligent automation in physical, neuronal and information systems. Our new generation of Industry 4.0 aims at an increased flexibility in manufacturing and mass customization, at improved quality and enhanced productivity. There are smart technologies, like smart community, cloud implementation, 5G technology, machine-to-machine measurement, smart cities, IoT gateway systems, cyber-physical systems and security (CSPs), augmented and virtual reality, healthcare and social applications, improvements in logistics and production, in agricultural electrification with intelligent extensions to manufacturing of crops, milk diaries, drug discovery and information resource management, facilitating and supporting the creation, use and application of those intelligent and innovative smart technologies. They are employed in the fields of computer science, computer engineering, information and communication, neuroscience and medicine, creativity and recreation. The respect and responsibility to all humans and all the creations are of premium importance. For reasons like these, this edited collection of original research chapters has been initiated, prepared and completed.

Classical, heuristic, nature-inspired and holistic optimization techniques are increasingly well-known and accepted as key technologies in research areas of engineering, science and technology, and in practice, in solving immense challenges worldwide, in support of individual fulfillment and social relations of high quality, and in getting prepared for a better future. Contemporary trends visible in the application on problems of science, classical and emerging technologies are becoming prolonged also into the areas of economy, finance, life quality, including health, hotel, tourism and travel industries across the globe. This monograph and compendium has asked for, gathered and refined best and high-level papers, selected on a wide and rich variety of research and application themes. Well-known and newly created methods of classical, heuristic, nature-inspired and holistic optimization have been employed to analyze, handle and resolve some of the very complex and hard problems in the areas of modern business, industries and economies.

This book, *Handbook of Research on Smart Technology Models for Business and Industry*, has welcomed submissions by interested authors from business, management, industries and economies, and from social, environment and developmental sciences. It has invited for premium chapters from the research fields of mathematics, data mining, analytics, machine learning, artificial intelligence, information technology and computer science which are related to our sectors of application. The book chapters reflect and provide new developments and contributions of the current and growing body of knowledge about the areas of the problems studies, about their underlying sectors of human life, of exchange of information and finance, of trade and advancements in education, culture, freedom, peace and empathy. This book will hopefully become a help for investigators, researchers, scientists, decision makers, managers, supervisors, executives, medical doctors, patients, consumers, employees, travelers, tourists, hoteliers, agents, directors, financiers, artists, modelers and philosophers, as well as anyone who is curious and interested in business, industries, economies and in the world in which we are living.

The contributing authors of the chapters of this book are experienced in science and practice, they are from all over the world, they use and refine the deep model-based means of mathematics and the less model-based, so-called heuristic, smart or intelligent algorithms, rooting in engineering, in computer science and applied mathematics. The latter ones are also named as model-free; they are mathematically less rigorous, released from the firmness of calculus and combinatorics, to integrate nature-, e.g., bio-inspired approaches to efficiently overcome hard problems. The rise of these techniques from artificial intelligence occurred along with to the powerful contributions of mathematics and physics which have often been model-based. Nowadays, labeled as Big-Data, Data Mining, Statistical Learning, Machine Learning, Neuroscience, Remote Sensing, Metaheuristics and Matheuristics, and fostered by Operational Research, both the model-free and the model-based streams of academic cultures meet, communicate and interact in various centers of research excellence, at important congresses, in remote forums and in premium projects and science agendas in all over the world, to overthrow misunderstandings and wrong perceptions between these two academic traditions, and to benefit from synergies, to jointly strengthen scientific progress and to provide common contributions to the solution of due and overdue real-life challenges of us all. Those immense problems do exist in every field of the modern world and in academia, in the business world and the industries, and their dynamic and mutual interrelations with engineering, social sciences, life and human sciences, collaboration with the developing and emerging countries, for higher qualities in living conditions, social and human relations, freedom, peace and empathy, for being trained and ready for today and tomorrow. From the variety of these subjects and realizing them all to their best, the editors have chosen and richly elaborated the areas of Smart Technology Models for Business and Industry.

To all the respected authors of these valuable 19 chapters, I convey my appreciation and gratitude for having shared their enthusiasm, dedication, knowledge, care, assessment, wisdom and vision with the scientific and applied community, and with humankind. I express my thanks to the publishing house

IGI Global, and to the editors, Prof. Dr. J. Joshua Thomas (UOW Malaysia, KDU Penang University College, Malaysia), Prof. Dr. Ugo Fiore (University of Naples Parthenope, Italy), Prof. Dr. Dr. Gilberto Pérez Lechuga (Universidad Autónoma del Estado de Hidalgo, México), Prof. Dr. Valeriy Kharchenko (Federal Scientific Agroengineering Center VIM, Russia), and Prof. Dr. Pandian Vasant (University Technology Petronas, Malaysia), for having provided the chance for many experts to make known and public best pieces of their achievements, contributions, works and lifeworks. I express to them my gratefulness for having ensured a premium monograph and compendium of a high academic, applied, and social significance and promise to our world.

Now, I wish you all a lot of joy in browsing through, studying deeply, getting inspired and gaining a lot from this interesting book. Yes, I hope that the blessings will be high, personally, societally, for humankind and for every creature.

Sincerely yours,

Gerhard-Wilhelm Weber
Poznan University of Technology, Poland & Middle East Technical University, Turkey
May 2020

Gerhard-Wilhelm Weber *is a Professor at Poznan University of Technology, Poznan, Poland, at Faculty of Engineering Management, in the Chair of Marketing and Economic Engineering. His research is on data mining, analytics, artificial intelligence, machine learning, mathematics, operational research, finance, economics, optimization and optimal control, neuro-, bio- and earth-sciences, medicine and development; he is involved in the organization of scientific life internationally. He received Diploma and Doctorate in Mathematics, and Economics / Business Administration, at RWTH Aachen, and Habilitation at TU Darmstadt (Germany). He replaced Professorships at University of Cologne, and TU Chemnitz, Germany. At Institute of Applied Mathematics, Middle East Technical University, Ankara, Turkey, he was a Professor in Financial Mathematics and Scientific Computing, and Assistant to the Director, and has been a member of five further graduate schools, institutes and departments of METU. G.-W. Weber has affiliations at Universities of Siegen (Germany), Federation University (Ballarat, Australia), University of Aveiro (Portugal), University of North Sumatra (Medan, Indonesia), Malaysia University of Technology, Chinese University of Hong Kong, KTO Karatay University (Konya, Turkey), Vidyasagar University (Midnapore, India), Mazandaran University of Science and Technology (Babol, Iran), Istinye University (Istanbul, Turkey), Georgian International Academy of Sciences, at EURO (Association of European OR Societies) where he is "Advisor to EURO Conferences" and IFORS (International Federation of OR Societies), where he is member in many national OR societies and working groups, at Pacific Optimization Research Activity Group, etc. G.-W. Weber has supervised many MSc. and PhD. students, authored and edited numerous books and articles, and given many presentations from a diversity of areas, in theory, methods and practice. He has been a member of many international editorial, special issue and award boards; he participated at numerous research projects; G.-W. Weber received various recognitions by students, universities, conferences and scientific organizations, nationally and internationally.*

Preface

The effect of information technology (IT) is building a modern world which is not completely understood yet. It is hard to estimate the magnitude and pace of cultural, lifestyle, and social changes already experienced in daily life without knowing the technological driving forces behind it. This book presents the contributions featuring the latest research and development in the different information engineering. The smart technology (ST), designed to connect people, organizations and smart things, promises business to fundamentally change how we live and interact and it can redefine a wide range of industries.

Given the substantial interest in these new concepts, which have the potential to radically change where we live, how we work and how we interact with each other and with organizations, there is a lack of understanding of smart technology model's emergence will impact businesses. Businesses that succeed in adapting their extant smart business models to the new technological possibilities have considerable opportunities to innovate and with high competitive capabilities. For businesses, it is therefore important to understand the extent to which smart things will transform existing business models, and as a part of this, how value creation in such industries will be influenced by the rise of the smart technology models .

In this book, among these, we particularly focus on smart technology models for business and industry. Smart technology also presents major challenges for businesses, including the advancement of interoperability between systems, dealing with established business partners that do not comply with new technologies, transactions, contractual and security threats, loss of control and privacy concerns related to data explosion collected and used by businesses and their smart things. Hence, it is considered as the basis for new generation smart technology intelligence. It is closely related with artificial intelligence in the sense that financial paradigm in bitcoin, elecronification of agricultural systems, 5G technolgy catalyst for industrial Internet of Things (IoT's) perform better particularly under high amount of obervations.

On the other hand, optimization is a very well-known area with a vast number of applications, from route finding problems to smart grids, hydropower energy systems and maintenance schedules in plants. Under the high amount of wireless data, the membraine computing mixed new emergent technology solutions and earlier heuristics fall short for the big data in manufacturing systems, that benefits mutiplier effect as operating. Information technologies achieve strength in combination and improve efficiency at all levels in industries. Therefore, new scientific smart models are emerged with the foundation of cloud wireless network, sensors, analytic tools and people in ever more intricate and resourceful ways.

OBJECTIVE OF THE BOOK

The book includes a description of a series of technological methods and a number of theoretical means of discussion on emergent smart technologies, Industry 4.0 applications, IoT in 5G network theoritical arcitectures, smart business management models, prediction models, renewable energy models, innovative signal theoris and implementation are available and in use. We aim to provide resouce for the audience. Being dual important aspects in smart technology business models to impove transactions in financial benefits, and integrated artificial intelligance models in industry. Another objective of the book is to present a variety of domain applications using smart technology model. Additionally, this book to provide a reference and guidence for the use of smart technology models in varies domains and it could be expanded to mutiple industry relavant operations in future.

In terms of smart technologies for business and industry models, the book covers a wide range of smart business paradigm in cryptocurrency security, operational modes in power plants, Intelligent management of powergrid energy resources, distribution of energy transition in consumers profiling, cloud computing model for forest animals monitoring, data hiding process models, smart deep neural network model on automated essay scoring with natural language projective models, Wind energy location analysis for a state, methodology to calculate solar resources analysis, smart forcast destination management for tourists, deep neural network (DNN) model for pressure tripping in oil and gas industry, modelling forcast on various machine learing models and its comparisions. There are optimization models for industries are addressed with meta-heuristics approaches, Genetic, partical swarm algorithm techniques on supply chain design and it extends to varies domain applications.

TARGET AUDIENCE

The book intends to provide useful information for audience with various background and with various data analytics expertise level. The book includes chapters offering overview of smart technology models in both business and industry implications, which would be useful for audience from any background. The details for specific smart models and their adaptation in various domains are also available, which would address the audience with data models background or with domain knowledge on areas, such as geology, power systems, healtcare, agriculture or natural language processing.

The book includes 20 chapters, presenting recent advancements on Smart technology models for Business and Industry with a variety of models, algorithms and application domains.

In Chapter 1, Bitcoin network along with its security and financial implications are discussed. This chapter offers an in-depth investigation of the Bitcoin network along with its security and financial implications. All the components of the network are described along with details of the cryptographic foundations of the system. The chapter also describes the security and privacy implications of Bitcoin and proposes methods to minimize their effect. Finally, the chapter discusses the regulatory and financial implications of the Bitcoin network to show how countries around the world are reacting to the widespread use of cryptocurrencies.

Chapter 2 focuses on the application of the electrical currents is discussed from the point of view of plants cultivation improvement in the horticulture. With underlining that the electricity is an abiotic stress stimulant, the electricity use ways classification is given. The application of the electric currents and other similar influences can – directly or indirectly – affect plants causing series of physiological and biochemical reactions. The authors offers an effective technique development of the plants electric stimulation, there are many aspects almost impossible to be taken into account within one experiment.

In Chapter 3, the potential of the Internet of Things (IoT) in the context of Fifth-Generation Networks (5G) is envisioned as suggesting several beneficial opportunities for companies, industries, and users to exploit this technology's applications. The authors focus on establishes how the IoT works, considering its 5G-architecture. The emphasis is on the infrastructural characteristic in terms of transmission power, frequency, speed, security, localization, device's lifetime and others. Additionally, the chapter illustrates what the IoT entails; discussing its workability and efficiency. Furthermore, it highlights a range of newly distinguishing features that would give it much-touted success in comparison with other technologies.

Chapter 4 develops principles for the formation of the optimal composition of the energy complex from all the given power plants based on renewable energy sources for an autonomous consumer, taking into account the variable energy loads of the consumer, changing climatic conditions and the possibility of using local fuel and energy resources. The authors provide solutions in addition to the optimal configuration of the power complex, it is also necessary to solve the problem of optimizing the joint operation of different types of power plants from the selected optimal configuration, i.e. it is necessary to determine the optimal modes of operation of power plants and the optimal share of their participation in providing consumers at every moment in time with numerical model.

Chapter 5 is on another important business cluster wit the help of wireless mesh network, The authors device in all the existing schemes, whenever any data is newly cached or any data is evicted then the information needs to be broadcasted to all the Mesh Access Points (MAPs). As a result, communication cost increases. In this chapter, a cluster and sub-cluster based cooperative caching scheme is designed to minimize the communication cost. Clusters and sub-clusters are formed by taking advantage of static position of Mesh routers. Cluster-heads (CHs) are selected using minimum spanning tree and minimum weighted vertex cover algorithm. Then the clusters and sub-clusters are formed. CHs and other Maps keep different cache information tables and based on those, cache decision is taken. Client mobility is also considered.

In Chapter 6, the authors develop a web-based automated text scoring (ATS) system which can grade essays and check for spelling errors. The main reasoning behind this work is to alleviate the labour-intensive marking of essays and ensures equality in scoring for high-stakes exams like TOEFL. The researcher had performed a detailed investigation on deep learning techniques used in the field of ATS and developed a recurrent neural network model that can score essays in an end-to-end approach. Using the developed deep learning model, a web application was also developed to showcase the process of ATS by letting the web application to communicate with the trained model. The evaluation metrics chosen to evaluate the model are the quadratic weighted kappa (QWK) score and the trained model can achieve 0.6 in QWK score.

Chapter 7 is on another important domain, The prediction of stock prices has always been very challenging problem for investors. Using machine learning techniques to predict stock prices is also one of the most favourite topics for academicians working in this domain. This book chapter discusses five supervised learning techniques and two unsupervised learning techniques to solve the problem of stock price prediction and has compared the performances of all the algorithms. Among the supervised learning techniques, Long Short Term Memory (LSTM) algorithm performed better than the others whereas among the unsupervised learning techniques, Restricted Boltzmann Machine (RBM) performed better. RBM is found to be performing even better than LSTM.

Chapter 8 is about the problem of Partially energy shortage is possible to decrease by the construction of sizeable grid-connected offshore wind farms. As the most perspective region for offshore wind energy development considered water area near the west coast of Myanmar and theoretical offshore wind energy potential of that water area at 10 and 100 m height defined. Eight prospective construction sites selected and assumed wind turbine model chosen. The authors describes the methodic for determining the optimal composition of the wind farms complex, consisting of several offshore wind farms located at a considerable distance from each other in areas with significant wind regime asynchrony. To illustrate this methodic, the optimal composition of complex with an installed capacity of 47.6 MW and located in Myanmar west coast water also defined.

In Chapter 9, the problem of handling today the transition to a cleaner, more reliable and affordable energy system where power grids are becoming less centralised, more flexible and digitalised, and where not only power utilities, but also consumers are playing a significant role. Distributed Energy Systems (DES) constitute a key element in such transition, with decentralised renewable energy generation near the consumption points, energy storage, electric vehicles and energy management systems, with the potential to ensure continuous supply and achieve higher efficiency, while reducing costs and adverse environmental impacts.

In Chapter 10, the aim of the present study is to review and to verify some existing empirical models of the global solar radiation and its components for the climatic conditions of Kyrgyzstan as well as to estimate the applicability of Meteonorm database model for the available solar radiation in the territory of Kyrgyzstan. However, in most cases, obtaining data on measurements of solar radiation is connected with difficulties due to financial or technical restrictions. Often, ground-based measurements of solar radiation are either not carried out at all, or only the value of the global horizontal intensity of solar radiation is measured.

In Chapter 11, tourism destination is place where tourism demand and supply meet. Destination is often the main reason why people travel. The authors scope to continue previous conducted research in Croatian coastal town Šibenik and build additional scenarios using technological advancements and available data to build a cornerstone for data-driven Destination Management System. The results of two experiments suggest that usage of simple techniques can be widely adopted in search for sustainable management of the destination. In second part of our research we were able to combine data before outbreak of Coronavirus COVID-19 and during its early growth phase in Croatia to show what devastating impact it has on tourist arrivals is short period of time and demonstrates in real time how fragile is tourism demand.

In Chapter 12, a large-scale application of machine learning tools in drilling hydrocarbon well is still exploratory in stage. This chapter presents a brief review of various applied research in drilling operations using machine learning (ML) tools and develop a deep neural network (DNN) model for predicting the downhole pressure surges while tripping. Tripping in or out drill-string/casing with a certain speed from the wellbore will results in downhole pressure surges. This surges could result in well integrity or well control problems which can be avoided if pressure imbalances are predicted before this operation is engaged. Existing analytical models focus on forecasting the pressure imbalance but requires cumbersome numerical analysis.

In Chapter 13, the era of Information Technology revolution has endeavored mankind with enormous technological comforts. Apart from human health care, Cloud Computing technology can be exploited for the welfare of endangered wild life animals that are currently at verge of facing extension in near future.International Union for Conservation of Nature (IUCN) has declared a huge number of wild life species into red list falling into 'critically endangered', 'endangered' and 'vulnerable' species because of pathogenic diseases, poaching, habitat degradation, and climate change. The authors aim to propose a cloud computing based health care model to preserve, sustain and protect the global wildlife heritage by identifying and providing treatment to diseased animals, monitoring their health conditions periodically and protection against poaching and hunting.

Chapter 14 discusses the efficiency of Support Vector Regression (SVR), Artificial Neural Networks (ANNs) and Structural Vector Autoregressive (SVAR) models in terms of in-sample forecasting of portfolio inflows (PIs). The authors showed results that SVR has the best modelling performance when compared to others. In determining factors that affect allocation of PIs into South Africa based on SVAR, 69% of the variation was explained by pull factors while 9% was explained by push factor. Hence, SVR model is more accurate than ANNs. This chapter therefore recommends that banking sector particularly RMB should use machine learning technique in modelling PIs for better financial solution.

In Chapter 15, the authors focus on the concept of a multireservoir systems in hydropower introduces the function of multiple units simultaneously to reach the peak requirements. The reservoir optimized operation is a complex, extremely nonlinear, high dimensional and multimodal task. The options that can be evaluated manually are generally limited in numbers, which made it difficult to identify the most appropriate option and should be taken into account while making decisions. Presently, for solving the optimization problems in multireservoir system many modern heuristic stochastic search algorithms were established. By connecting metaheuristic algorithms, the decision options can be identified to make the most suitable utilization of the scarce resources, best natural results for a given allotment can be attained also the best trade-offs between contending goals.

In Chapter 16, the authors witness and participate in the development of advanced technologies. New wireless resonant electric-power system for different power consumers are considered including stationary single-conductor transmission lines and single-trolley and noncontact high-frequency electric transport, using non-metal conductive media. The trends of the future development and application of wireless resonant systems for electric power transmission are described. The authors present chapter describes the method and means for electric power transmission without metal wires. In this case, several components of conventional transmission lines, such as metal wires, insulators and cables, are excluded, which results in considerable reduction of the equipment cost.

Chapter 17 focuses on the implementation of Box Behnken Design (BBD) to increase permeate flux of rectangular sheet membrane. Box Behnken Design (BBD) was used to optimize the membrane operation by predicting the optimum conditions. The factors such as operating pressure, feed velocity, and pore size selected as the input of the model. The study illustrates the optimum conditions of operating pressure, feed velocity and pore size, which found to be 14.5Pa, 0.179 m/s and 0.59μm respectively. Analysis of variance used to identify the significant terms in the model equation. The effect of input parameters on the model output evaluated using Pareto analysis. It shows that operating pressure is the most significant parameter in the developed model. The BBD predicted results follow the actual results with high accuracy.

In Chapter 18, securing data transmitted over the internet becomes a challenging issue caused in digitization and networking over the past decade. Data hiding schemes have been adopted to protect digital media content which involves confidential data such as text, video, audio, images and compression coding. A good reversible data hiding scheme is characterized by the possession of attributes like reversible, imperceptible, high payload capacity and robustness. The author finally count for the ability to sustain the secret payload against both intentional and unintentional attacks, it has been observed that all the proposed algorithms are more robust and reversible against various attacks in lower bit error rate and higher normalization coefficient.

Chapter 19 discusses numerous optimization problems have been solved by metaheuristics as they showed comprehensive improvements to solve these intractable optimization problems. Complex problems like supply chain design problems need strategic decisions and metaheuristics can intensify the decisions while designing supply chain network. In this chapter authors have introduced how nature memetic algorithm e.g. Genetic algorithm and particle swarm algorithms are implemented to solve supply chain network design problem. A discussion about the recent research in this field shows an important direction to the future research.

We strive to include a rich range of recent developments in smart technology in both algorithms and techniques as well as technological applications in the aspects of intelligent and proposals for new models for business and industry. We assume that the book would be a valuable tool for a wide audience, including researchers on computer technology. It provides useful cases to cities and counties, ministries, and other organizations that involve a variety of agricultural, energy, and healthcare areas. Improvements in learning and optimization tasks will be of particular interest to IT companies regarding the processing of natural language, information models and mangment of smart factories.

J. Joshua Thomas
UOW Malaysia KDU Penang University College, Malaysia

Ugo Fiore
University of Naples Parthenope, Italy

Gilberto Perez Lechuga
Autonomous University of Hidalgo State, Mexico

Valeriy Kharchenko
Federal Agroengineering Centre VIM, Russia

Pandian Vasant
Universiti Teknologi PETRONAS, Malaysia

Acknowledgment

The editors would like to appreciate the contributions of all the people involved in this book's development activity. They particularly wish to thank each one of the writers for their contributions. Our sincere thanks go out to the authors of the chapters who contributed the time and expertise to this book. Special thanks should be given to the people who, besides the preparations and were active in the evaluation process. That book would not have become a reality without invaluable support from all of them.

My gratitude to the members of the editorial advisory board who have done a lot to make the book high quality. Particularly the critical position of the Editorial Board member, Professor Gerhard-Wilhelm Weber, Professor Gilberto Pérez Lechuga for excellent forward and precious assistance. The editors are confident that this book will certainly be useful to readers, because, among other things, it gives them the opportunity to be acquainted with the results of research that has published. It would be reference the enormous role that IGI Global publishing house has played in order for this information to publish.

Finally, many thanks the editors would like to express to the entire staff of IGI Global publishing houses particularly Josh Christ who helped the editors and authors in preparing the manuscript of this book professionally and extremely friendly, and have made a lot to make this book the edition of exceptional quality. Not least, I want to thank our Lord and Saviour Jesus Christ who gave life and strength to those who worked throughout the book process to achieve the book during the Covid-19 period.

J. Joshua Thomas
UOW Malaysia KDU Penang University College, Malaysia

Ugo Fiore
University of Naples Parthenope, Italy

Gilberto Perez Lechuga
Autonomous University of Hidalgo State, Mexico

Valeriy Kharchenko
Federal Agroengineering Centre VIM, Russia

Pandian Vasant
Universiti Teknologi PETRONAS, Malaysia

Chapter 1
Decentralized Cryptocurrency Security and Financial Implications:
The Bitcoin Paradigm

Amany Mohammed Alshawi
https://orcid.org/0000-0002-9406-9900
King Abdulaziz City for Science and Technology, Saudi Arabia

ABSTRACT

The worldwide growth in electronic commerce has led to the emergence of digital currencies, which have been gaining public and research attention as an alternative method of payment. In recent years, Bitcoin has become a global currency with a multibillion USD market value. This is a result of the many attractive features of the currency including its distributed algorithms, cryptography, and incentive-driven performance. This chapter offers an in-depth investigation of the Bitcoin network along with its security and financial implications. All the components of the network are described along with details of the cryptographic foundations of the system. The chapter also describes the security and privacy implications of Bitcoin and proposes methods to minimize their effect. Finally, the chapter discusses the regulatory and financial implications of the Bitcoin network to show how countries around the world are reacting to the widespread use of cryptocurrencies.

INTRODUCTION

Digital cryptocurrencies, such as Bitcoin, pose regulation challenges to financial and governmental institutions due to the decentralized open source nature of their creation and exchange. First put into circulation in January 2009, Bitcoin is one of the first online currencies to provide peer-to-peer, open-access tracking instruments for monetary transactions without external backing from a central bank (Nakamoto, 2008). Digital currencies based on a mathematics-backed payment protocol, such as Bitcoin, have also exhibited extreme upward and downward swings in their valuation; for example, Bitcoin grew

DOI: 10.4018/978-1-7998-3645-2.ch001

in value to reach a record high value of more than 20,000 USD in October 2017. The starting value of the Bitcoin was only 0.0009 USD when it was first sold in 2009 (Beigel, 2019).

Cryptographic transaction protocols establish distributed networks of trust that are independent of social support structures for the verification of identity (Chait, 2015). Unlike conventional payment platforms (e.g., PayPal), digital cryptocurrencies (e.g., Bitcoin) can be less secure because of the increasing customization, centralization, and regulation of their transactions. Although the original design of Bitcoin aims at a fully decentralized payment system, recent events have revealed the true limits of the decentralization of the Bitcoin network. One is that a considerable number of centralized services currently host Bitcoin and control a large share of the Bitcoin market. Also, a number of Bitcoin developers hold privileged rights in the conflict resolution and maintenance of the official client version. These entities can decide the fate of the entire Bitcoin network by bypassing the rights and computing powers of the majority of other network users (Gervais, Karame, Capkun, & Capkun, 2014).

This has spawned a host of alternative cryptocurrency platforms, such as Ethereum (Youssefzadeh, 2014), which has created tension in the Bitcoin community between those who support its integration with existing institutions and those who resist its mediation by human judgment (Lustig & Nardi, 2015). Since digital cryptocurrencies enable direct money exchanges between different agents by eliminating intermediaries, they also create novel, legal possibilities of bargained-for exchanges (Fairfield, 2014). Virtual currencies such as Bitcoin are not inherently resistant to abuse, for instance, money laundering, financial fraud, or tax evasion. The illegal use of digital currencies can occur because they exist outside of established international institutional frameworks that intermediate the delivery of financial services (Abadi & Brunnermeier 2018; Pflaum & Hateley, 2014). As their computational models and verification protocols limit existing cryptographic currencies, virtual monetary transactions may involve inherent security risks (Kumaresan & Bentov, 2014).

At present, the value of Bitcoin, which already serves as a means of monetary exchange, can be predicted through transaction network characteristics, the cryptocurrency transaction volume, and the digital money supply. From a law enforcement and financial perspective, bringing Bitcoin into wider circulation may demand radical changes to the domain of international monetary control (Jeans, 2015). In recent years, digital cryptocurrencies have attracted an increasing amount of public attention. This is because they represent a novel kind of financial instrument that involves both risks, such as cybercrime facilitation, and financial opportunities derived from their innovative nature (Rehman et. al., 2019 and Huhtinen, 2014). This chapter presents the implications the tradeoff between distributed, decentralized, and deregulated foundations of digital cryptocurrencies (e.g., Bitcoin) and their cryptographic security has for their status as a means of unmediated digital monetary exchange (Spethoven, 2019)

The first part of this chapter describes the main Bitcoin system components and the way they interact to complete a single transaction. Special emphasis is placed on the steps involved in verifying all transactions. Then the chapter explains the algorithmic foundation of Bitcoin, including the steps involved to utilize hash functions and the elliptic curve protocol. The chapter then discusses the security implications of cryptocurrencies, in addition to privacy and anonymity issues. Then, it explains cryptocurrencies and open-access cryptographic platforms. This section presents details of how cryptocurrency transactions are verified and kept anonymous. The last section of the chapter describes the regulatory, security, and financial implications of cryptocurrencies in detail. Finally, the discussion section highlights the most important aspects of the chapter and summarizes all observations.

BACKGROUND

Foundation of Bitcoin

Bitcoin is a digital decentralized currency that was introduced in 2008 by Nakamoto (2008), and a proof-of-concept client was created in 2009. Bitcoin enables instant global payments without the central financial authority of traditional payment systems. Instead, the whole Bitcoin network acts as the financial authority, using cryptography to control the creation and transfer of money. This section describes the main components of the Bitcoin network and how they interact to create and transfer funds.

Bitcoin Components

1. Wallets
   A Bitcoin wallet is a digital case that is used to store digital cryptocurrency as a value. It enables users to always have their Bitcoins in their possession. A Bitcoin wallet provides services to users that make it possible to make different kinds of payments for free. As in physical wallets, Bitcoin wallets must be secured and used only by authorized people (Youssefzadeh, 2014).

2. Transaction
   "A transaction is a transfer of value between Bitcoin wallets which gets included in the block chain" (Huhtinen, 2014). Bitcoin wallets contain a secret piece of information known as a private key, which is used to sign transactions to prove they were initiated by the owner of the wallet. The signature also prevents the transaction from being changed by anybody once it has been issued. All transactions are broadcasted between users, and usually begin to be confirmed by the network in the following ten minutes, using a process known as mining.

3. Mining
   "Mining is a distributed consensus system that is used to confirm transactions by including them in the block chain" (Lustig & Nardi, 2015). It implements a sequential order in the blockchain and maintains the neutrality of the network. Mining also allows different computers to agree on a single state of the system. In order to confirm a transaction, it must be added to a block that applies strict cryptographic conditions to be verified by the network. The same conditions prevent previous blocks from being modified, as this will void all the following blocks. Mining is the basis for the competition to prevent any user from simply adding new blocks consecutively in the block chain. Consequently, no user can control what is added to the blockchain or replace units of the blockchain to reverse their payments (Lustig & Nardi, 2015; Youssefzadeh, 2014).

4. Blockchain
   The blockchain is represented as a public ledger for recording the cryptocurrency transactions on which Bitcoin is based. It represents a system that registers transitions from one state to another according to the transactions it processes. With each transaction, the Bitcoin system checks the condition of all unspent digital coins, the ownership status of the transacting parties, and the total number of all crypto-coins in circulation. This is essential to reject fraudulent transactions with forged or stolen coins (Huhtinen, 2014).

5. TimeStamp Server
   A TimeStamp server is used for proving that specific data exist at a certain time. It takes the hash of a block of items to be timestamped and then publishes the hash. Each timestamp also includes the previous timestamp in its hash (Huhtinen, 2014).

6. Proof-of-Work
   The Bitcoin network must prevent miners from hashing unauthorized transaction blocks, or all the Bitcoins would be mined in minutes. The Bitcoin protocol makes this possible by using proof-of-work. The proof-of-work system is used to implement a timestamp server based on cryptographic hashing. The system utilizes a double hashing, fixed-size block header using the SHA-256 algorithm and a specific target number larger than 256 bits. Only hashes satisfying these conditions will be accepted by the system. The miner has to append a random number to the new block and make sure the hash value of the new block begins with a series of zeros, which cannot be created by brute-force attacks. During the process of block chaining, verification is performed using the previous user's public key, and miners use their private key to sign any new block they create (Huhtinen, 2014; Sompolinsky & Zohar, 2015).

How the Bitcoin System Works

Bitcoins are created with the help of a computer encryption algorithm that discovers hash numbers in the process of its operation. These hash numbers represent the result of continuously solving mathematical functions that finds unique data blocks, which constitute the basic units of this cryptocurrency. This highly secure cryptographic algorithm requires that every next unit of Bitcoin currency must be computed according to a mathematical formula that includes data blocks from every previous currency unit. Furthermore, this cryptographic currency network timestamps all the transactions it processes. This makes these digital currency transactions computationally irreversible and transparent since the digital identities of transacting entities are public within the network (Chait, 2015). When the person running this mathematical algorithm on his/her computing equipment finds a numerical sequence that conforms to the requirements of this cryptocurrency-creation process, individual ownership of this block of data can be registered (Abadi & Brunnermeier 2018; Grushack, 2014). Figure 1 illustrates Bitcoin's most important building blocks.

Algorithmic Foundations of Bitcoin as a Cryptocurrency

Given that each new block of digital currency is derived directly from its previous unit, every user in this network maintains records of every cryptocurrency unit being created and every transaction. This ensures the authenticity of digital currency units and the integrity of cryptocurrency transactions. Moreover, those who discover new Bitcoin units are automatically rewarded gradually decreasing amounts of Bitcoins in recognition of currency inflation and network maintenance costs. Those who create new blocks of this cryptocurrency are directly contributing to the development of the distributed network it is based upon (Buterin 2014; Chait, 2015; Grushack, 2014).

Figure 1. Bitcoin building blocks (Mercator Research Group, 2015)

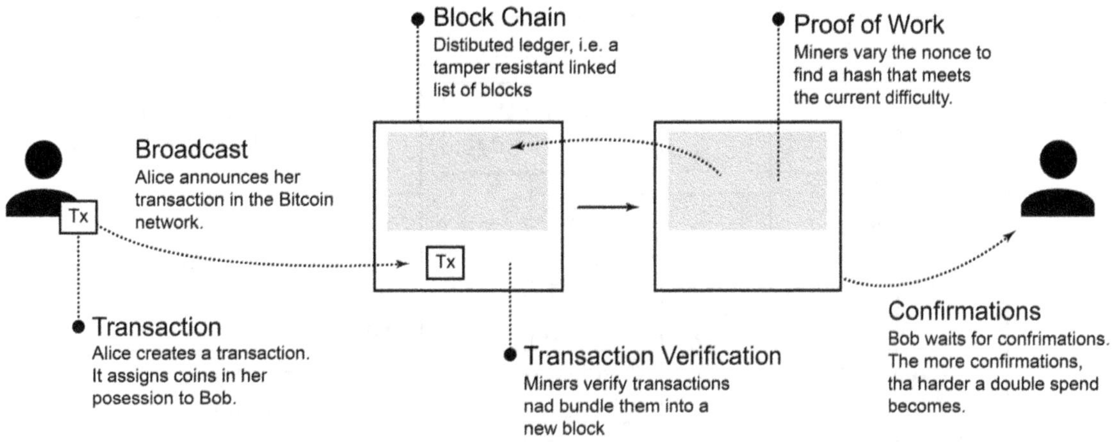

Additionally, the ease of discovering Bitcoins is algorithmically regulated in order to manage both the competition to find new Bitcoins and the pace with which new Bitcoins are put into circulation. This process has led to an exponential rise in the algorithmically determined difficulty level of discovering new Bitcoins over recent years. This also reflects the tremendous computing power that the Bitcoin network has accumulated by pooling the processing power of thousands of computers to perform highly complex and costly calculations (Grushack, 2014).

Driven by both the financial profit that Bitcoin discovery involves, and its growing popularity, this rising computational difficulty led to the predominant use of Application Specific Integrated Circuits (ASICs). These enable users to perform many calculations, which can match the computing capabilities currently required to find the next data block of Bitcoins. Despite increasing cost and energy efficiency of the Bitcoin discovery or mining, this development attracted commercial operators (Youssefzadeh, 2014). Unlike the initial use of Central and Graphic Processing Units, the predominant use of ASICs significantly raises the equipment investment requirements for engaging in Bitcoin mining. These requirements also decrease the competition between the agents that generate Bitcoins (Grushack, 2014).

In addition, the development of Bitcoin as a widely circulated digital currency depends on its cryptographic foundations, including both openly and privately accessible components. Transactions in Bitcoin require a public 34-bit alphanumeric key, serving as an individual digital-currency wallet, and a private key. The private key is algorithmically derived from the public key and used as a digital signature for transaction verification (Grushack, 2014; Youssefzadeh, 2014). The information related to Bitcoin transactions are recorded and verified by the public ledgers of the cryptocurrency network. This procedure verifies the existence of Bitcoin units that are exchanged in this network. These algorithmic foundations also verify the digital identities of the transacting parties by creating a computationally unique record of each transaction (Grushack, 2014). Figure 2 describes the process used for verification.

There is a continued need for blockchain computation, as it involves not only Bitcoin discovery, but also transaction verification. The authorization of digital currency transactions requires their inclusion in the public legers and data blocks that constitute digital currency units (Youssefzadeh, 2014). However, the process of handling Bitcoins is not inherently secure because the computer equipment, where digital wallet applications are stored, can be hacked. The risks of using Bitcoin pertain to the degree

Figure 2. Bitcoin transaction verification (Nakamoto, 2008)

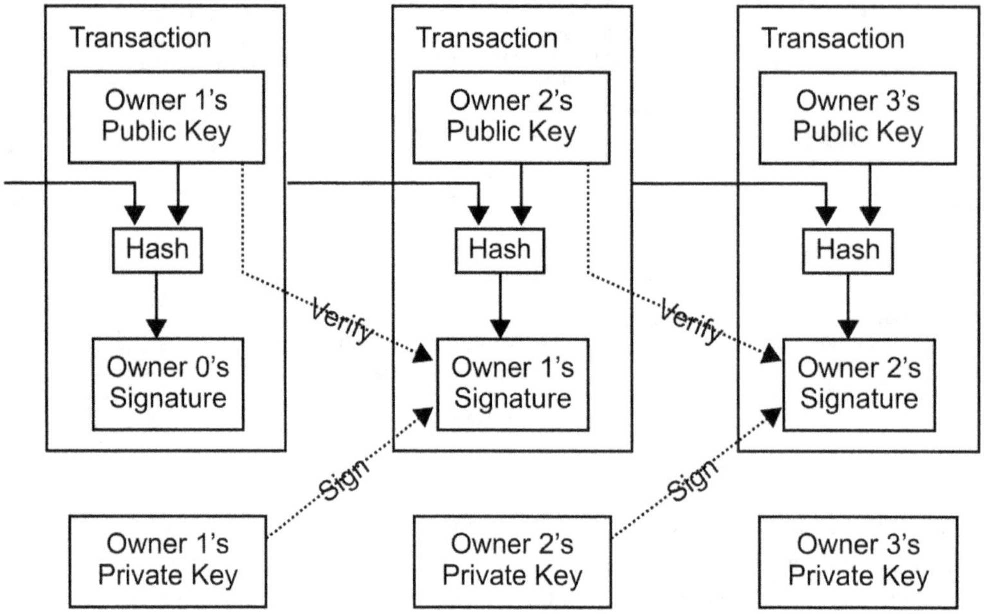

to which the computers used for transactions are protected from hacking attacks. Given that Bitcoin represents a virtual currency solely based on algorithmic information about its units, it is also possible to lose Bitcoins when the private keys encoding particular Bitcoin units cannot be found. This could result from the loss of relevant information on offline media or data erasure due to computer equipment damage (Grushack, 2014).

CRYPTOCURRENCY ISSUES

Security

The relative security and robustness of cryptocurrencies as a means of monetary exchange are derived from the decentralized and distributed design of the computer networks they are based on. In these cryptocurrency networks, every node contains identical information that enables the verification of cryptocurrency transactions. Every alteration of cryptocurrency block chains requires that all nodes of the corresponding networks update their public ledgers to remain synchronized (Youssefzadeh, 2014). Further, over time, the declining rate of cryptocurrency discovery, growing hardware requirements, and diminishing profits may weaken cryptocurrencies security (Andolfatto, 2018; Grushack, 2014; Sompolinsky & Zohar, 2015). Even though a considerable number of measures are taken to ensure the security of cryptocurrency transactions, there remain threats and vulnerabilities that must be analyzed and researched.

Bitcoin Wallets

In any Bitcoin application, the user wallet holds a public/private key pair, which most closely corresponds to a user's account. These keys must be protected at all times. There are multiple types of Bitcoin wallets including: software, hardware, brain, paper, and online wallets. Software wallets are one of the most common types that are required to use a local running Bitcoin. The Bitcoin developer team releases a reference implementation of the Bitcoin protocol (bitcoin.org). It is a full client, which processes the whole block chain.

A hardware wallet utilizes a separate device that usually operates offline. Since the device is not directly connected to any network, it is much harder for an attacker to gain access to it (Bamert, Decker, Wattenhofer, & Welten, 2014). A paper wallet stores the keys holding coins as a physical document, similar to cash money. A brain wallet stores keys in the user's mind by requiring him/her to memorize a passphrase. The passphrase is converted into a private key, from which the public key and the Bitcoin address are derived. To avoid dictionary or brute-force attacks; the phrase must be sufficiently long and have a high level of entropy.

Multi-signature transactions can be used to increase the security of Bitcoin wallets. For example, BitGo (bitgo.com) offers online wallets with 2-of-3 multi-signature transactions. In the same manner, threshold signatures can be used to impede theft and add two-factor security to wallets (Abadi & Brunnermeier 2018; Goldfeder, Bonneau, Felten, Kroll, & Narayanan, 2014; Goldfeder, Gennaro, & Kalodner, 2015). The idea comes directly from secret sharing (Shamir, 1979): the secret, in this case the private key, is split into shares. Any subset equal to or greater than a predefined threshold is able to reconstruct the secret, but any subset that is smaller gains no information about the secret. The main property of threshold signatures is that the key is never revealed. Participants directly construct a signature. This way, similar to multi-signatures, threshold signatures can be used to require m-of-n shares in order to sign a transaction (Goldfeder et al., 2014; Goldfeder et al., 2015).

In order to secure transactions, the Bitcoin protocol makes extensive use of elliptic curve cryptography (Koblitz, 1987; Miller, 1985). For transaction signatures, the elliptic curve digital signature algorithm (ECDSA) as standardized by NIST (Gallagher & Kerry, 2013) is used, parametrized by the secp256k1 curve defined by Brown (2010). For example, consider the standard P2PKH transaction script. The user needs to provide both his/her public key and signature to prove ownership. To provide a signature, the user chooses a per-signature random value, which must be kept secret and must not be used for more than one transaction. Otherwise, the secret key can be computed from the signature. Even partially bit-wise equal random values suffice to derive the private key (Wang, Wu, & Tan, 2006).

Double Spending

Bitcoin was originally suited for slow-pay monetary transactions, where the seller delivers the goods only after the network confirms the validity of the transaction. Now, it is possible to use Bitcoin for fast-pay scenarios, where fast delivery of the goods is expected after payment. Examples include ATM cash withdrawals, vending machines, and other point-of-sale operations. Recent research has shown that double-spending attacks on fast payments are possible, and they are less costly to the attacker than originally believed (Andrychowicz, Dziembowski, Malinowski, & Mazurek, 2015b; Bamert, Decker, Elsen, Wattenhofer, & Welton, 2015; Karame et al., 2012; Karame, Androulaki, Roeschlin, Gervais, & Capkun, 2015; Meirs, Garman, Green, & Rubin, 2013).

Double-spending attacks result in businesses losing money, with costs being passed on to customers. In some cases, businesses may limit the amount for fast-pay transactions, or even decline the use of Bitcoins. Researchers have proposed a number of countermeasures to minimize these attacks (Bamert et al., 2015; Karame et al., 2012; Karame et al., 2015). Examples of these countermeasures include communicating double-spending warnings between peers and blocking incoming connection requests.

Transaction Malleability

Transaction malleability is a bug in the Bitcoin protocol that changes the transaction identifier without voiding the transaction itself. As stated previously, each transaction includes references to previous transactions (inputs) with a corresponding redeem script (scriptSig) and specifies one or multiple destinations (outputs). Each transaction can be uniquely identified by its identifier (TXID), which is a hash of the transaction data, including the redeem script(s). The signature in a script, however, does not cover the same data as the hash, which forms the TXID. In particular, it does not cover the redeem script. The idea of this design is that signing the script would indicate signing the signature itself, which is noticeably impossible. Instead, during signature generation, Bitcoin replaces all redeem scripts with a dummy script consisting of a single OP_0 (pushing an empty string onto the stack). However, this exchange is not utilized when calculating the TXID.

A transaction-malleability attack can be considered a form of double spending (Decker & Wattenhofer, 2014). However, the attacker is the receiving, not the spending, party. Malleable transaction must be complete and the exchange must check for TXIDs only. Researchers have designed a malleability-resilient refund transaction, which does not require protocol modifications (Andrychowicz, Dziembowski, Malinowski, & Mazurek, 2015a). It is clear that an exchange is a point of failure in a highly distributed environment. As a result, additional methods are utilized, like the use of cloud providers, to protect from distributed denial of service (DDoS) attacks (Vasek, Thornton, & Moore, 2014).

Pooled Mining

It is always desirable that Bitcoin maintains a case in which no single entity controls more than half of the computational power. However, if a single user succeeds in being the first to verify a block and finds a valid nonce, the payoff is always a significant reward. The probability of this scenario is very low, so users group together to create mining pools where multiple miners contribute to the block generation. Every miner explores parts of the nonce space for a valid value. If anyone in the group succeeds, they will split the profit with the group (Rosenfeld, 2011).

Mining pools raise significant security concerns since they combine the computational power of miners who should be working individually. A look at the blockchain reveals that multiple consecutive blocks are mined by one mining pool group. Big mining pools such as GHash.io approached the critical threshold of 51% of the hash rate of the network (Orecchini F. et al., 2019 and Eyal & Sirer, 2014). If one mining pool group cannot reach the critical threshold, they partner with other groups to achieve their goal. However, rogue miners can achieve an unfairly high share by attacking the mining process and weakening opponents to gain higher rewards (Vasek et al., 2014).

There are a number of solutions to overcome the issue of pooled mining. Miners know their interest lies in keeping the distributed Bitcoin system intact. Consequently, miners need to move between mining pools to reallocate the computing power. Experiences utilizing this solution proved successful, as calls by the Bitcoin community to switch pools were perceived and the mining pools groups themselves started to support this movement (Hajdarbegovic, 2014).

Cryptographic Security

Bitcoin is based on the 256-bit Secure Hashing Algorithm (SHA-256), which is widely used for information encryption due to its computational security and robustness against efforts to tamper with the resultant cryptographic keys (Chait, 2015; Grushack, 2014). This encryption algorithm translates the original alphanumeric sequence into an encoded message by adjoining additional bits of information to the initial sequence of data. This creates a series of 512-bit extended messages and stores encryption-related coding information in a related sequence of 32-bit messages. Afterwards, the data are encrypted using a primary number based algorithm that concatenates the original information to the encoded data. It then records the subsequent transformation of the data based on the SHA-256 encryption protocol (BitLegal, 2016; Hendrickson, Hogan, & Luther, 2016).

The result of these algorithmic data manipulations are mathematically unique blocks of encrypted information, also referred to as keys. The keys correspond to certain arbitrary conditions of the digital currency protocol requirements. These cryptographic requirements are designed to significantly increase the computational difficulty of tampering with publicly circulated data sequences (Chait, 2015; Grushack, 2014).

All parts of the encoded data sequences that carry the information related to digital currencies and their transactions are algorithmically connected to randomly encrypted blocks of control data. Any unauthorized changes to the encrypted information will result in the rejection of these data blocks to ensure the computational security of the transactions. These cryptographic keys need to be simultaneously present for a transaction to take place. The simultaneous presence of these encrypted data blocks is the only way to verify the authenticity of digital currency transactions (Youssefzadeh, 2014).

The algorithm that is used with Bitcoins to verify the authenticity of transactions is the Elliptic Curve Digital Signature Algorithm (ECDSA). The ECDSA creates a public key from a prime number q; an elliptic curve $E \mod q$; a random point G of an arbitrary prime order p; and a point Q that is a function of dG, where d is the private key of the Bitcoin user (Youssefzadeh, 2014). Given that the cryptographic key calculations involve a single large prime number p, the base point G in compressed hexadecimal form, a reduction of a number *mod q* to an integer between 0 and q-1, and a mathematical mapping from q to p, the resultant private key d is a random number *modulo p* that can serve as a secure digital signature (Pflaum & Hateley, 2014). By computing Q as $G*d$, it is possible to define a public key using Q, p, E, and G, in the process of which a cryptographic signature is created, based on the SHA-256 algorithm that uses the number *modulo p* (Youssefzadeh, 2014).

The result of that operation is a 64-bit integer, z. To ensure the cryptographic security of this digital signature, for every private key, a random number k ranging from 0 to p-1 is generated. Based on this random number k, a function kG is computed. Afterward, the following equations are used to calculate the values r and s, which are used to check the validity of the digital signature (Youssefzadeh, 2014):

$$r = \phi\left(x\left(k\left(G\right)\right)\right) \in F_p \text{ and } s = \frac{z + rd}{k} \in F_p$$

Given that, according to the above, r is a product of the function $kG \bmod p$, and s equals $(z+rd)/k \bmod p$, it is possible to validate the digital signature without direct access to the private key. This will ensure the security of the cryptographic algorithm. The validation of the authenticity of this digital signature is performed by computing the value C as it is defined in the following formula (Youssefzadeh, 2014):

$$C = \frac{z}{s}G + \frac{r}{s}Q \in E(F_q)$$

According to this cryptographic algorithm, because the digital signature is valid if $\phi(x(C)) = r$, it is possible to indicate the following proof of validity of the digital signature. Provided that the cryptographic keys r and s are valid, it is possible to assume that $s = (z + rd)/k$ so that $k = (z + rd)/s$. Therefore, according to Youssefzadeh:

$$kG = \frac{z + rd}{s}G = \frac{z}{s}G + \frac{rd}{s}G = \frac{z}{s}G + \frac{r}{s}Q = C$$

The importance of recalculating a random value k for every transaction that is digitally signed also follows from this relation. Otherwise, it would be possible to derive the private key from the digital signature components that use identical k values. It is important to note that the cryptocurrency platform Ethereum uses the same ECDSA encryption algorithm that is used for Bitcoin transactions.

Another layer of cryptographic security is related to the multi-level data structure of Bitcoin. The security keys comprise the transaction timestamp and the cryptographic key of the previous data block. Additionally, the security keys contain the root key of the information tree related to the transactions performed with this unit of digital currency. This cryptographic approach employs the Merkle tree method of relating data blocks to each other within hierarchically organized information flows. This approach ensures that malicious attempts to conduct counterfeit transactions within the Bitcoin network are detected (Buterin, 2014).

Privacy and Anonymity

User anonymity is the major factor ensuring the privacy of any online transaction. The Bitcoin network is considered to be anonymous if an attacker is not able to identify the payer or payee within the Bitcoin user community (Fabian, Ermakova, & Sander, 2016). However, there is a belief that Bitcoin is fully anonymous, which is a result of a misunderstanding of the structure of the currency (Brito & Castillo, 2013). Actually, anonymity was not a major design objective of the Bitcoin system (Reid & Harrigan, 2011). All Bitcoin transactions are openly visible in the block chain, but instead of the user's name, his/her public address appears as a pseudonym (Ober, Katzenbeisser, & Hamacher, 2013).

Some researchers stated that unlinkability is a measure to quantify privacy in Bitcoin (Androulaki, Karame, Roeschlin, Scherer, & Capkun, 2013; Ren et al., 2018). Privacy can be increased by linking less personal data to the pseudonym, and each user should use multiple pseudonyms for his own identity. According to Nakamoto (2008), some linking cannot be avoided since multi-input transactions belong to the same owner. The risk of anonymity occurs when the owner of one public key is known, in which case linking could result in exposing other transactions that belong to the same user.

Reid and Harrigan argued that the fundamental limits of anonymity stem from the blockchain technology itself. The publicly available ledger of all confirmed transactions is necessary to avoid double spending since there is no central authority. All addresses used as an input for one transaction can be mapped to one individual, as a user must retain all of the addresses for multi-input transactions. An attacker might assemble all such addresses belonging to the same user into a single user network (Kaminsky, 2011; Meiklejohn et al., 2013; Ober et al., 2013; Ron & Shamir, 2013).

The Bitcoin network might also be de-anonymized through mapping of Bitcoin addresses to Internet Protocol (IP) addresses. It works by associating an IP address with a particular Bitcoin address obtained from the block chain. Knowing which device a Bitcoin address originated from is clearly a breach of user anonymity. Inspired by Kaminsky (2011), Koshy, Koshy, and McDaniel (2014) analyzed the risks of mapping Bitcoin addresses directly to IP data. They built a Bitcoin client to collect the data of 5.6 million transactions.

Bitcoin experts agree on a number of best practices to maintain user privacy in the Bitcoin network. One major factor is the secure storage of private keys. The user wallet is usually kept in a device that communicates with the rest of the Bitcoin network through the Internet (Franco, 2015). This storage method is known as an online, Web, or hot wallet. Since these wallets are vulnerable to online attacks, users are advised to keep only small amounts of Bitcoins in their web wallet (Barski & Wilmer, 2015; Franco, 2015). A Web wallet is less secure because an external Internet provider manages the funds (Franco, 2015; Hong & Kim, 2019).

A safer way to store Bitcoins is in cold storage, which is an offline device that is not accessible via the Internet and is recommended for large amounts of Bitcoins (Barski & Wilmer, 2015). Cold storage methods include flash drives, paper wallets, and hardware wallets. Although hardware wallets cost more, experts expect them to become the dominant method of Bitcoin storage in the future (Antonopoulos, 2015).

CRYPTOCURRENCIES AND OPEN-ACCESS CRYPTOGRAPHIC PLATFORMS

The idea of decentralized cryptocurrencies was proposed in the 1980s and 1990s. However, those efforts failed because of their reliance on centralized intermediaries and trusted servers to secure their networks. Owing to the invention of Bitcoin, a technical solution has been found that enables a decentralized consensus between at least three-quarters of cryptocurrency network participants about the authenticity of its transactions. At the same time, this solution renders digital currency services inherently vulnerable to takeover attempts by external agents with computer processing power that exceeds 51% of the computational capacity of a cryptocurrency network. This can enable digital currency double spending, prevent transaction confirmation, and block digital coin discovery (Buterin, 2014; Gruber, 2013; Rahouti, Xiong, & Ghani, 2018).

The decentralized cryptocurrency network manages the results of this process. The network utilizes digital signatures to verify coin creation and transactions. In addition, the signatures guarantee integrity through the simultaneous and distributed processing of public ledgers. Also referred to as block chains, these public ledgers record digital currency operations and transactions (Chait, 2015; Grushack, 2014). The mathematical problems serve as means of computationally creating basic units of digital currencies, or blocks. The mathematical properties of the algorithms used for issuing digital currency units progressively decrease the amount of crypto-coins that can be created. This sets an absolute limit on the

Figure 3. Conceptual overview of a Bitcoin transaction (BitLegal, 2016)

number of digital coins that can be generated (De Filippi, 2014; Grushack, 2014; Spethoven, 2019). For blocks of data to be accepted by the cryptocurrency network, complex calculations need to be performed.

The cryptographic algorithms and peer-to-peer communication technologies of digital currencies enable both the security of monetary transactions and the anonymity of their participants. This is possible to achieve without requiring that prior trust relations exist between transacting parties (Chait, 2015; Nakamoto, 2008). However, cryptocurrency transactions are not entirely anonymous, due to the nature of the cryptographic protocols they are based on. It is possible to record, verify, and trace all transactions to the network addresses and specific user identities involved (Brito & Castillo, 2013; Jeans, 2015; Grushack, 2014). Although cryptocurrency protocols do not specifically incorporate anonymization features, various digital money-laundering services, such as BitMixer, can hide the identity of cryptocurrency users.

The initial response to digital cryptocurrencies, created independent of the conventional financial system, has been skeptical (Grinberg, 2012). The main reason for this is that digital currencies are not intrinsically redeemable for or convertible into gold, services, or commodities. Many governments do not recognize digital cryptocurrencies as legal tender, since they seek to maintain their authority over the creation of money by fiat or law. However, at present, most national currencies are not redeemable for gold either (De Filippi, 2014; Grushack, 2014).

The reward for creating a single block of the cryptographic data on which Bitcoin is based is 25 Bitcoins. This reward is expected to decrease by half every four years, in which approximately every 210,000 cryptographic blocks are mined. This process will be repeated until the maximum amount of these digital currency units is generated (Youssefzadeh, 2014). These limitations may motivate the development of alternative digital cryptocurrencies with different cryptographic foundations. There is always a need to balance cryptographic security with a decentralized operation. This resulted in Bitcoin becoming the digital currency that has been used the longest (Gruber, 2013).

DISCUSSION

Although there was a low possibility of cryptocurrencies gaining wide acceptance, since 2012, the value of Bitcoin has risen dramatically to over 200 USD. Bitcoin is not only accepted as a form of payment by some online retailers, but also serves as a prototype for other cryptocurrencies, such as Peercoin and Litecoin (Adolfatto, 2018; De Filippi, 2014; Grushack, 2014; Hong & Kim, 2019; Jeans, 2015). Alternative systems of online currency exchange, such as Bitstamp, provide opportunities for the acquisition and resale of digital currencies (Karlstrom, 2014). Currently a large number of intermediaries, such as BitPay, convert cryptocurrencies into fiat currencies at low fees. This resulted in a growing number of merchants accepting payments in cryptocurrencies. This increasing acceptance stems from the fact that the intermediation of digital currency transactions minimizes the risks associated with the high rate of fluctuation in value (De Filippi, 2014; Grushack, 2014).

Cryptocurrency transactions involve a higher level of financial risk than intermediated payments in traditional currencies via wire transfer, PayPal, or credit card. This is because cryptocurrency transactions cannot be reversed. Such transactions can only be voided or refunded by the individual who receives the funds. Despite value fluctuations, decentralized cryptocurrencies have reached an overall market capitalization in excess of 7 billion USD and gained growing acceptance as Bitcoin automatic teller machines (ATMs) are deployed in the United States, Canada, and Germany (De Filippi, 2014; Grushack, 2014; Jeans, 2015). By September 2016, the daily transaction value was over 100 billion USD, and the total market value of all Bitcoins in circulation was about 14 billion USD (Coindesk, 2017; Hong & Kim, 2019).

Despite the growing acceptance of cryptocurrencies, the difficulty of tracking the identity of users poses law enforcement and government challenges. Since anonymized websites accepting transactions in Bitcoins can facilitate the illegal trade of drugs or weapons, some countries have raised the issue of banning cryptocurrencies. Cryptocurrencies can be used for financial fraud, illegal transactions, and money-laundering operations. Owing to the degree of anonymity that cryptocurrencies offer, law enforcement agencies find it difficult to identify criminals operating via encrypted online networks. Given that cryptocurrency transactions do not require the disclosure of the identity of the transacting parties, digital currencies can also assist in tax evasion (Adolfatto, 2018; Chait, 2015; Gruber, 2013).

Since governments have no control of cryptocurrencies, their decentralized circulation may pose a threat to the implementation of national financial and economic policies (Twomey, 2013). Digital cryptocurrencies are impervious to governmental interventions to their market valuation. This is because their underlying cryptographic protocols determine the overall amount of digital money allowed to be in circulation (De Filippi, 2014; Grushack, 2014). For example, in total, there will be no more than 21 million Bitcoins in circulation after all of these digital currency units are discovered (Youssefzadeh, 2014). In countries such as Brazil, the recognition and regulation of cryptocurrencies for monetary transactions may contribute to their credibility, trading volumes, and use as financial assets (Carstens, 2019; De Filippi, 2014; Luther, 2013; Pedro, 2016).

Some countries, such as Canada, Germany, Singapore, China, and India have restricted cryptocurrencies to commodities that can be exchanged without recognizing them as legal tender. This has precluded the circulation of these cryptocurrencies, even though individuals retain their rights to purchase or sell them. Russia's central bank has discouraged local financial and other institutions from processing transactions in digital cryptocurrencies. However, India, Thailand, and Indonesia have taken no steps to explicitly prohibit transactions in cryptocurrencies such as Bitcoin. In contrast, numerous countries have prohibited foreign exchange transactions involving digital cryptocurrencies for fear of uncontrolled flows of capital. The European Union's legislation does not recognize digital cryptocurrencies as a valid form of monetary exchange. This prevents financial institutions from issuing digital currencies due to the risks associated with trading in them (De Filippi, 2014; Pflaum & Hateley, 2014; Wu & Xu, 2019).

Nevertheless, certain countries, such as Israel and Singapore, are already promoting guidelines for the taxation of cryptocurrency transactions. The financial regulatory frameworks, such as those planned for Australia and Canada, are aimed at preventing tax evasion and money laundering. In the United States, profits and losses from cryptocurrency transactions are considered capital gains for taxation purposes. European Union countries are not prevented from introducing their own cryptocurrency transactions frameworks, such as value-added taxation. However, apart from Denmark, European countries have yet to develop national frameworks to organize digital currencies. In recent years, the United Kingdom and Germany have made efforts to propose special tax treatment of cryptocurrency income, charges, and transactions (Abadi & Brunnermeier, 2018; De Filippi, 2014; Jeans, 2015; Pflaum & Hateley, 2014; Wu & Xu, 2019).

Because of a lack of regulation for exchange platforms and financial institutions dealing in cryptocurrencies, consumers using digital currencies continue to be exposed to significant risks (De Filippi, 2014; Grushack, 2014). In recent years, online cryptocurrency platforms have experienced losses in excess of 375 million USD or filed for bankruptcy because of hacker attacks. The risks involved with digital cryptocurrency trading frequently result from improper security configurations, negligent digital transaction management, and insufficient security measures (Grushack, 2014; Moore & Christin, 2013).

National or international regulatory frameworks can mitigate the risks associated with using digital cryptocurrencies, digital cryptocurrency networks, and their platforms. However, the Bitcoin Foundation may be capable of intermediary action vis-à-vis regulatory, legislative, and public bodies. Non-profit organizations can coordinate, promote, and guide the development of cryptocurrency protocols in view of the economic benefits they can bring. With the increasing acceptance of digital payment systems in many countries, digital currency platforms for money transfer, mobile payment, and remittance transactions are gaining ground around the world (Kim, 2018; Wu & Xu, 2019; Youssefzadeh, 2014).

Cryptocurrencies pose unique regulatory, security, and cryptographic challenges (Gruber, 2013), the nature of which stems from their open-source, distributed, and decentralized foundations (Fairfield, 2014). Striking a balance between the cryptographic security of digital currencies and their regulatory oversight by legal and governmental institutions is essential. This will result in calling not only for technical solutions, such as more secure encryption algorithms, but also innovative frameworks for their internal and external oversight (Gruber, 2013). Digital cryptocurrencies lack the regulatory mechanisms that conventional currencies backed by governments are associated with (Jeans, 2015; Youssefzadeh, 2014). Governmental authorities seeking to prevent financial fraud, illicit activities, or tax evasion will continue to monitor digital cryptocurrency transactions (Gruber, 2013; Youssefzadeh, 2014). Bitcoin remains an algorithmic prototype for alternative cryptocurrencies that typically utilize its blockchain infrastructure for their transactions (Buterin, 2014; Karlstrom, 2014). At the same time, Bitcoin and other cryptocurrencies remain far from posing a competitive threat to credit card companies that habitually process almost 2,000 transactions per second (Sompolinsky & Zohar, 2014).

The decentralized nature of cryptocurrency networks may also be significantly reduced, as the volume of transactions constantly increases because of the growing size of cryptocurrency blocks of information. Large-scale currency-mining and transaction-processing operators may take advantage of these decentralized networks. Digital currency users increasingly tend to operate with truncated blocks of cryptographic information, which decreases the security of these distributed networks (Buterin, 2014; Koblitz & Menezes, 2016; Youssefzadeh, 2014). As opposed to traditional financial intermediaries, digital currencies do not necessarily represent a more secure means of monetary exchange, especially due to multiple recorded cases of breaches of trust (Barski & Wilmer, 2015; Lustig & Nardi, 2015). Thus, this research indicates that cryptocurrencies in general and Bitcoin in particular can be expected to become ever more embedded in financial regulation frameworks rather than displacing conventional, centralized means of monetary exchange (Karlstrom, 2014; Kim, 2018; Koblitz & Menezes, 2016).

As for the role of user trust in the Bitcoin network, it is clear that most aspects of trust are related to the nature of digital cryptocurrencies. This includes payment methods and other considerations related to the current legal concerns. These concerns stem from the fact that digital cryptocurrencies are not regulated, but at the same time, they are not banned in most countries around the world. The influence of the legal status may change fundamentally, and this would cause a significant change in consumer trust (Hong & Kim, 2019; Mercator Research Group, 2015; Tschorsch & Scheuermann, 2016).

Currently, cryptocurrencies are taxable property in the United States, Germany, and many other countries. Unlike most financial entities that were originally created in a specific country, with a specific regulatory framework, digital currencies exist entirely online. Trust changes based on different aspects of digital currencies' transactions. For example, it is somewhat risky for a business making a Bitcoin payment to another business or a consumer making a purchase with Bitcoins (Andolfatto, 2018; Mercator Research Group, 2015; Spethoven, 2019; Tschorsch & Scheuermann, 2016; Zarifis, Cheng, Dimitrious, & Efthymious, 2015; Zarifis, Efthymiou, Cheng, & Demetriou, 2014; Zohar, 2015).

FUTURE RESEARCH DIRECTIONS

With much research on the investment returns of Bitcoin, there is little knowledge on the true "intrinsic value" of Bitcoin and/or other cryptocurrencies (Luther, 2018). For example, does the Lightning Network that allows for greatly increased transaction speed with a reduction in cost and its implementation reflect on the value and price of cryptocurrencies (Mendling et al., 2018). Other fundamentals of cryptocurrencies include their decentralized nature; their ability to send monetary value over the Internet (Internet of Value); their potential to be more stable and acceptable than fiat currencies; their transparency and security against fraud because of blockchain; that they are an investment tool; that they have a public ledger; and other related value propositions.

Related to the above—do cryptocurrencies have an intrinsic value at all (Berentsen & Schär, 2018)? What are the arguments for the long-term viability and outlook for cryptocurrencies, and what are the arguments for the inevitable failure of cryptocurrencies? Are the cryptocurrencies destined for a bumpy ride through pricing bubbles and turbulence, or will the price be dictated by the so-called "fundamentals" (Corbet, Lucey, Urquhart, & Yarovaya, 2018)?

Considering Bitcoin's ten-year existence, what are the traditional trading patterns of securities that can predict the price of cryptocurrencies (Hong, 2017)? Do cryptocurrencies actually follow their own pattern of value behavior, or can these be analyzed as other securities? This area needs thorough investigation; as practitioners tend to emphasize comparisons to gold, shares, and other stores of value, this research is still in its early stage.

From the exploratory analyses carried out throughout this chapter, we can clearly see that the only cryptocurrency mentioned systematically is Bitcoin. With a reduction in the stringency of filter parameters, Ethereum, Ripple, XRP, Litecoin, and a small number of others are also part of the field.

This leads to the conclusion that other cryptocurrencies are severely under researched, despite amounting to almost half of the entire market capitalization of all cryptocurrencies. For example, XRP cryptocurrency market capitalization was third, after Ethereum, as of July 2019 (CoinMarketCap, 2019), yet received no considerable attention in the literature.

Furthermore, there is little research comparing the influences and consequent dynamics of various cryptocurrencies. Cryptocurrencies have diverse programming chains. For example, while Bitcoin blockchain stores information related to all past transactions, Ethereum is built on Ether, which, on top of storing information on accounts and transactions, also executes smart contracts through stored programming logic (Chen, 2018). Cryptocoins are generally not considered competitors; rather they share a co-existence and various capabilities that allow for the use of blockchains (Adhami, Giudici, & Martinazzi, 2018). There needs to be further research into the fundamentals of these coins, in order to inform and educate businesses about their use.

CONCLUSION

This chapter presented details of the technical specifications of the Bitcoin algorithms, specifically in terms of security-related aspects. It describes the major system components and how they interact to produce a complete financial transaction. The chapter also sheds light on the extent to which countries around the world are adopting the paradigm. However, there are no clear solutions for Bitcoin security risks, such as the price volatility or technical vulnerabilities of the currency. Institutions that are seeking to participate in the Bitcoin economy must take into consideration a wide range of risk factors. Bitcoin and other crypto-currencies remain far from posing a competitive threat to credit card companies that process almost 2,000 transactions per second. Digital currency networks can process only several transactions per second. Overall, to compete with traditional monetary systems, Bitcoin requires more efforts to enhance its prospects based on its technical implementations.

REFERENCES

Abadi, J., & Brunnermeier, M. (2018). *Blockchain economics* (Working paper). Princeton University.

Adhami, J., Giudici, G., & Martinazzi, S. (2018). Why do businesses go crypto? An empirical analysis of initial coin offerings. *Journal of Economics and Business*, 20–32.

Andolfatto, D. (2018). Blockchain: What it is, what it does, and why you probably don't need one. *Federal Reserve Bank of St. Louis Review, 100*(2).

Androulaki, E., Karame, G. O., Roeschlin, M., Scherer, T., & Capkun, S. (2013). Evaluating user privacy in Bitcoin. In A. R. Sadeghi (Ed.), *Proceedings of the 17th International Conference on Financial Cryptography and Data Security, LNCS 7859* (pp. 34-51). Springer.

Andrychowicz, M., Dziembowski, S., Malinowski, D., & Mazurek, L. (2015a). On the malleability of bitcoin transactions. In M. Brenner, N. Christin, B. Johnson, & K. Rohloff (Eds.), Lecture notes in computer science: Vol. 8976. *Financial cryptography and data security (FC 2015)* (pp. 1–18). Springer. doi:10.1007/978-3-662-48051-9_1

Andrychowicz, M., Dziembowski, S., Malinowski, D., & Mazurek, U. (2015b). On the malleability of bitcoin transactions. *Proceedings of the 2015 Conference on Financial Cryptography and Data Security (FC 2015)*. doi: 10.1007/978-3-662-48051-9_1

Antonopoulos, A. M. (2015). *Mastering Bitcoin*. O'Reilly Media.

Bamert, T., Decker, C., Elsen, L., Wattenhofer, R., & Welton, S. (2015). Have a snack, pay with bitcoins. *Proceedings of the 13th IEEE International Conference on Peer-to-Peer Computing*. doi: 10.1109/P2P.2013.6688717

Bamert, T., Decker, C., Wattenhofer, R., & Welten, S. (2014). Bluewallet: The secure bitcoin wallet. In *Proceedings of the 10th International Workshop Security Trust Management*, (pp. 65-80). Academic Press.

Barski, C., & Wilmer, C. (2015). *Bitcoin for the befuddled*. No Starch Press.

Beigel, O. (2019). *Bitcoin historical price & events*. Retrieved from https://99bitcoins.com/bitcoin/historical-price/

Berentsen, A., & Schär, F. (2018). *A short introduction to the world of cryptocurrencies*. Academic Press.

BitLegal. (2016). *BitCoin legal status around the world*. Retrieved May 9, 2016 from http://bitlegal.io/; http://money.cnn.com/interactive/technology/where-is-bitcoin-legal/

Bos, J. W., Halderman, J. A., Heninger, N., Moore, J., Naehrig, M., & Wustrow, E. (2014). Elliptic curve cryptography in practice. *Proceedings of the 18th International Conference on Financial Cryptography and Data Security (FC'14)*, 157-175.

Brito, J., & Castillo, A. (2013). *Bitcoin: A primer for policymakers*. Mercatus Center and George Mason University.

Brown, D. R. L. (2010). *SEC 2: Recommended elliptic curve domain parameters, standards for efficient cryptography group (SECG)*. Retrieved March 22, 2017 from http://www.secg.org/sec2-v2.pdf

Buterin, V. (2014). *A next-generation smart contract and decentralized application platform* (White paper). Toronto: Ethereum.

Carstens, A. (2019). Money in the digital age: What role for central banks? In Central Bank of Ireland. Whitaker Lecture.

Chait, G. (2015). *Distributed crypto identity as a mechanism for legal empowerment of the poor and stimulating local economic development* (Working paper). Sidney: Whythawk.

Chen, Y. (2018). Blockchain tokens and the potential democratization of entrepreneurship and innovation. *Business Horizons*, *61*(4), 567–575. doi:10.1016/j.bushor.2018.03.006

Coindesk. (2017). *Bitcoin - Daily number of transactions*. Retrieved January 10, 2017 from http://www.coindesk.com/data/bitcoin-daily-transactions/

CoinMarketCap. (2019). *Top 100 cryptocurrencies by market capitalization*. Retrieved August 8, 2019 from https://coinmarketcap.com/

Corbet, S., Lucey, B., Urquhart, A., & Yarovaya, L. (2018). *Cryptocurrencies as a financial asset: A systematic analysis*. International Institute of Financial Analytics Journal.

De Filippi, P. (2014). Bitcoin: A regulatory nightmare to a libertarian dream. *Internet Policy Review*, *3*(2), 43. doi:10.14763/2014.2.286

Decker, C., & Wattenhofer, R. (2014). Bitcoin transaction malleability and MtGox. *Proceedings of the 19th European Symposium for Computer Security Research (ESORICS'14)*. Retrieved July 10, 2016 from https://link.springer.com/chapter/10.1007/978-3-319-11212-1_18

Eyal, I., & Sirer, E. G. (2014). *How to disincentivize large Bitcoin mining pools*. Retrieved January 6, 2017 from http://hackingdistributed.com/2014/06/18/how-to-disincentivize-large-bitcoin-mining-pools/

Fabian, B., Ermakova, T., & Sander, U. (2016). Anonymity in Bitcoin – The users' perspective. *37th International Conference on Information Systems*.

Fairfield, J. (2014). Smart contracts, Bitcoin bots, and consumer protection. *Washington and Lee Law Review Online*, *71*, 35–50.

Franco, P. (2015). *Understanding Bitcoin: Cryptography, engineering and economics*. John Wiley & Sons.

Gallagher, P., & Kerry, C. (2013). *Federal Information Processing Standards (FIPS) Publication 186-4: Digital Signature Standard (DSS)*. Retrieved March 20, 2017 from http://nvlpubs.nist.gov/nistpubs/FIPS/NIST.FIPS.186-4.pdf

Gervais, A., Karame, G. O., Capkun, V., & Capkun, S. (2014). Is Bitcoin a decentralized currency? *IEEE Security and Privacy*, *12*(3), 54–60. doi:10.1109/MSP.2014.49

Goldfeder, S., Bonneau, J., Felten, E. W., Kroll, J. A., & Narayanan, A. (2014). Securing bitcoin wallets via threshold signatures. *International Technical Review*. Retrieved June 20, 2016 from http://www.cs.princeton.edu/~stevenag/bitcoin_ threshold_signatures.pdf

Goldfeder, S., Gennaro, R., & Kalodner, H. (2015). *Securing bitcoin wallets via a new DSA/ECDSA threshold signature scheme* (Technical report). Retrieved February 5, 2017 from http://stevengoldfeder.com/papers/threshold_sigs.pdf

Grinberg, R. (2012). Bitcoin: An innovative alternative digital currency. *Hastings Science & Technology Law Journal*, *4*, 159–174.

Gruber, S. (2013). Trust, identity and disclosure: Are bitcoin exchanges the next virtual havens for money laundering and tax evasion. *Quinnipiac Law Review*, *32*, 135–208.

Grushack, J. (2014). *Currency 3.0: Examining digital crypto currency markets* (Doctoral dissertation). Department of Computer Science and Economics, Union College, Schenectady, NY.

Hajdarbegovic, N. (2014). *Bitcoin miners ditch Ghash.io pool over fears of 51% attack, Coindesk*. Retrieved December 12, 2016 from http://www.coindesk.com/bitcoin-miners-ditch-ghash-io-pool-51-attack/

Hendrickson, J., Hogan, T., & Luther, W. (2016). The political economy of bitcoin. *Economic Inquiry*, *54*(2), 925–939. doi:10.1111/ecin.12291

Hong, K. H. (2017). Bitcoin as an alternative investment vehicle. *Information Technology Management*, *18*(4), 265–275. doi:10.100710799-016-0264-6

Hong, S., & Kim, H. (2019). Analysis of Bitcoin exchange using relationship of transactions and addresses. In *Proceedings of 21st International Conference on Advanced Communication Technology (ICACT)* (pp. 67-70). 10.23919/ICACT.2019.8701992

Huhtinen, T. (2014). *Bitcoin as a monetary system: Examining attention and attendance* (Thesis). Department of Finance, School of Business, Aalto University, Aalto, PA.

Jeans, E. (2015). Funny money or the fall of fiat: Bitcoin and forward-facing virtual currency regulation. *Journal on Telecommunications & High Technology Law*, *13*, 99–171.

Kaminsky, D. (2011). Black ops of TCP/IP. In *Proceedings of Black Hat Technical Security Conference*. Retrieved November 15, 2016 from http://de.slideshare.net/dakami/black-ops-of-tcpip-2011-black-hat-usa-2011

Karame, G. O., Androulaki, E., & Capkun, S. (2012). Double-spending attacks on fast payments in bitcoin. *Proceedings of the 2012 ACM Conference on Computer and Communications Security (CCS'12)*. doi: 10.1145/2382196.2382292

Karame, G. O., Androulaki, E., Roeschlin, M., Gervais, A., & Capkun, S. (2015). Misbehavior in bitcoin: A study of double-spending and accountability. *ACM Transactions on Information and System Security*, *18*(1), 1–32. Advance online publication. doi:10.1145/2732196

Karlstrom, H. (2014). Do libertarians dream of electric coins? The material embeddedness of Bitcoin. *Distinktion: Scandinavian Journal of Social Theory*, *15*(1), 23–36. doi:10.1080/1600910X.2013.870083

Kim, S. (2018). A novel Bitcoin mining scheme based on the multi-leader multi-follower Stackelberg Game Model. *IEEE Access: Practical Innovations, Open Solutions*, *1*(6), 48902–48912. doi:10.1109/ACCESS.2018.2867631

Koblitz, N. (1987). Elliptic curve cryptosystems. *Mathematics of Computation*, *48*(177), 203–209. doi:10.1090/S0025-5718-1987-0866109-5

Koblitz, N., & Menezes, A. (2016). Cryptocash, cryptocurrencies, and cryptocontracts. *Designs, Codes and Cryptography*, *78*(1), 87–102. doi:10.100710623-015-0148-5

Koshy, P., Koshy, D., & McDaniel, P. (2014). An analysis of anonymity in Bitcoin using P2P network traffic. In N. Christin & R. Safavi-Naini (Eds.), *Proceedings of the 18th International Conference on Financial Cryptography and Data Security, LNCS 8437* (pp. 469-485). Springer.

Kumaresan, R., & Bentov, I. (2014). How to use bitcoin to incentivize correct computations. *Proceedings of the 2014 ACM SIGSAC Conference on Computer and Communications Security*, 30-41. 10.1145/2660267.2660380

Lischke, M., & Fabian, B. (2016). Analyzing the Bitcoin network: The first four years. *Future Internet, 8*(1), 7. doi:10.3390/fi8010007

Lustig, C., & Nardi, B. (2015). Algorithmic authority: The case of Bitcoin. *48th Hawaii International Conference on System Sciences IEEE*, 743-752.

Luther, W. (2013). *Cryptocurrencies, network effects, and switching costs* (Working paper). Arlington, VA: Mercatus Center and George Mason University.

Luther, W. (2018). Is bitcoin intrinsically worthless? *J. Priv. Enterprise*, (33), 31-45.

Meiklejohn, S., Pomarole, M., Jordan, G., Levchenko, K., McCoy, D., Voelker, G. M., & Savage, S. (2013). A fistful of Bitcoins: Characterizing payments among men with no names. *Proceedings of the 2013 Conference on Internet Measurement (IMC '13)*. 10.1145/2504730.2504747

Meirs, I., Garman, C., Green, M., & Rubin, A. D. (2013). Zerocoin: Anonymous distributed e-cash from bitcoin. *Proceedings of the 2013 IEEE Symposium on Security and Privacy (S&P'13)*. 10.1109/SP.2013.34

Mendling, J., van der Aalst, W., vom Brocke, J., Cabanillas, C., Weidlich, M., Weske, M., ... Zhu, L. (2018). Blockchains for business process management - Challenges and opportunities. *ACM Transactions on Management Information Systems, 9*(1), 1–16. doi:10.1145/3183367

Mercator Research Group. (2015). *Understanding Bitcoin: Blockchain applications beyond payments*. Retrieved December 24, 2016 from https://www.mercatoradvisorygroup.com/Reports/Understanding-Bitcoin--Blockchain-Applications-Beyond-Payments/

Miller, V. S. (1985). Use of elliptic curves in cryptography. *Proceedings of the Fifth Conference of Advanced Cytology*. doi: 10.1007/3-540-39799-X_31

Moore, T., & Christin, N. (2013). Beware the middleman: Empirical analysis of Bitcoin-exchange risk. In A. R. Sadeghi (Ed.), *Financial Cryptography and Data Security: 17th International Conference Revised* (selected papers) (FC 2013) (pp. 25-33). Okinawa, Japan: Springer.

Nakamoto, S. (2008). *Bitcoin: A peer-to-peer electronic cash system* (White paper). Bitcoin Project. Retrieved February 15, 2017 from www.bitcoin.org

Ober, M., Katzenbeisser, S., & Hamacher, K. (2013). Structure and anonymity of the Bitcoin transaction graph. *Future Internet, 5*(2), 237–250. doi:10.3390/fi5020237

Orecchini, F., Santiangeli, A., Zuccari, F., Pieroni, A., & Suppa, T. (2019). Blockchain Technology in Smart City: A New Opportunity for Smart Environment and Smart Mobility. In P. Vasant, I. Zelinka, & G. W. Weber (Eds.), *Intelligent Computing & Optimization. ICO 2018. Advances in Intelligent Systems and Computing* (Vol. 866). Springer. doi:10.1007/978-3-030-00979-3_36

Pedro, F. (2016). *Understanding Bitcoin, Cryptography, Engineering and Economics*. TJ International Ltd.

Pflaum, I., & Hateley, E. (2014). A bit of a problem: National and extraterritorial regulation of virtual currency in the age of financial disintermediation. *Georgetown Journal of International Law*, *45*, 1169–1215.

Rahouti, M., Xiong, K., & Ghani, N. (2018). Bitcoin concepts, threats, and machine-learning security solutions. *IEEE Access: Practical Innovations, Open Solutions*, *3*(6), 67189–67205. doi:10.1109/ACCESS.2018.2874539

Rehman H.., Yafi E., Nazir M., Mustafa K. (2019). Security Assurance Against Cybercrime Ransomware. In *Intelligent Computing & Optimization. ICO 2018. Advances in Intelligent Systems and Computing*, (vol. 866). Springer.

Reid, F., & Harrigan, M. (2011). An analysis of anonymity in the Bitcoin system. In Y. Altshuler, Y. Elovici, A. B. Cremers, N. Aharony, & A. Pentland (Eds.), Security and privacy in social networks (pp. 197-273). New York: Springer. doi:10.1109/PASSAT/SocialCom.2011.79

Ren, Z., Cong, K., Aerts, T., de Jonge, B., Morais, A., & Erkin, Z. (2018). A scale-out blockchain for value transfer with spontaneous sharding. In *Proceedings of Crypto Valley Conference on Blockchain Technology* (vol. 3, pp. 1-10). 10.1109/CVCBT.2018.00006

Ron, D., & Shamir, A. (2013). Quantitative analysis of the full Bitcoin transaction graph. In A. R. Sadeghi (Ed.), *Proceedings of the 17th International Conference on Financial Cryptography and Data Security, LNCS 7859* (pp. 6-24). Springer. 10.1007/978-3-642-39884-1_2

Rosenfeld, M. (2011). Analysis of bitcoin pooled mining reward systems. *Computing Research Repository* (Technical report). Retrieved May 22, 2016 from https://arxiv.org/pdf/1112.4980.pdf

Shamir, A. (1979). How to share a secret. *Communications of the ACM*, *22*(11), 612–613. doi:10.1145/359168.359176

Sompolinsky, Y., & Zohar, A. (2015). Secure high-rate transaction processing in Bitcoin. *Proceedings of Financial Cryptography and Data Security: 19th International Conference*. doi: 10.1007/978-3-662-47854-7_32

Spethoven, A. (2019). Theory and reality of cryptocurrency governance. *Journal of Economic Issues*, *53*(2), 385–393. doi:10.1080/00213624.2019.1594518

Tschorsch, F., & Scheuermann, B. (2016). Bitcoin and beyond: A technical survey on decentralized digital currencies. *IEEE Communications Surveys and Tutorials*, *18*(3), 2084–2123. doi:10.1109/COMST.2016.2535718

Twomey, P. (2013). Halting a shift in the paradigm: The need for Bitcoin regulation. *Trinity College Law Review*, *16*, 67–90.

Vasek, M., Thornton, M., & Moore, T. (2014). Empirical analysis of denial-of-service attacks in the bitcoin ecosystem. In R. Böhme, M. Brenner, T. Moore, & M. Smith (Eds.), Lecture Notes in Computer Science: Vol. 8438. *Financial Cryptography and Data Security. FC 2014* (pp. 57–71). Springer. doi:10.1007/978-3-662-44774-1_5

Wang, H., Wu, A., & Tan, X. (2006). A new secure authentication scheme based threshold ecdsa for wireless sensor network. *Security Management*. Advance online publication. doi:10.1007/978-3-662-45472-5_11

Wu, Y., Luo, A., & Xu, D. (2019). Forensic analysis of Bitcoin transactions. In *Proceedings of IEEE International Conference on Intelligence and Security Informatics (ISI)* (pp. 167-169). Shenzhen, China: IEEE.

Youssefzadeh, S. B. (2014). *Future of Payment Platforms* (Thesis). Department of Electrical Engineering, California Polytechnic State University, San Luis Obispo, CA.

Zarifis, A., Cheng, X., Dimitrious, S., & Efthymious, L. (2015). Trust in digital currency enabled transactions model. *Proceedings of the 9th Mediterranean Conference on Information Systems (MCISE)*. Retrieved March 10, 2017 from http://www.icsd.aegean.gr/group/data.php?usrnm=eloukis&data=1b

Zarifis, A., Efthymiou, L., Cheng, X., & Demetriou, S. (2014). Consumer trust in digital currency enabled transactions. In W. Abramowicz & A. Kokkinaki (Eds.), *Business Information Systems Workshops (BIS 2014). Lecture Notes in Business Information Processing* (Vol. 183, pp. 241–254). Springer International Publishing. doi:10.1007/978-3-319-11460-6_21

Zohar, A. (2015). Bitcoin: Under the hood. *Communications of the ACM*, *58*(9), 104–113. doi:10.1145/2701411

ADDITIONAL READING

Buy Bitcoin Worldwide. (2018). Bitcoin price history chart. Retrieved January 31, 2020, from https://www.buybitcoinworldwide.com/price/

Coin Market Cap. (2016). Crypto-currencies market capitalizations. Retrieved May 20, 2018, from http:// coinmarketcap.com/

Governance, I. T. (2018). Software capability maturity model (CMM). Retrieved May 20, 2019, from https://www.itgovernance.co.uk/capability-maturity-model

Remit, C. (2017). 8 key factors that affect foreign exchange rates. Retrieved May 20, 2018, from https://www. compareremit.com/ money-transfer-guide/key-factors-affecting-currency-exchange-rates/

Ross, A. (2016). 12 cryptocurrency alternatives to bitcoin. Retrieved March 6, 2020, from http://www.bankrate.com/ finance/investing/cryptocurrency-alternatives-to-bitcoin-1.aspx

Source Forge Homepage. (2018). Retrieved March 6, 2020, from www.sourceforge.net

Umm-e-Laila Zahoor, A., Mehboob, K., & Natha, S. (2017). Comparison of open source maturity models. *Procedia Computer Science*, *111*, 348–354. doi:10.1016/j.procs.2017.06.033

Van Alstyne, M. (2014). Why bitcoin has value. *Communications of the ACM*, *57*(5), 30–32. doi:10.1145/2594288

Villamena, V. (2018). Everything you need to know about bitcoin and your taxes. Retrieved May 20, 2019, from https://www.cnbc.com/2018/02/21/everything-you-need-to-know-about-bitcoin-and-your-taxes.html

KEY TERMS AND DEFINITIONS

Bitcoin: A type of digital currency in which a record of transactions is maintained, and new units of currency are generated by the computational solution of mathematical problems, and that operates independently of a central bank.

Blockchain: A system in which a record of transactions made in Bitcoin or another cryptocurrency is maintained across several computers that are linked in a peer-to-peer network.

Cryptocurrency: A digital currency in which encryption techniques are used to regulate the generation of units of currency and verify the transfer of funds, operating independently of a central bank.

Cryptographic Algorithm: The means of altering data from a readable form (also known as plaintext) to a protected form (also known as ciphertext), and back to the readable form. Changing plaintext to ciphertext is known as encryption, whereas changing ciphertext to plaintext is known as decryption.

Cryptographic Security: The study of secure communications techniques that allow only the sender and intended recipient of a message to view its contents. Data are encrypted using a secret key, and then both the encoded message and secret key are sent to the recipient for decryption.

Mining: The process of adding transactions to the large distributed public ledger of existing transactions, known as the blockchain.

Transaction: A Bitcoin transaction is a transfer of Bitcoin value that is broadcast to the network and collected into blocks. Standard transaction outputs nominate addresses, and the redemption of any future inputs requires a relevant signature.

Wallet: A blockchain wallet is a digital wallet that allows users to manage Bitcoin and Ether. A blockchain wallet is provided by blockchain technology.

Chapter 2
Technique of Plant Electrical Stimulation by Weak Electric Currents

Alexander Smirnov
https://orcid.org/0000-0002-9236-2281
Federal Scientific Agroengineering Center VIM, Russia

Yuri Proshkin
Federal Scientific Agroengineering Center VIM, Russia

Alexander Sokolov
Federal Scientific Agroengineering Center VIM, Russia

Sergey Kachan
Federal Scientific Agroengineering Center VIM, Russia

ABSTRACT

The impact of electricity is described on plant growth and vegetation. The application of the electrical currents is discussed from the point of view of plant cultivation improvement in horticulture. Underlining that the electricity is an abiotic stress stimulant, the electricity use ways classification is given. The application of the electric currents and other similar influences can—directly or indirectly—affect plants causing a series of physiological and biochemical reactions. This technique enables the yield optimization and the fruits quality improvement by regulating the intensity and duration of the exposure according to different types and kinds of vegetables. In the area of an effective technique development of the plant electric stimulation, there are many aspects almost impossible to be taken into account within one experiment. An option for solving such problems is the compilation of the experimental databases, the introduction of the smart control systems, and the management of the technological processes of plant electric stimulation.

DOI: 10.4018/978-1-7998-3645-2.ch002

INTRODUCTION

The use of the electricity can – to a certain extent – stimulate the plants growth (Sokolov, 2018). This little-known technique called electroculture is able to accelerate the growth rates, increase the yields, improve the fruits quality and protect the plants from diseases, insects and frost (Ishikawa et al., 1990). Also, the electroculture application can reduce requirements for fertilizers or pesticides. There exist several approaches to the electroculture; they include exposure of weak and strong electric fields, magnetic fields, direct and alternating electric currents. Besides, the electricity can be applied to seeds, plants, soil, water or nutrient solutions. In particular, the application of an electrical field can – directly or indirectly – affect plants launching series of physiological and biochemical reactions. (Vallverdu-Queralt et al., 2013).

The first results of the electricity effect on plants, both positive and negative, were obtained by scientists as early as in the XVIII century. The experiments for the electricity influence studying on the plant life were conducted by the French physicist Jean-Antoine Nollet, the French abbot Bertolon, the Italian researcher Gardini, etc. (Gordeev et al., 1991). In the XIX century, some inactivity was observed until the 1880-ies, when the level of laboratory equipment provision for scientific works grew quite high. In the first half of the XX century, it was noted that the electric stimulation accelerated the plants growth and increased the yields; however, with some plants, failures were encountered. Conclusions were made that the exposure effectiveness is influenced by weather conditions, time of year as well as soil fertility. From the end of the XX century to the present time, the effects of the external electric field have been revealed on intermolecular bonds and on physiological & biochemical processes inside of the plant organism. In Russia, the experiments were conducted in the area of the seeds electric stimulation; also the acceleration possibility was studied of the scion increment process under the electric exposure. In the area of an effective technique development of the plants electric stimulation, there are many aspects almost impossible to be taken into account within one experiment. Those include: electric parameters of the plant, its surface and volume resistance, dielectric losses, active resistance of separate tissues as well as a set of such external factors as soil conductivity, atmospheric electricity, temperature, pressure, humidity, etc. An option for solving such problems is the compilation of the experimental databases, the introduction of the smart control systems and the management of the technological processes of the plants electric stimulation.

BACKGROUND

The living plant behavior under external electric influences can be attributed to its amazing properties. Inside of the plants, bioelectric potentials are generated and they themselves are exposed to the effects of the atmospheric electricity. In plants, signals are distributed; for example, the root "informs" the stem when damaged, then the irritation is transmitted to the leaves; and while responding, all of them together change their root activity. Mainly, the damaged part becomes electrically negative in the range of 20-120mV depending on the type of the plant and other reasons (Gordeev et al., 1991).

There exists a potential difference between different organs of plants, for example, between separate parts of a stem, a root, a flower, etc. (Opritov et al., 1991). In different trees, between the electrode in the soil and the one in the stem, the potential difference reaches 0.1V - 0.7V (Kholmansky et al., 2016).

Plants are not indifferent to the electricity; this was established a long ago by scientific and amateur experiments both successful and ended in failure. In 1747, the French physicist Jean-Antoine Nollet published his results on germination accelerating of seeds exposed to an electrical field. In 1783, the French abbot Bertolon informed on the negative result obtaining in his experiment on the electricity effect. He concluded that the electricity results in delays of the plants germination and growth. The reason for his failure is similar to the results of the experiments of the Italian researcher Gardini, who screened his plants from the atmospheric electricity with a metal wire grid (Gordeev et al., 1991).

As early as in the XIX century, after series of failures, an experiment was conducted with a plant (corn) placed under a dome made of wire. An electrode was inserted into this cage, from the tip of which the charge flowed into the atmosphere. After 10 days, the corn under the hood grew up to 17 cm, though in natural conditions it grew up to 8 cm (Gordeev et al., 1991).

Since the late 1900-ies, the researchers started experiments with the atmospheric electricity. In a vine sprout at a height of 0.8m, a platinum needle with copper wire was inserted, while the second wire at the bottom of the plant went into the ground. After that the atmospheric electric current flew through the plant, in its berries, the glucose content got twice higher and the acid content twice less. Also, a decrease in the number of ash elements in the wood was recorded, while in the leaves, on the contrary, an increase was observed. In his book "Electrical life" (1883), the French writer Albert Robida wrote that in the future as to stimulate the seedlings growth, fields would be subjected to the electrical treatment.

In 1904, the physicist Lemstrom from Finland noticed that the intense growth of vegetation during the short Arctic summer is due to the peculiarity of the electrical conditions of the atmosphere in the northern polar regions. He conducted the following experiment: at a height of 0.5m from the ground, he placed a metal grid and applied a high voltage to it. He was amazed with the effect: the field crops accelerated their growth by 100%, while the garden crops by 120%. In his gardens, the strawberries ripening in an electric field began on the 26th day, while in natural conditions it happens on the 54th day. Also, the Finnish physicist found that the effect of the electric field on plants depends directly on the fertility of the used soil.

In England, experiments were conducted in the 1920-ies. The scientist Blackman hung a network of wires at a height of 2.1m and supplied the voltage from 40 to 80 kV; the current was within 0.5-1mA. The 3 hours-long exposures were repeated twice a day for 20 days. The experiments with field crops resulted in an increase in the dry weight of the above-ground mass of the crops; however, with some cultures, the experiments turned out to be unsuccessful. From the experiments conducted by Blackman and scientists of his school, the dependence got known of the influence of weather conditions and time of year on the efficiency of electrical effects on plants; also, these researchers proved that the current intensity should not exceed 0.1·10-10A.

In Russia, I. V. Michurin subjected ridges with plants to electricity action and it gave amazing results. He found out that in order to achieve the effect, the voltage should not exceed two volts (Gordeev et al., 1991).

The Russian agronomist N.N. Speshnev was engaged in researches in the area of the electroculture; in his experiments, he electrified not only the soil but also the ground layer of the air. This way, he obtained the growth of yields for barley, oats, roots of radish and carrots. N. N. Speshnev promoted the use of the electroculture among the population, but his attempts were vain.

At the turn of the XIX and XX centuries, E. Pilsudsky developed his method of the electroculture. He dag zinc and iron sheets into the ground at the opposite ends of the field so that the currents in each pair went towards each other. In his experiments, the galvanic currents were equal to 0.05-0.25V. Weekly for a day, the wires connecting the sheets were disconnected. The yield of beet in Pilsudsky's experiments increased thrice and the sugar content in the beet roots increased by 12%.

In the 1930s, the laboratory experiments conducted by N. A. Artemyev and his group revealed the harmful effect of the strong ionization of the atmosphere on their plants. In particular, when the ions concentration decreased 10 times, the yield of cucumbers increased by half.

In 1924, Z.I. Zhurbitsky noticed that nitrogen and phosphorus are absorbed by plants 1.3 times more than potassium at the usual positive potential of the atmosphere (130V/m); but at the negative potential of the atmosphere (500V/m), the nitrogen and the phosphorus were absorbed 2.5 times less.

The scientists from the Institute of Plants Physiology found out that the yield of oats, barley, cucumbers and corn increased up to 50% with the potential difference growth up to 90V between the plants and the atmosphere. The opposite effect expressed in the plants growth reduction by half was obtained under a metal grid covered the plants from the atmospheric electricity (Gordeev et al., 1991).

In 1986 at the K.A. Timiryazev Moscow Agricultural Academy, the researchers observed the effect of the electrical stimulation at low soil moisture. The experimental (electrified) plants showed no signs of a lack of the moisture, though the control plants (grown in the ground with the normal moisture) were the most developed ones. The worst look had the plants with the low soil moisture and absence of the electric stimulation.

The scientists from the Volgograd State Agrarian University conducted their experiments on the effect of the electric current on survival of plant scions. The positive effect was obtained not less than twice and, besides, the faster process of adnation was observed. (Baev et al., 2014).

Other Russian scientists studied the effect of the electromagnetic influence on the rice seeds collectible samples. The increased germination energy of the seeds and the growth processes stimulation were noticed. All the studied batches reacted positively to the electrical treatment, though with different degrees of the reactivity. (Yudaev et al., 2015).

Interesting are designs by V.N. Pocheevsky, who applied – to his plants – the electrical stimulation from the "free" (cheap) natural current source. In his garden, the electrical stimulation of the plants was done with the converted electricity of the earth itself. The author claimed that he obtained "abundant" yields without additional fertilizing and was proud with the simplicity and visual beauty of the developed by him device. (Pocheevsky, 2016).

Many diverse versions of the electrical stimulation have been proposed by scientists of the world. In the Figure 1, one of the electrical stimulation versions with a current source is given, where the negative pole is attached to the upper leaves of the plant, while the positive pole to the roots.

Also, on the vital activity of plants, the weak electric currents act favorably, when they are passed through the soil. A lot of experiments on the soil electrification have been carried out all over the world. Notably, the soil is a very complex object, the electrical conductivity of which depends on a number of factors such as humidity, temperature, chemical composition, density, structure, etc. (Gureeva, 2018).

Many scientists have noted that any electrical stimulation in the summer is ineffective, while in the same experiments in the winter, the effectiveness was high. Also, the dissimilarity of the electricity effect on seedlings at different times of day was observed. This speaks for the atmospheric natural electric level influence on the life of plants (Chen et al., 2016).

In more detail below are discussed the plant reactions caused by strong/weak electric and magnetic fields/electric currents. The data collected from various sources have been analyzed. The analysis revealed that, the use of the electrophysical effects causes the plants growth acceleration, the increase of their number of leaves, their wet and dry mass as well as the content of minerals and valuable bioactive compounds. The similar results were obtained by both domestic and foreign researchers.

Figure 1. Electrical stimulation of plants

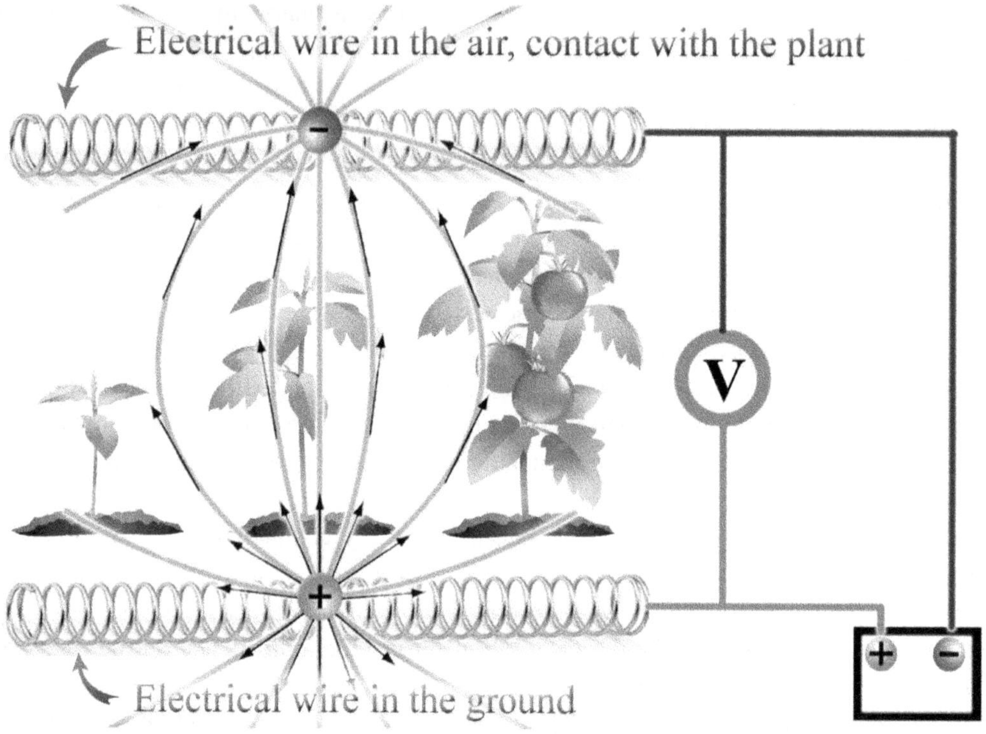

It has been found that in contrast to the strong pulsed electric field causing a cell membrane destruction, the treatment with a weak pulsed electrical field can cause changes at the level of a plant cell accompanied by such stress responses as the secondary metabolites accumulation, etc. (Dannehl, 2018). Yeh and his co-authors proved the ability of the pulsed electric field (10V/m, 50 Hz) to act as an abiotic elicitor for the secondary metabolites formation in suspension-based cultures of the Chinese yew-tree. In comparison with the control plant samples, the scientists found an increased intracellular accumulation of taxuyunnanine (the biologically active secondary metabolite) by 30% without a biomass loss. The Chinese scientists came to the conclusion that the pulsed electric field enhances the secondary metabolite production without harmful effects on the plant cell growth (Dannehl, 2018). Guderjan et al. (2005) studied the influence of a weak pulsed electrical field on stress responses of corn and soy plants. They demonstrated that in comparison with their control plant samples, after the plants treatment with the pulsed electric field (0.6kV/cm) changing the reversible cell permeability, the level of phytosterol in corn embryos and the amount of isoflavonoids in soy plants increased by 32.4% and 20.5%, respectively. However, a stronger field (up to 7.3kV/cm) resulted in a decrease in phytosterol yield in the maize embryos, which shows convincingly that only the samples exposed to a weak pulsed electric field react with the increased metabolic activity.

Lebedev et al. (1977) found out that ultra-weak magnetic induction equal to 10 nT applied to barley seedlings reduced the fresh mass of both sprouts and roots and their dry mass by 12, 35, 19 and 48%, respectively, as compared to the plants grown in the natural geomagnetic field. In addition, for many plant species, the use of the weak magnetic induction causes a reduction in cell reproduction in the meristem of plant roots (Dannehl, 2018).

In the other experiments, electrodes were used for passing the intermittent direct current (IDC) through the roots of radish and watercress plants. Compared to untreated plants, the highest total phenol content was observed in radish roots treated with 1000 mA for one hour per day: the increase by 46% was obtained (Dannehl et al., 2012). While studying radish tubers, the same scientists found that the phenolic compounds content and their associated antioxidant activity were less affected. These values were increased by 12.8% and 9.1%, respectively, as compared to the values obtained for the control plants (Dannehl et al., 2009). The content of anthocyanins in the radish tubers turned out to be increased by 28% under the same treatment conditions. In contrast, the phenolic compounds in radish sprouts remained unaffected by the IDC treatment. While the IDC equal to 1000 mA was sufficient for increasing secondary metabolites in radish, as to increase the total phenol content and the associated antioxidant activity in watercress sprouts, the researchers had to use the IDC of 1400 mA at a current 583 $\mu a/cm^2$. With the application of this stronger IDC, both contents were increased by 16% and 18%, respectively (Dannehl et al., 2012). Also, the contents of other human-friendly compounds such as chlorophyll, proteins, calcium, magnesium and iron were increased by 17%, 33%, 23%, 11% and 76%, respectively.

A significant increase in plant growth (by 30%) and potassium and calcium uptake was obtained on tomato plants when the direct current (DC) did not exceed 15μa per plant and the negative pole was placed on the ground (Black et al., 1971). It was assumed that the mediated by weak DC plant responses were caused by stimulation of active ion pumps or changes in growth-regulating compounds. Ward (1996) studied the varying effect of DC intensities on hydroponically growing tomato plants, where electrical contacts were placed on upper leaves of each plant and into the nutrient solution. He showed that the electric current of 125 μa applied downwards increased the fresh weight of the offshoots by 13.3% and the dry weight of the offshoots by 22%. However, supply of the electric current of 125 μa applied upwards resulted in the improved absorption of calcium (by 16.6%), potassium (by 3.4%) and magnesium (by 12.4%) as compared to the control plants. Gogo et al (2016) found no differences in terms of such plant characteristics as plant height, number of leaves, stem diameter, and number of primary seedlings when the upward oriented DC current of 73 and 127 μa was applied on the Solanum Africanum plants throughout 12 days (8 hours per day). However, they showed that the current of 127 μa resulted in significant increase of both the wet mass and dry weight of the leaves – by 14.4% and 24.4%, respectively. The lower DC current (73 μa) increased the number of marketable leaves by 55.3%; notably, at the same time, the number of non-market leaves was reduced by 77.2%. The similar results for non-market leaves were observed at plants exposed to the current of 127 μA. It is interesting that the treatment with the direct current of 127 μa was insufficient for a content increase of lutein, K-carotene, total carotenoids, chlorophyll A or B in the leaves. In contrast, the same metabolites were increased in stems of Solanum Africanum by 20%, 30%, 22%, 22% and 22%, respectively. However, the content of hemicellulose in the leaves increased by 40% and in the stems by 38.7%. Rathore et al (1988) showed that when a much lower direct current (2 μa) of any polarity was applied between an electrode in the tobacco callus and another electrode in the cultural medium, in comparison with the control plants, this current increased the wet mass (by 34%), dry mass (by 22%) and sprouts regeneration per callus. As it was recently shown, a DC of 30 to 100 mA was sufficient for inducing the biosynthesis of the secondary metabolites in intact roots or cell suspension cultures of peas, fenugreek (trigonella), cask medical biofilm cells and chickpeas (Kaimoyo et al., 2008). For example, in pea roots, the content of the phenolic compound (+) pisatin (phytoalexin) was 13 times higher than in roots of the untreated plants when the experimental plants were treated with the direct currents between 30-199 mA. The analysis revealed the similar differences for naringenin in cask medical biofilm cells suspensions, though to a lesser extent.

The scientists concluded that the electricity induces the biosynthesis of the secondary metabolites in a wide range of plant species without an adverse effect on the physiological functions of cells or tissues (Kaimoyo et al., 2008).

Based on this review, it has been concluded that, as it was defined above, the electricity can be seen as an affecting plants abiotic stress stimulant. If to consider the matter in this context, one can say that many reports revealed more advantages than disadvantages for plants when strong or weak electric fields, magnetic fields or electric currents were applied to them.

The electrodes attached to the bodies of the plants and placed into their nutrient medium allow a weak electric current to flow through callus tissues and whole plants. It is difficult to introduce this method into practice as each installation must maintain the same electric current; however the plants have different electrical conductivity, surface&volume resistance, dielectric losses, active resistance of separate tissues, etc. That is why at the present time, these methods are used for basic researches and not on practice.

The application of the electric current passing through plants causes increase in plant growth, leaf number, wet and dry mass and content of minerals and such valuable bioactive compounds as carotenoids, phenolic compounds and chlorophyll. From this point of view, the high correlations were found between the applied electric currents and the stress-related compounds.

Also taking all the used types of the electrical stimulation in consideration, the conclusion can be made that the reaction intensity of these plants depends on a specific installation and a treatment mode. However, it is difficult to compare the observed effects of electricity on plants due to the fact that often in the cited articles, the experimental details were incomplete including the insufficient descriptions of the equipment and the treatment conditions.

MOTIVATION

In 1953, Breazeale et al. stated that plants absorb the mineral nutrients in the form of ions; the ions carry electrical charges; the absorption of the ions is a purely electrical phenomenon and must involve a current flow unless neutral vapors absorption takes place. Some ions can be accumulated against the electrochemical gradient. But the phenomenon of the passive ion absorption consists in the ions movement along (and not against) the electrochemical gradient. One should assume that this latter type of absorption would be easily modified by an electric current applying to the accumulation system. By analogy with the polarograph, Breazeale argued that the ions absorption could be changed by applying a variety of currents depending on the applied voltage. It seems more likely that the applied electric current can affect not only the mechanism of the ion accumulation but also the internal activity of auxin or gibberellin. It is quite clear, however, that the currents, which far exceed the natural physiological electrical phenomena or opposite to them in their polarity, are harmful for plants. (Black et al., 1971).

For the penetration into a plant, the current must pass through the epidermal tissue of the roots and then pass through the apoplast or simplast or both. The current must be carried by the ions along electrochemical gradients that will guide the ions in their penetration through cell walls/membranes or along cytoplasmic passages.

it has been found out that, in contrast to the strong pulsed electric field causing a cell membrane destruction, a plant treatment with a weak pulsed electric field can result in changes at the level of a plant cell that are accompanied by such stress reactions as the secondary metabolites accumulation (Dannehl, 2018).

Other studies showed that with tomatoes (Solanum lycopersicum L.), the DC intensities of 100 to 500 mA at different exposure times from 15 to 60 min resulted in the accumulation of carotenoids & phenolic compounds and the increased antioxidant activity (Dannehl et al., 2011). It is expected that the electric fields accelerate mass transfer by affecting the permeability properties of the cell membranes (Ade-Omowaye et al., 2002). However, scrupulous studies on tall-growing agricultural plants such as cucumber have not been conducted yet.

DEVELOPMENT OF PLANT ELECTRICAL STIMULATION DEVICE

The developed device of electrostimulation of plants cultivated in both open and protected soils is primarily intended for research purposes, for studying vegetation effects of plants growing in both open and protected soils in the case of their electrostimulation.

There has been known the method of yields increase & plants vegetation acceleration consisting in the application of a constant electric current to the soil, which is the root zone of the plants. In the root zone, the process is carried out of electrolysis of the available in the soil mineral salts, the rate of flow of which is maintained in direct proportion to the intensity of photosynthesis (Gorshenin et al., 1994). The device is powered by a solar battery and the electrodes act only on the root system.

Also, there has been known the other plant life electrostimulation method. It consists in plants growing *in vitro*. The electrically conductive test tube for plants growing has a metal tip and a stopper. The tube is mounted on a tripod so that the metal tip touches the metal base of the tripod, to which the conductor from the plus terminal of the battery is connected. In order to stop the current supply, a switch is used. The current supply is controlled by a current regulator including registering devices for the current and voltage. The current supply is set with use of a time relay. The electrical stimulation starts when a slice of a plant meristem is placed into a nutrient solution, so that the electrical conductor of the plug touches the mirror of the nutrient solution. The plug with an electrical conductor is connected to the negative terminal of the battery. After reaching the required level of vegetation, the plant is brought to the open ground (Samarin et al., 2013).

The device implemented this action principle includes the conductor connecting the minus and plus terminals of the power supply source with the plant, the current supply regulator, the current and voltage registration devices, the potentiometer and the nutrient medium for the plant cultivation, into which the plus electrode is plunged.

The disadvantages of the known devices are a fall in yield and an elongation of the plants vegetation period.

Technically, the developed device is aimed at increasing the yield and reducing the vegetation period of plants by the way of the constant electric current application to the plant's root system and stem.

The engineering task is solved due to the following technical design. The plants electrical stimulation device for both open and protected soils consists of a) the conductor connecting negative and positive terminals of the power supply source with the plant; b) the regulator of the power supply made in the form of potentiometer; c) the device of the amperage and voltage registration; and d) the nutrient medium for plant growth connected with the positive electrode. The plants electrical stimulation device equipped with a LED indicator, a clip-clamp with carbonaceous conductive felt. The current supply regulator is made in the form of a chip connected to a potentiometer. The power supply source is made in the form

Figure 2. Diagram of plants electrical stimulation device

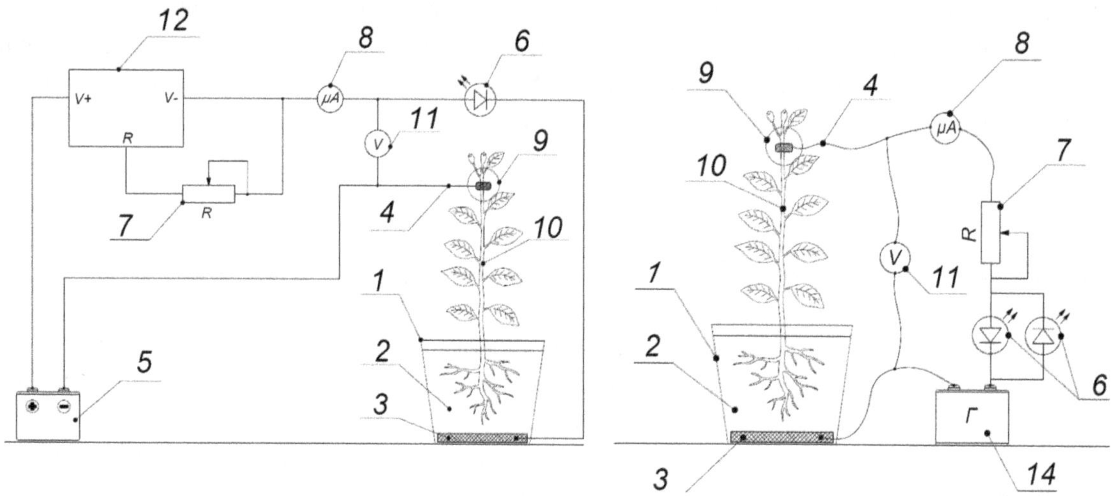

of a DC battery or a signal generator of a certain shape. In the Figure 2, the diagram of the device is represented, while the Figure 3 shows the clip-clamp.

The plants electrical stimulation device for both open and protected soils includes as follows: the container 1 for the nutrient medium 2, the carbonaceous electrode 3, the electrical conductor 4, the power supply source, which is the DC battery 5, the LED indicator 6, the potentiometer 7, microammeter 8, the combined clip-clamp 9, the plant 10, the voltmeter 11, the current supply regulator made in the form of the chip 12 connected with the potentiometer 7 and the carbonaceous conductive felt 13.

Into the container 1, the soil is put or other nutrient medium 2. Into the lower part of the soil 2, there placed the carbonaceous electrode 3 connected by the electrical conductor 4 with the LED indicator 6, the microammeter 8, the chip 12, the potentiometer 7 and with the plus terminal of the battery 5. The minus terminal of the power supply source 5 is connected with the voltmeter and upper part of the plant 10. The negative contact of the voltmeter 11 is connected with the clip-clamp 9 placed on the upper part of the plant 10, while the positive contact is situated between the microammeter 8 and the LED indicator 6.

Figure 3. Clip-clamp for plants connection with the current source

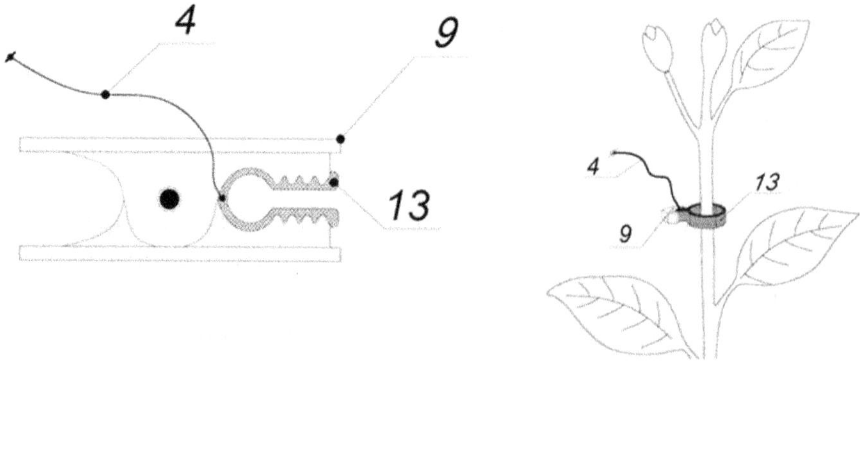

Adjustment of the microcurrent supplied to the plant 10 is carried out by means of the potentiometer 7 with a rotating mechanism. The latter rotates clockwise for the resistance reduction and counterclockwise for the resistance increasing, thereby increasing or decreasing the microcurrents passed through the plant depending on the selected parameters needed for the growth of the plant 10 intensification.

The chip 12 LM334 with its preset parameters provides stabilization of the microcurrent. The functional features of the chip 12 allow to hold the initially configured microcurrents regardless of the resistance changes in a) the stem of the plant 10, b) the transient contact resistance of the clip-clamp 9, and c) the electrode 3. The factors affecting changes in the physical-chemical properties of the plant 10 affect also the resistance of the plant.

As in the power supply circuit, the LED indicator 6 is included, the latter provides with indication of the presence of voltage and microcurrent in the circuit, so it is possible to detect a poor contact, a circuit breakage or overdrying of the soil. The overdrying of the soil increases the resistance, which can result in the circuit breakage.

The combined clip-clamp 9 is a plastic clamp supplemented with carbonaceous conductive felt. The attachment of the clip-clamp 9 can be done by two ways depending on the thickness of the plant stem and its meristematic tissues.

In both cases, the clip-clamp 9 is attached to the very top of the plant 10. The first method is used in the case of a hard and thick stem and is based on the classical clamp designed in such way that on its internal part, the clip-clamp 9 has the carbonaceous electro-conductive felt. The second method is applicable in the case of a sensitive stem of the plant 10. The clinging by the clip-clamp 9 is performed by way of throwing – onto the plant stem – a loop (collar) made of a soft carbonaceous electro-conductive felt non-injurious to the plant tissues.

The power supply source can be made in the form of a signal generator 14 of a certain shape. The device provides with the pulse characteristics specified by the generator 14.

The power supply 5 or the generator 14 provides every plant with the necessary electrical potential featured with the specified parameters (constant, variable, sine, meander, pulse, etc.) as every plant is characterized with a number of chemical electrical properties. The resistance of the stem of the plant 10 influences results of the electric microcurrent application to the plant, for example, the microcurrent can vary from 0.1 to 15 μa when the potential is applied to the upper and lower parts of the plant. Every plant culture needs an individual approach to the level of microcurrent and voltage and to the signal form. The supply of the voltage and the microcurrent from the power supply source 5 or the generator 14 allows us to influence the biologically active meristems of the cell structure of the plant and therefore to accelerate its growth.

The device design ensures the stabilization of the current passing through the plant, regardless of changes in the circuit resistive characteristics – this is reached due to using the chip 12 LM334 (figure 4) and the potentiometer 7.

In the developed device, there are 11 independent output channels. Its main technical specifications are represented in the Table 1.

Figure 4. Schematic diagram of the device LM334

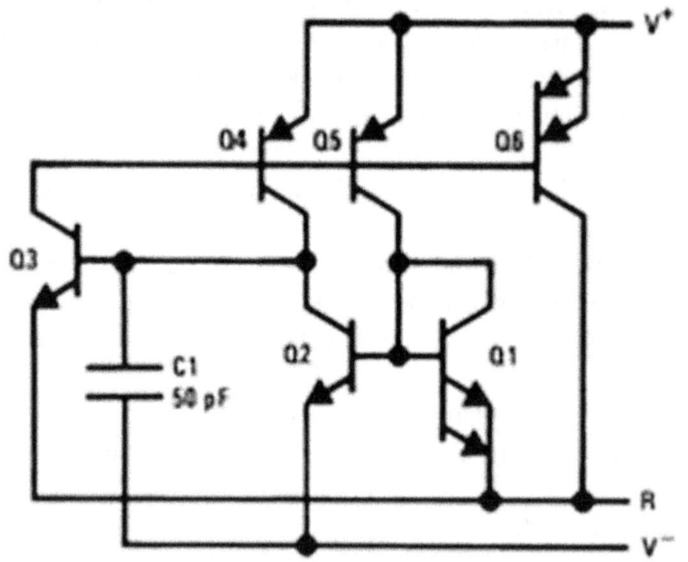

Table 1. Technical specifications of device of plants electro-physical stimulation with weak constant currents

Parameter	Value
Supply voltage, V DC	1-40
Stimulation current changes control range, μA (DC)	13-65
Working temperature, °C	0-30

EXPERIMENTAL RESEARCH

The laboratory studies of the effect of weak DC currents on the growth of cucumber seedlings were carried out. As the object of the research, there was chosen the parthenocarpic cucumber hybrid for photoculture. The subject of the research was the morphological parameters set of the cucumber grown up under an artificial irradiation and under the weak DC currents action.

The research was carried out in the testing laboratory of the Federal Scientific Agro-Engineering Center (Moscow, Russian Federation). For studying, there was chosen the smooth-fruited parthenocarpic hybrid of the cucumber Ermak F1 selected by the company "Gavrish" and intended for cultivation in industrial greenhouses in winter and spring. The cucumbers are grown up in a phyto-chamber on hydroponics. The appearance of the phyto-chamber is shown in the Figure 5. The nutrient solutions were prepared on the basis of deionized water in line with the traditional for cucumbers formulations.

Figure 5. Phyto-chamber with cucumber seedlings and electrical stimulation device

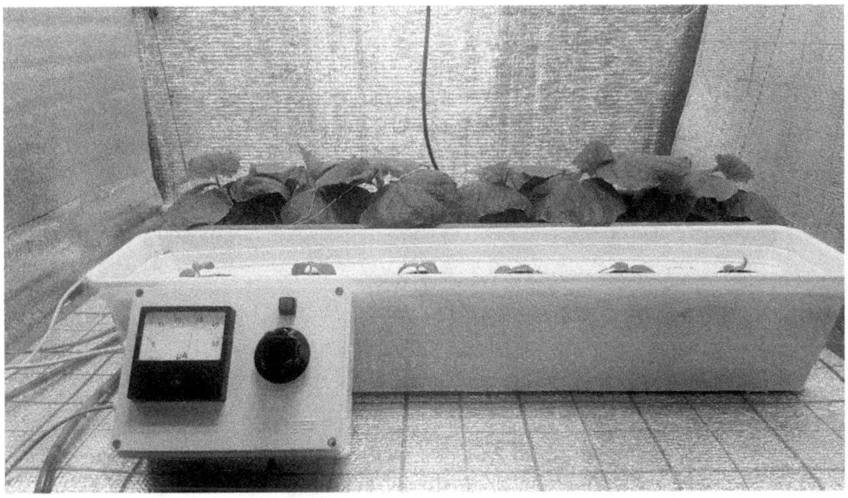

The cucumber plants were exposed continuously to DC 10, 20 and 30mA all over the time of their vegetation and fruiting. Untreated plants serve as the control samples. Each experiment involved 3 cucumber plants.

For stabilization of the current, the reference voltage and current sources were used of the model LM334, which allowed to adjust the current in the set limits. Thus, the output current intensity did not depend on the input voltage.

For the injury to plants minimization, the connection of the wire was carried out with aid of a special clip-clamp, which was a clamp with a conductive material (Figure 6, 7). The second wire was connected to a general electrode installed in the ground.

The negative electrode was attached to the plant in concord with the theory of electroosmosis. Besides, this order was chosen because our experience prompted that always use of the opposite polarity has been associated with tissue damage near the electrode (the damage was observed after 12 hours of exposure) (Figure 7). Also in our tests, the damage took place at the current 30μA and more during 62 hours regardless of the polarity of the applied voltage. Every 6 days, the upper negative electrode was removed closer to the plant growth point.

As the radiation sources in the phyto-chambers, the LED phyto-emitters were used. The spectral characteristics of the studied irradiators were chosen based on the previous studies (Mironyuk et al., 2020; Smirnov et al., 2017, 2018). The ratio of the fluxes Blue:Green:Red+Far-Red (% of total flow) was 14:31:55+14. The irradiance in all phytotrons was set as ~55 Wt/m^2 at 18-hour-long photoperiod and the air temperature was set 21°C/16°C (day/night). For measuring PAR-irradiancy and analyzing the radiation spectrum, the spectrophotometer TKA-Spectrum was used. In order to control heating, ventilation, carbon dioxide enrichment, humidity, illumination and environmental conditions, the climatic computer was used.

The plant growth, the number of leaves, the leaf area, the yields and the characteristics of the harvested fruits were evaluated.

Figure 6. Attachment of clip-clamp with electro-conductive material to plant

Figure 7. Tissue damage in area of contact with electrode in case of polarity change

The method of finding out the content of the mass fraction of moisture (W,%) and dry residue (X,%) in the cucumber fruits consisted in 6 hours-long drying of a 5-gram fruit weighed portion down to a constant mass at the temperature of 70°C. The measuring was repeated thrice. Drying of the samples was carried out in metal weighing cups with use of the quartz sand (10 g) and a glass stick for mixing of the samples with the sand. The samples uptake for the analysis was carried out from the middle part of the fruit by cutting a slice onto cubes about 3 mm in size. The mass of the combined sample made 120g.

The mass fraction of solids in cucumber fruits (X, %) was calculated by the formula:

$$X = 100\left(1 - \frac{m_1 - m_2}{m_1 - m_3} K\right) \quad (1)$$

where m_1 is mass of a cup with a lid, a stick, sand and the sample before drying, g;
m_2 is mass of the cup with the lid, the stick, the sand and the sample after drying, g;
m_3 is mass of the cup with the lid, the stick and the sand, g;
K is correction factor equal to 1.

The mass fraction of moisture in the product (W, %) was calculated by the formula:

$$W=100-X \quad (2)$$

After conducting the experiments, the results were processed in the software environment Python. The platform Python supports structural, object-oriented, functional, imperative, and aspect-oriented programming. Using the ready-made libraries "numpy" and "pandas" as well as the modules "preprocessing", "metrics", "PolynomialFeatures" and "LinearRegression" of the library "sklearn", the regressions estimation was carried out.

The main elements of the program code are presented below.

1. Data normalization
```
df['factor1'] = preprocessing.scale(df['factor1'])
df['factor2'] = preprocessing.scale(df['factor2'])
df['factor3'] = preprocessing.scale(df['factor3'])
```

2. Average result calculation
```
df = df.assign(rez=pd.Series(df.iloc[:,3:6].mean(axis=1)).values)
```

3. Input data structuring
```
factor = df[['factor1','factor2','factor3']]
rez = df['rez']
```

4. Matrix transformations
```
poly = PolynomialFeatures(degree=3, interaction_only=True)
factor_ = poly.fit_transform(factor)
print(factor_)
```

5. Self-learning
```
lm = LinearRegression()
model = lm.fit(factor_, rez)
```

6. Verification of results adequacy

```
predict = model.predict(factor_)
print(metrics.mean_squared_error(rez,predict))
```

The results of the experiments indicating the arithmetic average and the standard deviation are presented in the Tables 2 and 3.

Table 2. Results of cucumber fruits analysis

Number of sample (at processing)	Fruit mass, g	Fruit length, cm	Fruit thickness, cm	Mass fraction of moisture, %	Mass fraction of solids	
					%	g
1 (10μA)	193,8 (±19,8)	19,3 (±1,0)	4,6 (±0,21)	97,3 (±0,2)	2,7 (±0,1)	0,137 (±0,002)
2 (20μA)	246,3 (±25,9)	21,3 (±1,5)	4,6 (±0,23)	96,7 (±0,1)	3,3 (±0,2)	0,165 (±0,005)
3 (30μA)	172,3 (±21,3)	20,2 (±1,4)	3,6 (±0,20)	97,1 (±0,3)	2,9 (±0,2)	0,147 (±0,003)
4 (control)	202,6 (±27,4)	20,2 (±1,5)	4,0 (±0,25)	97,1 (±0,1)	2,9 (±0,1)	0,147 (±0,005)

Table 3. Morphological parameters of cucumber plant

Number of sample (at processing)	Stem length, mm	Nodes number, pieces	Wet mass of leaf, g	Yield of one plant, g
1 (10μA)	7200 (±325)	87 (±3)	10,3 (±1,4)	3421 (±265)
2 (20μA)	7300 (±298)	97 (±4)	16,9 (±2,1)	3009 (±280)
3 (30μA)	7750 (±424)	80 (±2)	7,3 (±1,8)	2337 (±302)
4 (control)	7470 (±476)	81(±3)	11,9 (±1,3)	2900 (±368)

EXPERIMENTAL RESULTS

The cucumber fruits exposed to the action of the weak direct current had an abnormal curvature, which can be attributed to the potassium absorption blocking from the nutrient solution (Figure 8). With a lack of potassium, the cucumbers are deformed: the fruit gets a narrowing at the base (at the attachment area of the pedicel) and, on the contrary, thickening at its top (neck). As a result, the green vegetables acquire a claviform or pear-like shape. In addition to the deformation, the potassium starvation is reflected in the appearance of the entire plant (its leaf apparatus): the old lower leaves of the cucumbers have yellowness at the edges (in the center, they remain green), i.e. the marginal chlorosis appears at them. Along with it, the younger leaves are reduced in size, have a dull color and a cup-like shape (Figure 9).

As for the control samples, the cucumber fruits had the correct shape and no chlorosis on the leaves was observed.

Figure 8. Appearance of cucumber fruits

Figure 9. Cucumber leaf exposed to current of 30 μA

A weak correlation was revealed between the intensity of the current passed through the plant, its yield and the length of the stem. Thus, with the current growth, proportionally and simultaneously, there was observed a length increase of internodes and a decrease in yield. Nevertheless, in comparison with the control plants, these changes were insignificant.

It is known for the tomato that the applied weak currents tend to stimulate the potassium absorption and this is not proportional to the current intensity, nor has it anything to do with the electrodes polarity (Dannehl et al., 2011). Often, a high increase of the potassium content in the roots is associated with a slow increase of the same in leaves or a stem growth and vice versa.

The potassium enters to the plant in the form of the ion K+. Unlike the calcium, the potassium reduces the viscosity of the protoplasm, increases its water content making higher the proteins hydration. This feature of the potassium action can be easily seen due to the fact that in its salts, the plasmolysis has a convex shape and the protoplasm detaches easily from the cell walls. The potassium salts are soluble and participate in regulation of the osmotic potential of cells. In particular, K+ is very important in the course of the leaves stomatal mechanism regulation. Also, the root pressure is largely dependent on the K+ presence in the root exudate.

Then, studies were carried out using 2^3-factorial planning with combinations of three independent variables: energy exposure of the PAR, the magnitude of the electric current through the plant and the level of mineral nutrition (electrical conductivity of the nutrient solution). The following variables were selected as dependent variables: the mass of the aerial part of the plant, the number of leaves and the height of the cucumber plant. Each experiment was carried out in triplicate. The planning matrix is presented in the Tables 4 and 5.

Table 4. The matrix conditions of each treatment

Experiment No.	Energy exposure PAR, W/m²	Electric current through the plant, μA	EC of the nutrient solution, dS m⁻¹
1	30	20	2.5
2	55	20	2.5
3	30	0	2.5
4	55	0	2.5
5	30	20	1.5
6	55	20	1.5
7	30	0	1.5
8	55	0	1.5

Table 5. Planning matrix for processing results

Experiment No.	Factors			Interactions				The results of the experiments		
	x_1	x_2	x_3	x_1x_2	x_1x_3	x_2x_3	$x_1x_2x_3$	Mass of the aerial parts, g Y_1	Plant height, cm Y_2	Number of leaves Y_3
1	1	1	1	1	1	1	1	32	28	6
2	-1	1	1	-1	-1	1	-1	36	26,3	6,3
3	1	-1	1	-1	1	-1	-1	37,3	38	6
4	-1	-1	1	1	-1	-1	1	38,7	29	6
5	1	1	-1	1	-1	-1	-1	35	33	6
6	-1	1	-1	-1	1	-1	1	35	26,5	6
7	1	-1	-1	-1	-1	1	1	35,3	35	6
8	-1	-1	-1	1	1	1	-1	31,3	24,6	6

Regression equations in coded variables:

For the mass of the aerial parts of the plant:

$$y=35{,}1-0{,}18x_1-0{,}58x_2+0{,}93x_3-0{,}83x_1x_2-1{,}18x_1x_3-1{,}43x_2x_3 \qquad (3)$$

For plant height:

$$y=30+3{,}45x_1-1{,}6x_2+0{,}28x_3-1{,}4x_1x_2-0{,}78x_1x_3-1{,}58x_2x_3 \quad (4)$$

For the number of leaves:

$$y=6{,}04-0{,}04x_1+0{,}04x_2-0{,}04x_3-0{,}04x_1x_2-0{,}04x_1x_3-0{,}04x_2x_3 \quad (5)$$

According to the equations, it can be seen that the factor x_1, the level of energy irradiation, has the greatest influence on the height of plants. For the mass of the aerial parts of the plant, the conductivity factor of the nutrient solution had the greatest influence. With an increase in the level of nutrition, the mass of the aerial part of the plant increased. None of the three factors had a significant effect on the number of leaves. Electrical stimulation had the greatest effect on plant height. At an electric current of 20 μA, the height of the plants was less than in the absence of external influences. This effect can be used to prevent elongation of cucumber seedlings in low light.

Other studies show that the ion density increase as a result of the stress exposure may lead to the higher nutrient availability in the plant cells (Dannehl, 2018). This phenomenon can result in an improvement of a number of the physiological processes including the photosynthesis, whereby more photosynthetic products can be accumulated in the leaves, which ends with a higher leaf area index. The increase of the mineral elements content (e.g. Mg and Zn) in the DC-treated plants is accompanied with more intensive formation of the chlorophyll (green pigment), so it prevents leaves yellowing (chlorosis) and leads to a decrease of non-edible leaves mass, as reported in the study.

It is possible that the electrical stimulation of ion uptake is not associated with an increased passive cation uptake and electroosmosis, nor is it related to the reduction potential of each cation (Black et al., 1971). Most likely, this is due to the stimulation of the active ion absorption apparatus, (such as the K+ anion pump on the root surface) or to the redistribution of substances regulating plant growth under an electric field action inside of the living tissue. In this case, the great variability of results from one test to another is not surprising and can be explained by the natural variety of the endogenous factors as well as significant variations in root and ground surface resistance.

CONCLUSION

It was found that the weak electric currents of 10÷30μa do not affect the cucumber growth processes and yield significantly. The cucumbers exposed to the weak direct currents were featured with unusual curvature of the fruits, which can be attributed to potassium absorption blocking from the nutrient solution. If to conduct a micronutrient analysis of cucumber leaves and fruits, it can help to identify the cause of this effect.

It is also necessary to select scrupulously the DC intensity and the duration of its exposure suitable for different types and kinds of vegetables in order to optimize the yield and the quality.

In spite of the fact that researches on the electric currents action on the plants growth and development were begun long ago, much in the mechanism of the influence remains not clear yet. This study confirms the complexity of the aspects of the plants electrical stimulation.

Thus, in the area of the plants electrical stimulation, there is a lot of contributing factors that should be taken into account, though doing it within one experiment is almost impossible. Those factors are as follows: soil conductivity, atmospheric electricity, temperature, time of year, time of day, electrical parameters of the treated organism itself, its dielectric conductivity, dielectric loss, specific surface and volume resistance, the active resistance of separate tissues and organs, body condition, metabolic activity in a particular period of time, etc.

FUTURE RESEARCH DIRECTIONS

In the scientific works direction related to the electrical stimulation of plants, a lot of unresolved issues remains. Their implementation will require the large-scale researches, the experiments, the results databases maintenance, the analysis algorithms development and the forecasting approaches development of crops growth and vegetation. All of this will be necessary to be worked through in order to form a knowledge base for the implementation of the technological processes control. Also, a gradual transition will be needed to an intelligent system of control and management of the technological processes of the electrical stimulation of the plants and their root system.

The solution of such problems should lead to dependencies identification between the plants electrical stimulation and their yield. The system parameters should be substantiated for the electrical stimulation of specific types of plants with the purpose of obtaining certain consumer products' quality and creation of the plants stimulation system able to produce the desired qualities of the plant products. In addition, it is necessary to study in detail the effectiveness of such technologies from the economic point of view.

REFERENCES

Ade-Omowaye, B. I. O., Rastogi, N. K., Angersbach, A., & Knorr, D. (2002). Osmotic dehydration of bell peppers: Influence of high intensity electric field pulses and elevated temperature treatment. *Journal of Food Engineering*, *54*(1), 35–43. doi:10.1016/S0260-8774(01)00183-2

Baev V.I., Petrukhin V.A. (2014) Kachestvennoye opisaniye protsessa prirastaniya privivok rasteniy pri elektricheskoy stimulyatsii [A qualitative description of the process of growth of plant grafts during electrical stimulation]. *Izvestiya Nizhnevolzhskogo agrouniversitetskogo kompleksa: Nauka i vyssheye professional'noye obrazovaniye, 4*(36), 241–245.

Black, J. D., Forsyth, F. R., Fensom, D. S., & Ross, R. B. (1971). Electrical stimulation and its effects on growth and ion accumulation in tomato plants. *Canadian Journal of Botany*, *49*(10), 1809–1815. doi:10.1139/b71-255

Breazeale, E. L., & McGeorge, W. T. (1953). Cation uptake by plants as affected by an applied potential. *Soil Science*, *75*(6), 443–448. doi:10.1097/00010694-195306000-00004

Chen, Y., Zhao, D.J., Wang, Z.Y., Wang, Z.Y., Tang, G.L., & Huang, L. (2016). Plant Electrical Signal Classification Based on Waveform Similarity. *Algorithms, 9*(4), 70. doi:10.3390/a9040070

Dannehl, D. (2018). Effects of electricity on plant responses. *Scientia Horticulturae, 234*, 382–392. doi:10.1016/j.scienta.2018.02.007

Dannehl, D., Huyskens-Keil, S., Eichholz, I., Ulrichs, C., & Schmidt, U. (2009). Effects of intermittent-direct-electric-current (IDC) on polyphenols and antioxidant activity in radish (Raphanus sativus L.) during growth. *Journal of Applied Botany and Food Quality*, *83*, 54–59.

Dannehl, D., Huyskens-Keil, S., Eichholz, I., Ulrichs, C., & Schmidt, U. (2011). Effects of direct-electric-current on secondary plant compounds and antioxidant activity in harvested tomato fruits (Solanum lycopersicon L.). *Food Chemistry*, *126*(1), 157–165. doi:10.1016/j.foodchem.2010.10.092

Dannehl, D., Huyskens-Keil, S., Wendorf, D., Ulrichs, C., & Schmidt, U. (2012). Influence of intermittent-direct-electric-current (IDC) on phytochemical compounds in garden cress during growth. *Food Chemistry*, *131*(1), 239–246. doi:10.1016/j.foodchem.2011.08.069

Gogo, E. O., Huyskens-Keil, S., Krimlowski, A., Ulrichs, C., Schmidt, U., Opiyo, A., & Dannehl, D. (2016). Impact of direct-electric-current on growth and bioactive compounds of African nightshade (Solanum scabrum Mill.) plants. *Journal of Applied Botany and Food Quality*, *89*, 60–67.

Gordeev, A. M., & Sheshnev, V. B. (1991). *Elektrichestvo v zhizni rasteniy* [Electricity in plant life]. Nauka.

Gorshenin, V.A., Adamovich, B.A., & Adamovich, A. B. (1994) *Sposob povysheniya urozhaynosti i snizheniya srokov vegetatsii rasteniy* [Method for increasing yield and reducing growing period of plants]. Patent RF, 2058717.

Guderjan, M., Topfl, S., Angersbach, A., & Knorr, D. (2005). Impact of pulsed electric field treatment on the recovery and quality of plant oils. *Journal of Food Engineering*, *67*(3), 281–287. doi:10.1016/j.jfoodeng.2004.04.029

Gureeva, M. A. (2018). The effects of electrical stimulation on plants through the root system. *Exact Science.*, *21*, 18–21.

Ishikawa, H., & Evans, M. L. (1990). Electrotropism of maize roots. *Plant Physiology*, *94*(3), 913–918. doi:10.1104/pp.94.3.913 PMID:11537481

Kaimoyo, E., Farag, M. A., Sumner, L. W., Wasmann, C., Cuello, J. L., & VanEtten, H. (2008). Sub-lethal levels of electric current elicit the biosynthesis of plant secondary metabolites. *Biotechnology Progress*, *24*(2), 377–384. doi:10.1021/bp0703329 PMID:18331050

Kholmansky A.S., Kozhevnikov Yu.M. (2016) Mekhanizm generatsii elektrokhimicheskogo potentsiala dereva [The mechanism of generation of the electrochemical potential of a tree]. *Izvestiya vysshikh uchebnykh zavedeniy. Lesnoy zhurnal, 5*(353), 73–80.

Lebedev, S. I., Baranskiy, P. I., Litvinenko, L. G., & Shiyan, L. T. (1977). Barley growth in superweak magnetic field. *Electronic Treatment of Materials*, *3*, 71–73.

Mironyuk, S. S., Smirnov, A., Sokolov, A., & Proshkin, Y. (2020) Optimization of spectral composition and energy economy effectiveness of phyto-irradiators with use of digital technologies. In Handbook of Research on Energy-Saving Technologies for Environmentally-Friendly Agricultural Development. IGI-Global.

Opritov, V. A., Pyatygin, S. S., & Retivin, V. G. (1991). *Bioelektrogenez u vysshikh rasteniy* [Bioelectrogenesis in higher plants]. Nauka.

Pocheevsky V. (2016) Ovoshchi na elektrichestve [Vegetables on electricity]. *Priusadebnoye khozyaystvo, 5*, 32–35.

Rathore, K. S., Hodges, T. K., & Robinson, K. R. (1988). A refined technique to apply electrical currents to callus-cultures. *Plant Physiology, 88*(3), 515–517. doi:10.1104/pp.88.3.515 PMID:16666337

Samarin, G.N. (2013). *Sposob elektrostimulyatsii zhiznedeyatel'nosti rasteniy* [Method electrostimulation of plant life]. Patent RF, 2555449.

Smirnov, A., Kholmanskiy, A., & Ukhanova, V. (2018). Optimization of lighting spectrum of greenhouse vegetables by using light emitting diodes. *International Journal of Research in Pharmacy and Biosciences., 5*(4), 11–17.

Smirnov, A.A., & Kholmansky, A.S. (2017). Zavisimost' fotosinteza pigmentov i produktivnosti tomata ot spektral'nogo sostava obluchatelya [Dependence of pigments photosynthesis and productivity of tomato on spectral composition of irradiator]. *Nauchnaya zhizn, 10*, 14-19.

Sokolov A.V. (2018) Analiz i istoriya razvitiya elektricheskikh stimulyatsiy rasteniy [Analysis and history of the development of electrical stimulation of plants]. *Innovatsii v sel'skom khozyaystve, 3*(28), 87–93.

Vallverdu-Queralt, A., Oms-Oliu, G., Odriozola-Serrano, I., Lamuela-Raventos, R. M., Martm-Belloso, O., & Elez-Martinez, P. (2013). Metabolite profiling of phenolic and carotenoid contents in tomatoes after moderate-intensity pulsed electric field treatments. *Food Chemistry, 136*(1), 199–205. doi:10.1016/j.foodchem.2012.07.108 PMID:23017413

Ward, R. G. (1996). The influence of electric currents on the growth of tomato plants. *Acta Physiologiae Plantarum, 18*, 121–127.

Yudaev, I. V., & Nosova, T. A. (2015) Vliyaniye predposevnoy elektrostimulyatsii na semyana risa s nizkimi posevnymi kachestvami [The effect of pre-sowing electrical stimulation on rice seeds with low sowing qualities]. In *Metody i tekhnicheskiye sredstva povysheniya effektivnosti ispol'zovaniya elektrooborudovaniya v promyshlennosti i sel'skom khozyaystve 80-ya* [Methods and technical means to improve the efficiency of use of electrical equipment in industry and agriculture 80th scientific-practical conference] (pp. 464–468). Academic Press.

Chapter 3
IoT Limitations and Concerns Relative to 5G Architecture

Ohood Saud Althobaiti
https://orcid.org/0000-0002-0483-1798
King's College London, UK & Taif University, Saudi Arabia

ABSTRACT

Several current computer science applications, implemented within specific paradigms, work at different levels to solve various challenges facing particular sectors. The potential of the internet of things (IoT) in the context of fifth-generation networks (5G) is envisioned as suggesting several beneficial opportunities for companies, industries, and users to exploit this technology's applications. This chapter establishes how the IoT works, considering its 5G architecture. The emphasis is on the infrastructural characteristic in terms of transmission power, frequency, speed, security, localization, device lifetime, and others. Additionally, the chapter illustrates what the IoT entails, discussing its workability and efficiency. Furthermore, it highlights a range of newly distinguishing features that would give it much-touted success in comparison with other technologies. It also presents research issues and challenges.

INTRODUCTION

Many telco operators are already immersed in the deployment of the next generation of mobile networks, 5G. 5G networks not only offer enhancements and additional features compared to 4G and its predecessors, but it also disrupts the network infrastructure, enabling the development of innovative use cases and applications. 5G offers higher connection speeds up to 10 Gbps (ITU, 2017), lower latency and delay i.e. less than 10 ms (Raftopoulou et al., 2019), faster response times, very high capacity, the capability of supporting a hundred times more devices per cell (Marzetta, 2010), higher reliability of the communications, ubiquitous connectivity, and lower battery requirements for devices (Lauridsen et al., 2016).

All of these advancements are possible thanks to the introduction of a set of innovative schemes such as, (i) a brand new architecture that standardizes radio access networks and pushes cloud capabilities to the edge of the network (C-RAN); (ii) new high-frequency bands to support a higher throughput (Wang et al., 2014); (iii) massive Multiple-input multiple-output (MIMO) to allow the transmission and recep-

DOI: 10.4018/978-1-7998-3645-2.ch003

tion of more than one signal concurrently over the same channel; and (iv) network slicing to provision dedicated virtual networks for specific applications and customers (Zhang et al., 2017).

One of the fields that will benefit more from the realization of 5G networks is the IoT. The limitations in terms of delay, throughput, and device density of current mobile networks are preventing many IoT applications to become a reality. Fortunately, all these limitations are solved in 5G networks. The low latency and high throughput of 5G networks will enable all the promises of industry 4.0 and e-health applications especially amid the critical experiencing Corona Virus Disease 2019 (COVID-19) (Bai et al., 2020); they will enable the deployment of smart cities with a high density of devices and sensors; the ubiquity of the connectivity and its high capacity will enable the promise of "full-remote" work for everyone. Moreover, 5G's brand new architecture will better existing applications such as media streaming, and it will the door to the realization of brand-new mobile applications with high computational requirements based on Virtual Reality (VR)/Augmented Reality (AR) and Artificial Intelligence (AI).

All of the aforementioned applications will have an increasing impact in our post-COVID society, where automation will become significantly important (Ting et al., 2020); full-remote work will become the new normal; mobile networks, due to their distributed nature, will become a source of anonymous data collection and processing to prevent and avoid future pandemics (Zhai et al., 2020); and our daily basis will be increasingly digital (Porpiglia et al., 2020) with many of our daily tasks and social interactions supported by VR/AR and media communication. Despite its clear advantages, many have stated that the frequency bands used in 5G networks can spread COVID-19 (Davies, 2020), but the World Health Organization (WHO) quickly released a statement denying this because it is propagated through breathing droplets (WHO, 2020).

IoT can be simply defined as being a technology which connects an electronic device on and off the internet. Several devices can be connected to the internet so long as it is determined to have the capabilities and approaches to the connection. Example of these devices includes washing machines, cell phones, lamps, wearable devices, headphones, coffee makers, automated cars, and other products. Due to these connections, the technology creates people to people, people to things, and things to things relationships; and approximately 26 billion devices are expected to access the internet by the year 2020 under the influence of the growth of IoT technology (Weber, 2010).

IoT is applicable in both the domestic and industrial sectors. For instance, it is applicable in such domestic activities as enhanced learning, e-health, and assisted living. Additionally, in industrial activities, the technology will affect business activities by playing a leading role in logistics, manufacturing, the management of business processes, advanced and intelligent products and goods, transportation, and the automation of significant services (Atzori et al., 2010). However, its impact on the economic growth is significant when applied effectively in key areas requiring faster service and product delivery. Besides other people criticizing IoT's adverse effects, they can see its impact on development as it is widespread and most people are adopting it for faster processing. Some of those who oppose it are quite worried about its attendant risks such as data privacy breaches, security issues, and trust issues. Although sensitive issues might result in serious consequences if compromised, technology cannot advance without precautions and smart techniques of protecting and promoting security being implemented. These technical and social issues need to be collectively addressed in order to include everyone's concerns for effective and widespread acceptance of IoT. Security is a necessity for any information system, and it must be addressed with immediate technologies which protect both the system and the data being transmitted via the devices. Wireless networks face various challenges such as distributed denial of service (DDoS) attacks (Bandyopadhyay & Sen, 2011), malware infection, and other forms of attacks which compromise their functionality and interoperability.

CORE IOT APPLICATIONS IN URBAN SCENARIOS

The Internet of Things 'intends' to connect and revolutionize the world through the seamless connection of heterogeneous smart devices (Akpakwu et al., 2018). The interaction between the fifth-generation networks and the Internet of Things is probably the most awaited technology geared to changing how people and smart devices interact, share information, and communicate. With its numerous applications—ones such as a smart city for creating situational awareness; a smart home in health assistance and energy monitoring; a smart grid for monitoring infrastructure and energy; and smart farming in the agricultural sector—it is evident that IoT has the potential for transforming techniques and modes of interaction which are supported by intelligent systems. The interconnection of "things" by IoT has enabled machine-to-machine (M2M) communication. M2M is a unique mode of heterogeneous device data communication without the need for intervention by humans. The revolution of M2M, mobile, and Internet technologies is termed as being the first phase of IoT applications. Moreover, there are anticipations of IoT bridging numerous techniques for enabling new applications with the connection with physical objects in support of making intelligent and informed decisions. Depending on the different application scenarios, IoT is believed to positively influence and impact the environment; to enhance global economic growth; and to transform communication models. According to Al-Fuqaha et al. (2015), the latest developments in RFID, communication technologies, smart sensors, and Internet protocols are the main enablers of IoT. This section highlights the main IoT application scenarios where IoT application requirements are classified into Consumer IoT (cIoT) and Industrial IoT (iIoT).

One of the benefits of cIoT is its ability to improve the quality of life and improve the human experience (Miraz et al., 2018). Another is that iIoT assists organizations in performing repetitive and dangerous actions; ones which are difficult for a human to accomplish. Moreover, the iIoT also helps in managing repetitive services efficiently (Palattella et al. 2016). cIoT and iIoT share the same attributes with regard to software and hardware platforms. In the main requirements, iIoT and cIoT architecture assists in collecting data, storing data, analyzing data, and sharing results. The architecture of iIoT and cIoT also meet strict security requirements protecting the data during the communication transfer. Figure 1 reveals the security architecture of data distribution service (DDS) for both iIoT and cIoT (Soroush et al., 2016).

However, as is revealed in Figure 2, cIoT is characterized by low data level security level— whereas iIoT, by contrast, is characterized by high-level security systems and requires a high data rate, low latency, bandwidth efficiency, and fault tolerance more than cIoT requires (DDS Foundation, 2019).

The architecture of iIoT demands high modularity, scalability, and interoperability at different levels, involving data interoperability and technical interoperability. However, it does not demand mobility features, since its nodes are fixed. Moreover, the Clot hardware requirements are designed for a limited lifetime, low cost, maintenance. However, the iIoT hardware supports is based on M2M communication for process monitoring and communication. The integration of iIot is also achieved through the hardwired internet (Palattella et al. 2016). Many cIoT communications and iIoT communications' requirements satisfy reliability and timeliness constraints for information exchange and correct, safe processes control. Thus, the communication network of both iIoT and cIoT is engineered with robust protections against losses; stringent deadlines limiting delays; and resilient safety protections against damages. The iIoT

Figure 1. Security architecture of iIoT and cIoT

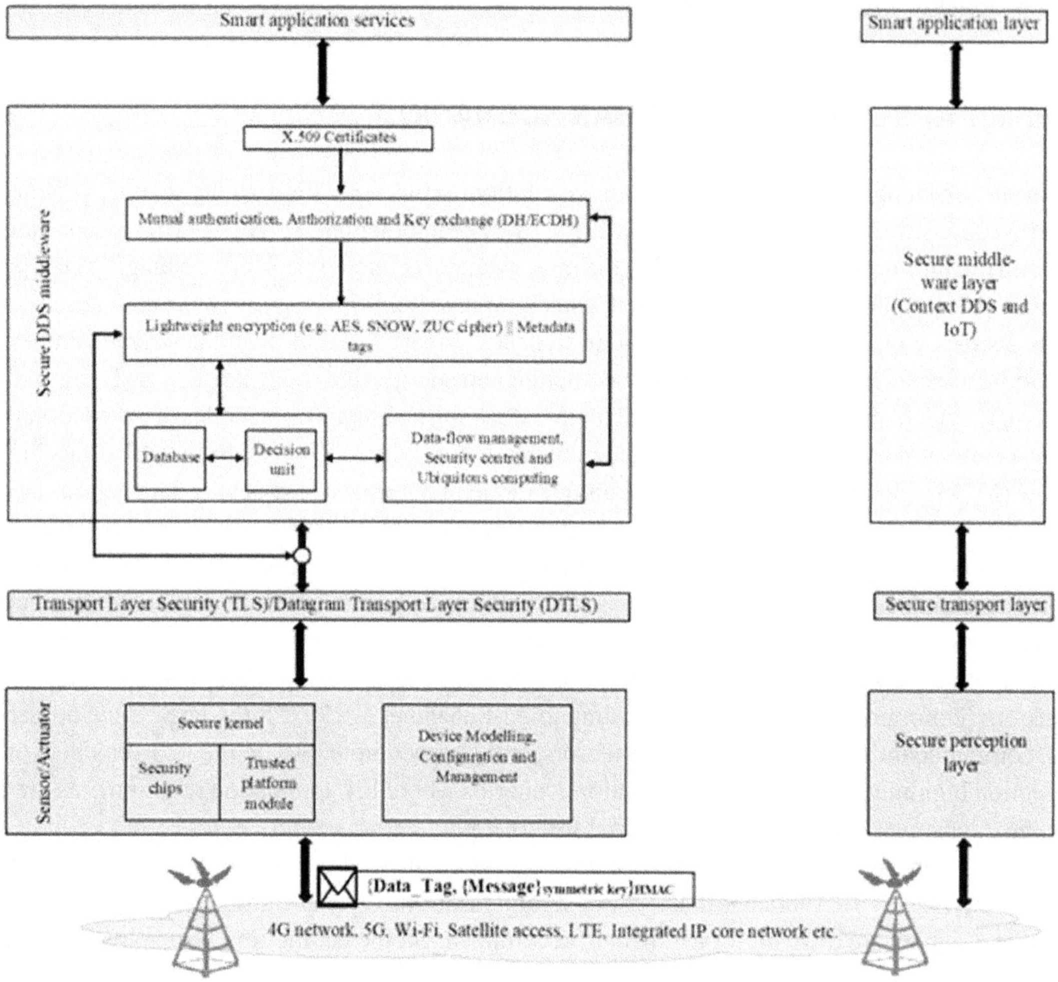

and cIoT are also characterized by their use of 3G and 4G technologies. However, 3GPP LTE (long-term evolution) is the most appealing of technologies for the IoT connectivity landscape (Palattella et al. 2016). To reduce costs, the iIoT is increasingly using cloud-based support—employing AI in analyzing electrical fingerprints. Moreover, the requirements of iIoT equipment are as highly demanding as its maintenance is complex.

Both technologies also share the scalability, friendliness to IP ecosystem, and constrained devices in lean protocol stack implementations. However, there are differences in iIoT and cIoT with regard to their specific communication requirements. For example, the fundamental differences in the iIoT and cIoT requirements are privacy, reliability, and QoS (latency; throughput; etc.). Machine-to-user communication is a typical aspect of cIoT communications and is characterized by low power consumption and ease of installation, of maintenance, and of integration (Miraz et al., 2018). The iIoT paradigm assumes a high-power consumption requirement. By contrast, the demand for sensor-rich smartphones, smartwatches, and health tracking systems (all of which are a part of cloT) encourages and enhances the development of high-power efficiency products. The cloT paradigm has IEEE lower layers (IEEE 802.15.4e TSCH

Figure 2. Requirements iIoT and cIoT

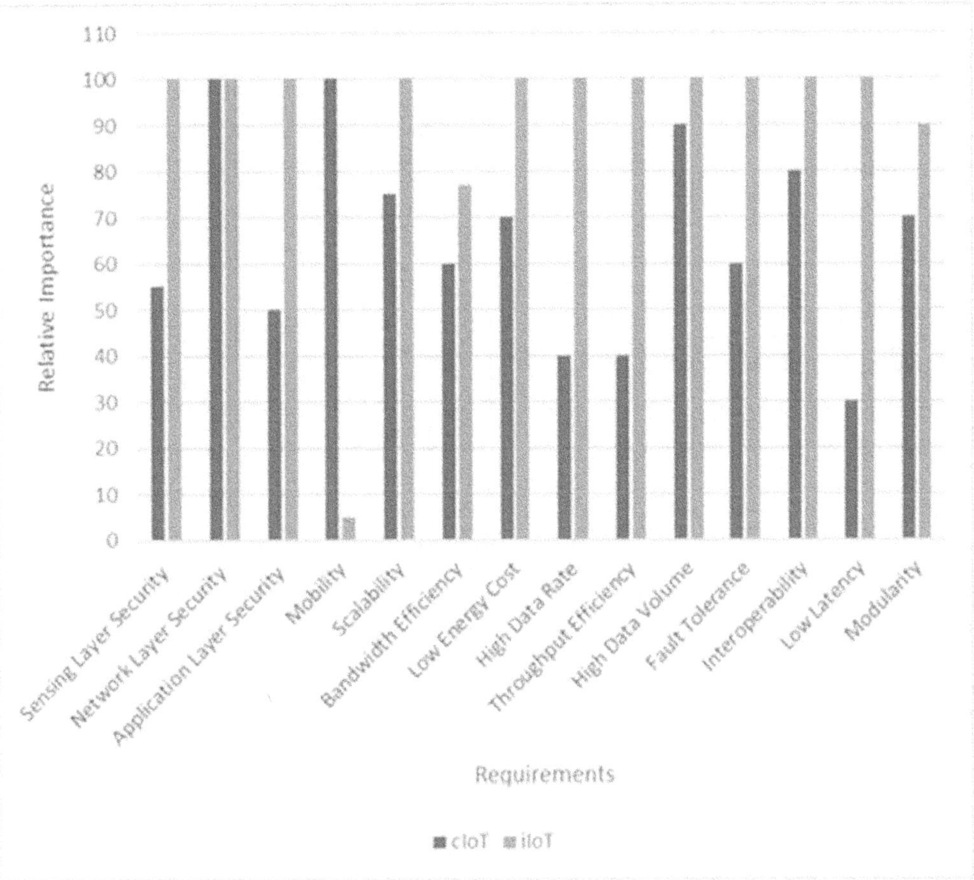

and IEEE 802.15.4 PHY). By contrast, the iIoT paradigm combines IEEE lower layers (IEEE 802.15.4e TSCH and IEEE 802.15.4 PHY) with IETF (Internet Engineering Task Force) higher layers (6LoWPAN) characterized by high reliability, low latency, and ultra-low jitter (Palattella et al. 2016).

Both technologies iIoT and cIoT share some common challenges and communication requirements. The iIoT and cIoT requirements are comprised of the IEEE802.15.4-2006 MAC layer(s) that make its single channel unreliable. Moreover, Bluetooth has a shorter communication range between 1 and 100 m (Miraz et al., 2018). In addition, Wi-Fi suffers from poor roaming support and poor mobility. Wi-Fi cannot guarantee QoS, influenced by large interference because of the unlicensed 2.4 GHz band. Low-Power Wide Area (LPWA) technology is characterized by iIoT and cIoT. The downside of LPWA its its unlicensed long-range communication spectrum. Furthermore, both iIoT and cIoT face similar challenges in terms of exposing the consumer- and industry-sensitive data to a third party. As both iIoT and cIoT applications are increasingly facing the data breach risks in the present IT environment, the applications face the risks of being compromised by the third parties. The data leak is one of the serious challenges in both iIoT and cIoT applications since they both store private and industry data (Palattella et al. 2016). As both applications require several nodes for the exchange of information, the attackers can penetrate the information systems and get access to sensitive data leading to potential vulnerabilities, system attacks, and privacy leaks. However, security systems of both applications are different. While iIoT

Table 1. Comparison between cIoT and iIoT

cIoT	iIoT
Associated with the advent of 5G communications that reduce end-to-end latency, increases data rate, and improve communication coverage.	Data management, middleware, and application programming interfaces are the iIoT key design architecture (Gardašević et al., 2017).
The cIoT protocol consists of IEEE lower layers (IEEE 802.15.4e TSCH and IEEE 802.15.4 PHY)	iIoT combines IEEE lower layers (IEEE 802.15.4e TSCH and IEEE 802.15.4 PHY), as well as IETF higher layers (6LoWPAN).
Low Data-Level Security	High Data-Level Security
Focused on improving the quality of life. For example, the smart home application enables people to manage climate control systems, open their garage automatically upon their arrival at home, prepare coffee, and control security (Kubilay & Bingöl, 2014).	The major aim of applications is to grow the world's economy (Al-Fuqaha et al., 2015).
Low Latency	High Latency
Low data rates	High data rates
Consumer devices	Machine/systems
Consumer-centric	Industry-centric
Machine-to-User communications	Machine-to-Machine communications
Faces challenges in developing suitable responses to customer demands and availability issues.	Applications in most industries are required to grow proportionally in relation to customer requirements and market needs.
Low energy consumption	Significant energy consumption

requires high-level security systems for the protection of industry data, the security of cIoT applications is minimal for the protection of consumer data. Table 1 summarizes the classification of iIoT and cIoT.

There are numerous applications and examples of IoT, and it is expected that the trend for developing these applications will increase more rapidly. This expectation is due to the technology's wide range of applications in various fields; its great impact; and due to various projects (recently recorded as being underway) focusing on IoT implementations. Figure 3 presents a worldwide statistic for the established base of IoT form 2014 and 2020 by category (Statista Research Department, 2020).

Smart City

This concept is the modern approach of advancing technology among cities, where the naming varies according to a group or individual understanding which results in other names. Examples include: knowledge city, digital city, ubiquitous city, intelligent city, and sustainable city. All these definitions are based on the meaning of the word "smart," with different literature materials using the names interchangeably (Cocchia, 2014). A smart city is viewed as being a ubiquitous service application which is aimed at improving the living standards and quality of life in the city by conveniently availing required information to the residents at any moment (Gea, 2013). Multiple devices are interconnected in this scenario where smart technologies are used to offer the needed services such as transportation, utilities, health, homes and buildings, and the government (Vasant, 2019). Challenges facing most of the traditional cities such as traffic jams will be no more as smart cars and intelligent systems are in continuous communication

Figure 3. IoT devices installed base by type (Statista Research Department, 2020)

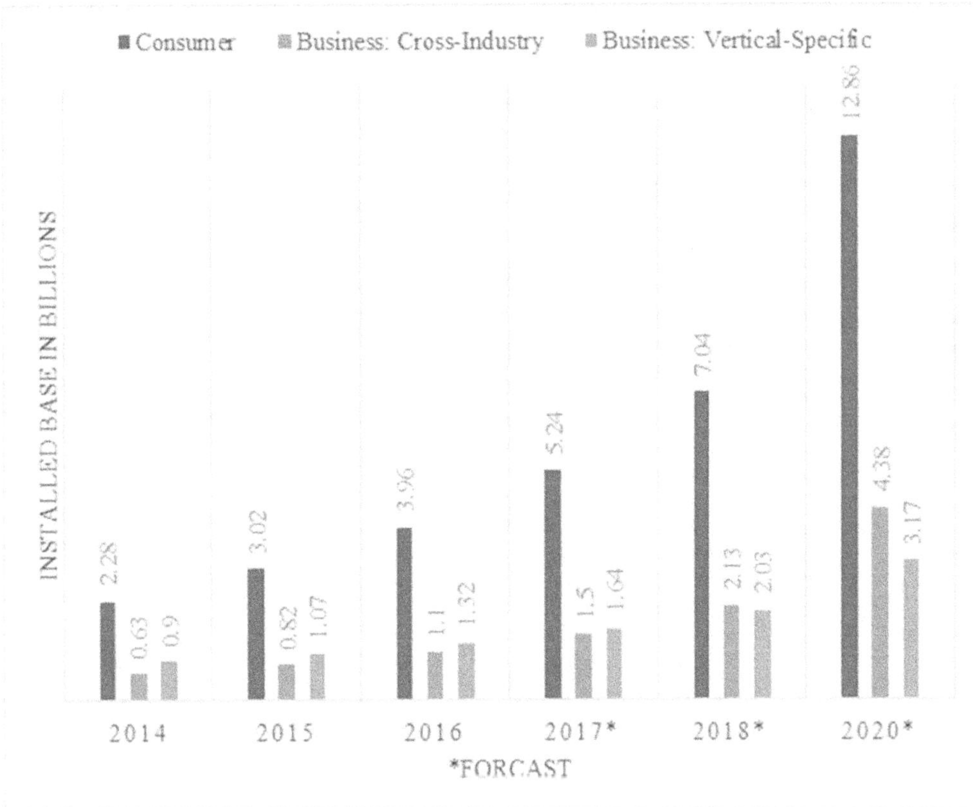

where cars get the information, and they can determine appropriate routes without jams and specifies the right time for travels.

A smart city comprises wireless technologies that connect all devices building within the city. Figure 4 displays the connectedness of smart infrastructure in the smart city. The sensors from each application detect and collect different information gathered together and transmitted to be stored and processed in the cloud through the Gateway.

According to Hancke et al. (2012), Smart cities have multiple sensing requirements due to the multiple applications integrated with the smart city application. These applications include smart surveillance, infrastructures, transportation, services, buildings, healthcare, and water and electricity distribution. In all these sensitive areas shaping smart cities, sensing is at the central part for their operations since it monitors the applications while acting on their own intelligently. Sensors are essential in increasing the efficiency of the utilization of resources based on the collected data with the sensors to reduce the measuring energy consumption, reducing costs, and implementing intelligent systems such as the intelligent transportation systems (Hancke et al., 2012). Thus, in this case, real-time sensing capabilities are required in smart cities to enhance its performance. Moreover, Smart cities require critical computing requirements such as cloud computing technology—which helps in the collecting, storing, and processing bulk of information simultaneously. A reliable and efficient cloud computing services are crucial in this sector to help users to upload important files and data into their computer programs or mobile applications from any location, provided there is internet access.

Figure 4. Connected smart city environment

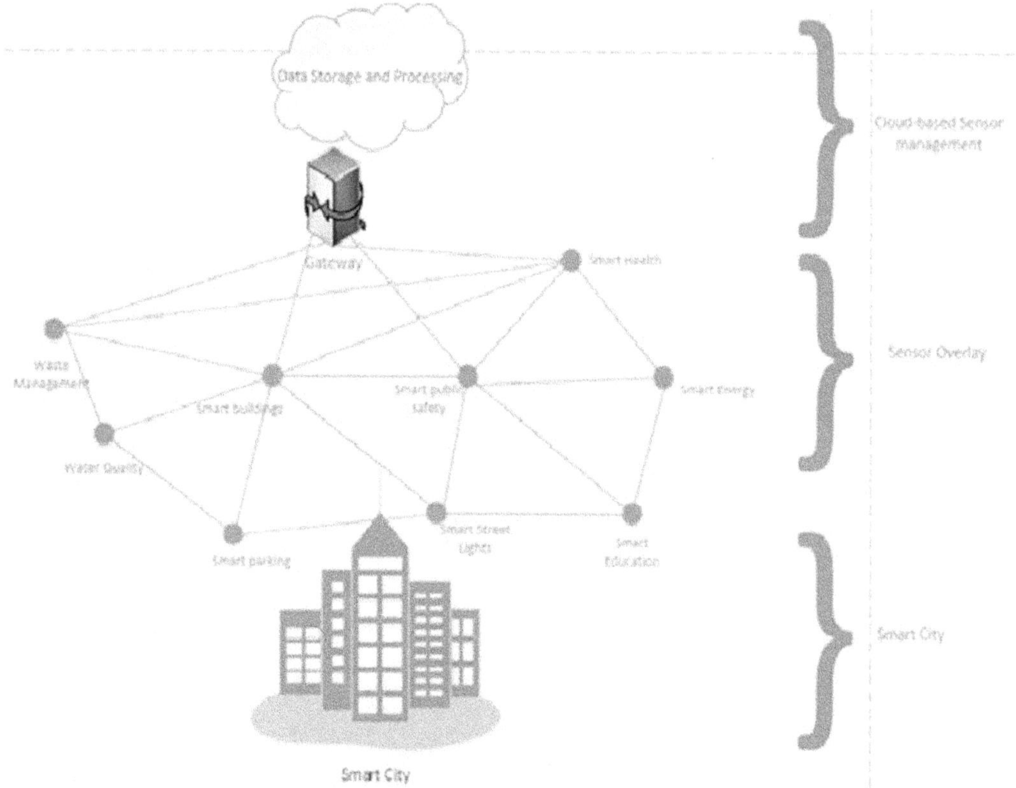

Due to the ubiquitous nature of this application, cloud computing is noted to be a major requirement (Hancke et al., 2012). On the other hand, a reliable, efficient, and real-time communication channel must be established in order to ensure devices continuously communicate—since there are multiple scenarios involving the constant exchange of information to support smart infrastructures from providing users with information, especially in such critical areas as healthcare and transportation. Examples of such communications include the two-way communications between consumers, systems operators, and the consumer loads and resources for advanced metering infrastructures, which indicate that smart cities must solve the issue of communication between the nodes to enhance their application efficiency. The most recommended communication is the wireless technology for connecting multiple devices, with communication standards being critical requirements (Hancke et al., 2012).

Smart Home

The idea of a smart home is shaped by the HVAC (heating, ventilation, and air conditioning) monitoring systems and thermostats. Smart homes and healthcare applications assure careful management of patient medications in the right time and with the right medicines by the applications keeping track of the patient records; and this is aimed at establishing a significant economic impact (Al-Fuqaha et al., 2015). IoT-based health services such as m-Health (mobile health) and Telecare are crucial in enabling disease or infection prevention, medical wellness, diagnosis, treatment, and management services, which are delivered via the electronic devices. By 2025, it is expected that this application will improve the economy by economic growth of up to $2.5 trillion per annum and that the estimated annual economic impact of IoT will range between $2.7 and $6.2 trillion (Al-Fuqaha et al., 2015). Moreover, smart homes are categorized under the application layer of IoT architecture. The application layer is vital in offering customer-requested services such as measurements of air humidity and temperature for each customer who is demanding this crucial information. High performing smart devices are applied at this level to meet specific consumer requirements. Other services available in smart homes include the ability to automatically lower the blinds of the upstairs windows and close other windows based on the analysis of the weather forecast. However, this application needs to have regular internal and external environment interactions. The regular interaction is associated with the differences in appliances available for internal and external environments such as domestic home appliances which are Internet-connected unlike the external appliances such as smart grids (Komninos et al., 2014). Most of the devices operating in the smart home make decisions autonomously based on sensor signals which enable the coordination of the monitoring and control of home appliances.

Through the utilization of a smart application from a user's device such as a smartphone, he/she can control house devices in ways such as adjusting lighting, regulating temperature, opening doors, heating/cooling the air, and directing specific applications to perform certain tasks (such as directing the coffee pot to start brewing). Figure 5 shows a clear illustration of some of the Smart home features integrated and controlled by a smart application.

Figure 5. Smart home scenario

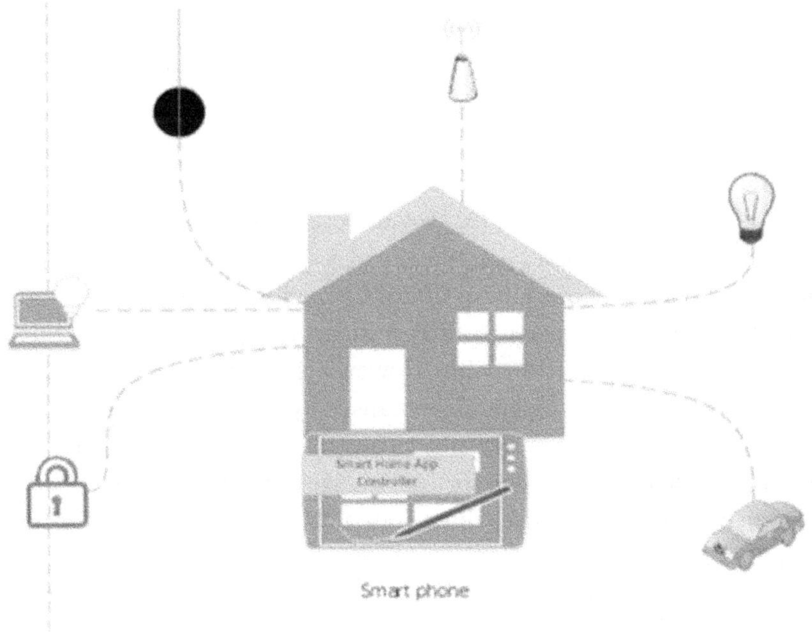

Smart home applications entail several regularly interacting objects—including furniture, smart objects, home appliances such as lighting, and temperature monitoring systems. Such objects require a distributed sensing infrastructure to help in efficiently monitoring the Smart home's constant change of state among the objects (Surie et al., 2008). Moreover, it is recommended that in an intelligent home environment, information processing ought to be integrated with all the daily objects to provide better services beyond the initial purpose, properties, and characteristics. Sensor information from such objects is vital in helping recognize user situations and activities. The ZigBee communication standard is mostly applied in the wireless sensor network in smart home applications. Wireless sensor networks are required for providing functionalities that individual objects may fail to provide where middleware is included (Surie et al., 2008) for protecting the devices and applications against intrusions. Moreover, simple devices, such as smartphones, tablets, and PCs are required for implementing the smart home environment. However, the computing requirements for this application entails ubiquitous and real-time processing and storage systems, including the cloud computing services that require high-speed broadband networking capabilities. Cloud computing also provides an application programming interface for enables users to configure the various services within the smart home using web browser. Furthermore, smart homes applications require accurate, timely, and relevant data or services where it is suggested that the context-aware computing is most applicable since it provides information based on "*where you are, who you are with, and what resources are nearby*" (Hui et al., 2017). Moreover, smart home applications require a secure environment of operation to avoid compromise of the home database containing critical information through such attacks as a malware infection.

Smart Grid

Smart grids use IoT technology applications for enhancing and improving the energy consumption in buildings and houses (Yan et al., 2013). IoT application scenarios in smart grids are essential in helping suppliers of power to manage and control resources for providing power proportionally according to the increase in the population. For instance, IoT is applied in smart grids to interconnect numerous network devices of the providers of energy. Hence, these devices gather, analyses, regulate, manage, and monitor the consumption of energy—which, in turn, provides crucial information for energy which helps in improving the efficiency of service delivery and quality of services, as well as reducing potential failures. Additionally, due to the availability of millions of devices and consumers in smart grids, security and reliability are crucial. According to Yan et al. (2013), smart grids are capable of eliminating electricity blackouts by improving reliability and quality. However, some challenges in intelligent grids are of great concern such as the increase in vulnerabilities caused by remote monitoring, automation, and the interconnection of supervised entities. Cyber-attacks are also on the rise, targeting smart power grids by mainly tampering with their transient stability on voltage (Liu et al., 2015). Hence, IoT application designers not only need to develop devices which are reliable but also safe and secure operating with the latest security protocols.

Due to the several functions assigned in devices of the smart grid including data collection, analysis, regulation, management, and monitoring of energy consumption rates, the application requires an extensive computing power to provide efficient results. Moreover, due to the heterogeneous nature of the smart grid involving multiple energy sources and users, the smart grids require efficient computing devices such as the mobile agent that was proposed by Bera et al. (2015) as the most suitable for supporting the power system under the cloud computing platform. Moreover, Loia et al. (2015) stated that future smart grids should be shaped with bio-inspired and cooperative computing that operates under the basis of the self-organizing sensor networks. Finally, the smart devices and applications within the smart grid including smart meters they should have efficient functionalities to realize digital management and make intelligent decisions which are communicated to the entire smart grid system in helping resource allocation and asset optimization (Brunekreeft et al., 2015).

KEY PERFORMANCE INDICATORS (KPIS): RESEARCH ISSUES AND CHALLENGES IN IOT

Zero Complexity

Zero complexity is a technical requirement in machine type communication devices. As one of the 3GPP cellular systems, these devices have been designed for data use and human voice and whether classified as cIoT or iIoT, zero complexity requirements towards enabling match in technology migration with backward compatibility. As the device complexity is being virtually reduced to zero, it will significantly impact device costs due to the virtual absence of silicon costs. The 3GPP project has identified specific features which can be removed in Machine-Type Communication (MTC) devices to reduce their complexity while maintaining their computing power. KPIs relative to achieving this mission are a restriction of the maximum requirements of IoT applications such as the peak data rate, limitation of device capability by the LTE into a single receive RF chain, and support for half-duplex operations and the reduction of sustainable data bandwidth. According to Palattella et al. (2016), the latest versions of LTE releases, especially in the 12 release's reduction specifications have been closed by the 3GPP, and it is expected that new versions will also be handled. Moreover, Akpakwu et al. (2018), stated that one of the requirements of IoT design requirements is low device cost where it implies that as IoT connectivity is proposed to offer services on ARPU (very low average revenue per user) resulting in low revenue generation, unlike mobile broadband subscriptions. Hence device complexity reduction is crucial in enabling mass-market and massive-volume applications and IoT implementations.

Long Battery Lifetime

Most of the IoT devices will be operating on batteries located in remote areas where the charging cannot be economically feasible or possible to charge. Additionally, besides battery technology evolution, total available energy is likely not to change, and the physical size batteries for IoT devices will continue to be smaller due to the miniaturization of the devices. This calls for the increased energy efficiency in the communication module which will enable IoT device batters to last for decades. Although the battery life of ten years is feasible in LTE Rel-12 and LPWA technologies for infrequent data transmissions, 5G faces challenge maintaining a battery life beyond one decade for increase frequent data transmissions (Palattella et al. 2016). The Release-13 of 3GPP standardization introduced critical features aimed at enhancing the massive deployment of smart services and devices such as smart grids and smart cities. These features include eMTC for improving communication of available cellular technologies such as LTE and GSM networks more efficiently and EC-GSM-IoT. The primary purpose of introducing these standardizations is to ensure IoT applications over an extensive coverage, backward compatibility, and an efficient battery with a long lifetime. For instance, eMTC standardization is designed to support massive IoT deployment and coverage while keeping the battery life to approximately a decade with a 5watt-hour battery system for efficient utilization.

Pervasive Coverage

Both iIoT and cIoT applications require increased levels of coverage such as in smart metering. All devices within the network need to access information shared across various IoT devices suitable for most business models. Additionally, the increased costs incurred when improving the coverage levels by increasing the number of base stations, 3GPP has stipulated some improved coverage MTC and low-complexity devices—which, in turn, facilitate IoT uptake scalability. Pallattela et al. stated that the improvement in coverage could be achieved by repeating information with further stipulations provided by the Release 12 of 3GPP. Whitmore et al. (2015) stated that coverage of IoT enabled business models to have a good limitation which is in correspondence with little management literature on the coverage of IoT, also described as "the coverage of IoT driven business models is scant" (Whitmore et al., 2015).

User Identification and Control

MTC user identification and control requirement entails SIM-integrated low-cost MTC devices. However, due to reasons of complexity, scalability, and configuration, there will be no integrations of SIM with future MTC devices. Every SIM card is equipped with an international mobile subscriber identity (IMSI), belong to each subscriber, and it includes details of the SIM profiles and MTC services as well as containing a direct link to HLR (Home location register). Hence, based on the subscriber details obtained from the profile, most operators can support MTC services. Some of these MTC services include optimal routing with dedicated APN (access point name), optimal data packet size (Palattella et al. 2016). The operator has full control of the MTC subscription through the IMSI for every subscriber within the network. A key indicator of improvement in control of these MTC devices involves new MTC categories of LTE EU by 3GPP, which will enable operators to establish restrictions of MTC devices from the access of roaming devices which will affect performance and allow efficient management for those devices not authorized by the operator to access the network.

Availability and Reliability

Availability in IoT applications is essential in ensuring effective utilization of IoT resources. For instance, MTC devices corresponding to applications with critical infrastructure features of control and connection to wireless networks requires high availability of wireless network connectivity. 5G is a promising network architecture available with minimal downtimes with low energy consumption and accessed by low-cost devices such as in MTC applications. There are guaranteed availability and very high-reliability connectivity, and service and end-to-end delays appear in a magnitude ranging between several milliseconds. This assures an operation which is reliable and uninterrupted for mission-critical applications (Mavromoustakis et al., 2016).

Mobility

The mobility patterns of users serve as the primary determinant of the chances of a Device-to-Device-based (D2D-based) interaction and connection—since, by nature, their connection cannot easily be approximated. Thus, the potential mobility of all devices involved in the connection results in increased opportunistic contacts. Moreover, in a study conducted by Militano et al. (2015) on "device-to-device communications for the 5G Internet of Things," it was stated that user mobility effects have a high impact on the system performance when native support for D2D communication is integrated into the architecture of 5G systems. Some of the parameters of mobility which determines the link performance of each D2D include throughput, duration, and length. These parameters also define the overall performance of the D2D system. Also, Militano et al. (2015) reported that the type of application which runs on the tops of D2D links also affects the overall system performance. 5G mobile networks for the Internet of Things are also expected to enable the mobility of vehicles in emergency scenarios which will improve traffic efficiency and reduce the number of road accidents for eV2X ("enhancement of vehicle to everything") vehicles such as ambulances and trucks (Akpakwu et al., 2018).

Scalability

Sometimes it is difficult to design or develop a scalable platform for IoT in a quick environment, as doing so is a complicated and complex task. Most business models operating with remote devices across the globe face challenges in the management of these devices due to scalability challenges. Some of the advantages of ensuring scalable IoT connectivity include making it easy to implement security updates among devices, thus leading to improved performance which is optimized in most use cases of IoT and traffic patterns. A scalable IoT application needs the network to be aligned with the device longevity, allow systems to be easily expanded and increase the lifetime of the devices. Since the 5G mobile network is noted as being the most promising network for enabling the Internet of Things, some service requirements such as low latency and scalability which will efficiently allow Massive to critical IoT applications. IoT scalability is defined as the ability to introduce new functions, applications, and heterogeneous devices focused on serving end users without any compromise of the current service provisions and quality (Akpakwu et al., 2018). Also, scalability and other factors such as security and privacy, interoperability and heterogeneity are termed as the major challenges which confront the vision of IoT, and they were proposed to be considered in future investigations.

Interoperability

Interoperability is identified as one of the features which hinders the IoT vision, as noted by Akpakwu et al. (2018) although there is no one-fit-for-all solution, this factor can be evaluated to determine 5G Internet of Things efficiency. Moreover, interoperability is a significant issue in applications of MTC, since these devices are supplied by different vendors. Hence, they lack standardization—which, in turn, limits heterogeneous device interactions. The seamless end-to-end interoperability among different network technologies is a major challenge faced when determining MTC applications for IoT which involves the connecting of everything, anywhere, and at any time. Upon successful addressing of interoperability issues, multiple device connectivity across various networks of communication will be enabled. Akpakwu et al. (2018) argued that perhaps this might lead to the realization of the concept of IoT that allows connection of heterogeneous devices through communication technologies for disseminating, communicating, and collecting critical information in relation with other smart applications or devices. This challenge of interoperability among heterogeneous devices needs to be addressed to ensure consumer demands are adequately satisfied. Akpakwu et al. (2018) reported that IoT heterogeneous nature is the major contributor to issues of interoperability—as it introduces numerous challenges of integration such as discovery mechanisms, standard practices, and service descriptions that must be addressed adequately in order to ensure the creation of an interoperable environment for IoT heterogeneous networks.

Security Management and Trust

Security and trust are the most crucial factor that must be considered in most IoT applications, as they carry sensitive information transmitted across the 5G mobile networks. The encryption algorithms used need to be reliable and safe in order for them to be considered for the sharing of information. For instance, the 3GPP security framework was developed for GSM with advanced encryption features of the air interface to prevent eavesdropping. A robust security mechanism must authenticate every user or device within the network. Some of the other security mechanisms suitable for enhancing security management and trust include mutual authentication, integrity protection of signaling, advanced key management systems, and the use of state-of-the-art encryption algorithms. The security framework has access to SIM card information which contains subscriber credentials which are used to the user or device authentication (Palattella et al. 2016). The trust of subscribers in the 3GPP systems is taken into account through random assignment of temporal identifies which makes the device and user tracking more complicated. Information shared within devices should never be altered or intercepted by any malicious user or device. Hence, designers of applications need to consider security as a firm requirement with support networks for providing an IoT system which is reliable with improved quality of services. Additionally, the Social Internet of Things is designed with a specific level of trust which enables them to communicate with each node, which is a trusted friend (Mavromoustakis et al., 2016). Hence, the isolation of nodes which are malicious can be easily achieved to ensure improved trustworthiness. Moreover, flexibility and security are some of the open issues that need to be handled during the gateway design which connects IoT objects and networks. Privacy issues such as application misuse are also a significant challenge in smart objects (Khan et al., 2017).

IOT COMMUNICATION TECHNOLOGIES

Low-Power Wide Area Network (LPWAN)

Modern technology has witnessed tremendous growth in the number of devices being connected to low-power wireless communication. Moreover, actuators and sensors are simple devices which are being rapidly connected, and all these are classified under the LPWAN communication technologies. The main aim of developing these technologies is to offer information transmission efficiently while using low power, low cost, and low bandwidth among various devices. According to Mekki et al. (2018), approximately 50 billion devices are likely to access the internet via radio communications by 2020. The also stated that the increased growth of IoT adoption in most markets had made LPWAN a widespread, long-range, low-rate radio communication technology. Among the mostly developed LPWAN technologies that compete in the market for large-scale deployment, they include Narrowband Internet of Things (NB-IoT), LTE-M, Sigfox, and LoRa, having numerous technical differences. LTE-M is also an LPWAN technology, which is part of the 3GPP standard release 13 that is characterized by reduced costs and complexity, and lower power consumption; and it covers an in-depth area reaching even to challenging areas such as deep inside buildings.

LPWAN technologies emerged after multiple implementations of various communication technologies such as ZigBee and Bluetooth radio technologies operating over short ranges, which proved to be ineffective and which increased IoT applications' requirements which demanded low-power-consuming communication technologies (Mekki et al., 2018). Cellular communications such as 2G, 3G, and 4G were developed to be solutions for larger coverage, but they were also ineffective due to excessive device energy consumption (Mekki et al., 2018). Moreover, the increased adoption of LPWAN technology in industries and research communities is influenced by the long range, low-power and low-cost characteristics of communication. Its extended range of coverage covers 10 to 40 kilometers in rural regions and 1 to 5km and urban regions, respectively. Also, its battery life is 10+ years; hence, it is inexpensive and highly energy efficient. The operation cost 1€ per device per year, while the value of the chipset is below 2€ (Raza et al., 2017). Mekki et al. (2018) reported that most of the recent experimental investigations on the indoor and outdoor environment performance of LPWAN show that the technology is most suitable for IoT applications transmitting small amounts of data over a long distance. Based on IoT application requirement factors, Sigfox, LoRa, NB-IoT, and LTE-M possess some similarities and differences, which have been noted in Table 2. The similarities and differences also indicate how far the technologies have achieved IoT KPIs.

All of the four technologies of LPWAN are applicable in most application domains—including manufacturing automation, smart farming in the agricultural field, electrical metering, and pallet tracking for logistics. NB-IoT provides a high-quality service while SigFox, LTE-M and LoRa are of low cost with an extended battery lifetime suitable for sensors in manufacturing automation. Due to the unavailability of LTE cellular coverage across many farms, NB-IoT might not be used in future. On the other hand, LoRa, LTE-M and Sigfox are still ideal for smart farming (Mekki et al., 2018). However, technical differences and IoT factors are the main determinants of their feasibility of being applied in specific domains (Mekki et al., 2018). Additionally, the differences between the technologies indicate that no particular technology can be used in all IoT applications. Hence NB-IoT and LTE-M fit in retail point of sale terminals application while Sigfox and LoRa are applied in the smart building domain. Sigfox and LoRa fit in smart building domain since the devices used to require long battery life and low costs. Moreover, the above differences and some of the similarities indicate that LPWAN communication technology is designed with low energy consumption and low-cost model for most devices helping realize some of the IoT KPIs. Also, LTE-M is suitable for alarm or warning systems in smart homes, metering application domains (Petajajarvi et al., 2015). Star topology is most adopted in LPWAN technologies, especially LoRaWAN networks where base stations communicate between the network server and sensor nodes (Petäjäjärvi et al., 2017). In this type of topology, the sensor nodes and base stations communicate through wireless channels which utilize the LoRaWAN physical layer. The connection between the central server and gateways require an IP-based network to be managed. Star topology in LPWAN enables the base station to carry all the complexity which keeps the end devices simpler—hence lowering energy consumed and lowering costs spent on the devices (Li et al., 2017).

Table 2. Comparison between, Sigfox, LoRa, NB-IoT, and LTE-M

Criteria	Sigfox	LoRa	NB-IoT	LTE-M
Bandwidth	100 Hz	125 kHz and 250 kHz (Mekki et al., 2018)	180 kHz	Has the highest (1.4–20 MHz)
Transmitted power	20 dBm	Up to 27 dBm	17 on average	17 on average
Frequency	The frequency is unlicensed ISM bands, where it operates on different ranges in different locations: 868MHz, 433MHz, and 915MHz in Europe, Asia and North America, respectively (Mekki et al., 2018).	The frequency is unlicensed ISM bands, where it operates on different ranges in different locations: 870MHz, 433MHz, and 920MHz in Europe, Asia, and North America, respectively.	Operates on licensed 3GPP frequency bands (200 kHz used).	Operates on a licensed spectrum (1.8 GHz).
Transmission rates	Maximum: 100 bps	Maximum: 50kbps	Maximum: 250 kbps	Has the highest (1Mbps for UL and DL)
Mobility	No	Yes	Yes	Yes
Interference immunity	Very high	Very high	Low	Low
Latency	Medium	Medium	Low	Low (Bhatoolaul et al., 2017)
Payload size	Maximum: 8bytes (DL) and 12bytes (UL) (Mekki et al., 2018)	240bytes	Up to 1600 bytes	100 bytes–1000 bytes
Security	Not maintained	Uses 128-AES	Inherits LTE-security features through attached SIM-card	Supports mutual authentication and encryption through attached LTE-M SIM-card
Localization	RSS (Received Signal Strength)	TDOA (Time difference of arrival)/ AoA (Angle of arrival) + RSS	RSS	RSS
Mode of communication	Asynchronous communication	Asynchronous communication	Synchronous communication	Synchronous communication
Energy efficiency	Operates on sleep mode; hence, only a low amount of energy is consumed	Very high	Medium high	Medium
Device lifetime	Devices have a long lifetime.	>10 years device's battery life.	Up to more than 10 years device's battery life.	5–10 years device's battery life.
Coverage capabilities	High	Low	Low	High
Deployment model	Actually deploys	Deployments raised rapidly	In-band LTE/ Standalone/ Guard-band LTE	In-band LTE/ Standalone

Narrowband Internet of Things (NB-IoT)

NB-IoT refers a communication technology which operates on Narrow Bands that have been specified by 3GPP Release 13. The technology was established purposely to cover a wide coverage area for IoT applications. NB-IoT, LTE and GSM (global system for mobile communication) operate interchangeably under frequency bands that are licensed—ranging from 700MHz, 900 MHz and 800 MHz, respectively. According to Wang et al. (2017), it is stated that NB-IoT occupies frequency bands corresponding to one resource block of LTE and GSM transmission, which is 200 KHz. Thus, this selection of the frequency bands gives possibilities of three operation modes which include operations such as guard band, in-band, and stand-alone. The Operations of Stand-alone occurs in scenarios utilizing the current GSM frequency bands (Figure 6).

Figure 6. Stand-alone NB-IoT deployment

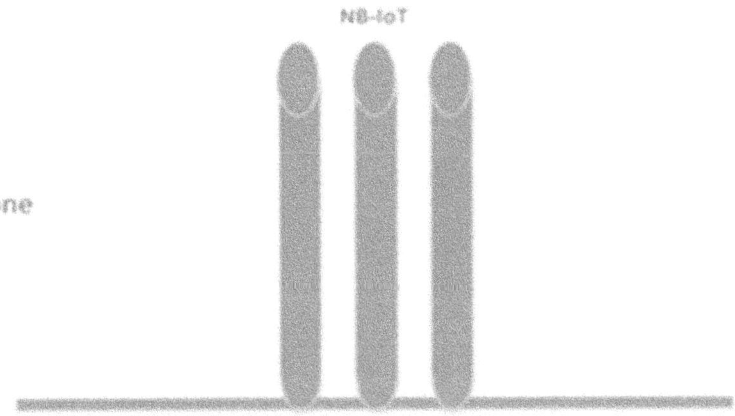

Figures 6, 7, and 8 show the different deployment models of operation.

The Guard-band operation uses resource blocks that are unused in the guard band of LTE carriers (Figure 7).

And In-band operation uses a single LTE carrier resource block (Figure 8).

These deployment options are required to be transparent to UE (User Equipment) upon turning it on for the first time, and NB-IoT carrier search. NB-IoT is focused on providing flexibility in deployment, which will allow operators to establish NB-IoT with the utilization of the available spectrum in small portions. This technology's primary design target is to operate on ultra-low-end IoT applications (Wang et al., 2017).

As NB-IoT technology is being implemented on LTE infrastructure that is well-established; and from the protocol stack perspective, it can be viewed as being an emerging interface. NB-IoT supports up to 100,000 devices per cell in connection with the Internet; and it has the potential of improving its scalability through the addition of NB-IoT carriers (Mekki et al., 2018). Wang et al. (2017) stated that NB-IoT uses FDMA (frequency division multiple access), OFDMA (orthogonal FDMA), and QPSK (quadrature phase-shift keying) for uplink, downlink, and modulation, respectively with a limited data rate of 20kbps for uplink and 200 kbps for downlink. The minimum system bandwidth required by NB-

Figure 7. Guard-band NB-IoT deployment

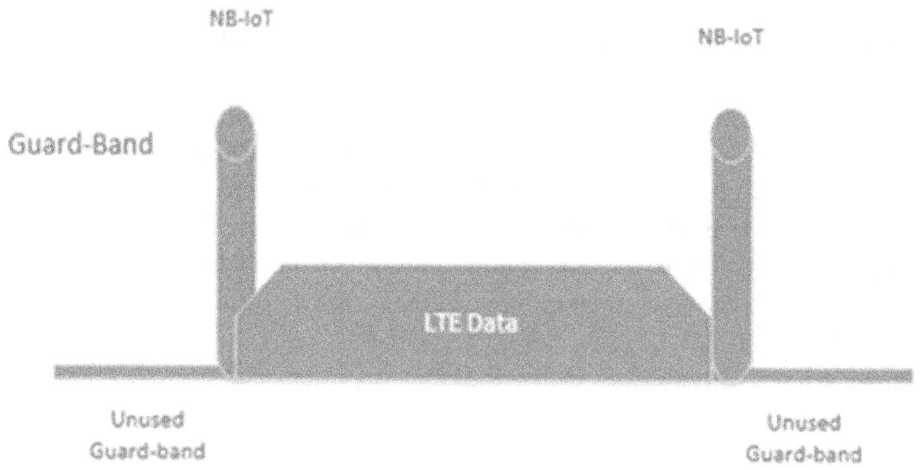

Figure 8. In-Band NB-IoT deployment

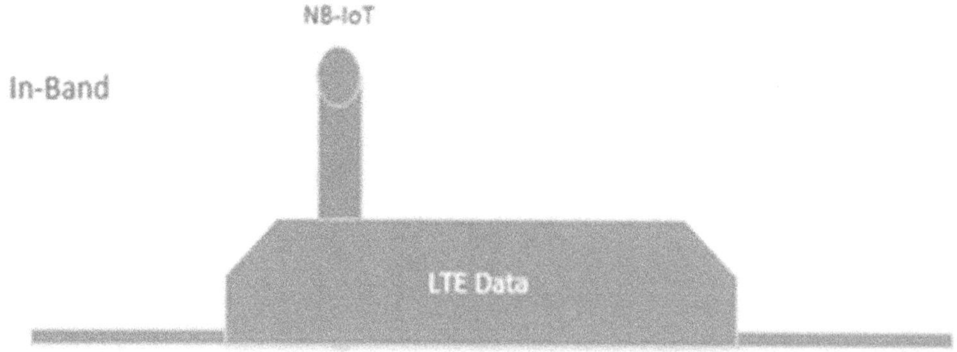

IoT is 180 kHz operating both uplink and downlink. Also, Mekki et al. (2018) stated that NB-IoT could achieve the battery lifetime IoT key performance indicator of up to a decade when it operates on the average transmission of 200 bytes per day.

According to Wang et al. (2017), NB-IoT has no full backward compatibility with 3GPP devices. NB-IoT is also characterized by the extensive use of LTE design features such as channel coding, interleaving, rate matching, single-carrier FDMA (SC-FDMA), and OFDMA. These features are advantageous, as they reduce the time taken in developing full specifications for NB-IoT. More researchers suggest that there are expectations of reducing the time taken to develop products of NB-IoT for the existing LTE software and equipment suppliers. Moreover, NB-IoT has a downlink transmission scheme similar to LTE (for instance, slow speed: 0.5 ms; sub frame: 1 ms; frame duration: 10 ms). On the other hand, the uplink transmission also indicates similarity to LTE including 1 ms sub frame, 0.5 ms slot, and 15 kHz subcarrier spacing. Even the NB-IoT physical channel design entirely relies on LTE legacy. During downlink, NB-IoT offers three signals and three physical channels which include NPSS (Narrowband Primary Synchronization Signal), NSSS (Narrowband Secondary Synchronization Signal), NPBCH (Narrowband Physical Broadcast Channel), NPDCCH (Narrowband Physical Downlink Control Channel), and NPDSCH (Narrowband Physical Downlink Shared Channel). Moreover, during the uplink, two channels are presented which include NPRACH (Narrowband Physical Random-Access Channel) and Narrowband Physical Uplink Shared Channel (NPUSCH). However, its physical channels slightly differ with the LTE in that its signals and physical channels, after some time, get multiplexed.

NB-IoT Resource Mapping

NB-IoT resource mapping has been developed strategically to enable effective performance, coexistence with LTE in case the deployment happens in the LTE carrier. For instance, LTE signals orthogonality has been preserved through the avoidance of NB-IoT signals mapping to recently used resource elements by signals of legacy LTE. Below is an illustration of resource mapping with each column, showing the resource elements in a single symbol of OFDM (Wang et al., 2017).

The Figure 9 resource mapping example contains 12 resources elements for each symbol of OFDM with 12 subcarriers which correspond to the resource elements.

Figure 9. NPDCCH/NPDSCH Stand-alone or Guard-band (Wang et al., 2017)

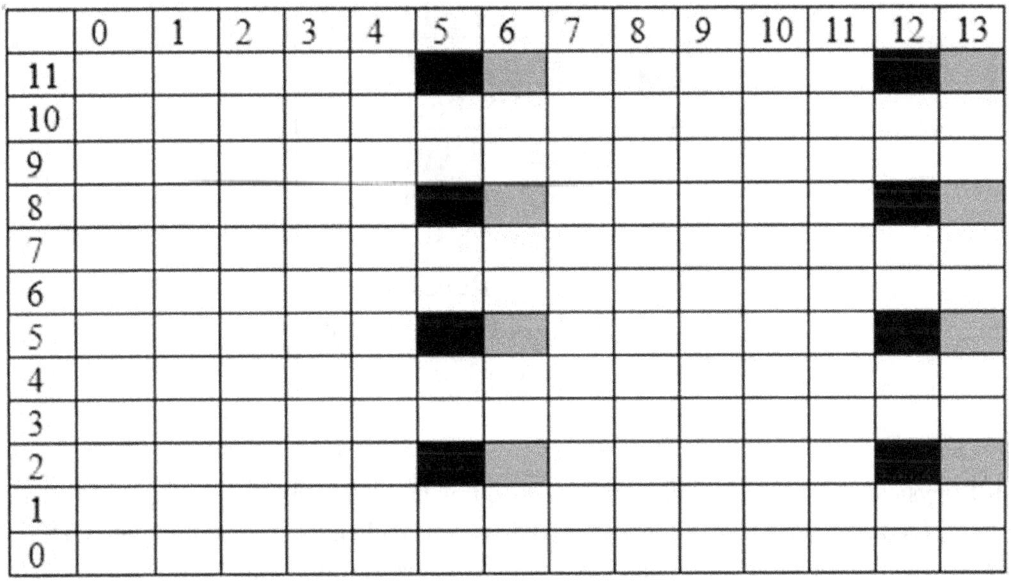

Moreover, in Figure 9 (which indicates guard-band and stand-alone deployments), all the resource elements of LTE need not be protected; hence NPDSCH, NPDCCH, and NRS have the advantages of using all element resources in one pair of PRB (Physical resource blocks) (Wang et al., 2017). On the other hand, Figure 10 shows in-band deployment; whereas NPDCCH/NPDSCH cannot be resource mapped onto LTE CRS and LTE PDCCH.

Additionally, NB-IoT signals have not been mapped onto the already-used elements of resources in legacy LTE signals which help in preserving the orthogonally of LTE signals. Also, the design of NB-IoT considers enabling UE to learn the different modes of deployment (stand-alone, guard-band, or, in-band) and the identity of cells through initial acquisition in NB-IoT and LTE (Wang et al., 2017).

Figure 10. NPDCCH/NPDSCH In-band (Wang et al., 2017)
Key for Figure 10. NB-IoT resource mapping examples

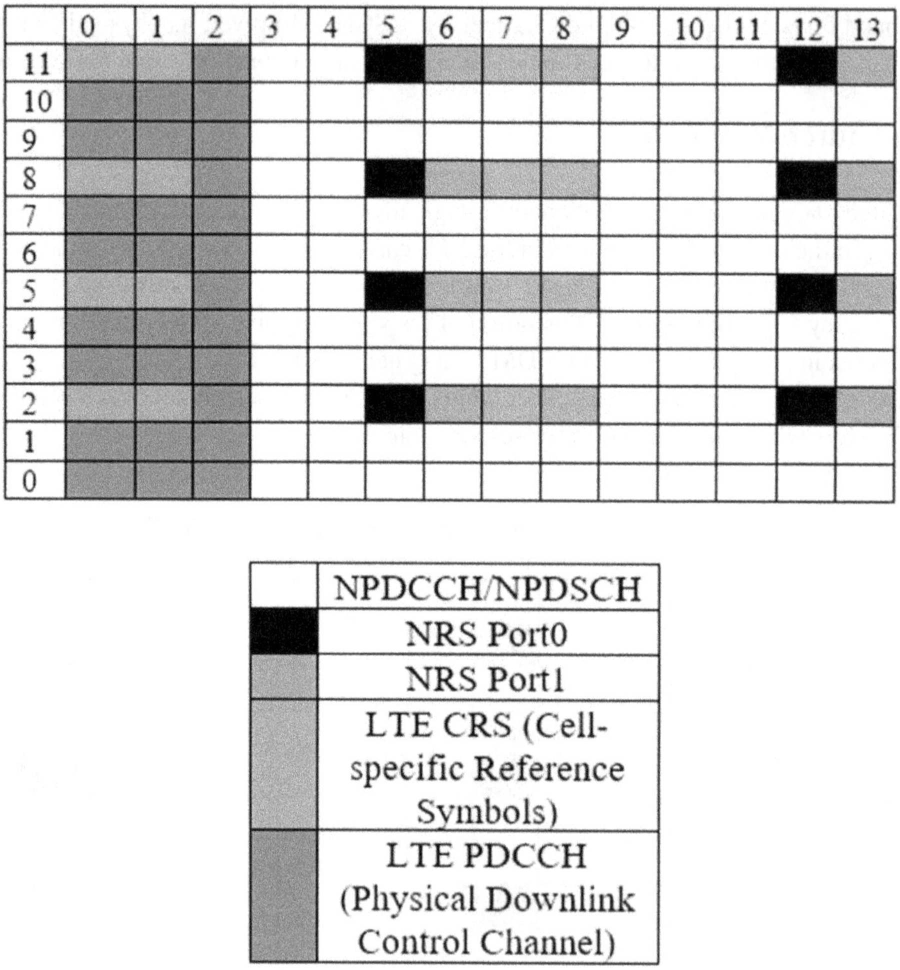

NB-IoT physical channels and signals of both downlink and uplink modes have specific relationships with many LTE counterparts. For instance, the NPSS physical channel relationship with LTE is characterized by all cells sharing one NPSS and has a new sequence fitting in one PRB in LTE. On the other hand, NPBCH is associated with 640ms Transmission Time Interval in LTE. NSSS have similarities to a new sequence which fits into a single PRB, and it also offers the lowest/least significant 3 bits of the number of the system frame. However, this differs from LTE SSS which does not provide the similar bits. Also, there is an overlap of LTE SSS (Secondary Synchronization Signal) with the six middle PRBs (Wang et al., 2017). NPDSCH has a single version of redundancy, and it uses TBCC (Tail-Biting Convolutional Code); whereas LTE counterparts contain more versions of redundancy, and they use Turbo Code. While NPDSCH uses a single QPSK, there are multiple higher order modulations used on LTE counterparts. Lastly, NPDSCH has the support of single-layer transmission, whereas there are several spatial-multiplexing layer transmissions supported in LTE (Wang et al., 2017).

According to Mekki et al. (2018), 3GPP's current plan suggests that they will extend mobility, multicast services, and localization methods for NB-IoT. Also, some additional technical details will, hence, be enhanced in order to ensure the effective application of NB-IoT technology. NB-IoT technology does not require any gateways, since it is a potentially less expensive option of communication technology, although other infrastructures have gateways which aggregate data from the sensors in order to facilitate communication with the central servers. This indicates that all sensor information is sent directly to the central server. As a result, companies such as Vodafone, Ericsson, Huawei, and Qualcomm are actively engaged in putting more effort into the commercialization of NB-IoT (Lucero, 2016). According to Lucero (2016), it is stated that these companies and others which operate in the mobile industry are developing NB-IoT in 3GPP with the aim of providing a better technology suitable for most IoT applications which require long battery life with low cost. Huawei's "Cellular IoT" technology, which was its original development, is termed as being the basis of NB-IoT technology in the global mobile industry which is later supported by the 3GPP process (Lucero, 2016). Besides the benefits of NB-IoT of increasing device battery life and reducing device costs, it also extends the connection range by 20 dB, which enables basement device connectivity (such as sump pumps lying a few feet into the ground; and smart electric meters).

NB-IoT has accomplished some of IoT KPIs such as lowering device complexity through UE implementation designs. One redundancy version support in downlink; the use of single antenna UE; no requirements of measurements of connected mode mobility; and low sampling rates are some of the features characterizing NB-IoT low device complexity (Wang et al., 2017). NB-IoT design allows latency which is less than 10s, thus making it suitable for applications such as sending alarm signals. Performance of NB-IoT use cases is defined by the following requirements, coverage, battery lifetime, data rates, and coverage range. It is also proposed that IoT traffic annual growth is expected to rise to 23% from 2015 to 2023—thus needing NB-IoT design in order to effectively incorporate all the performance metrics and good capacity which can support the high growth rates (Wang et al., 2017). Since NB-IoT is also classified as LPWAN communication technology, it supports star topology. This topology simplifies network maintenance and deployment operations (Adelantado et al., 2017). Firstly, NB-IoT application domains include electric metering because of its frequent communication and high data rates required. Secondly, it is applied in manufacturing automation because it offers a high quality of service and regular communications. Lastly, this technology is the best fit for being the retail point of sale terminal application, due to its low latency and long battery life.

CONCLUSION

In the industrial world, as with every other innovation, IoT and 5G networks face obstacles. The main problems include; the data contain various noise features. During communication, there may be high opportunities for interference since a variety of wireless network linked devices. In addition, the IoT security issues have still not been completely overcome, and although global enterprises continue to improve security models, it seems that the issue remains. To tackle emerging threats, IoT needs to consider specific conditions and security requirements. Furthermore, IoT's equipment must be controlled and improved to meet all the security needs, as it offers the first insight into the application of IoT methods, where most of the devices have very limited resources.

REFERENCES

Adelantado, F., Vilajosana, X., Tuset-Peiro, P., Martinez, B., Melià-Seguí, J., & Watteyn, T. (2017). Understanding the limits of LoRaWAN. *IEEE Communications Magazine*, *55*(9), 34–40. doi:10.1109/MCOM.2017.1600613

Akpakwu, G. A., Silva, B. J., Hancke, G. P., & Abu-Mahfouz, A. M. (2018). A survey on 5G networks for the Internet of Things: Communication technologies and challenges. *IEEE Access : Practical Innovations, Open Solutions*, *6*, 3619–3647. doi:10.1109/ACCESS.2017.2779844

Al-Fuqaha, A., Guizani, M., Mohammadi, M., Aledhari, M., & Ayyash, M. (2015). Internet of Things: A survey on enabling technologies, protocols, and applications. *IEEE Communications Surveys and Tutorials*, *17*(4), 2347–2376. doi:10.1109/COMST.2015.2444095

Atzori, L., Iera, A., & Morabito, G. (2010). The Internet of Things: A survey. *Computer Networks*, *54*(15), 2787–2805. doi:10.1016/j.comnet.2010.05.010

Bai, L., Yang, D., Wang, X., Tong, L., Zhu, X., Bai, C., & Powell, C. A. (2020). Chinese experts' consensus on the Internet of Things-aided diagnosis and treatment of coronavirus disease 2019. *Clinical eHealth.*

Bandyopadhyay, D., & Sen, J. (2011). Internet of Things: Applications and challenges in technology and standardization. *Wireless Personal Communications*, *58*(1), 49–69. doi:10.100711277-011-0288-5

Bera, S., Misra, S., & Rodrigues, J. (2015). Cloud computing applications for smart grid: A survey. *IEEE Transactions on Parallel and Distributed Systems*, *5*(5), 1477–1494. doi:10.1109/TPDS.2014.2321378

Bhatoolaul, D., Ghosh, A., Mangalvedhe, N., & Ratasuk, R. (2017). *LTE-M evolution towards 5G massive MTC. In 2017 IEEE Globecom Workshops (GC Workshops)*. IEEE.

Brunekreeft, G., Buchmann, M., Dänekas, C., Guo, X., Mayer, C., Merkel, M., Rehtanz, C., Göring, A., Herrmann, A., Kodali, R., Stadler, M., Uslar, M., Vogel, N., Luhmann, T., Menz, T., Müller, S.-U., & Recknagel, P. (2015). *China's way from conventional power grids towards smart grids. Regulatory Pathways for Smart Grid Development in China*. Springer Vieweg. doi:10.1007/978-3-658-08463-9

Cocchia, A. (2014). *Smart and digital city: A systematic literature review. In Smart city*. Springer.

Davies, J. (2020). 5G Conspiracy Theories; What They Are, Why They Are Wrong and What Can Be Done. *Telecoms.com*. https://telecoms.com/503845/5g-conspiracy-theories-what-they-are-why-they-are-wrong-and-what-can-be-done/

Foundation, D. D. S. (2019). *How does DDS compare to other IoT technologies?* Accessed April 23, 2019 from https://www.dds-foundation.org/features-benefits/

Gardašević, G., Veletić, M., Maletic, N., Vasiljevic, D., Radusinovic, I., Tomovic, S., & Radonjic, M. (2017). The IoT architectural framework, design issues and application domains. *Wireless Personal Communications*, *92*(1), 127–148. doi:10.100711277-016-3842-3

Gea, T., Paradells Aspas, J., Lamarca, M., & Roldan, D. (2013). Smart cities as an application of Internet of Things: Experiences and lessons learnt in Barcelona. In *2013 Seventh International Conference on Innovative Mobile and Internet Services in Ubiquitous Computing (IMIS)*. IEEE. 10.1109/IMIS.2013.158

Hancke, G., Silva, B., & Hancke, G. Jr. (2012). The role of advanced sensing in smart cities. *Sensors (Basel), 13*(1), 393–425. doi:10.3390130100393 PMID:23271603

Hui, T. K. L., Sherratt, R. S., & Sánchez, D. D. (2017). Major requirements for building smart homes in smart cities based on Internet of Things technologies. *Future Generation Computer Systems, 76*, 358–369. doi:10.1016/j.future.2016.10.026

ITU. (2017). *Minimum requirements related to technical performance for IMT-2020 radio interface (s)*. ITU.

Khan, A., Rehmani, M. H., & Rachedi, A. (2017). Cognitive-radio-based Internet of Things: Applications, architectures, spectrum related functionalities, and future research directions. *IEEE Wireless Communications, 24*(3), 17–25. doi:10.1109/MWC.2017.1600404

Komninos, N., Philippou, E., & Pitsillides, A. (2014). Survey in smart grid and smart home security: Issues, challenges and countermeasures. *IEEE Communications Surveys and Tutorials, 16*(4), 1933–1954. doi:10.1109/COMST.2014.2320093

Kubilay, T., & Bingöl, O. (2014). Web-based smart home automation: PLC-controlled implementation. *Acta Polytechnica Hungarica, 11*(3), 51–63.

Lauridsen, M., Berardinelli, G., Tavares, F. M., Frederiksen, F., & Mogensen, P. (2016, May). Sleep modes for enhanced battery life of 5G mobile terminals. In *2016 IEEE 83rd Vehicular Technology Conference (VTC Spring)* (pp. 1-6). IEEE.

Li, L., Jiuchun, R., & Zhu, Q. (2017). On the application of LoRa LPWAN technology in Sailing Monitoring System. In *2017 13th Annual Conference on Wireless On-demand Network Systems and Services (WONS)*.IEEE.

Liu, R., Vellaithurai, C., Biswas, S. S., Gamage, T., & Srivastava, A. (2015). Analyzing the cyber-physical impact of cyber events on the power grid. *IEEE Transactions on Smart Grid, 6*(5), 2444–2453. doi:10.1109/TSG.2015.2432013

Loia, V., Terzija, V., Vaccaro, A., & Wall, P. (2015). An affine-arithmetic-based consensus protocol for smart-grid computing in the presence of data uncertainties. *IEEE Transactions on Industrial Electronics, 62*(5), 2973–2982. doi:10.1109/TIE.2014.2363046

Lucero, S. (2016). *IoT platforms: Enabling the Internet of Things* [white paper].

Marzetta, T. L. (2010). Noncooperative cellular wireless with unlimited numbers of base station antennas. *IEEE Transactions on Wireless Communications, 9*(11), 3590–3600. doi:10.1109/TWC.2010.092810.091092

Mavromoustakis, A., Constandinos, X., Mastorakis, G., & Batalla, J. M. (Eds.). (2016). Internet of Things (IoT) in 5G Mobile Technologies: Vol. 8. *Modeling and Optimization in Science and Technologies*. Springer.

Mekki, K., Bajic, E., Chaxel, F., & Meyer, F. (2018). *A comparative study of LPWAN technologies for large-scale IoT deployment*. ICT Express.

Militano, L., Araniti, G., Condoluci, M., Farris, I., & Iera, A. (2015). Device-to-device communications for 5G Internet of Things. *EAI Endorsed Transactions on Internet of Things*, *15*(1), 1–15.

Miraz, M. H., Ali, M., Excell, P. S., & Picking, R. (2018). Internet of nano-things, things and everything: Future growth trends. *Future Internet*, *10*(8), 68. doi:10.3390/fi10080068

Palattella, M. R., Dohler, M., Grieco, A., Rizzo, G., Torsner, J., Engel, T., & Ladid, L. (2016). Internet of Things in the 5G era: Enablers, architecture, and business models. *IEEE Journal on Selected Areas in Communications*, *34*(3), 510–527. doi:10.1109/JSAC.2016.2525418

Petäjäjärvi, J., Mikhaylov, K., Yasmin, R., Hämäläinen, M., & Iinatti, J. (2017). Evaluation of LoRa LPWAN technology for indoor remote health and wellbeing monitoring. *International Journal of Wireless Information Networks*, *24*(2), 153–165. doi:10.100710776-017-0341-8

Petajajarvi, J., Pettissalo, M., Mikhaylov, K., Roivainen, A., & Hänninen, T. (2015). On the coverage of LPWANs: Range evaluation and channel attenuation model for LoRa technology. In *2015 14th International Conference on ITS Telecommunications (ITST)*. IEEE.

Porpiglia, F., Amparore, D., Autorino, R., Checcucci, E., Cooperberg, M. R., Ficarra, V., & Novara, G. (2020). Traditional and Virtual Congress Meetings During the COVID-19 Pandemic and the Post-COVID-19 Era: Is it Time to Change the Paradigm? *European Urology*

Raftopoulou, M., Jorguseski, L., & Litjens, R. (2019, June). Design and Assessment of Low-Latency Random Access Procedures in 5G Networks. In *2019 European Conference on Networks and Communications (EuCNC)* (pp. 406-411). IEEE.

Raza, U., Kulkarni, P., & Sooriyabandara, M. (2017). Low power wide area networks: An overview. *IEEE Communications Surveys and Tutorials*, *19*(2), 855–873. doi:10.1109/COMST.2017.2652320

Soroush, H., Arney, D., & Goldman, J. (2016). Toward a safe and secure medical Internet of Things. *IIC Journal of Innovation*, *2*(1), 4–18.

Statista Research Department. (2020). Internet of Things Units Installed Base by Category 2014-2020. *Statista*. https://www.statista.com/statistics/370350/internet-of-things-installed-base-by-category/

Surie, D., Laguionie, O., & Pederson, T. (2008). *Wireless sensor networking of everyday objects in a smart home environment. In Intelligent Sensors, Sensor Networks and Information Processing (ISSNIP)2008*. IEEE.

Ting, D. S. W., Carin, L., Dzau, V., & Wong, T. Y. (2020). Digital technology and COVID-19. *Nature Medicine*, 1–3. PMID:32284618

Vasant, P. (2019). *Intelligent Computing & Optimization* (Vol. 866). Springer. doi:10.1007/978-3-030-00979-3

Wang, Y., Li, J., Huang, L., Jing, Y., Georgakopoulos, A., & Demestichas, P. (2014). 5G mobile: Spectrum broadening to higher-frequency bands to support high data rates. *IEEE Vehicular Technology Magazine*, *9*(3), 39–46. doi:10.1109/MVT.2014.2333694

Wang, Y.-P. E., Lin, X., Adhikary, A., Grövlen, A., Sui, Y., Blankenship, Y., Bergman, J., & Razaghi, H. S. (2017). A primer on 3GPP narrowband Internet of Things. *IEEE Communications Magazine*, *55*(3), 117–123. doi:10.1109/MCOM.2017.1600510CM

Weber, R. H. (2010). Internet of Things–New security and privacy challenges. *Computer Law & Security Review*, *26*(1), 23–30. doi:10.1016/j.clsr.2009.11.008

Whitmore, A., Agarwal, A., & Xu, L. D. (2015). The Internet of Things—A survey of topics and trends. *Information Systems Frontiers*, *17*(2), 261–274. doi:10.100710796-014-9489-2

WHO. (2020). *Myth Busters*. https://www.who.int/emergencies/diseases/novel-coronavirus-2019/advice-for-public/myth-busters

Yan, Y., Qian, Y., Sharif, H., & Tipper, D. (2013). A survey on smart grid communication infrastructures: Motivations, requirements and challenges. *IEEE Communications Surveys and Tutorials*, *15*(1), 5–20. doi:10.1109/SURV.2012.021312.00034

Zhai, Y., Wang, Y., Zhang, M., Gittell, J. H., Jiang, S., Chen, B., ... Wang, X. (2020). From Isolation to Coordination: How Can Telemedicine Help Combat the COVID-19 Outbreak? *medRxiv*.

Zhang, H., Liu, N., Chu, X., Long, K., Aghvami, A. H., & Leung, V. C. (2017). Network slicing based 5G and future mobile networks: Mobility, resource management, and challenges. *IEEE Communications Magazine*, *55*(8), 138–145. doi:10.1109/MCOM.2017.1600940

Chapter 4
Calculation of Receipt of Renewable Energy Resources and Operation Modes of Power Plants

Baba Dzhabrailovich Babaev
Dagestan State University, Russia

Vladimir Panchenko
https://orcid.org/0000-0002-4689-843X
Russian University of Transport, Russia

Valeriy Vladimirovich Kharchenko
https://orcid.org/0000-0003-3725-2976
Federal Scientific Agroengineering Center VIM, Russia

ABSTRACT

The main objective of the work is to develop principles for the formation of the optimal composition of the energy complex from all the given power plants based on renewable energy sources for an autonomous consumer, taking into account the variable energy loads of the consumer, changing climatic conditions and the possibility of using local fuel and energy resources. As a result of solving this optimization problem, in addition to the optimal configuration of the power complex, it is also necessary to solve the problem of optimizing the joint operation of different types of power plants from the selected optimal configuration, that is, it is necessary to determine the optimal modes of operation of power plants and the optimal share of their participation in providing consumers at every moment in time. A numerical method for analyzing and optimizing the parameters and operating mode of the energy complex with the most accurate consideration of the schedule of changes in consumer load and software that automates the solution of this optimization problem are also presented.

DOI: 10.4018/978-1-7998-3645-2.ch004

INTRODUCTION

Mastering energy sources has always been a way of human survival. And today, energy consumption remains one of the most important economic and social indicators that largely determine the standard of living of people. Therefore, it is often said that the world is ruled by energy.

According to experts, approximately 2/3 of the territory of Russia with a population of about 20 million people is deprived of centralized electricity supply. Providing consumers with electric and thermal energy is possible only with the help of small energy, i.e. where the construction of large power plants in some cases is impractical, in others it is not justified from an economic point of view, in the third it is impossible because of the lack of funds for the construction of expensive heating plants and the construction of power lines (Fortov & Popel, 2011).

In many such regions, autonomous installations, which run on imported liquid fuel or local energy resources, are used as a rule for energy supply. However, they are characterized by a high cost of fuel, due to the range of its transportation, the limited timing of seasonal delivery, multi-link supply chain. The transport component of the cost of fuel for such consumers can reach 70 - 80%, which leads to a high cost of electricity received from these types of plants (Fortov & Popel, 2011; Velkin, 2015; Babaev, 2003; OPEC Annual Statistical Bulletin – 2015; Makarov, Grigoryev & Mitrova, 2016; Alkhasov, 2010; Energy strategy of Russia for the period until 2035, 2014; On the development of renewable energy in Russia until 2030; Eleven EU countries have already fulfilled their res development plans for 2020).

Russian and foreign experience shows that the combined use of installations based on renewable energy sources in decentralized energy supply systems is an effective way of energy supply to consumers, both from an economic and environmental point of view. This method of energy supply will significantly reduce the consumption of liquid fuel, increase the reliability of the energy supply system and also reduce the cost of the generated unit of electricity (Velkin, 2015).

RENEWABLE ENERGY TRENDS IN THE WORLD AND IN INDIVIDUAL COUNTRIES

The main driving forces for the development of renewable energy technologies in the modern world are (Fortov & Popel, 2011; Velkin, 2015; Babaev, 2003; OPEC Annual Statistical Bulletin – 2015; Makarov, Grigoryev & Mitrova, 2016; Alkhasov, 2010; Energy strategy of Russia for the period until 2035; On the development of renewable energy in Russia until 2030; Eleven EU countries have already fulfilled their res development plans for 2020):

- renewable energy resources (solar energy (Kharchenko, Nikitin, Tikhonov, Panchenko & Vasant, 2019; Panchenko, 2018; Panchenko, 2019; Panchenko, Izmailov, Kharchenko & Lobachevskiy, 2020; Kharchenko, Panchenko, Tikhonov & Vasant, 2018; Panchenko, Kharchenko & Vasant, 2019; Panchenko & Kharchenko, 2019; Panchenko, Chirskiy & Kharchenko, 2019) wind energy, biomass energy, geothermal energy of small rivers, the energy of sea waves and tides, low-grade natural and waste heat (Babaev, Kharchenko, Panchenko & Vasant, 2019; Babaev, Kharchenko & Panchenko, 2019), etc.) are practically unlimited, many times exceed mankind's apparent energy needs and are constantly replenished;

- the desire of oil and gas importing countries to reduce dependence on energy imports, because unlike oil, gas, coal and uranium, renewable energy resources are more or less evenly distributed across countries and regions; they are not monopolized by a limited number of countries;
- increasing energy security and reducing political dependence on other countries;
- concern for the environment, since the use of renewable energy does not lead to significant environmental pollution and does not lead to a change in the thermal balance of the Earth;
- business development, creation of new jobs;
- the development of promising new technologies, the creation of a scientific and technical reserve for the export of technologies to other countries;
- the economic efficiency of renewable energy (in recent years).

The intensive development of renewable energy in most countries of the world takes place for environmental reasons, considerations of ensuring energy security, social motives, and, as a rule, is accompanied by significant state financial, legislative and political support.

EUROPEAN UNION LEGISLATIVE INITIATIVES FOR THE DEVELOPMENT OF RENEWABLE ENERGY MARKETS

It should be noted that in most leading countries the accelerated development of renewable energy sources is carried out with the defining state legislative and direct financial support. The most ambitious targets adopted in the European Union.

1. Introduced annual quotas for local energy companies for the production of electricity from renewable energy in the amount of 2 – 15% of total electricity sales. If the energy company does not produce and does not sell electricity from renewable energy sources, it should buy green certificates.
2. Funds to stimulate the development of the renewable energy market are taken from a special fund managed by the state through deductions from electricity tariffs and the sale of green certificates.
3. Each energy company must buy electricity from owners of generators using renewable energy sources at a fixed price for 20 years, for example, in Germany at a price of 0,57 Euro/kW h for 20 years with a capacity of less than 30 kW, in France – 0,295 Euro/kW h at a power of up to 450 kW.
4. No permission is required for the private owner to purchase, install and include in the energy system an electric generator using renewable energy sources.
5. Each owner of an electric generator using renewable energy is given the right to freely choose the energy company to which he sells or from whom he buys electricity.
6. The owner of an electric generator using renewable energy sources, when buying and installing a grant is given in the amount of 5 Euro/W, but not more than 40 – 75% of the cost of the generator.
7. Encouraged by cheap credit rates and tax incentives, the participation of a third party in financing the costs of the purchase, installation and operation of equipment using renewable energy sources.
8. The total amount of grants and tariffs is calculated in such a way that the owner is compensated for all expenses for the purchase and maintenance of an electric generator using renewable energy sources, and a profit of 200 – 300% of the bank interest on a contribution in the amount of funds spent is guaranteed for 20 years.

Sweden is the European leader by a large margin, where more than half, namely 53,9% of all energy consumed, is renewable energy (bioenergy, hydroelectric power station and wind energy). Finland (39,3%), Latvia (37,6%), Austria (33,0%), Denmark (30,8%) are also among the five leaders in the share of renewable energy sources in the energy balance. The lagging behind are Luxembourg and Malta (both 5%), the Netherlands (5,8%), Belgium (7,9%) and the United Kingdom (8,2%) [10].

The drum countries that have already fulfilled the renewable energy sources plan for 2020 are Bulgaria, the Czech Republic, Denmark, Estonia, Croatia, Italy, Lithuania, Hungary, Romania, Finland and Sweden.

For 2030, the European Union has already set a new target indicator "share of renewable energy sources in final energy consumption" - 27% (Eleven EU countries have already fulfilled their res development plans for 2020).

THE POTENTIAL OF RENEWABLE ENERGY SOURCES IN THE RUSSIAN FEDERATION, OBSTACLES TO DEVELOPMENT

The Russian Federation, being the largest country in the world, possesses significant resources of renewable energy sources: solar and wind energy, hydropower resources, geothermal energy and biomass energy. Due to the geographical location of the Russian Federation, its size, terrain and climate diversity, the types of renewable energy sources vary significantly from region to region. In fact, in all regions of the country there is one or two types of renewable energy sources, the operation of which can be economically justified, and in some regions there are all types of renewable energy sources. This distinguishes our country from many smaller countries, where, due to the uniformity of geographical conditions, only one type of renewable energy dominates or is present (Popel, Frid, Kolomiyets, Kiseleva & Terekhova, 2010).

Despite the fact that the Russian Federation has huge resources of wind, geothermal, solar energy, hydropower resources, biomass energy, today the role of renewable energy in the country is insignificant. At the moment, the share of renewable energy sources in its energy balance is slightly less than 1% and amounts to about 0,1% of the global indicator. This indicator has not changed over the past three years. And the prospects are not encouraging (Fortov & Popel, 2011; Makarov, Grigoryev & Mitrova, 2016).

This situation in the Russian Federation is associated with the presence of a complex set of factors that impede the development of renewable energy sources.

Legislative factors. The imperfection of the legislative base, its weakness, despite the adoption in recent years of a number of laws and regulations related to its inadequate study and a clear definition of the regulators that provide the mechanism for the supply and sale of electricity.

Economic factors. The high cost of renewable energy production and lower capacity compared to traditional sources, the lack of incentive measures for the production of energy equipment, the use of renewable energy (tariff, tax) and specific financial mechanisms for state support, the lack of funding from the budget, the lack of long-term priorities in the price ratio of the main types of traditional energy sources.

Administrative factors. These factors are associated with barriers to obtaining permits for the construction of alternative energy facilities and their connection to existing networks.

Personnel factors. Low qualifications of personnel and insufficiently developed and adjusted system in staff training, lack of professionalism in the field (local authorities) have a negative impact on the current situation in the field of development of alternative energy. According to experts, in 5 years the need for personnel in the Russian Federation will be 4500 – 5500 people.

Information and psychological factors. Lack of information on renewable energy opportunities, knowledge culture in this area.

In addition, the abundance of significant reserves of combustible minerals along with excess generating capacity in the electric power industry can be considered as other constraining factors for the development of renewable energy in the Russian Federation (Fortov & Popel, 2011). This forms in the minds of the public, authorities, and business circles a corresponding view of the development of energy based on renewable energy sources.

Thus, renewable energy in the Russian Federation at the moment can not compete with traditional energy in connection with the presence of these barriers. Therefore, an important task for our country is to overcome these barriers by developing and implementing development methods and supporting the use of renewable energy sources.

A positive factor for the development of renewable energy in the Russian Federation is the ongoing creation of a legislative framework.

The strategic objectives for the use of renewable energy and local fuels are:

1) Decrease in the growth rate of the total volume of fossil fuel consumption, preservation of its reserves for future generations;
2) Diversification of fuel and energy balances and increasing the level of energy security and reliability of energy supply of the constituent entities of the Russian Federation and Russia as a whole;
3) Reduction of costs for transmission and distribution of electric and thermal energy, as well as transportation of fuel and reduction of losses resulting from this;
4) Achieving comfortable working and living conditions of the population, reliable energy supply to local industry enterprises, primarily in remote and inaccessible areas that are not connected to public networks;
5) Bridging the world level in the production of high-tech equipment for renewable energy;
6) Reduction of greenhouse gas emissions and harmful emissions from the production of electric and thermal energy.

The state policy in the field of the use of renewable energy sources for the period until 2035 provides for the following measures:

1) Reimbursement of fees for technological connection to networks;
2) Removal of barriers when connecting renewable energy installations to public networks;
3) Subsidization of interest rates on loans raised for the development of production by organizations producing energy based on renewable energy sources;
4) Development of a set of measures for state support of industry and scientific institutions to provide the renewable energy industry with Russian equipment, components and advanced technologies,
5) Technology transfer and localization at Russian enterprises for the production of components for power plants operating on renewable energy sources.

The target vision for the development of renewable energy envisages an increase in the share of renewable energy sources by 2035 to 3,7% in the total volume of commissioned capacities of power plants and in the production of electricity – up to 2,2% of the total generation, which corresponds to electric energy production of 34 – 35 billion kW·h. To achieve the planned volumes of electricity production based on renewable energy sources in the period under review (until 2035), it is necessary to ensure the commissioning of generating capacities with a total installed capacity of over 18 GW. The development of renewable energy by 2035 should provide an increase in the average installed capacity utilization factor for wind farms up to 25%, microhydroelectric power plants – 50%, geothermal power plants – 75%, biomass thermal power plants – 50%, other renewable energy sources – at least 40% (Bezrukikh, Degtyarev, Elistratov et al., 2007; Babaev, 2015).

PROPOSED RENEWABLE ENERGY INCENTIVE MECHANISMS

To influence the direction of stimulating the use of renewable energy, the authors propose the following mechanisms that will allow the Russian Federation to successfully and inexpensively develop renewable energy-based energy without significant investments:

1) Compensation by the state to producers for a certain time the increased costs of energy received from renewable energy sources;
2) Providing free access to the electricity market;
3) Use of special green certificates for energy from renewable energy sources, aimed at selling to conscious consumers who are willing to pay a high price for renewable energy sources;
4) Mandatory quotas for the production (consumption) of electricity from renewable sources;
5) Non-discriminatory preferential connection to the electric network;
6) Introduction of taxes on emissions and environmental pollution by enterprises based on technologies for the use of fossil fuels, and the funds received from taxes to be used for the development of renewable energy sources;
7) Tax incentives for energy producers based on the use of renewable energy sources;
8) Financing research in the field of renewable energy both by the state and with the participation of private capital;
9) Establishing preferential loans for individual and industrial buyers of renewable energy systems, i.e., providing access to bank loans at low interest rates, as well as subsidizing investments for renewable energy sources;
10) Oblige energy companies to produce a certain share of energy from renewable sources or oblige energy system owners and large consumers to buy a certain percentage of energy or heat consumed from producers using renewable energy sources;
11) Accelerated depreciation of equipment to reduce the tax burden at the initial stage of the project and alleviate the high initial capital cost of renewable energy technologies;
12) Grants for research in the field of alternative energy;
13) Tendering for the implementation of projects in the field of renewable energy;
14) Development of educational programs of the population on possible options for using various energy sources, including renewable, principles of energy conservation, as well as awareness of the responsibility of each member of the society for improving the environmental situation in the area of residence and in the country;
15) Training and professional development of specialists engaged in the field of energy production based on renewable energy sources, organization of exhibitions, conferences, assistance in conducting advertising campaigns, presentations of energy-saving technologies.

Thus, the development of autonomous energy using renewable energy will be largely determined by the initiative of regional and local authorities, as well as private business.

An analysis of the literature (Fortov & Popel, 2011; Velkin, 2015; Popel, Frid, Kolomiyets, Kiseleva & Terekhova, 2010; Babaev, 2015; Elistratov, 2017; Elistratov & Aronova, 2012; Vissarionov, Deryugina, Kuznetsova & Malinin, 2008; Burmistrov, Vissarionov, Deryugina, et al., 2009; Babaev & Sheverdiev, 2017b; Babaev & Sheverdiev, 2018; Director & Maykov, 2010; Maykov, Director & Zaichenko, 2010; Maykov & Director, 2011; Director, Zaichenko & Maykov, 2012;. Director, Maykov & Ivanin, 2012), as well as the experience of Russia and foreign countries, shows that in decentralized regions efficient, modern and reliable energy supply can be ensured by creating efficient energy complexes based on renewable energy sources.

The systemic effects of the introduction of autonomous energy supply systems based on in these regions include the following (Elistratov, 2017):

Regional: improving energy security in remote regions of Russia by increasing self-sufficiency in fuel and energy resources;

Technological: reduction of energy losses for transportation and distribution of energy. Use of modern equipment and systems;

Social: improving the reliability and quality of energy supply to consumers;

Economic: reducing the volume of "long-distance" fuel, reducing the cost of energy at the end user, reducing state subsidies;

Ecological: conservation of nature, reduction of harmful emissions into the environment from operating diesel power plants and abandoned barrels.

To find the optimal configuration, parameters and operating modes of an autonomous energy complex based on renewable energy sources in real conditions, a methodology has been developed that represents the sequential implementation of 3 stages programmatically combined into a single multi-level and multi-criteria optimization space.

THE PROGRAM FOR CALCULATING RESOURCES OF RENEWABLE ENERGY SOURCES AND OPTIMIZING THE OPERATING MODES OF POWER PLANTS OF INTEGRATED ENERGY SUPPLY TO THE CONSUMER ACCORDING TO THE DYNAMICS OF ITS LOAD

When turn on the optimization program of the energy complex, a window appears (Figure 1), where there are schemes for connecting the consumer of electric and thermal energy, buttons for determining the resources of renewable energy sources and "Optimization".

When click the "Solar Energy Resource" button, a window appears, where you need to enter the current date, latitude and longitude of the area, maternity noon, the maximum solar radiation power per year characteristic of the area, cloud cover and albedo of the area where solar modules. When entering only these data, the program determines the optimal inclination angle of the receiving platform, the duration of the sunny day and the solar radiation power, taking into account the cloudiness, diffuse and reflected components on the horizontal and inclined (according to the optimal inclination angle) platform for each hour from sunrise to sunset, according to the methodology given in (Babaev & Sheverdiev, 2018).

And when enter the areas of solar modules and collectors, their efficiency, the program calculates the generation of electric and thermal energy by the hour in kW·h.

Figure 1. The window of the program for calculating resources of renewable energy sources and optimizing the operating modes of the integrated energy supply of a consumer according to the dynamics of its load

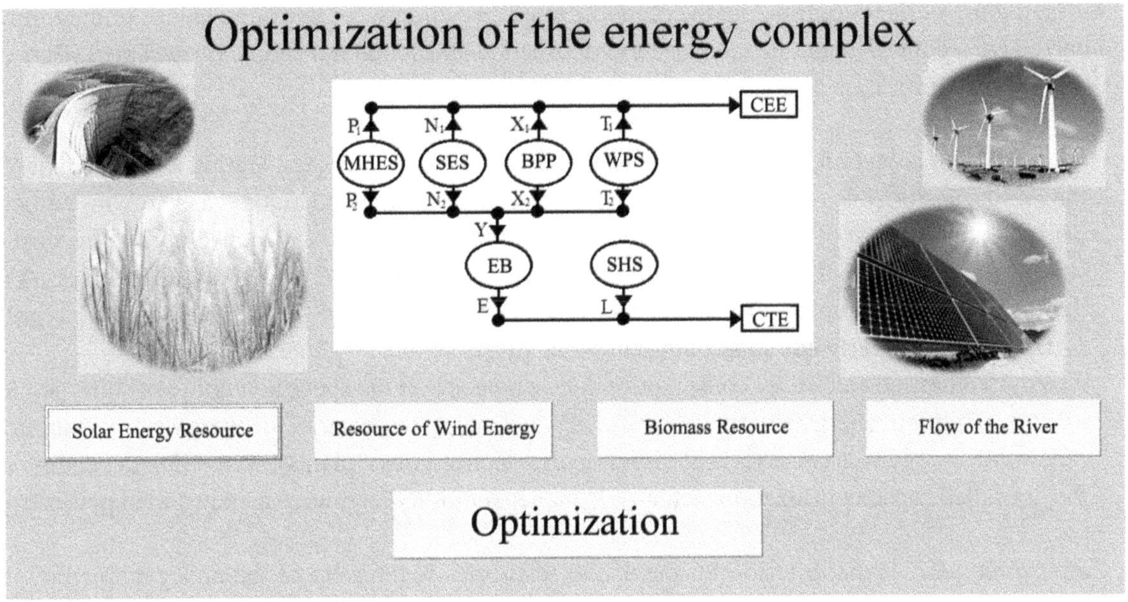

To determine the resource of wind energy in the window in Figure 1 press the button "Resource of Wind Energy". In this case, a window appears, where it is required to enter the height of the weather vane and the openness class at Milevsky local weather station. When you click the Milevsky Classification button, a classification table and their characteristics appear. At the same time, the program brings the wind speed according to the weather station to the conditions of an open area and to the indicated height of the location of the wind wheel.

To determine the biomass resource in the window in the Figure 1 presses the button "Biomass Resource". At the same time, a window appears, where it is required to enter the available farm waste, the number of domestic animals and their daily dry matter output, as well as plant dry waste, methane calorific value in MJ/m^3, since the resulting biogas is transferred to the content methane in it. Having indicated the efficiency of a biogas plant, the program calculates the generation of biogas plants in kW·h by clicking on the "Calculate" button.

Also, according to the well-known method (Velkin, 2015), the resource of local rivers and streams is determined by clicking on the button in the window in Figure 1 "Flow of the River". Having indicated the pressure and efficiency of the microhydroelectric power station in the window, the program calculates the generation of the microhydroelectric power station by the hours of the day.

DETERMINATION OF OPTIMIZATION CRITERIA FOR POWER PLANTS OF THE ENERGY COMPLEX

Before starting the task of optimizing the composition of an autonomous energy complex, it is necessary to select the criteria by which the optimization will be carried out. The selected optimal energy supply system using renewable energy sources should reliably, stably and at minimal cost provide consumers with electricity and heat.

As the main criteria for optimizing the composition of the energy complex, the values of the following interrelated technical, economic, energy, environmental and social indicators are accepted (Babaev, 2003; Babaev, 2016b):

a) ***technical and economic***: 1) reduced costs, which include the costs of operating a power plant (for the private sector, it is a waste of human labor on preparing firewood, dung, gas, etc.), investment, depreciation, the cost of the corresponding type of fuel in the area of its consumption, the cost of storage, loading, unloading for transport and transportation of fuel; 2) metal consumption, kg/kW; 3) durability (1/(equipment service life)); 4) the cost of building the system; 5) discounted payback; 6) 1/(effect of the early commissioning of the facility);
b) ***energy***: 1) fuel and electricity consumption for own needs; 2) the specific number of staff, people/kW; 3) stability of energy supply – the ratio of total energy to power of an energy storage unit (for renewable energy sources, except for microhydroelectric power plants); 4) a value depending on the "installed capacity utilization factor"; 5) "maneuverability" – time to gain installed power from 0;
c) ***environmental***: 1) the criterion for the use of territories is the ratio of the area occupied by the territory to the power of the power plant, ha/MW; 2) the criterion for the degree of environmental pollution – the content of ash, sulfur oxides, nitrogen and bacteria in 1 mm^3 of air or in emissions for the generation of 1 GJ of energy
d) ***social***: 1) satisfaction of the staff with work – the working conditions of the staff (from "better" - 0 to "worse" - 10); 2) the prestige of the work (from "prest" - 0, to "non-prest" - 10).

According to the above criteria, the optimality coefficient C_j corresponding to the electrical and thermal installations of the energy complex is determined. The calculation of the coefficient C_j of each j-th installation included in the energy complex is carried out by the "Spider" method. According to this method, the process of constructing a diagram in polar coordinates "Spider" is carried out (Figure 2).

On the axes of this diagram, we postpone the values of previously defined criteria indicators. These indicators are transformed in such a way that for the power plant under consideration, the smaller their values, the better. For example, the indicators "equipment service life", "effect of the facility's early commissioning" and "installed capacity utilization factor" are given as 1/(equipment service life), 1/(effect of the facility's early commissioning) etc.

We evaluate the power plants included in the power complex according to the following rule: *the optimal power plant for the area in question corresponds to the one whose diagram outlines the smallest area* (Figure 2).

THE MATHEMATICAL MODEL OF THE ENERGY COMPLEX

In the Figure 3 shows a diagram of an autonomous energy complex, which is a system of energy-generating installations using renewable energy sources – a wind power installation (wind turbine), a microhydroelectric power station, solar power systems and heat supply, a biogas power plant, an electric boiler, a consumer of electrical and thermal energy (Babaev & Sheverdiev, 2017b; Babaev & Sheverdiev, 2018).

Figure 2. "Spider" diagram

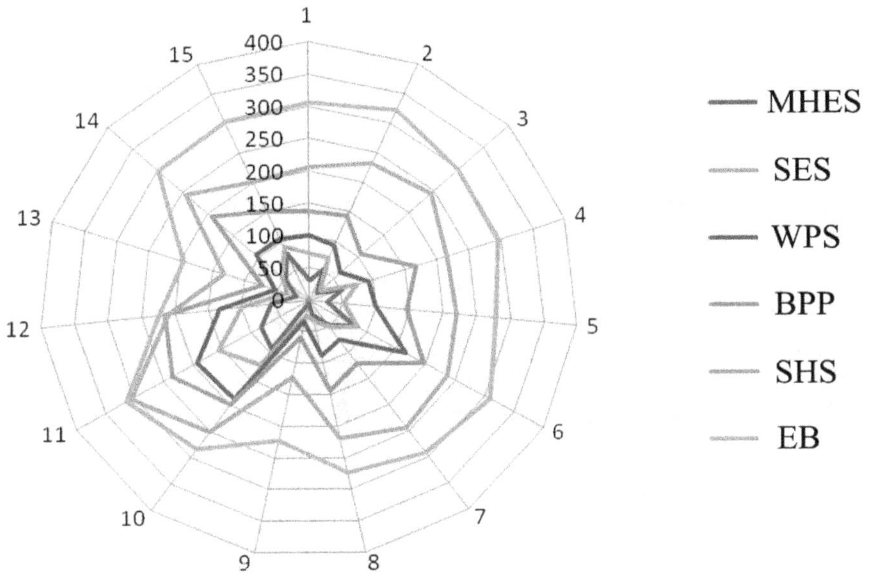

In order to solve the optimization problem, it is necessary to compile a mathematical model of the optimization object.

Since this power supply system of an autonomous consumer is closed, it can be argued that the sum of the capacities of electric and thermal energy sources must be equal to the sum of the capacities consumed by the consumer minus losses due to their transmission and various transformations. Based on this, it is possible to write the mathematical formulation of the considered model of the energy complex in the form of a system of equations for the balance of electric and thermal energy flows (Babaev & Sheverdiev, 2017b; Babaev & Sheverdiev, 2018).

The electric balance of the energy complex at each moment of time is written in the form:

$$P_1 + N_1 + X_1 + T_1 = Z, \tag{1}$$

where Z is the electrical load of the consumer at a given time.

Accordingly, the heat balance at each moment of time:

$$E + L = W, \tag{2}$$

where W is the thermal load of the consumer at a given time.

Additional ratios:

$$Y = P_2 + N_2 + X_2 + T_2 \tag{3}$$

$$E = k_{eb} \cdot (P_2 + N_2 + X_2 + T_2) = k_{eb} \cdot Y, \tag{4}$$

Figure 3. Scheme of an autonomous energy complex: P_1, N_1, X_1, T_1, P_2, N_2, X_2, T_2 – flows of electric energy at each moment of time from microhydroelectric power station, solar power system, biogas power plant, wind turbine, respectively; Y is the flow of electrical energy at each moment of time supplied to the electric boiler; E, L - thermal energy flows at each moment of time from electric boiler and solar power system for heat supply.

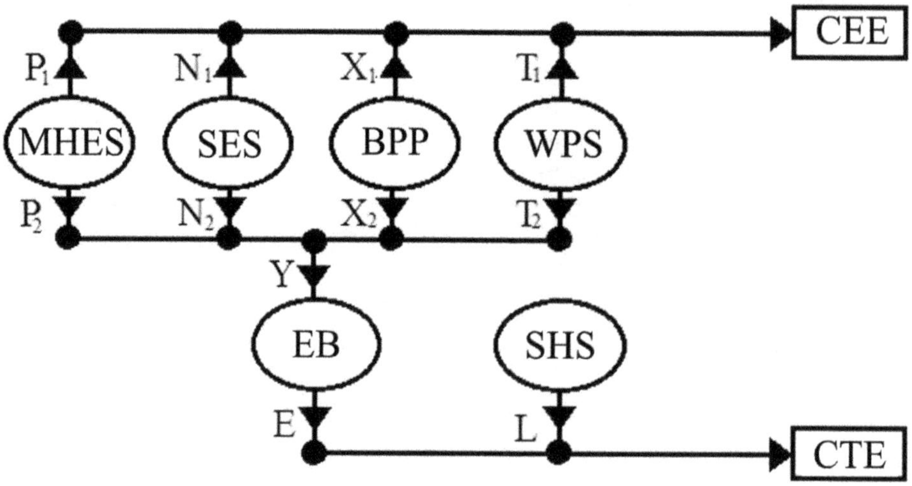

where k_{eb} is the efficiency of the electric boiler.

Taking into account formulas (3) and (4), the heat balance formula of the energy complex (2) will take the form:

$$k_{eb} \bullet Y + L = W \tag{5}$$

Thus, we obtained the system of equations (1) - (5), which represents the balance of electric and thermal energy of the considered energy complex for the time interval Δt_i, during which the consumer's electric and thermal loads are considered constant, i.e., $Z(t) = const$ and $W(t) = const$.

It is assumed that all power flows $P_1{}^3\,0$, $P_2{}^3\,0$, $N_1{}^3\,0$, $N_2{}^3\,0$, $X_1{}^3\,0$, $X_2{}^3\,0$, $T_1{}^3\,0$, $T_2{}^3\,0$, $E\,{}^3\,0$, $L\,{}^3\,0$.

Equality to zero means that the corresponding power plant, which is part of the energy complex, is not used for the territory in question due to lack of resources or is excluded due to the fact that it is not optimal (Babaev & Sheverdiev, 2017b; Babaev & Sheverdiev, 2018).

Introduce the objective function:

$$f_0 = \sum_i^{\tau} \Delta t_i \cdot \left(C_1 \cdot P_1^i + C_2 \cdot N_1^i + C_3 \cdot X_1^i + C_4 \cdot T_1^i \right) + \sum_i^{\tau} \Delta t_i \cdot \left(C_5 \cdot E^i + C_6 \cdot L^i \right), \tag{6}$$

where τ is the number of time intervals with a constant load; C_1, C_2, C_3, C_4, C_5, C_6 are the optimality coefficients corresponding to the "Spider" areas of the microhydroelectric power station, solar power system, biogas power plant, wind turbine, electric boiler and solar power system for heat supply diagrams respectively.

The optimization task is to determine the operating modes of power plants that are part of the energy complex (Figure 3), at each time Δt_i, at which the objective function (6) will take a minimum value.

NUMERICAL METHODS

The authors proposed algorithm for optimizing the composition of the energy complex (Figure 3) is carried out in 3 stages.

At the first stage, the consumer determines the electric and thermal loads, calculates the resources of local renewable energy sources and the possible generation of installations that are part of the energy complex (Figure 3) at each time Δt_i, during which the consumer's electrical and thermal loads are assumed to be constant. In this work, a time interval of 1 hour is taken as the time interval of stationarity. The calculation of local renewable energy sources is carried out according to the methods described in (Babaev, B.D. 2015). The calculation of the energy production of plants from the energy complex (Figure 3), taking into account the previously determined renewable energy resources at each hour *i*, is carried out according to the following formulas:

$$P_i = 9,81 \cdot Q_i \cdot \rho \cdot H \cdot \eta_{mhes} \cdot 10^{-3} \tag{7}$$

$$N_i = R_\beta^i \cdot S_{ses} \cdot \eta_{ses} \cdot 10^{-3} \tag{8}$$

$$T_i = \frac{V_{red\_i} \cdot \rho_{air} \cdot D^2 \cdot \pi}{8} \eta_{wps} \cdot n \tag{9}$$

$$X_i = \frac{V_{M\_i} \cdot G \cdot \eta_{bpp}}{3,6} \tag{10}$$

$$L_i = R_\beta^i \cdot S_{shs} \cdot \eta_{shs} \cdot 10^{-3} \tag{11}$$

$$E_i = \left(P_2^i + N_2^i + X_2^i + T_2^i\right) \cdot \eta_{eb} = Y_i \cdot \eta_{eb}, \tag{12}$$

where P_i, N_i, T_i, X_i, L_i, E_i – generation of microhydroelectric power station, solar power system, wind turbine, biogas power plant, solar power system for heat supply, electric boiler, respectively, in every hour, kW·h; η_{mhes}, η_{ses}, η_{wps}, η_{bpp}, η_{shs}, η_{eb} – effectiveness microhydroelectric power station, solar power system, wind turbine, biogas power plant, solar power system for heat supply, electric boiler, respectively; H – hydraulic head of microhydroelectric power station, m; S_{ses}, S_{shs} – square of solar power system and solar power system for heat supply, m^2; D – wind wheel diameter, m; n – number of wind power plants; G – calorific value of methane, MJ/m^3.

At the second stage, for each installation included in the energy complex circuit (Figure 3), the optimality coefficient C_j corresponding to the areas of the "Spider" diagrams (Figure 2) of the micro-hydroelectric power station, solar power system, biogas power plant, wind turbine, solar power system for heat supply, and electric boiler, respectively, is calculated.

The area of the "Spider" of the diagram of the j-th power plant corresponding to the optimization coefficient C_j is calculated as the sum of the areas of the triangles outlined by the "Spider" of the diagram (Figure 2), according to the formula (Babaev, 2016):

$$C_j = 0,5 \cdot \sin\left(\frac{2\pi}{K}\right) \cdot \left(k_1 \cdot k_2 + k_2 \cdot k_3 + \ldots + k_{k-1} \cdot k_k + k_k \cdot k_1\right) \tag{13}$$

where K is the number of input criteria; k_1, k_2, k_3, ..., k_k are the values of the optimization criteria (Babaev & Sheverdiev, 2017b; Babaev & Sheverdiev, 2018).

After that, the area of the power plants is compared, as a result of which all the elements of the circuit are arranged in chronological order of increase in the area of the "Spider" diagrams, that is, from the most optimal installation to not optimal for the particular area in question.

At the third stage, from the constructed hierarchy of power plants, starting with the first (most optimal) one, one selects the number of power plants that will provide the required electrical and thermal loads of the consumer for each time interval Δt_i, and the rest of the systems (less optimal in this period of time) are excluded. After calculating, therefore, the optimal combination of power plants, the optimal shares of power plants in the power system that are part of the selected optimal combination of power complex and their operating modes at each moment of time Δt_i, at which the objective function (6) will take a minimum value.

The power plants selected in this way and the shares of their participation in the energy supply to the consumer will correspond to the most optimal combination and operation modes of the energy complex.

To automate this process, software was developed in the Delphi 7 environment, which is based on the algorithm shown in the Figure 4 (Babaev & Sheverdiev, 2017a).

When entering the initial data, the program automatically determines the resources of renewable energy sources and the generation of power plants that are part of the energy complex, as well as the optimal combinations of power plants and their participation for the power supply of this consumer at any given time, taking into account many criteria by which optimization is carried out.

THE RESULTS OF TESTING SOFTWARE FOR OPTIMIZING THE COMPOSITION OF THE ENERGY COMPLEX BASED ON LOCAL RENEWABLE ENERGY SOURCES

To demonstrate the capabilities of the developed model and the efficiency of the optimization algorithm for the energy complex scheme (Figure 3), software was tested for the real conditions of the Republic of Dagestan – for an average village located in Akhtynsky district with coordinates of 41,5 ° north latitude and 47,7 ° east longitude.

Figure 4. The algorithm for solving the optimization task

1.0. Entering consumer data (electrical and thermal loads at any given time) - Z_i and W_i;
1.1. Enter data for the area in question;
1.2. Enter the efficiency values of power plants;
1.3. Enter optimization criteria.

↓

2.0. Calculation of renewable energy resources of the considered area;
2.1. Calculation of the generation of power plants at each point in time, taking into account their efficiency and available resources.

↓

3.0. Calculation for each power plant of the "Spider" area of the diagram according to the formula:

$$C_j = 0{,}5 \cdot \sin\left(\frac{2\pi}{K}\right) \cdot (k_1 \cdot k_2 + k_2 \cdot k_3 + \dots + k_{K-1} \cdot k_K + k_K \cdot k_1)$$

↓

4.1. Comparison of "Spider" areas of power plant diagrams;
4.2. Location of power plants in chronological order from $C_{j\ min}$ to $C_{j\ max}$

↓

5.0. The selection of the optimal combination of power plants and the calculation of the optimal shares of their participation in providing the consumer at each time interval.

↓

6.0. The conclusion of the optimal composition of the energy complex and the optimal participation shares of each power plant in providing the consumer at each time interval.

As the initial data, the electrical and thermal loads of this consumer at each time point of the day under consideration are set as well as other necessary data on the terrain.

And clicking in the window on the Figure 1 button "Optimization" it is necessary to enter the indicators given in (Babaev, 2003; Babaev, 2016b) for the various power plants used in it. By clicking on "Calculate cost", the cost price of the generated energy of each power plant is determined. When you click "Calculate Area", the area of the "Spider" diagrams for each power plant is determined according to the methodology described in (Babaev, 2016a; Babaev, 2016b) and a hierarchy of power plants is created to increase the area of the diagram, i.e. the most optimal efficient power plant is positioned 1 further as the efficiency decreases and the most inefficient power plant is positioned last.

To provide the consumer, the most efficient first power plant is first used, when there is a lack of energy from the first power plant, the second power plant is turned on to provide the consumer. If the consumer is not fully provided even when the second power plant is turned on, the third power plant is turned on, etc.

When you click "Optimization", data appears on the use of each power plant at every hour of the day and their share of participation in providing the consumer with energy. The optimal combinations of power plants and their shares for the complete provision of the consumer are given.

When you click the "Graphs" button in the program produces graphs showing the operation modes of each power plant (Figure 5).

Figure 5. The results of optimizing the composition of the energy complex and operating modes of power plants using a specific consumer as an example

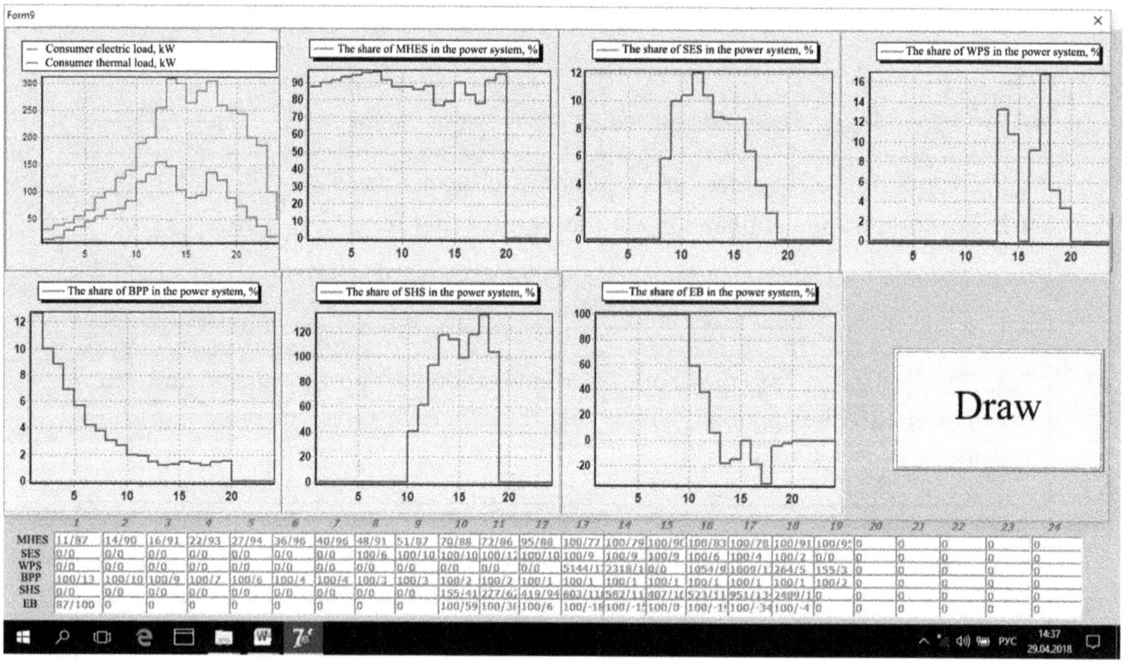

The optimization results are shown in the graphs of Figure 5. As a result of testing the software to optimize the composition of the energy complex, the optimal combinations of installations and their share in providing the consumer at each time point during the day were determined.

Thus, the principles of the formation of the optimal composition of the energy complex from all the listed renewable energy power plants for an autonomous consumer were developed taking into account the variable energy loads of the consumer, changing climatic conditions and the possibility of using local fuel and energy resources. The problem of optimizing the joint operation of different types of power plants from the selected optimal configuration was also solved, the optimal operating modes of power plants and the optimal share of their participation in providing the consumer at each time point were determined.

In addition, the principles of multicriteria optimization of combined power supply schemes for an autonomous consumer based on local energy sources using the Spider diagrams have been developed. A numerical method has been developed for analyzing and optimizing the parameters and operation mode of the energy complex with the most accurate allowance for the schedule of changes in consumer load and software that allows automating the solution of this optimization problem. The program has been tested for optimization power supply of a specific autonomous consumer.

REFERENCES

Alkhasov, A. B. (2010). *Vozobnovlyaemaya energetika* [Renewable energy]. FIZMATLIT.

Babaev, B.D. (2003). *Kriterii vybora al'ternativnyh kombinirovannyh sistem energosnabzheniya* [Criteria for the selection of alternative combined energy supply systems]. *Izv. vuzov. Sev. -Kavk. region. Tekhn. nauki*, 49 - 50.

Babaev, B. D. (2015). Resursy vozobnovlyaemyh istochnikov energii Respubliki Dagestan: Uchebno-spravochnoe posobie [Resources of renewable energy sources of the Republic of Dagestan: manual]. Mahachkala: Izdatel'stvo "Raduga".

Babaev, B.D. (2016a). Metodika sravnitel'nogo analiza variantov kombinirovannogo energosnabzheniya po neskol'kim kriteriyam [Technique for a comparative analysis of the options for combined energy supply according to several criteria]. *Energetik, 5*, 19 - 22.

Babaev, B. D. (2016b). *Razrabotka i issledovanie energosistem na osnove vozobnovlyaemyh istochnikov s fazoperekhodnym akkumulirovaniem tepla* [Development and study of energy systems based on renewable sources with phase transition heat storage]. Dis. dok. tekhn. nauk. Mahachkala.

Babaev, B.D., Kharchenko, V., Panchenko, V. & Vasant, P. (2019). Materials and Methods of Thermal Energy Storage in Power Supply Systems. *Renewable Energy and Power Supply Challenges for Rural Regions*, 115 – 135. DOI: . doi:10.4018/978-1-5225-9179-5.ch005

Babaev, B. D., Kharchenko, V. V., & Panchenko, V. (2019). Development and Research of Phase-Transition and Thermochemical Materials for Heat Accumulation. Handbook of Research on Smart Computing for Renewable Energy and Agro-Engineering, 1 – 26. Doi:10.4018/978-1-7998-1216-6.ch001

Babaev, B.D. & Sheverdiev, R.P. (2017a). Certificate of state registration of a computer program No. 2017663797. In *Optimizaciya sostava energokompleksa* [Optimization of the energy complex]. Registered in the Register of Computer Programs.

Babaev, B. D., & Sheverdiev, R. P. (2017b). Optimizaciya sistemy kompleksnogo energosnabzheniya avtonomnogo potrebitelya na osnove mestnyh vozobnovlyaemyh energoistochnikov [Optimization of the integrated energy supply system of an autonomous consumer based on local renewable energy sources]. In *Materialy V Mezhdunarodnoj konferencii "Vozobnovlyaemaya energetika: problemy i perspektivy". Tom 2. Mahachkala* [Materials of the V International Conference "Renewable Energy: Problems and Prospects". Volume 2. Makhachkala]. Academic Press.

Babaev, B.D. & Sheverdiev, R.P. (2018). Optimizaciya ugla naklona solnechnyh modulej v zavisimosti ot mesta raspolozheniya i avtomatizirovannyj raschet solnechnoj radiacii [Optimization of the angle of inclination of solar modules depending on the location and automated calculation of solar radiation]. *Energetik, 12*, 29 - 31.

Bezrukikh, P. P., Degtyarev, V. V., & Elistratov, V. V. (2007). *Spravochnik po resursam vozobnovlyaemyh istochnikov energii Rossii i mestnym vidam topliva/pokazateli po territoriyam* [Handbook of Russian renewable energy resources and local fuels/indicators by territory]. IAC Energiya.

Burmistrov, A. A., Vissarionov, V. I., & Deryugina, G. V. (2009). Metody rascheta resursov vozobnovlyaemyh istochnikov energii [Methods for calculating renewable energy resources]. Moskva: Izdatel'skij dom MEI.

Director, L.B. & Maykov, I.L. (2010). Reshenie zadach optimizacii slozhnyh energeticheskih sistem [The solution of problems of optimization of complex energy systems]. *Upravlenie bol'shimi sistemami, 28*, 274 – 292.

Director, L.B., Maykov, I.L. & Ivanin, O.A. (2012). Zadacha optimizacii avtonomnyh energeticheskih kompleksov v sostave lokal'nyh raspredelitel'nyh setej [The optimization problem of autonomous energy complexes as part of local distribution networks]. *Energotekhnologii i resursosberezhenie, 4*, 33 - 41.

Director, L. B., Zaichenko, V. M., & Maykov, I. L. (2012). Intellektual'nye sistemy upravleniya avtonomnymi energeticheskimi kompleksami v sostave lokal'nyh raspredelitel'nyh setej maloj energetiki [Intelligent control systems for autonomous energy complexes as part of local distribution networks of small energy]. *Izvestiya Rossiiskoi Akademii Nauk, 1*, 38 – 48.

Eleven EU countries have already fulfilled their res development plans for 2020. (n.d.). http://www.renen.ru/eleven-eu-countries-have-already-fulfilled-their-res-development-plans-for-2020/

Elistratov, E. E., & Aronova, E. S. (2012). Solnechnye energoustanovki. Ocenka postupleniya solnechnogo izlucheniya [Solar power plants. Assessment of solar radiation]. Izdatel'stvo Politekhnicheskogo universiteta.

Elistratov, V. V. (2017). Tekhnologicheskie i tekhniko-ekonomicheskie osobennosti energokompleksov na osnove VIE dlya slozhnyh prirodno-klimaticheskih uslovij [Technological and technical and economic features of energy complexes based on renewable energy for complex climatic conditions]. In *Materialy V Mezhdunarodnoj konferencii "Vozobnovlyaemaya energetika: problemy i perspektivy". Tom 1. Mahachkala* [Materials of the V International Conference "Renewable Energy: Problems and Prospects". Volume 1. Makhachkala]. Academic Press.

Energeticheskaya strategiya Rossii na period do 2035 goda [Energy strategy of Russia for the period until 2035]. (2014). Academic Press.

Fortov, V. E., & Popel, O. S. (2011). Energetika v sovremennom mire: Nauchnoe izdanie [Energy in the modern world: Scientific publication]. Dolgoprudnyj: Izdatel'skij Dom "Intellekt".

Kharchenko, V., Nikitin, B., Tikhonov, P., Panchenko, V., & Vasant, P. (2019). Evaluation of the Silicon Solar Cell Modules. Intelligent Computing & Optimization. *Advances in Intelligent Systems and Computing*, *866*, 328–336. doi:10.1007/978-3-030-00979-3_34

Kharchenko, V., Panchenko, V., Tikhonov, P. V., & Vasant, P. (2018). Cogenerative PV Thermal Modules of Different Design for Autonomous Heat and Electricity Supply. Handbook of Research on Renewable Energy and Electric Resources for Sustainable Rural Development, 86 – 119. Doi:10.4018/978-1-5225-3867-7.ch004

Makarov, A. A., Grigoryev, L. M., & Mitrova, T. A. (2016). Prognoz razvitiya energetiki mira i Rossii [Forecast of the development of the energy industry in the world and Russia]. Academic Press.

Maykov, I.L. & Director, L.B. (2011). Reshenie zadach optimizacii i upravleniya gibridnymi energeticheskimi kompleksami v strukture raspredelennoj generacii [Solving the problems of optimization and control of hybrid energy complexes in the structure of distributed generation]. *Upravlenie bol'shimi sistemami*, *35*, 250 – 264.

Maykov, I.L., Director, L.B. & Zaichenko, V.M. (2010). Reshenie zadach optimizacii energeticheskih sistem s neskol'kimi avtonomnymi energoustanovkami [Solving the problems of optimizing energy systems with several autonomous power plants]. *Upravlenie bol'shimi sistemami*, *31*, 110 – 129.

On the development of renewable energy in Russia until 2030. (n.d.). http://www.renen.ru/on-the-development-of-renewable-energy-in-russia-until-2030/

Panchenko, V. (2019). Photovoltaic solar modules for autonomous heat and power supply. *IOP Conference Series: Earth and Environmental Science*, 317, 9. 10.1088/1755-1315/317/1/012002

Panchenko, V., Chirskiy, S., & Kharchenko, V. V. (2019). Application of the Software System of Finite Element Analysis for the Simulation and Design Optimization of Solar Photovoltaic Thermal Modules. Handbook of Research on Smart Computing for Renewable Energy and Agro-Engineering, 106 – 131. Doi:10.4018/978-1-7998-1216-6.ch005

Panchenko, V., Izmailov, A., Kharchenko, V., & Lobachevskiy, Y. (2020). Photovoltaic Solar Modules of Different Types and Designs for Energy Supply. *International Journal of Energy Optimization and Engineering*, *9*(2), 74–94. doi:10.4018/IJEOE.2020040106

Panchenko, V., & Kharchenko, V. (2019). Development and Research of PVT Modules in Computer-Aided Design and Finite Element Analysis Systems. Handbook of Research on Energy-Saving Technologies for Environmentally-Friendly Agricultural Development, 314-342. Doi:10.4018/978-1-5225-9420-8.ch013

Panchenko, V., Kharchenko, V., & Vasant, P. (2019). Modeling of Solar Photovoltaic Thermal Modules. *Intelligent Computing & Optimization. Advances in Intelligent Systems and Computing*, *866*, 108–116. doi:10.1007/978-3-030-00979-3_11

Panchenko, V. A. (2018). Solar Roof Panels for Electric and Thermal Generation. *Applied Solar Energy, 54*(5), 350–353. doi:10.3103/S0003701X18050146

Popel, O. S., Frid, S. E., Kolomiyets, Yu. G., Kiseleva, S. V., & Terekhova, E. N. (2010). Atlas resursov solnechnoj energii na territorii Rossii, spravochnoe izdanie [Atlas of solar energy resources in Russia, reference publication]. Moskva: izd. OIVT RAN.

Velkin, V. I. (2015). *Metodologiya rascheta kompleksnyh sistem VIE dlya ispol'zovaniya na avtonomnyh ob"ektah* [Methodology for calculating complex renewable energy systems for use in autonomous facilities]. URFU.

Vissarionov, V. I., Deryugina, G. V., Kuznetsova, V. A., & Malinin, N. K. (2008). Solnechnaya energetika [Solar power]. Moskva: Izdatel'skij dom MEI.

KEY TERMS AND DEFINITIONS

Actinometrical Data: Results of long-term meteorological observations at weather stations processed and systematized by specialized organizations in the form of climate reference books and databases.

Geographical Coordinates: Angular values (latitude and longitude) that determine the position of objects on the earth's surface and on the map.

Heat Accumulator: Device for heat accumulation for the purpose of its further use.

Heat Storage Material: Material used for the accumulation of thermal energy and may be classified according to the material class, the way of accumulation and heat release as well as from the cyclic operation.

Heat Storage System: The system of devices intended for temporary storage of thermal energy by means of various heat-storage materials and including a number of auxiliary equipment including heat exchangers, pumps, pipelines, fittings, test equipment and automatic devices, as well as other devices serving and providing the work of the heat storage system.

Modeling: The study of objects on their models; building and studying models of real-life objects, processes or phenomena in order to obtain explanations of these phenomena, as well as to predict phenomena that interest the researcher.

Optimization: The process of maximizing profitable characteristics, ratios and minimizing costs.

Optimum Inclination Angle of the Receiving Surface: Inclination angle of the receiving surface relative to the horizon, which allows obtaining the maximum solar radiation flux on its surface for a given period of time.

Solar Battery: A combination of photoelectric converters (photocells) – semiconductor devices that directly convert solar energy into direct electric current, in contrast to solar collectors that which heat the coolant.

Chapter 5
Cooperative Caching in Wireless Mesh Networks

Abhishek Majumder
Tripura University, India

Swapna Nath
Tripura University, India

Sukanta Chakraborty
Tripura University, India

Malabika Sarkar
Tripura University, India

ABSTRACT

For reduction of gateway bottleneck effect, data access cost, and response delay, many cooperative caching schemes are available. But in all the existing schemes, whenever any data is newly cached or any data is evicted, the information needs to be broadcast to all the mesh access points (MAPs). As a result, communication cost increases. In this chapter, a cluster and sub-cluster-based cooperative caching scheme is designed to minimize the communication cost. Clusters and sub-clusters are formed by taking advantage of the static position of Mesh routers. Cluster-heads (CHs) are selected using minimum spanning tree and minimum weighted vertex cover algorithm. Then the clusters and sub-clusters are formed. CHs and other Maps keep different cache information tables, and based on those, cache decision is taken. Client mobility is also considered. The performance of the proposed scheme and some other existing schemes have been compared. It has been observed that in the proposed scheme, communication cost in terms of number of messages is reduced.

INTRODUCTION

The term 'Wireless Mesh Networks' (WMN) (Akyildiz et al., 2005; Odabasi et al., 2010; Sgora et al., 2009; Sichitiu, 2005) illustrate wireless networks in which every node can directly communicate with one or more peer nodes. The term 'mesh' is originally is used to demonstrate that all nodes are connected to all other nodes. In most modern mesh network, only a subset of nodes is connected to each other. A WMN is actually a communication network consisting of radio nodes. These nodes are arranged in a mesh topology. Here nodes send and receive data. The coverage area of the radio nodes serve as a single network. It can be called a mesh cloud. A mesh network is reliable and redundant. When one

DOI: 10.4018/978-1-7998-3645-2.ch005

node fails, the remaining nodes can still communicate with each other directly or through one or more intermediate nodes.

WMN is connected to the internet through the gateway (GW). As all the internet traffic flows through a few numbers of GWs, heavy congestion around these GWs creates a serious problem. The problem is termed as GW bottleneck problem. One solution of this problem is catching of data in some of the nodes of WMN. But the challenge is minimizing the distance between requesting client and the data serving node. It is more serious in WMNs than in wired networks. The access cost for a data item in a multi-hop network is calculated by the number of hops between the requesting node and the nearest data serving node. In WMN, when number of hops between requesting node and data serving node increases, contention among the mesh nodes also increases. Therefore, the MRs which are far from the GW suffer from long access latency and low throughput.

Data access cost and response delay can be minimized by reducing the distance between requesting node and data serving node i.e. by caching far distant requested data as close as possible to the requesting node. Gateway bottleneck problem can be reduced by caching popular data within the network i.e. by reducing number of requests made to GW for accessing data from the internet. In this chapter, a cooperative caching technique has been proposed to minimize the data access cost, to reduce the GW bottleneck effect and to minimize the response delay.

The chapter is organised as follows. Second section presents the motivation of the work. Third section gives a brief overview of existing cooperative caching scheme in WMN. Proposed scheme has been presented in next section. Assumptions and numerical analysis of the scheme has been done in the fifth section. Performance comparison of the proposed scheme with the existing schemes has been presented in the next section. The chapter is concluded in the seventh section. Future scope is presented in section eight.

MOTIVATION

In WMN there are only few GWs. If all the internet traffics are routed to the GWs, they become bottleneck of the network. For resolving the issue, the data catching can be done at different nodes of the WMN. But selection of nodes for data catching is a challenging issue because of data access cost. So there is a need to develop cooperative caching technique in WMN for reducing the data access cost.

RELATED WORK

This section discusses some of the existing schemes.

In group-based caching (Ting et al., 2007) there are five parts namely, group formation, cooperative caching table, cache control message, cache placement and replacement and search operation. In group formation phase every mobile host (MH) and its adjacent nodes which are one hop away make a group. Every MH maintains two cooperative catching tables namely, self table and group table. The self table is updated when placement of data is done in a MH. On the other hand, the group table is updated when MH receives the notification of catching status from the group member. Each MH periodically sends caching control message to exchange status with its group members. When a MH receives a data object it updates the cache space using cache placement and replacement process. When a MH wants to access a data item, it first searches its self-table and directly accesses data item if data item is locally cached. Otherwise, MH searches its group-table to know which group members cache that data and randomly sends request to any of them. If requesting MH cannot find data within its local cache as well as group members' cache, it will retrieve that data from the source node by data discovery process. In data discovery process the requester node sends the data request to the data source. When the data source MH receives the request, it replies with the data object.

Layer based Cooperative Caching in wireless P2P Network (Zhao et al., 2010; Zhao et al., 2008) illustrates the basic ideas of the three cooperative caching techniques namely, Cache-Path, Cache-Data and Hybrid-Cache. In Cache-Path, a node need not store data-path information of all passing data. It stores the data-path only when it is nearer to the caching node then the data source. In Cache-Data, a traditional rule is followed i.e., when a node receives data request for data D_i from the same node, it does not cache the data. This is because the cache-data perspective needs extra space to save the data which should be used carefully. In this scheme main focus is on design and carrying out of the cache-data scheme. For this purpose, an asymmetric approach is introduced for cooperative caching. The scheme considers layered design. In layered approach the cooperative cache layer is placed on top of the network layer. Asymmetric caching approach has three phases: forwarding the request message, determining the caching nodes and forwarding the data reply. In forwarding the request message phase, a request message is created and passed down to the cache layer. The original request message is enfolded by the cache layer with a new destination address which is next node to reach the server of that data. In the second phase when a request message arrives at the data server the caching nodes are decided by the cache manager on the forwarding path. Then the ids of these caching nodes are attached to cache list. Cache list is enfolded in the cache layer header. In forwarding the data reply phase, the reply message goes along a different path from the requesting message. In the case of replying message, the message proceeds to the cache layer via network layer of those nodes which want to cache that data. On the other hand, the data item only goes through the network layer of the intermediate nodes which don't want to cache that data.

Grid cluster based cooperative caching (Chauhan et al., 2010), scheme has four phases namely, cluster formation, global cluster cooperative caching, cache discovery process and consistency. In cluster formation phase the whole network is partitioned into predefined equal sized grid called cluster. Each cluster has a super node called Cache State Node (CSN) which is responsible for maintaining the Global Cache State (GCS) information of different clusters within the network domain. GCS contains list of data items along with TTL value stored in the network. When a node of any cluster caches or replaces data item, CSN of that cluster updates its GCS. Each node in the cluster maintains Cluster Cache State (CCS) to know the clustering information of all other nodes within the cluster. In global cluster cooperative (GCC) caching there are four cases of each request namely, local hit, cluster hit, remote hit and global hit. Local hit occurs when a node needs a data and observes it in the local cache. Cluster hit occurs when data is not available in the local cache but it is present within the cluster. Remote hit takes place when cluster miss occurs but data is present within the network CSN. Global hit occurs when data is not present in the network, it will be will be retrieved from the data source or server. Cache discovery process determines the data access path to a MH having requested data item or to the data source or server. When a data item is not present within the cluster of the requesting MH, the requester checks CSN for finding out the caching MH's id for that data item. After receiving the caching MH's id, the requesting MH sends a request message to the caching MH. At the time of sending reply, the caching MH sends data item to the requesting MH. For consistency, the GCC caching uses a simple weak consistency based on TTL, in which a MH considers a cached copy up-to-date if the TTL value is not expired.

In peer-to-peer cooperative caching (Chow et al., 2007), when a request is initiated for a data item in mobile host (MH), it first looks at its local cache. If data is not available in the local cache, the MH broadcasts a requesting message to all its adjacent peers. If data is available in its adjacent peers, data will be provided by one of the peers. If data item is not available in the adjacent peers, the requesting MH sends a request-message to Mobile Support Station (MSS). There are four probable cases for each data request namely, local cache hit, global cache hit, cache miss and access failure. If the required data item is found in the MH's local cache, local cache hit occurs. If requesting data item is not cached, the MH tries to fetch it from its peers. If the peers have the required item, a global cache hit occurs. If the MH fails to acquire a local cache hit or a global cache hit, it encounters a cache miss and has to make a contact with the MSS to obtain the item. If the MH fails to access the item from the MSS, because it is outside the service area of MSS, that is an access failure.

In Cluster based cooperative caching (Nkwe et. al., 2009) the system model considers a hierarchy of four levels namely, the internet, Internet Gateways (IGWs), Mesh Routers (MRs) and the Mesh Clients (MCs). Here the internet is considered as original data server. In this scheme, MRs are designed as Cluster Head (CH) taking advantages of their static positioning, good computational and energy resources. The local MNs will be the cluster member of the CH. The IGWs connect the MRs to the internet. Two lists are maintained in each MR namely, Cluster Cached List (CCL) and Network Cached List (NCL). CCL maintains the caching information within the cluster and NCL maintains the caching information within the network. When a data item is cached in a cluster member i.e MC, it will send notification to its CH i.e. MR. CH then update its CCL and sends notification to all other CHs to update their NCL. When a MC of a cluster wants to access a data item, it is searched in its local cache storage. On a hit, the data is returned to the user. On miss, MC sends a unicast request message to its CH. CH searches the data in the CCL, otherwise in NCL. On hit, CH forwards the request to the caching node. The node with the requested item will forward the requested data item to the requesting CH and after that CH forwards that data item to requesting MC. On miss, the request is sent to the original data source. Upon receiving the request the original data source replies with the requested data item.

In cooperative caching using extra storage (Denko et al., 2010), the WMN is divided into non-overlapping clusters. Each cluster has one MR which is designated as cluster head (CH). MCs are cluster members. All of the existing cache replacement algorithms replace a valid data item in the cache space, when there is no sufficient room for the newly arriving data item. The scheme allows to cache the valid but evicted data items in the MRs' expandable storage. Every valid data item that is replaced from the MC's local cache is sent to its corresponding CH for caching in the Cache Rescue database (CRDB), until its originally assigned TTL elapses. From the Cache Rescue storage, the MRs are able to serve clients' future requests. When the CRDB becomes full, it does not cache any longer, until there is enough space freed to accommodate this incoming data item. The structure of the CRDB is same as the local cache but greater in size. When a MC makes a request, it first looks up the requested data item in its local cache. On a miss, it then forwards the request to its CH to search from the CRDB. Since the CH is part of the cluster, that data item will not be cached by the MC if it is served from the CRDB. If the requested item is not found in the CRDB, it will be forwarded to the internet.

In hierarchical cooperative caching (Wu et al., 2010), there are three parts namely, network model, data access or caching model and data caching system. In the network model, Mesh Access Points (MAPs) compose a backbone network by connecting with each other through multi-hop wireless paths. Each mesh client associates with the nearest MAP through wireless links. A MAP and its local clients constitute a cell. In data access or caching model there are multiple items to be retrieved by mesh client nodes. A source node S_j serves a data item D_j. The data source node S_j may be a client or the gateway node. The data source is assumed to be known to all the nodes, including both MAPs and clients. The design of data caching system involves three parts: heretical cache architecture, data request or reply procedure and cache placement algorithm. In hierarchical cache architecture the cache system consists of two layers namely, MAP layer and mesh client layer. The nodes in both layers have local cache spare to store data cache copies. The upper layer dominates the system through a cache manager module. Cache manager is a logical module deployed at each MAP layer. In data request or reply procedure, three tables are maintained namely, local cache table, cell cache table and nearest cache table. Local cache table is maintained at each mesh client to indicate which data items are now cached locally by the client. Cell cache table is maintained at each MAP to indicate which data items are now cached in the cell. Nearest cache table is maintained at each MAP to indicate the nearest cached cell for each data item. When a MC has a request for data D_i which is not cached locally, MC sends a local request message to its local MAP and then waits for the data item to be delivered. After receiving the request message MAP will first check its cell cache table. If data is not present within the cell it will check nearest cache table, sends global request message to the nearest cached cell and then waits for the corresponding reply.

Different researchers have used data catching in wireless sensor (WSN) (Charan et al., 2019; Sureshkumar et al., 2016), mobile social networks (Wu et al., 2020), cognitive device-to-device networks (Chen et al., 2016), Ultra-Dense Cellular Networks (Lin et al., 2019), Social Wireless Networks (Wang et al., 2017), Mobile Ad-hoc Networks (Bhaskar et al., 2018), 5G wireless networks (Amentie et al., 2016) and Content-Centric Networks (Han et al., 2017) for enhancing the performance. But the chapter mainly concentrates on co-operative caching in WMN.

PROPOSED SCHEME

By taking advantage of the static nature of MAPs in WMN, clusters and sub-clusters are formed in the proposed scheme. Among all the MAPs present in the network, a subset of them is selected as CHs (Cluster Heads). The CHs in turn form individual cluster. All the MAPs form individual Sub-Cluster. The members of the Sub-Cluster are the local clients of the MAPs. The MAP itself is the Sub-Cluster Head of its Sub-Cluster.

In this scheme five hierarchy levels are present namely, the internet (first level), GW (second level), CHs (third level), MAPs (fourth level) and MCs (fifth level). Here only the CHs are able to take the caching and replacement decisions. All other MAPs follow the instructions given by their CH and accept the decisions taken by their CH. In this proposed approach, whenever any data is cached by any node, its Cluster-Head (CH) informs this to all other CHs only via predefined path. Here broadcasting is not required. As a result, caching notification cost will be reduced, which in turn will reduce the communication cost.

The proposed approach has seven parts namely, cluster-head (CH) selection, cluster formation, cluster architecture, hierarchical cache system architecture, messages and tables, algorithms and Mobility management.

Cluster-Head (CH) Selection

The cluster-head selection process has two phases namely, finding out the minimum spanning tree (MST) of the network graph and finding out the minimum weighted vertex cover of the MST.

Finding Out the Minimum Spanning Tree (MST) of the Network Graph (Router level)

In this phase to find the minimum spanning tree (MST), at first weights are assigned to the edges of the network graph of routers. Weight assignment is started from the Gateway node in hop count basis. For the nodes which are reachable via multiple paths from Gateway, minimum path in hop count is selected and weight is assigned according to that minimum distance. After weight assignment, minimum spanning tree is found out using the Kruskal algorithm.

Finding Out the Minimum Weighted Vertex Cover of the MST

Here after getting the MST, weights are assigned on vertices of MST based on minimum hop count from the Gateway. From the weighted graph, minimum weighted vertex cover is found out by using Support Ratio Algorithm (Balaji et al., 2010). By applying this Minimum weighted vertex cover algorithm on the MST of the router level network graph, finally CHs are selected.

Cluster Formation

After the selection of the CHs, all CHs will form their individual clusters. Here all the cluster will be non-overlapping. The steps of cluster formation are given below:

1. Mark all the CHs as selected.
2. Start with the GW which is in 1st level.
3. Under GW (CH), assign the unselected MAPs which are adjacent to the GW, as members of that cluster. Mark the members as selected.
4. Go to next level. Start with a random CH. For every CH, assign the unselected MAPs which are adjacent to that CH, as members of that cluster. Mark the members as selected.
5. Repeat step 4 until all the MAPs are selected i.e. all the levels are covered.

Cluster Architecture

The CH forms its cluster. The members of a cluster are the MAPs which are adjacent to the CH. Within a cluster, all the MAPs (including the CH itself) form their individual sub-cluster. The members of a sub-cluster are the local clients of the MAP and the MAP itself is the sub-cluster head.

Hierarchical Cache System Architecture

In the cache system of the proposed approach, there are three layers namely, the cluster-head (CH) layer, the mesh-access-point (MAP) layer and the mesh-client (MC) layer. The nodes in all the three layers have local cache space to store the cache copies of data. In every CH, there is a Super Cache Manager (SCM) module and in every MAP, there is a Cache Manager (CM) module.

All the CM contains cache related information such as cache copy tables, data request record etc. for its sub-cluster. All the CHs maintain a SCM containing cache related information such as cache copy tables, data request record, data access timestamp etc for the whole cluster. Based on such information, CHs decide whether a data should be cached locally within the cluster or not, where to cache, which data should be replaced when memory becomes full etc. On data request, CM or SCM can determine using the available information, where the requested data is present. According to that they can forward the request and then the requesting client will be able to access the data. Here the CMs can't take caching decisions themselves independently. They just follow the instructions given by their CH. The MCs also do not have any right to take caching decisions independently. They carry out the instructions given by their sub-cluster head i.e. MAP.

Messages and Tables

In the proposed scheme following message are used for data Request- Reply process,

i. LREQ(i, j): Local request message between client i and it's MAP for data D_j within a sub-cluster.
ii. LREP(j, i): Local reply message between MAP and client i. Data D_j is included in the message.
iii. CHREQ(i, j): Request message between MAP-I and its CH for data D_j within a cluster.
iv. CHREP(j, i): Reply message between CH and MAP-I. Data D_j is included in the message.
v. NREQ(M,j,N,P): Request message between CH-M and CH-N via path P for data D_j.
vi. NREP(N,j,M,P): Reply message between CH-N and requesting CH-M via path P. Data D_j is included in the message.
vii. GREQ(N,j S_j G P): Request message from CH-N to data server S_j of data D_j via GW-G. From node N to G, message will be sent via path P.
viii. GREP(G,j, S_j,N,P): Reply message from GW-G to CH-N via path P. D_j will be included in the message.

Following messages are used during data caching notification,

i. MAP_ADD(j,I,k): Caching notification message from caching MAP-I to its CH after caching D_j by replacing D_k.
ii. CH_ADD(j,I,k,P): Caching notification message from caching CH-I to all other CHs via path P after caching D_j by replacing D_k.

Following messages are used during Data Update

i. UPDATE_MAP(S_j,j,I): Update message between CH and it's caching MAP-I. The modified data copy will be included in the message.
ii. UPDATE_MC(S_j,j,i): Update message between MAP and its client x_i. The modified data copy will be included in the message.

iii. UPDATE (S_j,j,P, M): Update message from data-sever-CH to most distant caching CH-M (among all the CHs having matching path sequence) via path P. All the CH-ids having matching path will be included in the message. The modified data copy also will be included in the message.

Following messages are used during mobility management,

i. LEAVE(curMAP, newMAP, data_ids): CurMAP will hold the MAP-id in which the MC is currently registered. At the time of sending the LEAVE message, if the moving MC receives signal from any other MAP, that MAP-id will be recorded in the newMAP field. NewMAP will hold the MAP-id to which the MC wants to go i.e. MAP-id in which the MC will be registered in near future after leaving the current MAP. If newMAP field is blank or null, that means the MC will go to out of coverage area. Data_ids will hold the ids of data which are currently present in MC's local cache space.
ii. ADD (curMAP, prevMAP, data_ids): CurMAP will hold the MAP-id in which the MC has newly entered. PrevMAP will hold the MAP-id of the MAP from which it has come to the new sub-cluster. Data_ids will hold the ids of data which are currently present in MC's local cache space at the time of entering in a new sub-cluster. Here when MC enters in a new cluster from another cluster or from out of coverage area, data_ids field will contain some value only if the MC is a data server (because server data can't be evicted), otherwise it will be left blank or null.

Following tables are used in the proposed scheme,

i. At each client following table is used
   a. Local cache table [LCT]: (data id) Here the data ids of the data which are stored locally within the cache space of MC are listed.
ii. At each MAP following tables are used,
   a. Sub-Cluster cache table [SCCT] (data id, node id (client/MAP)): Here the data ids of the data which are stored locally within the sub-cluster, i.e. in the cache space of MCs or MAP are listed. The caching node id is also listed.
   b. Data request table [DTR] (data id, client id (requesting), no. of request): This table is used at the time of taking cache decisions. This table stores the data ids for which requests are coming. It also stores the client id from which request is coming and also the number of requests from the requesting client.
iii. At each CH following tables are maintained
   a. Cluster cache table [CCT] (client id, sub-cluster id (MAP id), data id, data size, timestamp, flag): This table keeps the data ids and their size which are stored locally within the cluster. The client id and its sub-cluster id at which the data is stored, is also recorded here. Timestamp means the time at which the data is accessed. This table keeps the record of last K timestamps for every cluster data. Flag is a binary variable here. If there is any server MC present within the cluster, flag value of that server data will be '1'. Otherwise for all cache copies of the data, flag value will be '0`.
   b. Network cache table [NCT] (data id, CH-id, update decision status): This table keeps the data ids of the data which are present within the network and also the CH-ids at which these data are stored (by caching or server data). The CH-id list is kept in ascending order of distance.

Initially the value of update decision status variable is set to '1'. This variable is required at the time of cache copy update on modification of the source data.

c. Route table [RT] (CH-id, path, distance, caching decision status): Every CH keeps this table in which all other CH-ids, their path (the shortest path via which they can be reached), and their distance in hop-count is stored. Initially the value of caching decision status variable is set to '1'. This variable is required at the time of sending caching notification to all the CHs.

d. Cache decision table [CDT] (data id, MAP id (requesting), no. of requesting client): This table keeps the record of data for which request is coming, from which sub-cluster the request is coming and the number of requesting clients in each requesting sub-cluster.

iv. At Gateway following table is maintained

a. Client position table [CPT] (client id, CH-id): This table keeps ids of all the clients which are present within the network and their CH-id in which they are currently registered. Always the updated location information of all the clients is kept in this table.

Algorithm

In the proposed approach of cooperative caching, the following algorithms are used,

i. Data request-reply algorithm
ii. Data caching algorithm
iii. Replacement algorithm
iv. Cache update algorithm

Data Request-Reply Algorithm

For handling data requests made by MCs and the replies for those requests, this algorithm is used.

When any MC x_i makes a request for any data item D_j which is not present locally within the cache space of MC itself, at first the data item is searched within the sub-cluster using the function Sub-cluster hit-miss checking. If the requested data D_j is found within the sub-cluster i.e. if cache hit occurs, data D_j is sent to requesting client x_i. If D_j is not found within the sub-cluster i.e. if cache miss occurs, D_j is searched within the cluster using the function Cluster hit-miss checking. If cache hit occurs, then data D_j is sent to requesting client x_i. If D_j is not found within the cluster i.e. if cache miss occurs, D_j is searched within the network using the function Network hit-miss checking. If cache hit occurs, data D_j is sent to requesting client x_i. If D_j is not found within the network i.e. if cache miss occurs, the function Gateway operation will be performed. Then data D_j is sent to requesting client x_i. At the end, i.e. after receiving data D_j the client x_i will access the data.

These functions are explained below,

Sub-Cluster Hit-Miss Checking Function

When any mesh client x_i makes a request for any data D_j, the data is searched in the local cache space of requesting client. If cache hit occurs, x_i can directly access the data from its cache space. Otherwise, request is sent to its MAP. MAP checks from SCCT whether the requested data is available within the sub-cluster. If cache hit occurs, the caching node id is checked. If the MAP itself is the caching node, the

MAP sends data reply to x_i. Otherwise, if some other client is the caching node, MAP sends request to that caching client and then the client sends data reply to MAP. At last MAP forwards this data reply to x_i. But if the requested data is not available within the sub-cluster, the function *Cluster hit-miss Checking* needs to be executed. The algorithmic steps of the function Sub-cluster hit-miss checking are as follows:

```
When any client x_i issues a request for data D_j, it checks whether the data is cached in the total cache space of MC-x_i or not.
If present,
        Requesting MC-x_i accesses the data D_j and stops processing
Else
        CM checks SCCT to find out whether D_j is available in sub-cluster
        If available,
                If D_j is present in MAP itself
                        MAP sends data reply of D_j to x_i
                Else
                        MAP sends request for D_j to corresponding caching client
                        Caching clients send data reply to x_i via MAP
                End if
                Requesting MC-x_i accesses the data D_j and stops processing
        Else
                cluster hit-miss checking ()
        End if
End if
End
```

Cluster Hit-Miss Checking Function

If the requested data D_j is not found within the sub-cluster of the requesting client x_i, at first SCM of CH checks in CCT to find out whether D_j is available within the cluster. If available, request will be sent to the caching node and accordingly that node will send data reply to x_i. But if the requested data is not available within the cluster, the function Network hit-miss checking needs to be executed. The algorithmic steps of the function cluster hit-miss checking are as follows:

```
SCM checks whether data D_j is available in CCT
If available
        SCM checks whether D_j have more than one entries in CCT
        If yes,
                Request is sent to nearest caching node
        Else
                send request for D_j to the particular caching node
```

```
                End if
                Requested node sends data reply to MC-x_i and MAP
                Requesting MC-x_i accesses D_j and stops processing
        Else
                 network hit-miss checking()
        End if
        End
```

Network Hit-Miss Checking Function

If the requested data D_j is not found within the Cluster of the requesting client x_i, SCM of CH Checks from NCT whether D_j is available within the network. If available, request is sent to nearest caching CH. The requested CH finds out the nearest caching node-id (MAP/MC) from its CCT. The requested CH retrieves the data from that node and sends data reply to requesting CH. The CH sends data reply to x_i and x_i accesses the data. But if the requested data is not available within the network, the function Gateway operation needs to be executed. The algorithmic steps of the function Network hit-miss checking are as follows:

```
        SCM checks NCT to checks whether D_j is available within the network
        If available,
                SCM sends data request to nearest caching CH
                Requested CH searches its CCT and selects the nearest caching node-id to servc D_j from its cluster
                Requested CH retrieves D_j from the selected node
                Requested CH sends data reply of D_j to requesting CH
                CH of x_i sends data reply to x_i via x_i's MAP
                x_i accesses D_j and stops processing
        Else
                 gateway operation ()
        End if
        End
```

Gateway Operation Function

If the requested data D_j is not available within the Network, request is sent to Gateway because in case of network cache miss, Gateway is considered as data source. Gateway retrieves the data from outside network and then sends data reply to requesting CH. CH sends the data reply to requesting client x_i. The algorithmic steps of the Gateway function are as follows:

```
        CH of x_i sends request for D_j to gateway
        Gateway retrieves D_j from data source S_j of D_j
        Gateway sends data reply to x_i's CH
        CH sends data reply to x_i via it's MAP
        x_i accesses data D_j and stops processing
        End
```

Data Caching Algorithm

When MC sends request to its MAP for data D_i which is not present within the sub-cluster, CM of MAP updates DRT i.e. CM keeps record of requested data id, requesting client id and number of request from the client. Then MAP forwards the request to CH. SCM of CH updates CDT i.e. SCM keeps the record of requested data id, requesting MAP id and number of requesting clients from that MAP. The time at which the first request was made and the time at which requested data arrives at CH, in between that time if any further request is issued within the cluster, requesting CM and the SCM keeps updating their DRT and CDT.

When D_i arrives at CH, SCM calculates the value of F_i X D, where, F_i = Request frequency of data D_i at CH, D = Distance between requesting client and serving node (client or MAP). If the value of F_i X D is greater than a threshold T, SCM decides to cache D_i. Then from CDT, SCM finds out the requesting MAP-id list for D_i and informs the selected MAPs to cache D_i if free space is available. Every CM finds out the client-id with maximum number of request for D_i from the DRT. If free space is available in the selected clients, D_i is cached. For the selected clients which have no space, their MAP checks if free space is available in the sub-cluster. If available, D_i is cached. After every caching, caching notification is sent to CH. If free space is not available within the sub-cluster also, SCM checks whether any sub-cluster has already cached the data. SCM can check it by observing the notification. If it has received any caching notification for D_i, that means D_i is cached, otherwise not cached. If not cached, SCM checks if free space is available in any sub-cluster within the cluster. If free space is available, D_i is cached in the free sub-cluster, otherwise D_i is cached using replacement algorithm. Then the required tables within the cluster are updated. After taking caching decision, the corresponding entries of CDT and DRT for D_i are reset.

Then new caching CH needs to inform all other CHs so that they can update their NCT. For this purpose, caching notification message will be sent to all other CHs. But individual notification will not be sent to all the CHs. Among all the CHs, a subset will be selected to which individual notification will be sent. The subset is selected by using the 'caching decision status' of RT. At first the most distant CH-id from the newly caching CH will be selected and its 'caching decision status' value will be incremented by 1 i.e. its value will become 2. Then 'caching decision status' value of all the CHs which lie in the same path of the selected most distant CH, will be decremented by 1 i.e. their value will become 0. This procedure continues until 'caching decision status' value of all CHs are changed from 1 to 2 or from 1 to 0. Ultimately individual notification is sent to only those CHs which have 'caching decision status' value 2. The ids of the CHs which lie in the same path of selected CHs, are included in the notification message. When the messages are sent towards the CHs having 'caching decision status' value 2, all other CHs that lie in the path will check their own id in the message. If id is matched, they will retrieve the content of the passing by message and then will forward it to next node. Otherwise, they will directly forward the message to next node without retrieving the content. This way number of messages is reduced. After sending caching notification, 'caching decision status' value for all the CHs in RT will be reset to 1.

The algorithmic steps for data caching is given below:

```
1.        Client issues a request for data Di which is not present in its sub-
cluster
2.        CM updates Data Request Table (DRT)
```

3. CM forwards the request to CH
4. SCM updates Cache Decision Table (CDT)
5. After receiving data D_i, SCM of requesting CH calculates the value of F_i x D
6. Check F_i x D > T
7. If no,
8. D_i will not be cached
9. Else
10. SCM decides to cache the data D_i
11. SCM finds out the requesting MAP-id list from CDT
12. Every CM of selected MAPs are informed to cache D_i if free space is available
13. Every CM of selected MAPs finds out the client-id with maximum number of requests for D_i from DRT
14. Check whether free space is available in selected clients
15. If yes,
16. Cache D_i
17. Send notification to CH
18. CH informs all other CHs and update required tables and then stop processing
19. Else
20. Check whether free space is available within the sub-cluster
21. If yes,
22. Cache D_i in any node (client or MAP) within the sub-cluster having sufficient space for D_i
23. Send notification to CH
24. CH informs all other CHs and updates required tables and then stop processing
25. Else
26. SCM of CH checks if any sub-cluster cached the data
27. If yes
28. no more caching will be done
29. Else
30. SCM checks if free space is available in that cluster
31. If yes,
32. Cache D_i in the free sub-cluster
33. Send notification to CH
34. CH informs all other CHs and updates required tables and then stop processing
35. Else,

```
                                        Cache Di within the clus-
ter using the replacement algorithm
                                        Send notification to CH
                                        CH informs all other CHs
and updates required tables and then stop processing
                                    End If
                               End If
                          End If
                     End If
                End If
                End
```

Replacement Algorithm

When replacement is required, SCM of CH checks CCT to find out which data are stored within the cluster and what their flag values are. Data having flag value '1' are source data i.e. the data of server node. Since source data can't be evicted, so only the cached data i.e. data having flag value '0' are considered at the time of replacement. SCM checks timestamps of all cached data from CCT and calculates their popularity (Chand et al., 2007).

Popularity

The access probability reflects the popularity of a data item for a host. An item with lower probability should be chosen for replacement. At a host, the access probability A_i for data item D_i is given as,

$$A_i = a_i / \sum_{k=1}^{N} a_k$$

Here a_i is the mean access rate of data item D_i. Based on sliding window method of last K access times a_i can be calculated. The access rate is calculated using the equation,

$$a_i = \frac{K}{t^c - t_i^K}$$

Here t^c is the current time and t_i^K is the timestamp of oldest access to item D_i.

Algorithm

The replacement algorithm is as follows

```
          SCM checks Cluster cache table (CCT) for data ids and their corre-
sponding flag values
          SCM selects the data ids which have flag value 'zero'
```

```
From CCT, SCM checks timestamps of selected data and computes their popularity
SCM selects data ids having minimum popularity
SCM checks whether more than one data have equal minimum popularity
If yes
        SCM checks from CCT, whether any of these data stored in multiple sub-cluster within the cluster
        If no,
                SCM checks distance of other serving nodes for those data from NCT
                Check whether distance is equal
                If equal,
                        SCM checks size of those data from CCT
                        Check whether size is equal
                        If no,
                                Replace the data having maximum size by new data
                        Else
                                Replace one of the data randomly by new data
                        End if
                Else
                        Replace the data having minimum distance
                End if
        Else
                Replace the data by new data
        End If
Else
        Replace the data having minimum popularity by new data
End If
Update required tables within the cluster
New caching node's CH informs all other CHs
All other CHs update their NCT and order the CH-id list and then stop processing
End
```

Cache Update Algorithm

If the data is modified in the server node and the server is MC, it sends update message to its CH. In the proposed scheme, there is no need to send individual update message to all caching CHs if they lie in the same path. Update message is sent to only the most distant CH. All other intermediate CHs can update their data by retrieving the data from the passing by update message.

The algorithm for cache update is given below,

```
1.      If server is MC
2.              MC sends UPDATE_MC(Sj,j,i) to its MAP
3.              MAP sends UPDATE_MAP(Sj,j,i) to its CH
4.      End If
5.      Server CH checks its NCT and finds out the ids of caching CHs for its data
6.      SCM checks Route table (RT) and finds out the path for all selected CHs
7.      SCM finds out the most distant CH id from CHs having update decision status value 1
8.      Increment decision status value of the most distant CH by '1' i.e. value will become '2'.
9.      Check whether some of the other CH ids having update decision status value '1' present in most distant CH's path
10.     If yes
11.             Decrement Update decision status values of the CHs present in the most distant CH's path by '1' i.e. value will become '0'
12.     End If
13.     SCM Checks in NCT, whether CH having Update decision status value '1' is present
14.     If yes
15.             go to step 7
16.     End If
17.     SCM sends UPDATE (Sj,j,P,M) to all CH-M having update decision status value '2'
18.     SCM of server CH sets all Update decision status values to '1' again
19.     During message passing via path P, every CH along with path P checks their own id in the message
20.     The destination CH-M and all other CHs whose id match, update data and stop processing
21.     End
```

Mobility Management

Since mesh clients are mobile in nature, whenever they move i.e. whenever they leave a sub-cluster or enter into a sub-cluster, the cache table information may become invalid and the requested data may not be delivered to the requesting client successfully. For solving these problems, mobility management is introduced here which will work as follows:

1. When an MC moves outward of a sub-cluster, at that time when the signal strength becomes close to the threshold T but still greater than T (T is the value of signal strength at which MC will be disconnected from current MAP), the MC will send a LEAVE (curMAP, newMAP, data_ids) message to its CH via the MC's MAP.

2. After receiving the LEAVE message from the moving MC, CH will check if there is any MAP-id present in the newMAP field.
3. If present, the CH will check whether that MAP belongs to the current CH's cluster. If yes, only the required tables within the cluster will be updated. When that MC enters into a new sub-cluster of that cluster, MC will send ADD (curMAP, prevMAP, data_ids) message to CH via new MAP. Then again, the required tables within the cluster will be updated.
4. In the LEAVE message, if the newMAP field is null or even if a MAP-id is present in the newMAP field but does not belong to the current cluster, the CH will cache all the data present in the MC's local cache within the cluster using cache placement/replacement algorithm.
5. If cache replacement is required and it is found that data of leaving MC has minimum access popularity, this data will not be cached. Otherwise, if a data is replaced which still have other copies present in the cluster, required tables within the cluster will be updated but no notification will be sent to other CHs. Notification will be sent to other CHs only when a data is replaced which does not have any other copy within the cluster i.e. if the evicted data does not exist within the cluster anymore [at that time other CHs need to update their NCT].
6. At the time of leaving a cluster, after caching, all the cache copies of the MC will be evicted from the MC's cache space.
7. When MC enters in a new sub-cluster of a new cluster, it will send ADD message to new CH via its MAP. Required tables within the cluster will be updated. At that time if MC is having any data (because server's data can't be evicted) which is not already present in the cluster, CH will send notification to all other CHs to update their NCT.
8. Whenever any MC leaves a cluster or newly enters into a cluster, in both the cases MC's current CH will send a notification to Gateway to update the current location of the MC (CPT). This way Gateway always keeps track of location information of all the MCs present in the network.
9. If any MC moves out of the cluster after issuing a data request, data reply will come to the requesting CH. CH will forward the message to Gateway to know the current location of the MC. Then that data reply will be forwarded and delivered to the requesting MC.

ASSUMPTION AND NUMERICAL ANALYSIS

This section presents the numerical analysis of the proposed scheme and existing hierarchical caching technique. In this mathematical analysis two performance metrics have been used namely, total communication cost and response delay. Total communication cost reflects the cost efficiency and response delay reflects the time efficiency. The definitions are given as follows,

- Total communication cost: It consists of the query cost (during data request and reply), cost of sending caching notification to all CHs, data update cost and cost of mobility management. Here the cost is calculated in terms of number of messages. More the number of messages, higher will be the communication cost. Total communication cost $\propto$ Number of messages transmitted or received.
- Response delay: This is the delay between sending of a data request and receiving the corresponding response. This is calculated in terms of distance (in hop-count) between requesting node and data serving node. More the distance is, higher will be the response delay.

The notations that are used in the numerical analysis are given as,

Γ = Total communication cost
Γ_Q =Query cost (during data request/reply) per query
Γ_C =Cost of sending caching notification to all CHs per caching
Γ_U =Cost of each data update
Γ_M =Cost of mobility management per movement of MC
Γ_{mc} =Query cost required when data is present in requesting MC itself
Γ_{sub} =Query cost required when data is not present in requesting MC but present in requesting MC's sub-cluster
Γ_{clus} =Query cost required when data is not present in requesting MC's sub-cluster but present in its cluster
Γ_{nw} =Query cost required when data is not present in requesting MC's Cluster but present within the network of MC
Γ_{gw} =Query cost required when data is not present in requesting MC's network
Γ_{leave} =Cost required when a MC leaves a sub-cluster
Γ_{add} =Cost required when a MC enters into a sub-cluster
R_Q = Query response delay per query
r_{mc} = Query response delay when data d is found in requesting MC's own cache space
r_{sub} = Query response delay when data d is not present in requesting MC's own cache space but present in sub-cluster of requesting MC
r_{clus} = Query response delay when data d is not present in sub-cluster of requesting MC but present in the cluster of that MC
r_{nw} = Query response delay when data d is not present in cluster of requesting MC but present within the network of that MC
r_{gw} = Query response delay when data d is not present in requesting MC's network
P_{cd} = Average probability that data d is cached in requesting client itself
P_{sd} = Average probability that data d is cached in requesting client`s sub-cluster
P_{cld} = Average probability that data d is cached in requesting client's cluster
P_{nd} = Average probability that data d is cached in requesting client's network
$1 - P_{nd}$ = Average probability that data d is not cached in requesting client's network
P_{ch} = Average probability that the mesh node (MC or MAP) is in the sub-cluster of CH
$1 - P_{ch}$ = Average probability that the mesh node (MC or MAP) is in the sub-cluster other than that of CH
P_r = Average probability that a data item is cached in a MAP
$1 - P_r$ = Average probability that a data item is cached in a MC
P_{move1} = Average probability that intra-sub-cluster node mobility will take place within same cluster
$1 - P_{move1}$ = Average probability that MC will move out of or move into a cluster while moving
P_{move2} = Average probability that some data will be deleted (will not exist in that cluster anymore) from the cluster or some data will be newly added to a cluster because of node mobility
$1 - P_{move2}$ = Average probability that no data will be deleted from a cluster and no data will be newly added to a cluster and if deleted, other copy of the same data will exist in that cluster. In all the cases, caching notification will be sent to all other CHs. So, this simply describes the probability that caching notification will be sent to all other CHs
η = Average caching rate of data objects
λ = Average rate of queries for data objects

μ = Average Update rate of data objects
σ = Average mobility rate of MCs
Ψ_c / Ψ_d = Cost required to transmit a control/data packet between two MRs
Ψ_c / Ψ_d = Cost required to transmit a control/data packet between a MC and its MR
τ_c/τ_d = Communication delay of transmitting a control/data packet between two MRs
τ_c/τ_d = Communication delay of transmitting a control/data Packet between a MC and its MR
δ = Communication delay when Gateway retrieves data from internet or another network
D = Average distance between requesting MC's CH and nearest caching CH of requested data
L = Average distance between an arbitrary CH and Gateway
m = Average number of caching CHs of certain data within the network
X = Average number of caching CHs to which individual update message needs to be sent (X<=m)
N = Total number of CHs within the network
n = Average number of CHs to which individual caching notification needs to be sent (n<=N-1)
Γ_{cell} = Query cost required when data is present within the cell of requesting client but not present in the cache space of requesting client itself
r_{cell} = Response delay when the requested data is present within the cell of requesting client but not present in the cache space of requesting client of itself
Dist = Average distance between requesting MAP and data serving MAP
L_1 = Average distance between an arbitrary MAP and Gateway
n_{edge} = Total number of edges present in the network graph of routers
m_1 = Average number of caching MAPs for any data

Performance Analysis of Proposed Scheme

This section analyses the performance of the proposed cooperative caching scheme in terms of the total communication cost and response delay.

Total communication cost can be calculated as,

$$\Gamma = \lambda.\Gamma_Q + \eta.\Gamma_C + \mu.\Gamma_U + \sigma.\Gamma_M \tag{1}$$

Query cost can be calculated as,

$$\Gamma_Q = P_{cd}.\Gamma_{mc} + (P_{sd} - P_{cd}).\Gamma_{sub} + (P_{cld} - P_{sd}).\Gamma_{clus} + (P_{nd} - P_{cld}).\Gamma_{nw} + (1 - P_{nd}).\Gamma_{gw} \tag{2}$$

The cost components are calculated as,

$$\Gamma_{mc} = 0 \tag{3}$$

$$\Gamma_{sub} = \begin{cases} \varphi_c + \varphi_d & \textit{If d is in requesting MC's router} \\ 2(\varphi_c + \varphi_d) & \textit{If d is in some other MC within the subcluster} \end{cases}$$

$$= P(\varphi_c + \varphi_d) + (1 - P_r)2(\varphi_c + \varphi_d)$$

$$=(2-P_r)(\varphi_c+\varphi_d) \tag{4}$$

$$\Gamma_{clus}\begin{Bmatrix}\varphi_c+\varphi_d+\Psi_c+\Psi_d\\ 2(\varphi_c+\varphi_d)+\Psi_c+\Psi_d\\ \varphi_c+\varphi_d+2(\Psi_c+\Psi_d)\\ 2(\varphi_c+\varphi_d+\Psi_c+\Psi_d)\end{Bmatrix}$$

If requesting MC's MAP itself is CH and d is found in some other MAP or if requesting MC's MAP is not CH and d is found in CH itself. If requesting MC's MAP itself is CH and d is found in some client of sub-cluster within the cluster or if requesting MC's MAP is not CH and d is found in some MC of CH's own sub-cluster. If requesting MC's MAP is not CH and d is found in some MAP (not CH) itself.

$$\begin{aligned}&=P_{ch}(1-P_{ch})P_r(\varphi_c+\Psi_c+\varphi_d+\Psi_d)+(1-P_{ch})P_{ch}P_r(\varphi_c+\Psi_c+\Psi_d+\varphi_d)+P_{ch}(1-P_{ch})(1-P_r)\\&(2(\varphi_c+\varphi_d)+\Psi_c+\Psi_d)+(1-P_{ch})P_{ch}(1-P_r)(2(\varphi_c+\varphi_d)+\Psi_c+\Psi_d)+(1-P_{ch})(1-P_{ch})\\&P_r(\varphi_c+\varphi_d+2(\Psi_c+\Psi_d))+(1-P_{ch})(1-P_{ch})(1-P_r)(2(\varphi_c+\varphi_d+\Psi_c+\Psi_d))\\&=(1-P_{ch}^2)(2-P_r)(\varphi_c+\varphi_d)+2(1-P_{ch})(\Psi_c+\Psi_d)\end{aligned} \tag{5}$$

$$\Gamma_{nw}\begin{Bmatrix}\varphi_c+\varphi_d+D(\Psi_c+\Psi_d)\\ 2(\varphi_c+\varphi_d)+D(\Psi_c+\Psi_d)\\ \varphi_c+\varphi_d+(D+1)(\Psi_c+\Psi_d)\\ 2(\varphi_c+\varphi_d)+(D+1)(\Psi_c+\Psi_d)\\ \varphi_c+\varphi_d+(D+2)(\Psi_c+\Psi_d)\\ 2(\varphi_c+\varphi_d)+(D+2)(\Psi_c+\Psi_d)\end{Bmatrix}$$

If requesting MC's MAP itself is CH and d is found in nearest CH i.e. MAP itself. If requesting MAP itself is CH and d is found in some MC of nearest CH's own sub cluster. If requesting MC's MAP is CH and d is found in some other MAP itself under the nearest CH's cluster or if requesting MC's MAP is not CH and d is found in some MC of nearest caching CH's own sub cluster. If requesting MC's MAP is not CH and d is found in some MAP (not CH) of nearest caching cluster itself. If requesting MC's MAP is not CH and d is found in some MC of a MAP (not CH) of nearest caching cluster itself

$$= P_{ch}P_{ch}P_r\left(\varphi_c+\varphi_d+D\left(\Psi_c+\Psi_d\right)\right)+P_{ch}P_{ch}\left(1-P_r\right)\left(2\left(\varphi_c+\varphi_d\right)+D\left(\Psi_c+\Psi_d\right)\right)+P_{ch}\left(1-P_{ch}\right)$$
$$P_r\left(\varphi_c+\varphi_d+\left(D+1\right)\left(\Psi_c+\Psi_d\right)\right)+\left(1-P_{ch}\right)P_{ch}P_r\left(\varphi_c+\varphi_d+\left(D+1\right)\left(\Psi_c+\Psi_d\right)\right)+\left(1-P_{ch}\right)$$
$$P_{ch}P_r\left(\varphi_c+\varphi_d+\left(D+1\right)\left(\Psi_c+\Psi_d\right)\right)+P_{ch}\left(1-P_{ch}\right)\left(1-P_r\right)\left(2\left(\varphi_c+\varphi_d\right)+\left(D+1\right)\left(\Psi_c+\Psi_d\right)\right)$$
$$+\left(1-P_{ch}\right)P_{ch}\left(1-P_r\right)\left(2\left(\varphi_c+\varphi_d\right)+\left(D+1\right)\left(\Psi_c+\Psi_d\right)\right)+\left(1-P_{ch}\right)\left(1-P_{ch}\right)$$
$$P_r\left(\varphi_c+\varphi_d+\left(D+2\right)\left(\Psi_c+\Psi_d\right)\right)+\left(1-P_{ch}\right)\left(1-P_{ch}\right)\left(1-P_r\right)\left(2\left(\varphi_c+\varphi_d\right)+\left(D+2\right)\left(\Psi_c+\Psi_d\right)\right)$$
$$=\left(2-P_r\right)\left(\varphi_c+\varphi_d\right)+\left\{D+2-2P_{ch}\right\}\left(\Psi_c+\Psi_d\right) \tag{6}$$

$$\Gamma_{gw}=\begin{cases}\varphi_c+\varphi_d+L\left(\Psi_c+\Psi_d\right) & \textit{if requesting MC's MAP itself is CH}\\ \varphi_c+\varphi_d+\left(L+1\right)\left(\Psi_c+\Psi_d\right) & \textit{if requesting MC's MAP is not CH}\end{cases}$$

$$=P_{ch}\left\{\varphi_c+\varphi_d+L\left(\Psi_c+\Psi_d\right)+\delta\right\}+\left(1-P_{ch}\right)\left\{\varphi_c+\varphi_d+\left(L+1\right)\left(\Psi_c+\Psi_d\right)\right\}$$
$$=\left(\varphi_c+\varphi_d\right)+\left(1-L-P_{ch}\right)\left(\Psi_c+\Psi_d\right) \tag{7}$$

Now substituting the values of Γ_{mc}, Γ_{sub}, Γ_{clus}, Γ_{nw}, and Γ_{gw}, query cost is calculated as,=

$$\Gamma_Q=P_{cd}.\Gamma_{mc}+\left(P_{sd}-P_{cd}\right).\Gamma_{sub}+\left(P_{cld}-P_{sd}\right).\Gamma_{clus}+\left(P_{nd}-P_{cld}\right).\Gamma_{nw}+\left(1-P_{nd}\right).\Gamma_{gw}$$
$$=\left(\varphi_c+\varphi_d\right)\left[1-\left(P_r-1\right)P_{nd}+\left(P_r-2\right)\left(P_{cd}-P_{cld}P_{ch}^2-P_{sd}P_{ch}^2\right)\right]$$
$$+\left(\Psi_c+\Psi_d\right)\left[D\left(P_{nd}-P_{cld}\right)+L\left(1-P_{nd}\right)+\left(2P_{sd}-P_{nd}-1\right)\left(P_{ch}-1\right)\right] \tag{8}$$

Caching notification cost can be calculated as,

$\Gamma_C=dnc_1.\Psi_c+dnc_2.\Psi_c+dnc_3.\Psi_c+\ldots+dnc_n.\Psi_c$ where n<=N-1

$$=\Psi_c\left(dnc_1+dnc_2+dnc_3+\ldots+dnc_n\right) \tag{9}$$

Here $dnc_1, dnc_2, dnc_3, \ldots, dnc_n$ are distance between the newly caching CH and $CH_1, CH_2, CH_3, \ldots, CH_n$ respectively. Here maximum value of n is N-1 because when all the (N-1) CHs are in different way, the caching CH needs to send (N-1) individual notifications.

Data update cost can be calculated as,

$\Gamma_u=ds_1.\Psi_d+ds_2.\Psi_d+ds_3.\Psi_d+\ldots+ds_x.\Psi_d$ where x<=m

$$=\Psi_d\left(ds_1+ds_2+ds_3+\ldots+ds_x\right) \tag{10}$$

Here $ds_1, ds_2, ds_3, \ldots, ds_n$ are distance between the server CH and caching $CH_1, CH_2, CH_3, \ldots, CH_x$ respectively. Here maximum value of x is m because when all the m caching CHs are in different way, the server CH needs to send m individual update message.

Mobility management cost can be calculated as,

$$\Gamma_M = \Gamma_{leave} + \Gamma_{add} \tag{11}$$

Now Γ_{leave} and Γ_{add} are calculated as,

$$\Gamma_{leave} = \begin{array}{c} \varphi_c \\ \varphi_c + \Psi_c \\ \varphi_c + L\Psi_c \\ (\varphi_c + \Psi_c) + L\Psi_c \\ \varphi_c + L\Psi_c + [d_1 + d_2 + \ldots + d_n]\Psi_c \\ (\varphi_c + \Psi_c) + L\Psi_c + [d_1 + d_2 + \ldots + d_n]\Psi_c \end{array}$$

When MC leaves sub-cluster (CH) and informs that it will go to some other sub-cluster within same cluster. When MC leaves sub-cluster (not CH) and informs that it will go to some other sub-cluster within same cluster. When MC leaves sub-cluster (CH) and informs that it will go to some other cluster or to out of coverage area and MC's data are cached in free space or by replacing some data which still have other copies present in that cluster. When MC leaves sub-cluster (not CH) and informs that it will go to some other cluster or out of coverage area and MC's data are cached in free space or by replacing some data which still have other copies present in that cluster. When MC's MAP is not CH and MC leaves a cluster and data of MC are cached within the cluster by replacing some data which does not exist within the cluster anymore.

$$\begin{aligned} &= P_{ch} P_{move1} (1 - P_{move2}) \varphi_c + (1 - P_{ch}) P_{move1} (1 - P_{move2})(\varphi_c + \Psi_c) + P_{ch} (1 - P_{move1})(1 - P_{move2}) \\ &(\varphi_c + L\Psi_c) + (1 - P_{ch})(1 - P_{move1})(1 - P_{move2})[(\varphi_c + \Psi_c) + L\Psi_c] + P_{ch}(1 - P_{move1}) P_{move2} \\ &[\varphi_c + L\Psi_c + \Psi_c (d_1 + d_2 + d_3 + \ldots + d_n)] + (1 - P_{ch})(1 - P_{move1}) \\ &P_{move2}[(\varphi_c + \Psi_c) + L\Psi_c + \Psi_c (d_1 + d_2 + d_3 + \ldots + d_n)] = \varphi_c [1 - P_{move1} P_{move2}] + \Psi_c \\ &[\{P_{ch} - (d_1 + d_2 + d_3 + \ldots + d_n)\} P_{move1} P_{move2} + (d_1 + d_2 + d_3 + \ldots + d_n) P_{move2} - L.P_{move1} - P_{ch} + L + 1] \end{aligned} \tag{12}$$

$$\Gamma_{add} = \left\{ \begin{array}{c} \varphi_c \\ \varphi_c + \Psi_c \\ \varphi_c + L\Psi_c \\ (\varphi_c + \Psi_c) + L\Psi_c \\ \varphi_c + L\Psi_c + [d_1 + d_2 + \ldots d_n]\Psi_c \\ (\varphi_c + \Psi_c) + L\Psi_c + [d_1 + d_2 + \ldots + d_n]\Psi_c \end{array} \right\}$$

When MC enters into a sub-cluster (CH) from another sub-cluster within same cluster. When MC enters into a sub-cluster (not CH) from another sub-cluster within same cluster. When MC enters into a sub-cluster (CH) of a new cluster and does not have any data or it has some data which is already present in the new cluster. When MC enters into a sub-cluster (not CH) of a new cluster and it does not have any data or it has some data which is already present in the new cluster. When MC enters into a sub-cluster (CH) of a new cluster and the MC has some data which was not already present in the cluster at past i.e. the data are newly added to that cluster. When MC enters into a sub-cluster (not CH) of a new cluster and the MC has some data which was not present in the cluster at past i.e. the data are newly added to that cluster.

$$= P_{ch}P_{move1}\left(1-P_{move2}\right)\varphi_c + \left(1-P_{ch}\right)P_{move1}\left(1-P_{move2}\right)\left(\varphi_c+\Psi_c\right) + P_{ch}\left(1-P_{move1}\right)\left(1-P_{move2}\right)$$
$$\left(\varphi_c + L\Psi_c\right) + \left(1-P_{ch}\right)\left(1-P_{move1}\right)\left(1-P_{move2}\right)\left[\left(\varphi_c+\Psi_c\right)+L\Psi_c\right] + P_{ch}\left(1-P_{move1}\right)P_{move2}$$
$$\left[\varphi_c + L\Psi_c + \Psi_c\left(d_1+d_2+d_3+\ldots+d_n\right)\right] + \left(1-P_{ch}\right)\left(1-P_{move1}\right)$$
$$P_{move2}\left[\left(\varphi_c+\Psi_c\right)+L\Psi_c+\Psi_c\left(d_1+d_2+d_3+\ldots+d_n\right)\right] = \varphi_c\left[1-P_{move1}P_{move2}\right]+\Psi_c$$
$$\left[\left\{P_{ch}-\left(d_1+d_2+d_3+\ldots+d_n\right)\right\}P_{move1}P_{move2}+\left(d_1+d_2+d_3+\ldots+d_n\right)P_{move2}-L.P_{move1}-P_{ch}+L+1\right] \quad (13)$$

Therefore, substituting the value of Γ_{leave} and Γ_{add} the cost of mobility management can be calculated as,

$$``_M = 2\varphi_c\left[1-P_{move1}P_{move2}\right]+2\Psi_c$$
$$\left[\left\{P_{ch}-\left(d_1+d_2+d_3+\ldots+d_n\right)\right\}P_{move1}P_{move2}+\left(d_1+d_2+d_3+\ldots+d_n\right)P_{move2}-L.P_{move1}-P_{ch}+L+1\right] \quad (14)$$

From (1) and (8), we get

$$\Gamma_Q = \lambda\left(\varphi_c+\varphi_d\right)\left[1-\left(P_r-1\right)P_{nd}+\left(P_r-2\right)\left(P_{cd}-P_{cld}P_{ch}^2-P_{sd}P_{ch}^2\right)\right]$$
$$+\lambda\left(\Psi_c+\Psi_d\right)\left[D\left(P_{nd}-P_{cld}\right)+L\left(1-P_{nd}\right)+\left(2P_{sd}-P_{nd}-1\right)\left(P_{ch}-1\right)\right] \quad (15)$$

From (1) and (9), we get

$$\eta\Gamma_c = \eta.\Psi_c.\left(dnc_1+dnc_2+dnc_3+\ldots+dnc_n\right) \quad (16)$$

From (1) and (4), we get,

$$\mu\Gamma_c = \mu.\Psi_d.\left(d_1+d_2+d_3+\ldots+d_x\right) \quad (17)$$

From (1) and (14), we get,

$$\sigma\Gamma_M = 2.\sigma.\varphi_c\left[1-P_{move1}P_{move2}\right]+2.\sigma.\Psi_c$$
$$\left[\left\{P_{ch}-\left(1+d_1+d_2+d_3+\ldots+d_n\right)\right\}P_{move1}P_{move2}+\left(d_1+d_2+d_3+\ldots+d_n\right)P_{move2}-L.P_{move1}-P_{ch}+L+1\right]$$

$$=2.\sigma.\varphi_c\left[1-P_{move1}P_{move2}\right]+2.\sigma.\Psi_c\left[(1-P_{ch})(1-P_{move1}P_{move2})+\left(1-P_{move1}\right)\left\{\left(d_1+d_2+d_3+\ldots+d_n\right)P_{move2}+L\right\}\right] \tag{18}$$

Now using (1), (15), (16), (17) and (18) the total cost of communication is calculated as,

$$\Gamma = \lambda\Gamma_Q+\eta\Gamma_c+\mu\Gamma_u+\sigma\Gamma_m$$
$$=\varphi_c\left[\lambda\left\{1-\left(P_r-1\right)P_{nd}+\left(P_r-2\right)\left(P_{cd}-P_{ch}^2\left(P_{cld}-P_{sd}\right)-P_{nd}\right)\right\}+2\sigma\left(1-P_{move1}P_{move2}\right)\right]$$
$$+\varphi_d\lambda\left[1-\left(P_r-1\right)P_{nd}+\left(P_r-2\right)\left(P_{cd}-P_{ch}^2P_{cld}-P_{ch}^2P_{nd}\right)\right]+\Psi_c\left[\lambda\left\{D\left(P_{nd}-P_{cld}\right)+L\left(1-P_{nd}\right)\right.\right.$$
$$\left.+\left(2P_{sd}-P_{nd}-1\right)\left(P_{ch}-1\right)\right\}+\eta\left(d_1+d_2+d_3+\ldots+d_3\right)+2\sigma\left\{\left(1-P_{ch}\right)\left(1-P_{move1}P_{move2}\right)\right.$$
$$\left.\left.+\left(1-P_{move1}\right)\left\{\left(d_1+d_2+d_3+\ldots+d_n\right)P_{move2}+L\right\}\right\}\right]+\Psi_d\left[\lambda\left\{D\left(P_{nd}-P_{cld}\right)+L\left(1-P_{nd}\right)\right.\right.$$
$$\left.\left.+\left(2P_{sd}-P_{nd}-1\right)\left(P_{ch}-1\right)\right\}\mu\left(d_1+d_2+d_3+\ldots+d_x\right)\right]$$

The response delay can be calculated as,

$$R_Q=P_{cd}.r_{mc}+\left(P_{sd}-P_{cd}\right).r_{sub}+\left(P_{cld}-P_{sd}\right).r_{clus}+\left(P_{nd}-P_{cld}\right).r_{nw}+\left(1-P_{nd}\right).r_{gw} \tag{20}$$

The components of the above mentioned equation can be calculated as,

$$r_{mc}=0 \tag{21}$$

$$r_{sub}=\begin{cases}\gamma_c+\gamma_d & \text{when } d \text{ is in MAP itself of requesting MC}\\ 2\left(\gamma_c+\gamma_d\right) & \text{when } d \text{ is in some MC of requesting MC's subcluster}\end{cases}$$

$$=P_r\left(\gamma_c+\gamma_d\right)+\left(1-P_r\right)\left(\gamma_c+\gamma_d\right)=\left(2-P_r\right)\left(\gamma_c+\gamma_d\right) \tag{22}$$

$$r_{clus}=\begin{cases}\gamma_c+\tau_c+\gamma_d+\tau_d & \text{If requesting MC's MAP itself is CH and d is found in some MAP itselt within cluster or requesting MC's MAP is not CH and d is present in CH itself}\\ 2\left(\gamma_c+\gamma_d\right)+\tau_c+\tau_d & \text{If requesting MC's MAP itself is CH and d is present in some CH of other sub-cluster within the cluster or if requesting MC's MAP is not CH and d is present in some MC of CH's sub-cluster}\\ \gamma_c+\gamma_d+2\left(\tau_c+\tau_d\right) & \text{If requesting MC's MAP is not CH and d is present in some MAP (not CH) itself within the cluster}\\ 2\left(\gamma_c+\gamma_d+\tau_c+\tau_d\right) & \text{If requesting MC's MAP is not CH and d is present in some MC of some other MAP (not CH) within the cluster}\end{cases}$$

$$\begin{aligned}&= P_{ch}(1-P_{ch})P_r(\tau_c+\gamma_c+\tau_d+\gamma_d)+(1-P_{ch})P_{ch}P_r(\tau_c+\gamma_c+\tau_d+\gamma_d)\\&+(1-P_{ch})P_{ch}(1-P_r)(2(\gamma_c+\gamma_d)+\tau_c+\tau_d)+(1-P_{ch})P_{ch}(1-P_r)(2(\gamma_c+\gamma_d)+\tau_c+\tau_d)\\&+(1-P_{ch})P_r(1-P_{ch})(\gamma_c+\gamma_d+2(\tau_c+\tau_d))+(1-P_{ch})(1-P_{ch})(1-P_r)(2(\gamma_c+\gamma_d+\tau_c+\tau_d))\\&=(1-P_{ch}^2)(2-P_r)(\gamma_c+\gamma_d)+2(1-P_{ch})(\tau_c+\tau_d)\end{aligned} \tag{23}$$

$$r_{nw}=\begin{cases}\gamma_c+\gamma_d+D(\tau_c+\tau_d) & \text{If requesting MC's MAP is CH and d is found in CH itself of nearest caching cluster}\\2(\gamma_c+\gamma_d)+D(\tau_c+\tau_d) & \text{If requesting MC's MAP is CH and d is found in some MC of CHs own sub cluster of nearest caching cluster}\\\gamma_c+\gamma_d+(D+1)(\tau_c+\tau_d) & \text{If requesting MC's MAP is CH and d is found in some MAP (not CH) itself of nearest caching cluster or if requesting MC's MAP is not CH and d is found in CH itself of nearest caching cluster}\\2(\gamma_c+\gamma_d)+(D+1)(\tau_c+\tau_d) & \text{If requesting MC's MAP is CH and d is found in some MCof MAP (not CH) in nearest caching cluster or if requesting MC's MAP is not CH and d is found in some MC of CH's own sub cluster of nearest caching cluster}\\\gamma_c+\gamma_d+(D+2)(\tau_c+\tau_d) & \text{If requesting MC's MAP is not CH and d is found in some MAP (not CH) itself of nearest caching cluster}\\2(\gamma_c+\gamma_d)+(D+2)(\tau_c+\tau_d) & \text{If requesting MC's MAP is not CH and d is found in some MC of a MAP (not CH) of the nearest caching cluster}\end{cases}$$

$$\begin{aligned}&= P_{ch}P_{ch}P_r(\gamma_c+\gamma_d+D(\tau_c+\tau_d))+P_{ch}P_{ch}(1-P_r)(2(\gamma_c+\gamma_d)+D(\tau_c+\tau_d))\\&+P_{ch}(1-P_{ch})P_r(\gamma_c+\gamma_d+(D+1)(\tau_c+\tau_d))+(1-P_{ch})P_{ch}P_r(\gamma_c+\gamma_d+(D+1)(\tau_c+\tau_d))\\&+P_{ch}(1-P_{ch})(1-P_r)(2(\gamma_c+\gamma_d)+(D+1)(\tau_c+\tau_d))+(1-P_{ch})P_{ch}(1-P_r)\\&(2(\gamma_c+\gamma_d)+(D+1)(\tau_c+\tau_d))+(1-P_{ch})(1-P_{ch})P_r(\gamma_c+\gamma_d+(D+2)(\tau_c+\tau_d))\\&+(1-P_{ch})(1-P_{ch})(1-P_r)(2(\gamma_c+\gamma_d)+(D+2)(\tau_c+\tau_d))\end{aligned}$$

$$=(2-P_r)(\gamma_c+\gamma_d)+\{D+2-2P_{ch}\}(\tau_c+\tau_d) \tag{24}$$

$$r_{gw}=\begin{cases}\gamma_c+\gamma_d+L(\tau_c+\tau_d)+\delta & \textit{if requesting MC's MAP is CH}\\\gamma_c+\gamma_d+(L+1)(\tau_c+\tau_d)+\delta & \textit{if requesting MC's MAP is not CH}\end{cases}$$

$$\begin{aligned}&= P_{ch}\{\gamma_c+\gamma_d+L(\tau_c+\tau_d)+\delta\}+(1-P_{ch})\{\gamma_c+\gamma_d+(L+1)(\tau_c+\tau_d)+\delta\}\\&=(\gamma_c+\gamma_d)+(1-L-P_{ch})(\tau_c+\tau_d)+\delta\end{aligned} \tag{25}$$

Now substituting the value of r_{mc}, r_{sub}, r_{clus}, r_{nw}, and r_{gw} in equation (20), response delay is calculated as,

$$\begin{aligned}R_Q &= P_{cd}.r_{mc}+(P_{sd}-P_{cd}).r_{sub}+(P_{cld}-P_{sd}).r_{clus}+(P_{nd}-P_{cld}).r_{nw}+(1-P_{nd}).r_{gw}\\&=(\gamma_c+\gamma_d)\left[1-P_{nd}(P_r-1)+(P_r-2)(P_{cd}-P_{cld}P_{ch}^2-P_{sd}P_{ch}^2)\right]\\&+(\tau_c+\tau_d)\left[D(P_{nd}-P_{cld})+L(1-P_{nd})+(2P_{sd}-P_{nd}-1)(P_{ch}-1)\right]+(1-P_{nd})\delta\end{aligned} \tag{26}$$

Performance Analysis of Hierarchical Cooperative Caching Scheme

In this section performance metrics of the existing hierarchical cooperative caching scheme (Wu et al., 2010) have been calculated.

Total communication cost consists of query cost, caching notification cost and data update cost. So, the total communication cost can be calculated as,

$$\Gamma = \lambda.\Gamma_Q + \eta.\Gamma_c + \mu.\Gamma_u \tag{27}$$

The query cost can be calculated as,

$$\Gamma_Q = P_{cd}.\Gamma_{mc} + (P_{sd} - P_{cd}).\Gamma_{cell} + (P_{nd} - P_{cld}).\Gamma_{nw} + (1 - P_{nd}).\Gamma_{gw} \tag{28}$$

The components of the above equation is as follows,

$$\Gamma_{mc} = 0 \tag{29}$$

$$\Gamma_{cell} = \begin{cases} \varphi_c + \varphi_d & \text{if } d \text{ is in requesting MC's MAP} \\ 2(\varphi_c + \varphi_d) & \text{if } d \text{ is in some other MC within the cell} \end{cases}$$

$$= P_r(\varphi_c + \varphi_d) + (1 - P_r)2(\varphi_c + \varphi_d) = (2 - P_r)(\varphi_c + \varphi_d) \tag{30}$$

$$\Gamma_{nw} = \begin{cases} \varphi_c + \varphi_d + Dist.(\Psi_c + \Psi_d) & \text{if } d \text{ is in MAP of nearest caching cell} \\ 2(\varphi_c + \varphi_d) + Dist.(\Psi_c + \Psi_d) & \text{if } d \text{ is in MC of nearest caching cell} \end{cases}$$

$$\begin{aligned} &= P_r\left[\varphi_c + \varphi_d + Dist.(\Psi_c + \Psi_d)\right] + (1 - P_r)\left[2(\varphi_c + \varphi_d) + Dist.(\Psi_c + \Psi_d)\right] \\ &= (2 - P_r)(\varphi_c + \varphi_d) + Dist.(\Psi_c + \Psi_d) \end{aligned} \tag{31}$$

$$\Gamma_{nw} = \varphi_c + \varphi_d + L_1(\Psi_c + \Psi_d) \tag{32}$$

Now by subtracting all the derived values in (28), query cost can be calculated as,

$$\Gamma_Q = (\varphi_c + \varphi_d)\left[(2 - P_r)(P_{nd} - P_{cd}) + (1 - P_{nd})\right] + (\Psi_c + \Psi_d)\left[Dist(P_{nd} - P_{sd}) + L_1(1 - P_{nd})\right] \tag{33}$$

Caching notification cost can be calculated as,

$$\Gamma_C = 2{,}n_{edge}.\Psi_c \tag{34}$$

Data update cost can be calculated as,

$$\Gamma_U = \Psi_d.(d_1 + d_2 + d_3 + \ldots + d_{m1}) \tag{35}$$

From (27), (33), (34) and (35), total communication is calculated as,

$$\begin{aligned}\Gamma = \lambda.\Gamma_Q + \eta.\Gamma_c + \mu.\Gamma_u &= (\varphi_c + \varphi_d)\left[\lambda\{(2-P_r)(P_{nd} - P_{cd}) + (1-P_{nd})\}\right] \\ &+\Psi_c\left[\lambda\{Dist(P_{nd} - P_{sd}) + L_1(1-P_{nd})\}2.\eta.n_{edge}\right] \\ &+\Psi_d\left[\lambda\{Dist(P_{nd} - P_{sd}) + L_1(1-P_{nd})\} + \mu(d_1 + d_2 + d_3 + \ldots + d_{m1})\right]\end{aligned} \tag{36}$$

The response delay can be calculated as,

$$R_Q = P_{cd}.r_{mc} + (P_{sd} - P_{cd}).r_{cell} + (P_{nd} - P_{sd}).r_{nw} + (1-P_{nd}).r_{gw} \tag{37}$$

The terms used in the above equation can be calculated as,

$$r_{mc} = 0 \tag{38}$$

$$r_{cell} = \begin{cases}\gamma_c + \gamma_d & if\ d\ is\ in\ requesting\ MC's\ MAP \\ 2(\gamma_c + \gamma_d) & if\ d\ is\ in\ some\ other\ MC\ within\ the\ cell\end{cases}$$

$$= P_r(\gamma_c + \gamma_d) + (1-P_r)2(\gamma_c + \gamma_d) = (2-P_r)(\gamma_c + \gamma_d) \tag{39}$$

$$\Gamma_{nw} = \begin{cases}\gamma_c + \gamma_d + Dist.(\tau_c + \tau_d) & if\ d\ is\ in\ MAP\ of\ nearest\ caching\ cell \\ 2(\gamma_c + \gamma_d) + Dist.(\tau_c + \tau_d) & if\ d\ is\ in\ MC\ of\ nearest\ caching\ cell\end{cases}$$

$$\begin{aligned}&= P_r\left[(\gamma_c + \gamma_d) + Dist.(\tau_c + \tau_d)\right] + (1-P_r)\left[2(\gamma_c + \gamma_d) + Dist.(\tau_c + \tau_d)\right] \\ &= (2-P_r)(\gamma_c + \gamma_d) + Dist.(\tau_c + \tau_d)\end{aligned} \tag{40}$$

$$\Gamma_{nw} = (\gamma_c + \gamma_d) + L_1(\tau_c + \tau_d) + \delta \tag{41}$$

Now by substituting all the derived values in (37), response delay can be calculated as,

$$R_Q = P_{cd}.r_{mc} + (P_{sd} - P_{cd}).r_{cell} + (P_{nd} - P_{sd}).r_{nw} + (1-P_{nd}).r_{gw} \tag{42}$$

$$(\gamma_c+\gamma_d)\left[(2-P_r)(P_{nd}-P_{cd})+(1-P_{nd})\right]+(\tau_c+\tau_d)\left[Dist(P_{nd}-P_{sd})+L_1(1-P_{nd})\right]+\delta(1-P_{nd}) \quad (43)$$

Total communication cost and response delay per query of hierarchical cooperative caching scheme has been represented using equation (36) and equation (43) respectively.

RESULTS AND DISCUSSION

To evaluate the performance of the proposed scheme, total communication cost of the proposed scheme is compared with the total communication cost of the hierarchical scheme. The default values of different parameters are given in table 1.

Table 1. Default values of the parameters used

P_{cd}=0.05	$L=L_1$=20	**σ=0.011**
P_{sd}=0.15	D=Dist=10	γ_c/γ_d=0.002/0.004
P_{cld}=0.3	$m=m_1$=5	τ_c/τ_d=0.001/0.002
P_{nd}=0.6	n_{edge}=100	Ψ_c/Ψ_d=0.003/0.006
P_{ch}=0.2	δ=0.007	φ_c/φ_d=0.004/0.008
P_r=0.6	λ=0.03	$(d_1+d_2+\ldots+d_n)=d$=90
P_{move1}=0.5	η=0.008	$(d_1+d_2+\ldots+d_x)=d'$=90
P_{mvoe2}=0.65	μ=0.005	

When η and μ are kept constant, total communication cost will vary with λ. The total communication cost of the proposed scheme both excluding mobility and including mobility is compared with that of hierarchical scheme.

From figure 1 it can be observed that, for equal value of λ total communication cost of proposed scheme is less than that of hierarchical scheme because hierarchical scheme's caching notification cost is greater than caching notification cost of proposed scheme. When the value of λ increases, total query cost of both the schemes increases. As a result, total communication cost for both the schemes increase.

Figure 2, shows the effect of average query rate on total communication cost when MC is mobile. Total communication cost of the proposed scheme is almost equal to the total communication cost of hierarchical scheme when a fixed mobility rate is considered. When value of λ increases, total communication cost of both the schemes increase for constant value of μ, η and σ.

Figure 3 shows that, when the value of λ and μ are kept constant, total communication cost will vary with η. The total communication cost of the proposed scheme both excluding mobility and including mobility is compared with the total communication cost of hierarchical scheme.

When value of η increases total caching notification cost increases for both the schemes. As a result, total communication cost of both the schemes increase. Since caching notification cost of hierarchical scheme is greater than that of the proposed scheme, total communication cost of hierarchical scheme is higher as compared to the proposed scheme when η increases.

Figure 1. Total communication cost (excluding mobility cost) vs. average query rate

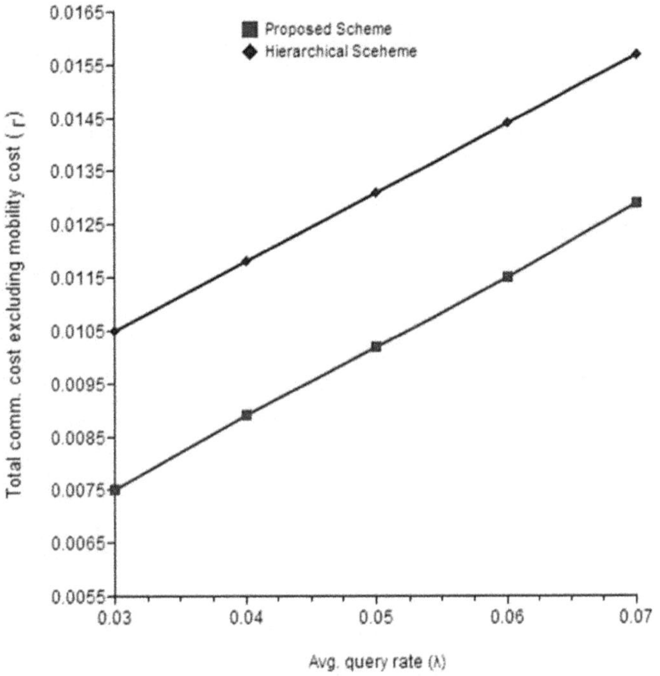

Figure 2. Total communication cost (including mobility cost) vs. average query rate

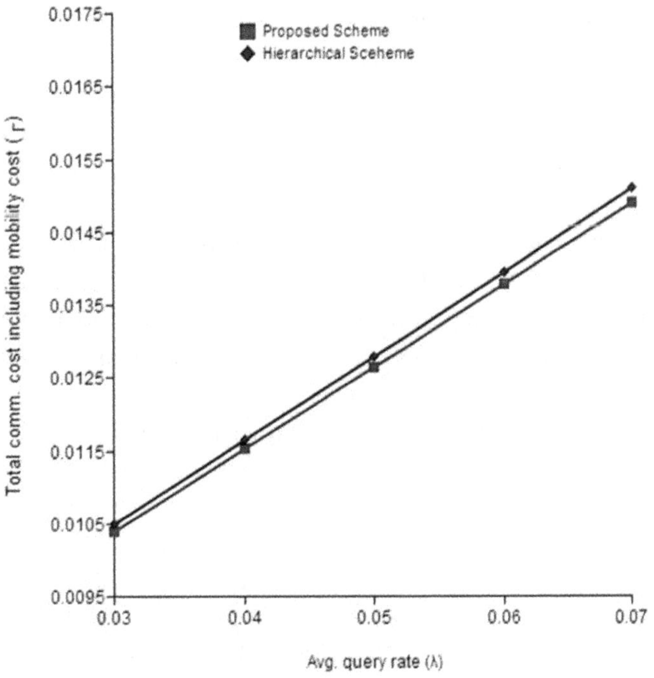

Figure 3. Total communication cost (excluding mobility cost) vs. average caching rate

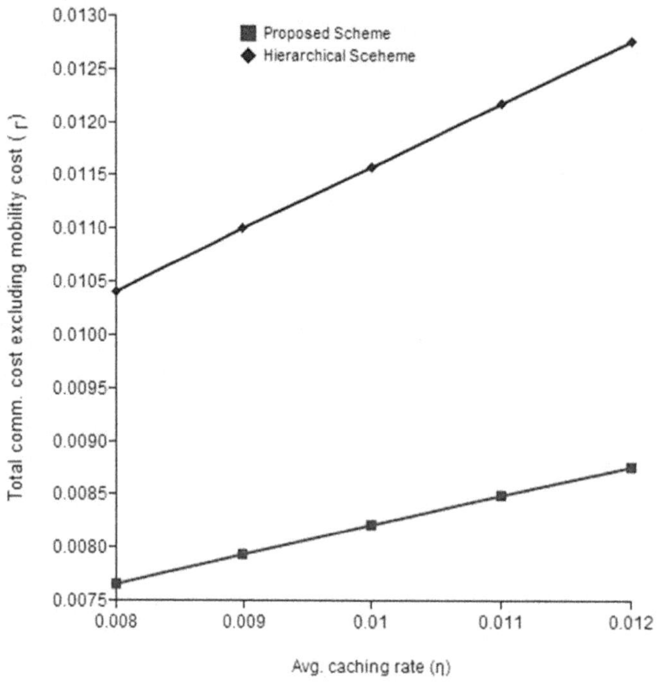

Figure 4 shows the effect of average caching cost on total communication cost when MCs are mobile. Total Communication cost of proposed scheme includes the mobility cost. When value of η increases total communication cost for both the schemes increase when λ and μ are kept constant.

Figure 5 and figure 6 show the effect of average update rate of data objects on total communication cost without and with mobility respectively. When the values of η and λ are kept constant, total communication cost will vary with μ. The total communication cost of proposed scheme both excluding mobility and including mobility is compared with the total communication cost of hierarchical scheme.

Here the worst case of proposed scheme is compared with hierarchical scheme when calculating the total update cost. In the worst-case scenario, update cost for both the schemes are equal. So, in figure 5 total communication cost for both the schemes increase equally if μ increases. For every value of μ, total communication cost of hierarchical scheme is greater than total communication cost of proposed scheme because caching notification cost of hierarchical scheme is greater than the caching notification cost of proposed scheme.

If node mobility is considered, total communication cost of proposed scheme increases. Now by keeping η, λ and σ as constant, if μ increases, total communication cost of both the schemes increases almost equally since update cost of both the schemes are considered as equal.

In all the above three cases, mobility rate is considered as fixed. But in practical situation, total communication cost varies with node mobility rate. In the hierarchical scheme, nodes don't have mobility i.e. they have considered the network as static. So, the impact of mobility rate on total communication cost of hierarchical scheme can't be shown. But in the proposed scheme since node mobility is considered, variation of total communication cost with mobility rate has been shown in figure 7.

Figure 4. Total communication cost (including mobility cost) vs. average caching rate

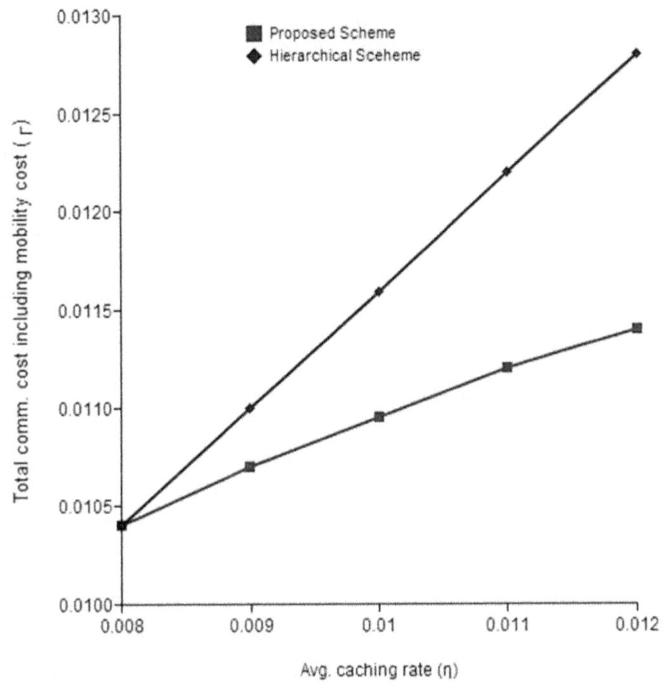

Figure 5. Total communication cost (excluding mobility cost) vs. average update rate of data objects

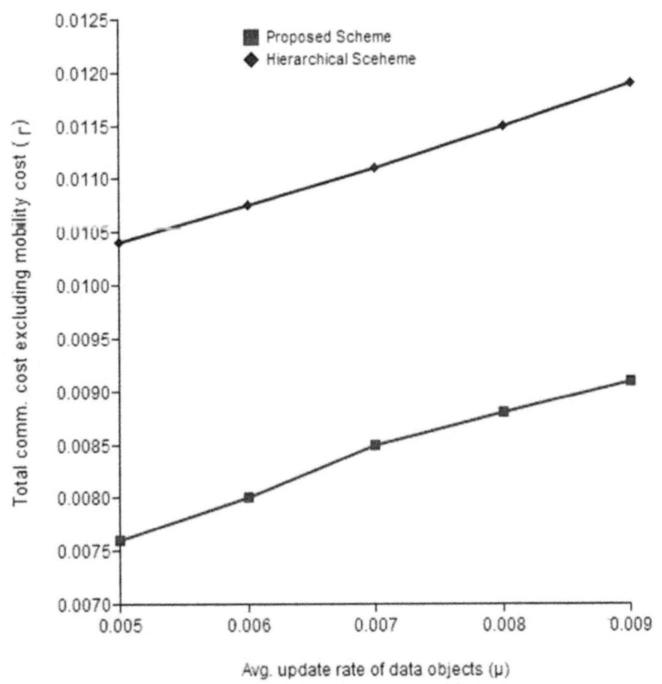

Figure 6. Total communication cost (including mobility cost) vs. average update rate of data objects

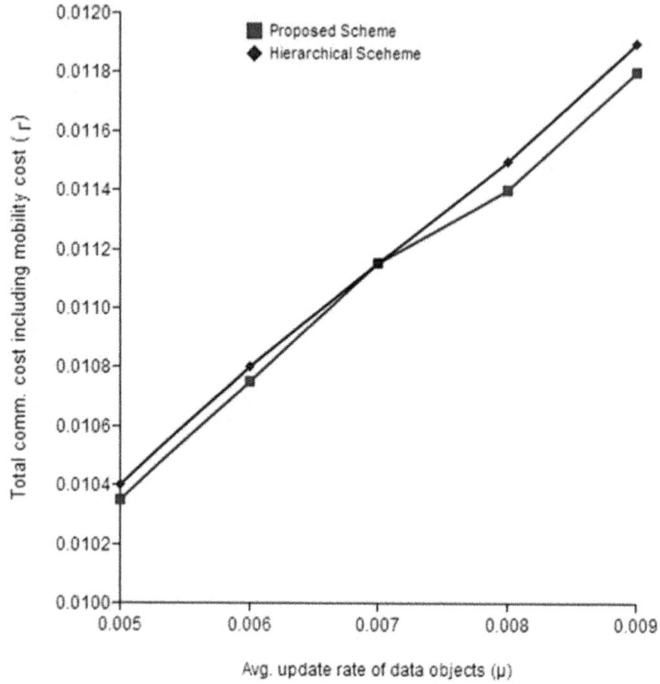

Figure 7. Total communication cost vs. Avg. mobility rate

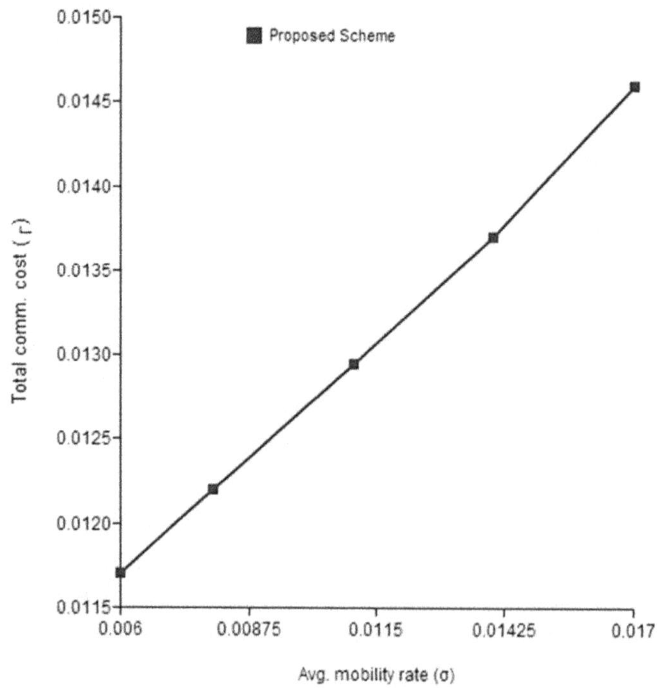

When average mobility rate of MCs increase, mobility management cost also increases. Therefore, total communication cost also increases with increased value of σ. This is demonstrated in figure 7.

CONCLUSION

In the proposed scheme, by taking advantage of the static nature of the MRs of WMN, clusters and sub-clusters are formed. Here different cache information tables are maintained only in the static MRs and cache decision is made in a hierarchical way. A CH first decides whether a data item should be cached within the cluster, and then determines which sub-cluster should cache the data, and finds out which node of the sub-cluster will cache it. Consistency is maintained by updating the cache copies on modification of the source data. To increase the cache hit ratio, an efficient replacement algorithm is designed, which takes into consideration different factors such as: popularity, distance, size etc. In the proposed scheme, the impact of client mobility on the caching system is also introduced. Communication cost is reduced by restricting the caching notification messages required to be sent.

FUTURE SCOPE

The client mobility affects the performance in terms of message cost. If mobility rate increases, total communication cost also increases. Therefore, mobility management is an issue to be further investigated to improve the performance. Moreover, to increase the cache hit further investigation can be carried out to design more efficient cache placement or replacement algorithm. Deep learning (Thomas et al., 2019) and optimization techniques (Vasant et al., 2019; Vasant et al., 2020) can be useful tools to achieve this objective. WMN with cache replacement policy can be applied in the field of agriculture and renewable energy (Kharchenko et al., 2018; Kharchenko et al., 2019a, Kharchenko et al., 2019b). Security (Lugovic et al., 2019) of cached data in WMN also remains as future work.

ACKNOWLEDGMENT

The authors would like to thank Mobile Computing Laboratory, Department of Computer Science & Engineering, Tripura University for providing the necessary infrastructure. They would also like to thank anonymous reviewers who have provided suggestions to enhance the quality of the chapter.

REFERENCES

Akyildiz, I. F., Wang, X., & Wang, W. (2005). Wireless mesh networks: A survey. *Computer Networks, 47*(4), 445–487. doi:10.1016/j.comnet.2004.12.001

Amentie, M. D., Sheng, M., Song, J., & Liu, J. (2016, October). Minimum delay guaranteed cooperative device-to-device caching in 5G wireless networks. In *8th International Conference on Wireless Communications & Signal Processing (WCSP)* (pp. 1-5). IEEE.

Balaji, S., Swaminathan, V., & Kannan, K. (2010). An effective algorithm for minimum weighted vertex cover problem. *International Journal of Mathematical, Computational, Physical, Electrical and Computer Engineering*, *4*(7), 770–774.

Bhaskar, B., Mahesh, D., Kumaraswamy, P., & Nagender, Y. (2018). Caching Strategies Based on Data Density Estimation in Mobile Ad-hoc Networks. *International Journal of Engineering & Technology*, *7*(1.8), 240-243.

Chand, N., Joshi, R. C., & Misra, M. (2007). Cooperative caching in mobile ad hoc networks based on data utility. *Mobile Information Systems*, *3*(1), 19–37. doi:10.1155/2007/193641

Charan, P., Usmani, T., Paulus, R., & Saeed, S. H. (2019). Performance of Distributed Energy Aware Routing (DEAR) Protocol with Cooperative Caching for Wireless Sensor Networks. *Wireless Sensor Network*, *11*(3), 35–45. doi:10.4236/wsn.2019.113003

Chauhan, N., Awasthi, L. K., Chand, N., Joshi, R. C., & Misra, M. (2010). Energy efficient cooperative caching in mobile ad hoc networks. *International Journal of Applied Engineering Research*, *1*(3), 360–371.

Chen, B., Shen, H., & Cao, X. (2016). Cooperative caching placement in cognitive device-to-device networks. *Journal of Communication*, *11*(1), 42–49.

Chow, C. Y., Leong, H. V., & Chan, A. T. (2007). GroCoca: Group-based peer-to-peer cooperative caching in mobile environment. *IEEE Journal on Selected Areas in Communications*, *25*(1), 179–191. doi:10.1109/JSAC.2007.070118

Denko, M. K., Nkwe, T., & Obaidat, M. S. (2010). Efficient cooperative caching with improved performance in wireless mesh networks. In *IEEE International Conference on Communications* (ICC) (pp. 1-5). IEEE. 10.1109/ICC.2010.5501921

Han, W., Wang, G., & Ying, P. (2017, December). Energy-Efficiency Aware Cooperative Caching Strategy for Content-Centric Networks. In *International Conference on Security, Privacy and Anonymity in Computation, Communication and Storage* (pp. 389-398). Springer. 10.1007/978-3-319-72395-2_36

Kharchenko, V., & Vasant, P. (2018). *Handbook of Research on Renewable Energy and Electric Resources for Sustainable Rural Development*. IGI Global. doi:10.4018/978-1-5225-3867-7

Kharchenko, V., & Vasant, P. (2019a). *Handbook of Research on Energy-Saving Technologies for Environmentally-Friendly Agricultural Development*. IGI Global.

Kharchenko, V., & Vasant, P. (2019b). *Handbook of Research on Smart Computing for Renewable Energy and Agro-Engineering*. IGI Global.

Lin, P., Song, Q., & Jamalipour, A. (2019). Multidimensional Cooperative Caching in CoMP-Integrated Ultra-Dense Cellular Networks. *IEEE Transactions on Wireless Communications*, *19*(3), 1977–1989. doi:10.1109/TWC.2019.2960329

Lugovic, S., Mrsic, L., & Korona, L. Z. (2019). Public WiFi Security Network Protocol Practices in Tourist Destination. In *International Symposium on Pervasive Systems, Algorithms and Networks* (pp. 321-332). Springer. 10.1007/978-3-030-30143-9_27

Nkwe, T. K., & Denko, M. K. (2009). Self-optimizing cooperative caching in autonomic wireless mesh networks. In *IEEE Symposium on Computers and Communications* (pp. 411-416). IEEE. 10.1109/ISCC.2009.5202375

Odabasi, S. D., & Zaim, A. H. (2010). A survey on wireless mesh networks, routing metrics and protocols. *International Journal of Electronics Mechanical and Mechatronics Engineering*, *2*(1), 92–104.

Sgora, A., Vergados, D. D., & Chatzimisios, P. (2009). IEEE 802.11s wireless mesh networks: Challenges and Perspectives. In *International Conference on Mobile Lightweight Wireless Systems* (pp. 263-271). Springer. 10.1007/978-3-642-03819-8_25

Sichitiu, M. L. (2005). Wireless mesh networks: opportunities and challenges. In *Proceedings of World Wireless Congress* (*vol. 2*, pp. 21). Academic Press.

Sureshkumar, A., & Ravindran, R. S. (2016). Swarm and fuzzy based cooperative caching framework to optimize energy consumption over multimedia wireless sensor networks. *Wireless Personal Communications*, *90*(2), 961–984. doi:10.100711277-016-3274-0

Thomas, J. J., Karagoz, P., Ahamed, B. B., & Vasant, P. (2019). *Deep Learning Techniques and Optimization Strategies in Big Data Analytics*. IGI Global.

Ting, Y. W., & Chang, Y. K. (2007, July). A novel cooperative caching scheme for wireless ad hoc networks: Group caching. In *International Conference on Networking, Architecture, and Storage* (pp. 62-68). IEEE. 10.1109/NAS.2007.4

Vasant, P., Zelinka, I., & Weber, G. W. (2019). *Intelligent Computing & Optimization*. Springer International Publishing. doi:10.1007/978-3-030-00979-3

Vasant, P., Zelinka, I., & Weber, G. W. (2020). *Intelligent Computing and Optimization*. Springer International Publishing. doi:10.1007/978-3-030-33585-4

Wang, R., Hajiaghajani, F., & Biswas, S. (2017, October). Incentive based cooperative content caching in Social Wireless Networks. In *IEEE 28th Annual International Symposium on Personal, Indoor, and Mobile Radio Communications (PIMRC)* (pp. 1-7). IEEE. 10.1109/PIMRC.2017.8292169

Wu, D., Liu, B., Yang, Q., & Wang, R. (2020). Social-aware cooperative caching mechanism in mobile social networks. *Journal of Network and Computer Applications*, *149*(102457), 1–12. doi:10.1016/j.jnca.2019.102457

Wu, W., & Huang, Y. (2010, December). Hierarchical cooperative data caching for wireless mesh networks. In *IEEE/IFIP 8th International Conference on Embedded and Ubiquitous Computing (EUC)* (pp. 289-296). IEEE. 10.1109/EUC.2010.48

Zhao, J., Zhang, P., & Cao, G. (2008). On cooperative caching in wireless P2P networks. In *The 28th International Conference on Distributed Computing Systems* (pp. 731-739). IEEE.

Zhao, J., Zhang, P., Cao, G., & Das, C. R. (2010). Cooperative caching in wireless p2p networks: Design, implementation, and evaluation. *IEEE Transactions on Parallel and Distributed Systems*, *21*(2), 229–241. doi:10.1109/TPDS.2009.50

ADDITIONAL READING

Abbas, N., Hajj, H., Sharafeddine, S., & Dawy, Z. (2018). Traffic offloading with channel allocation in cache-enabled ultra-dense wireless networks. *IEEE Transactions on Vehicular Technology*, *67*(9), 8723–8737. doi:10.1109/TVT.2018.2845135

Cvjetkovic, M., & Rakocevic, V. (2016, October). Alternative topology construction for cooperative data distribution in mobile ad hoc networks. In *8th International Congress on Ultra Modern Telecommunications and Control Systems and Workshops (ICUMT)* (pp. 171-176). IEEE. 10.1109/ICUMT.2016.7765352

Lepage, L., Sparks, B., & Tapia, B. (2019). Unstable Communication for Write-Back Caches. *International Journal of Software Systems Research and Methodology*, *2*(1), 1–6.

Mahajan, B. N., Mahajan, A. R., & Thomas, A. (2017, February). Co-operative push back streaming and routing in WSN. *In International Conference on Data Management, Analytics and Innovation (ICDMAI)* (pp. 306-311). IEEE. 10.1109/ICDMAI.2017.8073531

Parvez, I., Rahmati, A., Guvenc, I., Sarwat, A. I., & Dai, H. (2018). A survey on low latency towards 5G: RAN, core network and caching solutions. *IEEE Communications Surveys and Tutorials*, *20*(4), 3098–3130. doi:10.1109/COMST.2018.2841349

Yang, S., Chan, H. C., Lam, P. P., & Chong, P. H. (2017). MeshFS: A distributed file system for cloud-based wireless mesh network. *Journal of Systems and Software*, *131*, 201–217. doi:10.1016/j.jss.2017.06.020

KEY TERMS AND DEFINITIONS

Cluster Head: Cluster heads are the nodes which act as a leader of a cluster. Different techniques are used to form the cluster and to select the cluster head.

Cooperative Caching: Cooperative caching is the process of sharing and coordination of data among the nodes without going to the data source.

Gateway: Gateways are the nodes which connect the WMN to the external network. All the traffic to and from the external network goes through the gateway.

Mesh Access Point: These are the mesh nodes having access point capability. It can serve the clients as an access point forming a basic service set. At the same time, it can communicates with the WMN. By these nodes, clients can send and receive data packets to and from other nodes.

Mesh Client: Mesh clients are the mobile or static devices which are using the services of WMN.

Mesh Router: Mesh routers are the nodes which perform the routing activities in the WMN.

Mobile Host: These are the nodes which coordinate to cache the data.

Wireless Mesh Network: Wireless mesh network is a self-healing and self-organizing wireless network where the mesh nodes can communicate to each other through multiple path.

Chapter 6
Smart Computerized Essay Scoring Using Deep Neural Networks for Universities and Institutions

J. Joshua Thomas
https://orcid.org/0000-0002-9992-6094
UOW Malaysia KDU Penang University College, Malaysia

Lim Ting Wei
UOW Malaysia KDU Penang University College, Malaysia

Y. Bevish Jinila
Sathyabama Institute of Science and Technology, Chennai, India

R. Subhashini
https://orcid.org/0000-0001-5752-2588
Sathyabama Institute of Science and Technology, Chennai, India

ABSTRACT

This chapter develops a web-based automated text scoring (ATS) system that can grade essays and check for spelling errors. The main reason behind this work is to alleviate the labour-intensive marking of essays and ensures equality in scoring for high-stakes exams like TOEFL. The researcher had performed a detailed investigation on deep learning techniques used in the field of ATS and developed a recurrent neural network model that can score essays in an end-to-end approach. Using the developed deep learning model, a web application was also developed to showcase the process of ATS by letting the web application to communicate with the trained model. The model was trained using Keras framework and TensorFlow library and the web application was done using the Flask framework. This work is the LSTM network that can capture sequential dependencies. The evaluation metrics chosen to evaluate the model are the quadratic weighted kappa (QWK) score, and the trained model can achieve 0.6 in QWK score.

DOI: 10.4018/978-1-7998-3645-2.ch006

INTRODUCTION

This work introduces a Natural Language Processing (NLP)-related project which is based on deep learning models for the problem of Automated Text Scoring (ATS). The resulting product is a web application named EssayScore that allows users to upload essays to be scored by a deep learning model with instant feedback. The approach to the deep learning model used is Long-Short-Term-Memory (LSTM) units, which is a subset of Recurrent Neural Networks (RNN) that deals with sequential dependency. There are three major aims to be completed for this work. The first aim is to investigate the application of deep learning techniques on the problem of automatic text scoring (ATS). Besides, the second aim is to develop a recurrent neural network (RNN) model that can score essays without human intervention. The final aim is to develop a web application that can accept user essays as input and grade them using the trained RNN model. To achieve the aims stated above, 10 objectives are identified that can contribute to the three aims above. The first objective is to perform a comprehensive literature review on the automated text scoring using neural networks and any existing work done in this research area. This is done to understand how recurrent neural networks and long-short-term memory (LSTM) units work such as their strengths and weaknesses. Besides, on the technical side, the researcher is required to learn how to develop a deep learning model using Keras framework and TensorFlow to train an RNN model for the work. The functional and non-functional requirements of the web application are established before the development of the system began and commentary on these requirements from a user's perspective are made. Furthermore, a Software Requirement Specification (SRS) will also be written to document the requirements from a developer's perspective. Next, UML diagrams of the web application are also created during the designing phase such as the use case diagram, class diagram, and sequence diagram. On the documentation side, the researcher also produced a work report on the developed automated test scoring (ATS) system which includes the analysis, evaluation, and conclusion and recommendation. Lastly, the last objective is to critically evaluate the product and process in the report and suggest some improvements in the finished product.

An ATS system aims to act as a "second-grader" for high stake tests to reduce the burden of the teachers. Besides, EssayScore also acts as an effective learning tool for students to gain instant feedback on their essays by reducing the time taken for students to get back their essay scores. For example, a feedback loop such as the one proposed by Paruchuri (2019) in Figure 1 is an effective use case of the ATS system in facilitating the learning process of students.

EssayScore is mainly composed of two parts: the front-end of the application which handles user input and the backend which sends essays for grading. The deep learning model chosen will be a Long-Short-Term Memory model (LSTM), which is a kind of recurrent neural network (RNN) model that is useful in capturing temporal dependencies that are commonly present in essay texts. The deep learning model will be trained using Keras and TensorFlow and the deployment will be handled by TensorFlow Serving and Docker. The model will then act as the backend of the application which handles the scoring of the essay. The front-end is responsible for making REST API requests to the backend and populates the returned essay score to the user interface (UI).

Besides, EssayScore also provides spell checking feature for each uploaded essay as each word misspelled will be highlighted in red. The web application also provides recommendations for the misspelled word for users to choose from. Furthermore, uploaded essays and predicted scores will be stored in the Cloud Fire Store, a cloud-based storage and authentication platform from Google. Users of the web application will be able to view past submissions to see their learning progress.

Figure 1. Feedback Loop of an ATS system (Paruchuri, 2019)

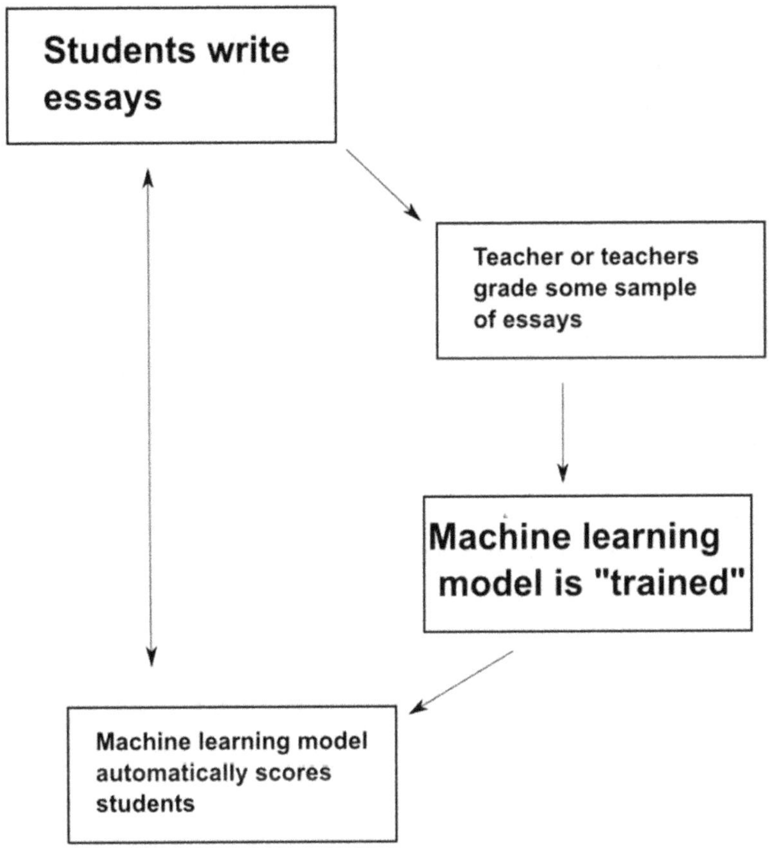

The deep learning architecture of the model will be based on works like Taghipour and Ng (2016) which utilizes word embedding to detect the semantic representations of the words in vector space. Each input essay will be treated as a sequence of word vectors through pre-processing with the NLTK tokenizer and the GloVe model. The architecture also includes a 1-Dimensional Convolution Neural Network (CNN) to extract n-gram features for embedding purposes. Different evaluation metrics such as mean squared error (MSE) and mean absolute error (MAE) is also used in determining the accuracy of the model. The model is trained through 50 epochs and the weights of the model that has the lowest MSE during the training phase are saved as the final model for deployment. Quadratic Weighted Kappa (QWK) is also used as the preferred metric to determine the inter-rater agreement between the deep learning model and the human grader.

The chapter organised in the following it has the background studies in three primitive areas, related works, main contribution of the chapter with problem statement continued with methodology and techniques, flow of the system design, training results, test cases, ethical consideration and further directions of the work.

BACKGROUND

In this section, three main areas of research will be discussed and justified, which are natural language processing (NLP), deep learning, and the type of platform chosen for the final product. Each of these areas will be further supported by at least two existing literature.

Natural Language Processing (NLP)

Natural Language Processing (NLP) is a research area in computer science that explores how computers can understand and interpret languages that are spoken or written by human beings (Chowdhury, 2003). NLP has considered one of the hardest problems in computer science as natural languages has a different meaning when placed in a different context. The common usage of slang, proper nouns, abbreviations, and acronyms also made the task of translating human language to machine-understandable equivalent a challenging task. The research of NLP began in the 1950s and is mainly based on machine learning and statistics. There are a lot of major tasks in natural language processing such as text summarization, text translation, named entity recognition (NEG), question answering, information retrieval, etc.

These tasks can be easily done by using Python's Natural Language Toolkit (NLTK) package by Stanford University (Arafat and Raihanuzzaman, 2016). NLTK is the ideal library for pre-processing a large corpus of text as it supports text classification, tokenization, parsing, and many more. This Python library will be used to pre-process the ASAP AES dataset before feeding it into the deep learning model for training. For this work, the research will be using the NLTK library for tokenization purposes using the Tokenizer class (NLTK 3.4.5 documentation, 2019).

Deep Learning

Throughout this sub-section, three relevant neural network architectures for the proposed system are discussed.

Artificial Neural Networks (ANN)

Artificial neural networks (ANN) have gained a steady surge of interest recently among machine learning practitioners. The application of ANNembeddings commonly known as deep learning enables computational models to learn the underlying representation of data through several levels of abstraction via multiple processing layers (LeCun, Bengio, and Hinton, 2015). Deep learning methods have produced state-of-the-art results in the various domains of artificial intelligence such as object recognition, natural language processing (NLP), image processing, and many more. Deep learning learns via a backpropagation algorithm that tweaks the parameters or weights in each layer to discover the intricate structures of data (LeCun, Bengio, and Hinton, 2015). ANNs can be further classified into two different groups, namely convolutional neural networks (CNN) and recurrent neural networks (RNN). CNN has brought lots of breakthroughs in image and speech processing while RNNs excel at sequential data such as texts and speech.

A typical neural network is typically made up of layers of nodes. The function of the node is to multiple the input from the data with a set of weights that signify the significance of the input in determining the output. All these products of inputs and weights will be passed through an activation function which

Figure 2. Architecture of an Artificial Neural Network (ANN) (Ahire, 2018)

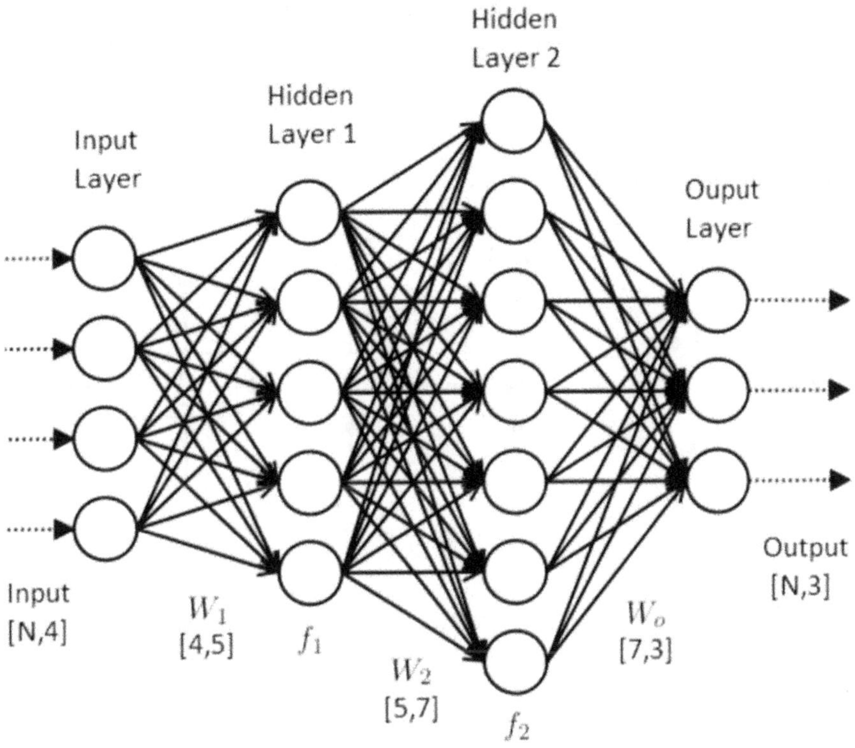

determines whether a neuron fire. An example is shown by Ahire (2018) in Figure 2 which shows an artificial neural network (ANN) with one input layer, one output layer, and two hidden layers.

Recurrent Neural Network (RNN)

Although ANNs are famed for their generalization power, they are essentially memoryless (Chollet, 2018). Each input through the network is processed independently and there is no state there is stored between inputs. However, to process a sequence or a time series of data points (such as a sample essay), the entire sequence must be passed into the network at the same time. In contrast, a recurrent neural network (RNN) processes sequences of data by iterating through the data points and maintaining a state that holds information relative to what the network has seen so far. It is a type of neural network that has an internal loop with states storing previous information. The state of RNN is reset between processing two different data sequences (e.g.: two separate essay samples). Looking back at the problem, the network must remember the words it has seen before to understand the context of the word in that essay.

RNN's ability to process vectors in the sequence is much more power compared to ANN which has a fixed number of layers from the get-go. Another advantage of RNN over vanilla neural networks is RNNs combine the input vector with their internal state vector with a fixed function to produce a new state vector (Karpathy, 2015). It is analogous to running a fixed program with certain inputs and internal variables. In other words, RNNs are essentially Turing complete in the sense that they can simulate any program with proper weight tuning. This is especially useful in the problem of ATS as RNNs can capture the local representation of the essay texts such as n-grams. Words are meaningless without context, and to understand the context one must have some information about the words that came before and after it. This is an ideal use case for RNNs as RNNs can store previous inputs that it has seen in its internal state, which can be passed down too many time steps as it affects the processing of each new example.

Figure 3. Illustration of an unrolled RNN (Olah, 2015)

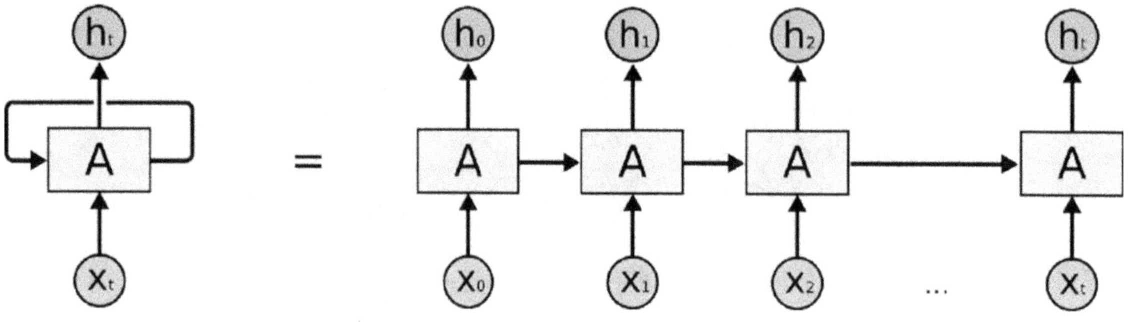

Figure 3 shows an illustration of a typical RNN that is unrolled across different time steps. The internal state is passed from one time-step to the next as input along with a new input at the new time step. This can be expressed mathematically as

$$h_t = \phi\left(W_{X_t} + U_{h_{t-1}}\right) \tag{1}$$

where the hidden state at time step *t* is h_t. h_t is a function of the input at the same time step x_t, multiplied by a weight matrix *W* added to the hidden state of previous time step h_{t-1} multiplied by its weight matrix *U*. Matrices *W* and *U* determine how much importance should be given to the present input and the past internal state. During the backpropagation process, the error generated by the output will be used to adjust the weight matrices until they reach an optimal level. The sum of the weighted input and hidden state is squashed by the activation function ϕ. This function can be either a logistic sigmoid function or a tanh function which is used to squash the value in the range of (0, 1). Since the feedback loop happens at every time step, each hidden state contains information of not only the previous hidden state but also all of those that preceded the previous hidden state.

Long Short-Term Memory (LSTM) units

However, in practice, RNNs are rarely used as a standalone network as they are too general to be of real use. Simple RNNs have a major problem: in theory, it should be able to retain information about inputs seen many timesteps before at any given time *t*, however, in practice, such long-term dependencies are impossible to learn. This problem is dubbed as the ***vanishing gradient problem***. This is due to the gradient becoming vanishingly small, preventing it to update the weights during backpropagation. Long-short-term-memory (LSTM) units are used to solve this vanishing gradient problem. It saves information for later use in its internal state, thus preventing older signals from gradually vanishing during processing.

The vanishing gradient problem has been extensively studied by Hochreiter and Schmidhuber (1997). Through years of research, they discovered a solution named Long Short-Term Memory (LSTM) to combat this problem. LSTMs are a subset of RNN that can learn long term dependencies and are widely used in NLP tasks. Instead of having a single layer, several layers are performing a variety of functions. Similar to how memory works in a computer, LSTM enables RNNs to remember their inputs over some time. LSTM units can be considered as a gated cell, where the cell decides whether to store or delete information based on the weight assigned to the information. Therefore, the network learns overtime which information should be remembered or forgotten.

There are three gates in an LSTM unit: input, forget, and output gate. These gates determine whether a new input is received, which information to delete or let it influence the output at the current time step. An illustration of an LSTM unit is shown below:

Figure 4. An illustration of an LSTM unit (Walia, 2017)

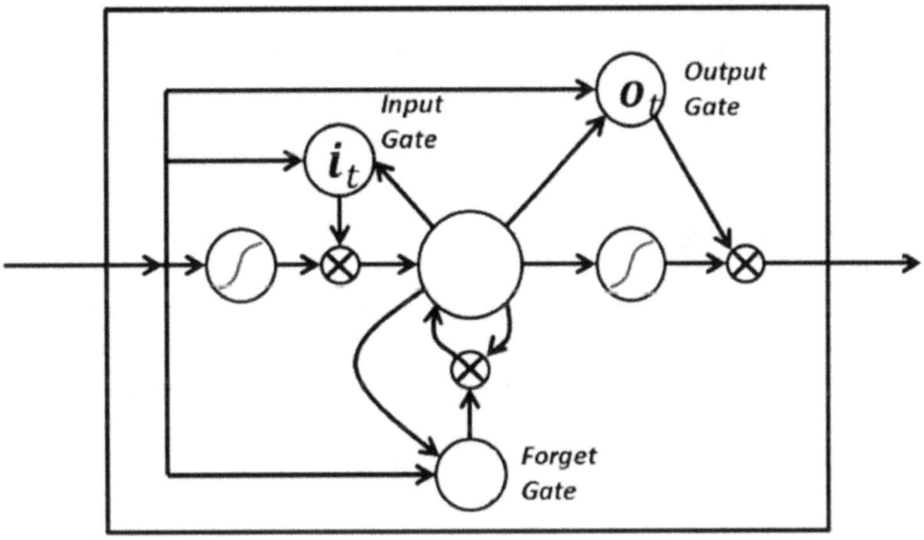

Figure 4 above can be described formally using mathematical notation. Based on the figure above, an LSTM unit calculates a hidden state H_t as

$$i_t = \sigma\left(x^t U^i + h_{t-1} W^i\right)$$

$$f_t = \sigma\left(x^t U^f + h_{t-1} W^f\right)$$

$$O_t = \sigma\left(x^t U^o + h_{t-1} W^o\right)$$

$$\bar{C}_t = tanh\left(x^t U^g + h_{t-1} W^g\right)$$

$$C_t = \sigma\left(f_t * C_{t-1} + i_t * \bar{C}_t\right)$$

$$h_t = \tan h(C_t) * O_t \quad (2)$$

where, *i, f, o* denotes input, forget, and output gates respectively. These three gates have the same equation but with different parameters (*W* is the recurrent connection connecting the previous state to compute the current state, *U* is the weight matrix applied to the inputs to the current hidden layer). The symbol σ denotes the sigmoid function where it squashes the value of these vectors in the range (0, 1). The input gate specifies how much of the current input is used to compute the current state by multiplying them elementwise with another vector. Next, the forget gate defines how much of the previous state must be forgotten. The output gate determines how much of the internal state is exposed to the next time step. *C* will be the internal memory of the unit and is an addition of the product of previous memory and the forget gate ($f_t * C_{t-1}$) and the product of the current hidden state and the input gate ($i_t * \underline{C}_t$).

RELATED WORKS

Automated Text Scoring (ATS)

Automated text scoring (ATS) has been a popular research area in the field of computer science back in the early 1960s. Automated text scoring (ATS) is the process of grading essays written by students without human intervention through the help of statistical and natural language processing (NLP) techniques (Alikaniotis, Yannakoudakis, and Rei, 2016). A typical ATS takes an essay as input written for a given topic and then assigns a numeric score to the essay that reflects the quality of the essay based on its organization, grammar, and content. However, predicting the score of an essay can be highly challenging as it requires knowledge on spelling, grammar, word meanings, discourse, and pragmatics (Dong, Zhang and Yang, 2017).

Besides, analyzing natural languages used in everyday human-to-human communication is a complicated problem for computers regardless of the medium, such as verbal, writing, or reading. Ambiguities are more than common in natural languages as the same text can have different interpretations depending on the context, making the task of grading and scoring essays a daunting undertaking.

ATS is typically considered as a **supervised** machine learning problem and is either a classification, regression, or rank preference task (Phandi, Chai and Ng, 2015). Most of the existing ATS system is based on several predefined features of the essay such as the length of essays, grammar, syntax, coherence, and the overall organization. (Zhao *et al.*, 2017). Another approach has emerged recently in solving ATS, which involves neural network models to discover the features automatically in an end-to-end fashion. With this approach, the process of extracting predefined features from the essay can be skipped and a direct prediction of scores from essay texts can be accomplished with less manual effort.

Word Embeddings

One of the ways to represent the vocabularies in a document is by using word embeddings. Arafat and Raihanuzzaman (2019) defined word embedding as "a collective term for models that learned to map a set of words or phrases in a vocabulary of vectors of numerical values". Word embeddings are the vector representations of a word that encodes their context in the corpus of texts. It can capture the context of a word in a document, semantic and syntactic similarity et cetera (Karani, 2018).

Word embeddings are better than the traditional one-hot encoding or bag of words (BOW) model as word embeddings are less sparse and therefore saves computation time. Furthermore, bag-of-words (BOW) models do not preserve the sequence of the text (Chollet, 2018). Hence, the general structure and meaning of a tokenized sentence are lost. Therefore, it is commonly used in shallow language processing models. Table 1 showed the advantages word embedding has over the traditional one-hot-encoding.

Table 1. Differences between one-hot encoding and word embeddings (Chollet, 2018)

One-hot encodings	Word embeddings
	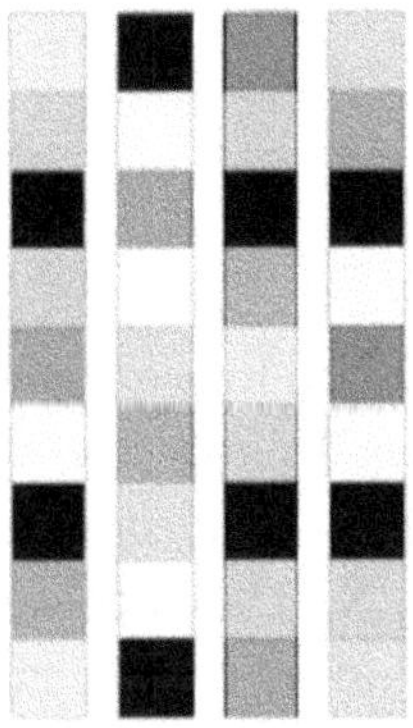
· Sparse	· Dense
· High-dimensional	· Lower-dimensional
· Hard-coded	· Learned from data

There are generally two-word embeddings that are being considered for EssayScore, which are Word2Vec and GloVe. The first word embedding Word2Vec was published by Mikolov et al. (2013) that uses neural networks to calculate word embeddings based on the words' context. There are two approaches to implement the network, continuous bag of words (CBOW), or continuous skip-gram model. The CBOW model takes the context of each word as the input and tries to perform prediction on the word corresponding to the context. On the other hand, the skip-gram model does the opposite by generating a probability distribution of words based on a word context. The difference in the neural network architecture between CBOW and skip-gram model is illustrated in Figure 5.

On the other hand, GloVe emphasizes on the fact that the frequency of words appearing together is vital information that is not captured by the Word2Vec model. As such, GloVe tries to build word embeddings in a way that the sum of these word vectors relates directly to the probability of these words appearing together in the corpus (Böhm, 2018).

Figure 5. CBOW and Skip-Gram Architectures (Bornstein, 2018)

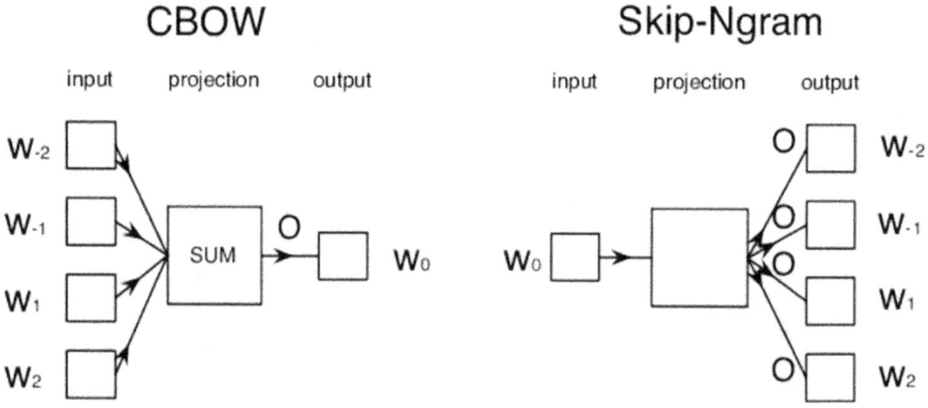

The author decided to use a pre-trained word embedding as opposed to manually training the word embedding. This is due to a few reasons. One reason is supported by Farahmand (2019)'s recommendation that pre-trained embeddings work better for small training set like the AES dataset that will be used in this work. Next, the decision to use pre-trained word embeddings such as GloVe or Word2Vec is because of the limited computational resources. By using pre-trained word embedding that is readily available, the researcher can cut down the training time and bring the model to the testing phase sooner by freezing the weights of the embedding layer during the training phase.

After some brief considerations, the author decided to use the GloVe model because it is very straightforward and easy to implement. Another reason is that GloVe imparts a more practical meaning into the word vectors produced by considering the relationships between word pairs and not words as suggested by Jiao (2018). Furthermore, GloVe also penalizes highly frequent word pairs such as stops words ("the", "an", etc) so that they will not dominate lesser frequent but important words. An example of a visualization of GloVe in action is shown in Figure 6. The figure shows a graph depicting the cosine similarity of the first 25 words in the corpus that is like the word 'happy' using Principal Component Analysis (PCA) and plotting module matplotlib.

Figure 6. Graph of cosine similarity of first 25 words from the word happy

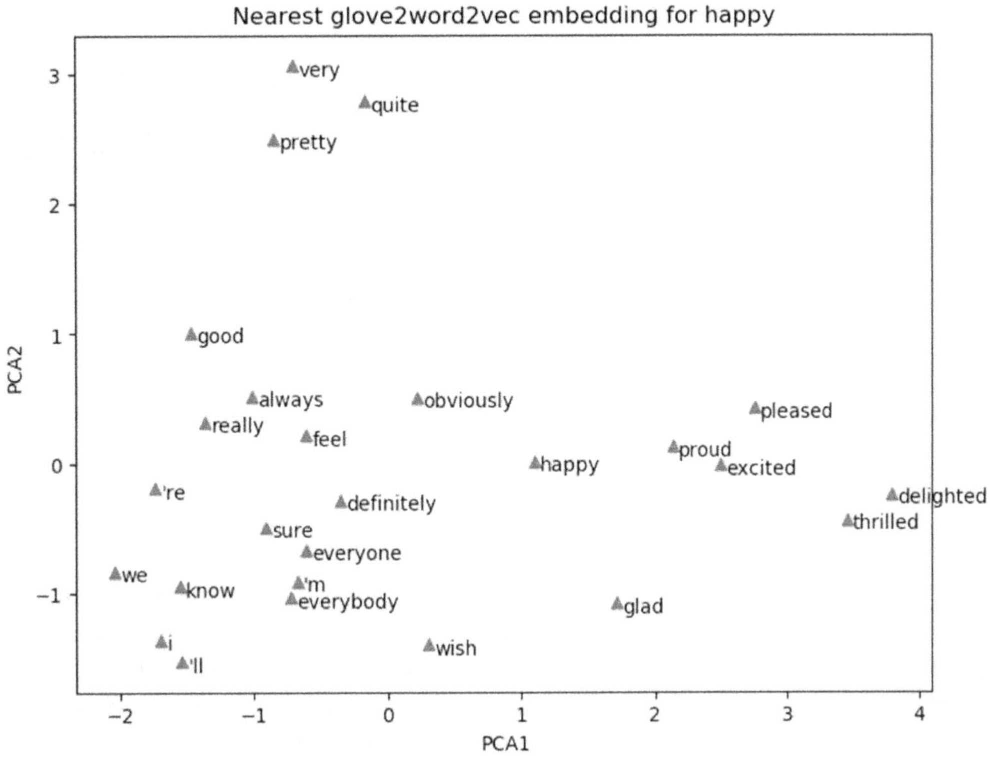

MAIN FOCUS OF THE CHAPTER

Problem Statement

The main reason that spurred the use of deep learning in automated text scoring (ATS) is because of the tedious and time-consuming task of feature engineering used in traditional machine learning models. ATS is typically considered as a supervised machine learning problem and is either a classification, regression, or rank preference task (Phandi, Chai and Ng, 2015). Therefore, most of the existing ATS systems are made using several predefined features of the essay through feature extraction. Examples of predefined features are the length of essays, grammar, syntax, coherence, and the overall organization (Zhao et al., 2017). However, the process of extracting these features can be difficult and time-consuming. Hence, the approach of solving ATS using deep learning techniques has gained traction in recent years, which involves the neural network to discover the features automatically in an end-to-end fashion. With this approach, the process of feature extraction can be skipped entirely and the prediction of scores can be done with less manual effort. Figure 7 by Heidmann (2019) illustrates this difference between machine learning and deep learning.

Figure 7. Differences between machine learning and deep learning (Heidmann, 2019)

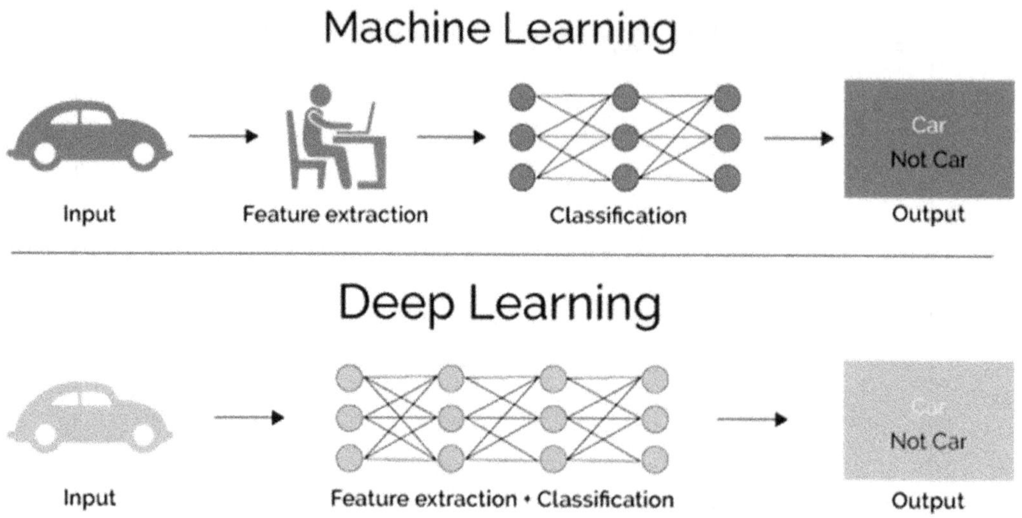

The ATS is created to solve the problem of a need to reduce the burden or teachers in marking essays and for students to gain immediate feedback for improvement. Essay writing is an indispensable part of assessing a student's language fluency in the academy. Students are generally given an essay topic to elaborate and marks are given based on the marking rubric that is predetermined. However, the sheer number of essays to score for teachers and the need for 'second graders' in high stake exams such as Test of English as a Foreign Language (TOEFL) is too overwhelming for the graders to handle. According to statistics by ETS (2013), there are around 655,000 students worldwide that took the GRE revised General Test, with each student having to answer two essay questions, producing a total of 1.3 million responses. Grading these essays can be time-consuming and this is where ATS can help to alleviate this burden.

Besides, ATS is also beneficial to the students as it serves as an in-classroom learning tool for students to practice their essay writing skills. Because of its ability to give instant feedback, ATS can reduce the time taken to receive feedback for each essay written, thereby facilitating the entire writing and learning process. In summary, the raison d'être of an ATS is to reduce the labor-intensive marking of essays and ensure human bias does not influence the grading outcome and ensures equality in scoring.

EXISTING SOLUTIONS/SYSTEMS TO THE PROBLEM

Work Essay Grade (PEG)

An extensive amount of research had been done on the problem of automated text scoring (ATS) since the 1960s. The first work related to essay scoring can be traced back to 1966 called Work Essay Grade (PEG) by Page (1966). PEG is considered the forerunner of automated text scoring, but the development of commercial ATS application does not begin until the 1990s when computers became cheaper and more pervasive. PEG grades essays by looking at its surface writing quality. It captures the intrinsic qualities that are in a person's writing styles and uses approximations of these variables, called *proxes* by the researchers, to measure these qualities. Specific qualities that are being captured include average word length, number of semicolons, and the rarity of words (SDLC Waterfall Model and Gagne, 2019a). These proxes are then being fed into a regression equation as predictors to compute beta weight (coefficients) which is then used to calculate a score for the essay.

For the training phase, PEG requires 100 to 400 sample essays. PEG had reported an impressive inter-rater agreement of 0.87 with 20 variables and proved that the score predicted by a computer is as well as a real human judge. However, it was criticized for not taking account of the semantic aspect of essays and placed more emphasis on the surface structures. Besides, it is also possible to trick the system by writing longer essays as PEG used indirect measures of writing skills.

Intelligent Essay Assessor (IEA)

Another AES system that is available in the market, Intelligent Essay Assessor (IEA), scores an essay using a method called Latent Semantic Analysis. It is a semantic text-analysis method created by psychologist Thomas Landauer assisted by Peter Foltz and Darrell Laham. LSA processes a corpus of text and represents the words that appear in a sentence, paragraph, or essay through statistical computations. It is defined as "a statistical model of word usage that permits comparisons of the semantic similarity between pieces of textual information" (Dikli, 2006). LSA has a high degree of correlation with human meaning similarities among words and texts. The basic idea of LSA is that the meaning of a passage is dependent on its words. Changing even only one of the words can result in an entirely different meaning in the passage. With this method, the semantic information is determined only by the co-occurrence of words in a large corpus of texts.

Using LSA as its backbone, IEA focuses on content related features rather than the surface features of the essay in contrast to the PEG system by Page (1966). IEA distinguishes itself from its contemporaries by only requiring 100 pre-scored essays to train compared to other AES systems. However, although IEA can evaluate semantic aspects of the essay, it cannot detect creativity and reflective thinking (Rudner and Gagne, 2019b).

Electronic Essay Rater (E-rater)

Finally, the last existing AES that will be discussed is the Electronic Essay Rater (E-rater). Developed by the Educational Testing Service (ETS), E-rater evaluates the quality of an essay by determining the linguistic features in the text (Dikli, 2006). E-rater employs natural language processing (NLP) methods that are used to identify lexical and syntactic cues in a text. NLP is considered a complex task to understand as it has several levels of processing. NLP comprises four categories of language tasks including speech recognition, syntactic analysis, discourse analysis, and information extraction and machine translation (Dikli, 2006). Currently, E-rater is used by ETS for scoring the Graduate Management Admissions Test (GMAT) AWA (Analytical Writing Assessment) (Dikli, 2006).

Dikli (2006) claims that the discrepancy rate between e-rater and human rater is less than 3 percent since the inception of the system. E-rater uses a corpus-based approach to build their model by examining sample essays from actual essay data. The system also assumes that good essays resemble other good essays, and the same applies to bad essays. Hence, the system will construct a vector model to determine the similarity of the essays to the ones found in the database to assign a score. The validity of the score is heavily dependent on the scoring of the sample essays in the database because the e-rater does not do any actual reading on the input essays.

In Figure 8Next, the second functional requirement is to assign a score to the uploaded essay in an end-to-end fashion. The architecture of the overall model will be based on previous works of Taghipour and Ng (2016). Their architecture uses a total of 5 layers starting with the lookup table layer, convolution layer, recurrent layer, meaning over time, and linear layer with Sigmoid activation function. The figure below shows the illustration of the network architecture included in their work:

Figure 8. An illustration of the architecture of the network (Taghipour and Ng, 2016)

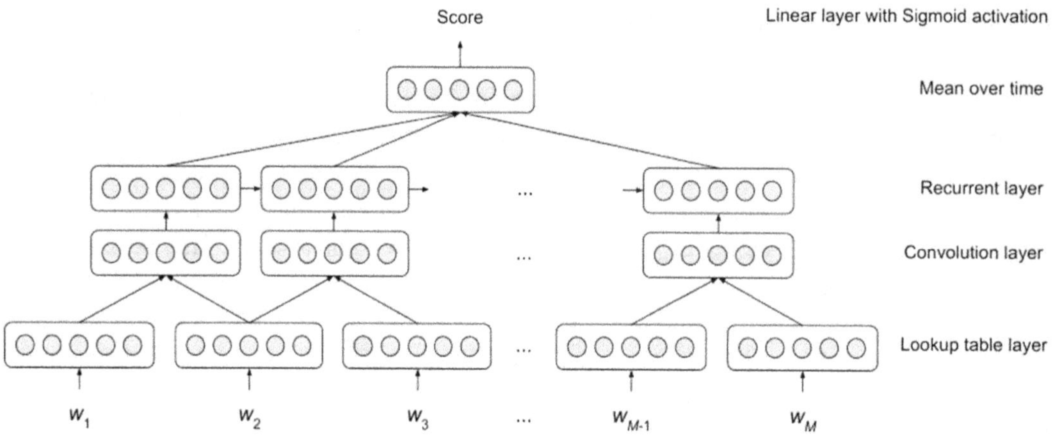

According to the authors of the network model, the model achieves an excellent result of 0.76 in Quadratic Weighted Kappa (QWK). As such, this network architecture is the most ideal for implementing EssayScore. The reason that the researcher is not implementing the best model from their work is because of resource constraints. Finally, the third requirement is to check the submitted essays for spelling errors using the *pyspellchecker* module. The justification of this requirement is because since the web application grades essays and assigns a score to them, it would be helpful to point out the spelling errors in the essay and suggests some of the improvements for the misspelled word.

Datasets

The proposed RNN model will be trained by using a dataset that is openly available on the internet. Specifically, the RNN model in EssayScore will be trained using the Automated Student Assessment Prize's (ASAP) Automated Essay Grading (AEG) dataset obtained from Kaggle (Kaggle, 2012). The ASAP dataset is sponsored by the William and Flora Hewlett Foundation in 2012 to create a fast, effective and affordable solution to automated grading of student-written essays. This dataset is popular among ATS researchers and is frequently used. Notable works that use this dataset include Alikaniotis, Yannakoudakis, and Rei (2016), Taghipour and Ng (2016), and Dong, Zhang, and Yang (2017).

The ASAP AEG dataset consists of approximately 13,000 essays, written across 8 prompts (i.e.: essay topics). These essays were written by students of class 7 to 10. Each of the essays was evaluated b 2 evaluators and only 6 out of the 8 prompts have overall scores. The last 2 prompts have their scores broken down for individual essay attributes such as content, organization, style et cetera. The main reason that this dataset is chosen for this work is that this dataset is publicly available on the Internet which eliminates any copyright infringement issues. The second reason is that this dataset originated from a Kaggle competition backed by William and Flora Hewlett Foundation. Hence, the validity of this dataset is assured. The high number of essays also helps to train a better network. Table 2 shows a statistical summary of the essays in the ASAP AEG dataset.

Table 2. Statistical Summary of ASAP AEG dataset

Set	Grade level	No. of Essays	Avg. Length	Max length	Main score	Max score	Mean score
1	8	1783	350	911	2	12	8
2	10	1800	350	118	1	6	3
3	10	1726	150	395	0	3	1
4	10	1772	150	383	0	3	1
5	8	1805	150	452	0	4	2
6	10	1800	150	489	0	4	2
7	7	1569	250	659	0	30	16
8	10	723	650	983	0	60	36

The topic of essay 1 deals with persuasive writing in which the writer needs to write a letter to a local newspaper outlet to persuade readers to agree with the writer's opinion on the effects the computer has on people. The rubric outlines a six-point scale that award scores based on an overall statement that captures the essence of the response. Below are the overall statements and their corresponding score point.

Score Point 1: An underdeveloped response that may take a position but offers no more than very minimal support.

Typical elements:

- It contains few or vague details.
- It is awkward and fragmented.
- It may be difficult to read and understand.
- May show no awareness of the audience.

Score Point 2: An under-developed response that may or may not take a position.

Typical elements:

- It contains only general reasons with unelaborated and/or list-like details.
- Shows little or no evidence of organization.
- May be awkward and confused or simplistic.
- May show little awareness of the audience.

Score Point 3: A minimally developed response that may take a position, but with inadequate support and details.

Typical elements:

- It has reasons with minimal elaboration and more general than specific details.
- Shows some organization.
- May be awkward in parts with few transitions.
- Shows some awareness of the audience.

Score Point 4: A somewhat-developed response that takes a position and provides adequate support.

Typical elements:

- Has adequately elaborated reasons with a mix of general and specific details.
- Shows satisfactory organization.
- May be somewhat fluent with some transitional language.
- Shows adequate awareness of the audience.

Score Point 5: A developed response that takes a clear position and provides reasonably persuasive support.

Typical elements:

- Has moderately well-elaborated reasons with mostly specific details.
- Exhibits generally strong organization.
- Maybe moderately fluent with transitional language throughout.
- May show a consistent awareness of the audience.

Score Point 6: A well-developed response that takes a clear and thoughtful position and provides persuasive support.

Typical elements:

- Has fully elaborated reasons with specific details.
- Exhibits strong organization.
- He is fluent and uses sophisticated transitional language.
- May show a heightened awareness of the audience.

The essays in the dataset are graded using the score points above. Essay A in Appendix F shows a bad essay that is graded 1 and Essay B shows an excellent essay that is graded 6. From the reasoning given, it is observed that essay A is undeveloped and brief. In contrast, essay B is well elaborated with good examples. The purpose of training the model is to find this relationship between the essay written and the given score and then make a prediction on new essays.

For the architecture of the deep learning model, the author had chosen LSTM units as the core structure of the network. The structure of the network can be seen in Figure 5 and is based on works like Alikaniotis, Yannakoudakis, and Rei (2016) and Taghipour and Ng (2016). The reason for choosing LSTM models over other deep learning models like FFNN or CNN is because of LSTM's ability to process sequential data such as text files. Another reason to justify the selection is because of LSTM's ability to solve the vanishing gradient problem as described in section 2.2.1.3.

Experimental Setup

The main resource required for the development phase would, of course, be a laptop with a minimum of 4 GB RAM and equipped with NVIDIA GPU. This is because the NVIDIA GPU will be used to train the RNN model in a reasonable period. Other alternatives include Google Colaboratory, a free Jupiter Notebook environment that provides cloud GPU and TPU resources, free-of-charge for up to 12 hours (Google Colaboratory, 2019). Secondly, out of a few selections of IDE such as Notepad++, PyCharm, Spyder, and Visual Studio Code, the VS Code is chosen as the preferred IDE for development. The justification for this selection is that the researcher had done previous development on Visual Studio Code before, so familiarity is a big plus in making this decision. Besides, VS Code supports intelligent code completion, in-built git version control extension, and many more custom extensions that can greatly shorten development time.

To train the model, the TensorFlow library has been chosen instead of other alternatives like Theano and CNTK. The reasoning is because the TensorFlow library has a large number of resources and communities supporting it. The chances of failing to implement the features are comparably lower compared to other machine learning libraries. Keras, a framework for implementing a deep-learning model built on top of the TensorFlow library will be used to develop the RNN model. The reasoning is because Kera's framework provides high-level abstractions for developing deep learning models, thereby facilitating the whole development process.

For the deployment of the trained model, there are a few alternatives that are being considered. The candidates are TensorFlow Serving and TensorFlow.js. TensorFlow.js is a library written in JavaScript for machine learning purposes while TensorFlow Serving is a serving system for serving machine learning models for production. The researcher had finally decided on using TensorFlow Serving as the preferred tool for deploying the trained model because of its ease of use. Furthermore, the model will be served as a RESTful API that can be called by any platform that supports HTTP requests. Therefore, TensorFlow Serving made it possible for future enhancements such as porting the web application to a mobile application.

General Workflow of the Application

The high-level overview of the application EssayScore is shown in Figure 9. The application generally operates in two different sections: the frontend and the backend. The frontend consists of the web application that is built using HTML, JavaScript, and CSS with Jinja. Jinja is a web template engine for the Python programming language which allows the developers to evaluate Python-like expressions in HTML codes. The reasoning for this is because using Jinja eases the process of populating data in the HTML by using templating methods. Next, a Flask micro-server is created as the middleman between the frontend and backend. The reasoning that Flask microframework is chosen instead of a more heavyweight framework like Django is because the functionality of this application is basic. Therefore, a simple microframework like Flask is enough for this purpose.

The Flask server will be communicating with two backend servers. One if it is the Firebase Authentication and Cloud Firestore. Here, all the authentication procedures are done, and essays are being uploaded to the Cloud Firestore. Cloud Firestore is chosen over the conventional SQL database because of its ease of use and zero setup time. Furthermore, Firebase and all its features are designed to be used out-of-the-box. Hence, this can save development and deployment time. On the other hand, the other

Figure 9. High-level overview of the proposed system

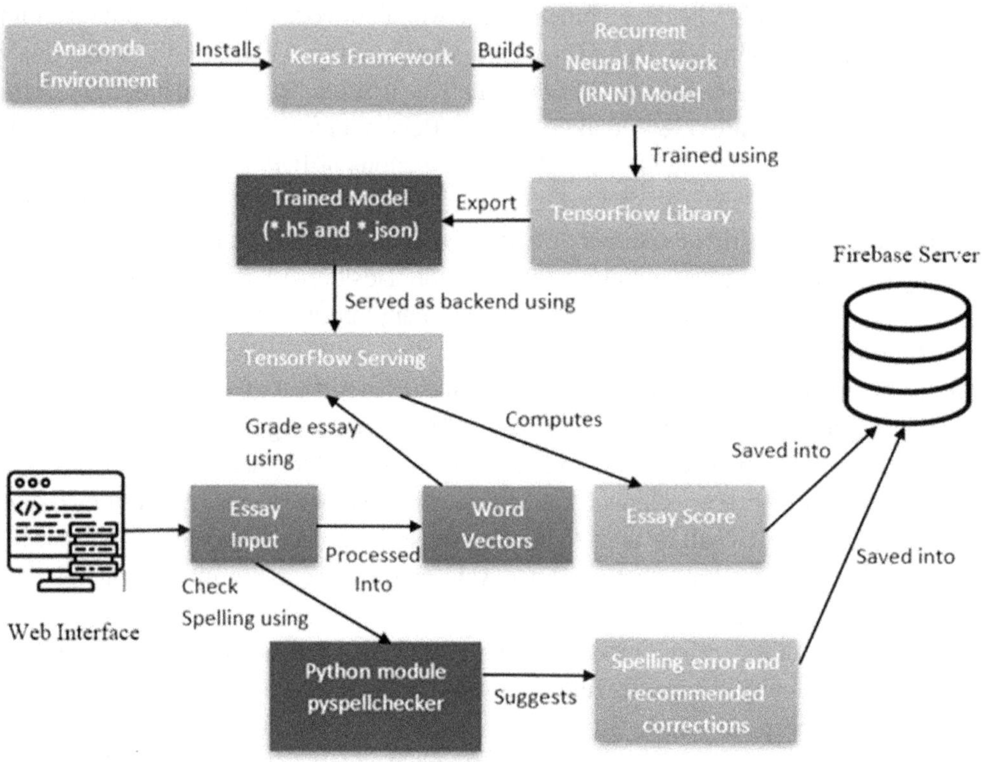

server is the RESTful API Server served through Docker that will be serving the trained model. The Flask server will be sending POST requests to these two servers and will be receiving JSON response from the backend. From here, the Flask server is responsible for processing the JSON response and populate the UI with the returned data.

Next, after the essay text is read from the file, the final scores for each essay test will be normalized in the range of [0, 1] using the method min-max normalization. This is done to standardize all the scores for each essay as each essay has a different minimum and maximum scores. The formula for min-max normalization is shown below:

$$z_i = \frac{x_i - \min(x)}{\max(x) - \min(x)} \tag{3}$$

where z_i represents the normalized essay score for the ith essay,

min(x) refers to the minimum score of each essay topic and

max(x) refers to the maximum score of each essay topic.

The function used to tokenize the essay texts into word vectors are shown in code snippet 3. In this function *tokenize()*, the NLTK Tokenizer is used to create a vocabulary index based on the word frequency of the corpus text. Every word in the text will be assigned to a unique integer value and 0 is reserved for padding. The lower the integer, the more frequent the word appears in the corpus. The helper function *texts_to_sequences()* is used to map an essay text to unique integer sequences based on the vocabulary index created by *fit_on_texts()*. Next, the vectors are padded to reach 500 essay length for each essay. Any essays with length greater than 500 are truncated. Finally, the label tensor is reshaped to (*num_essays,* 1) to fit the input shape of the neural network. After the pre-processing is done, the model training can begin.

```
Using TensorFlow backend.
INFO - 16:47:21: Begin tokenization of essays…
INFO - 16:47:23: Tokenization Done!
INFO - 16:47:23: Time Taken to tokenize essays: 2 seconds
INFO - 16:47:23: Tokenization Summary:
INFO - 16:47:23: Number of unique words found: 32156
INFO - 16:47:23: Number of essays used for tokenization: 8304
[1, 85, 368, 1, 97, 20, 11129, 155, 29, 6, 22, 610, 57, 5393, 79, 25, 29, 6,
22, 55, 76, 3480]
INFO - 16:47:24: Shape of data tensor: (8304, 500)
INFO - 16:47:24: Shape of label tensor: (8304, 1)
```

Model Creation and Training

The deep learning model created is based on the works of Taghipour and Ng (2016). As shown in Figure 10, the architecture of the neural network consists of 6 layers of neurons with each layer performing different tasks.

The first layer which is the **embedding layer** maps each sequence of essays in integers to a low-dimensional and learned continuous vector representations. It is useful because they can meaningfully represent discrete variables in the transformed space with lower dimensionality. Embedding layer is chosen here because it is a great alternative to one-hot encoding. The researcher used a technique call **transfer learning** which describes the training of a network from a pre-trained model. Here, the researcher used the finalized pre-trained GloVe model as the embedding layer to save computation time.

The second layer is the **convolution layer** which applies a 1-dimensional convolutional neural network on the dense word representation. It is a function that extracts feature vectors from n-grams of word vectors. This is done so that the network can capture local contextual dependencies in the essay to improve the performance of the system.

The third layer is the **recurrent layer** which generates a representation for the given essays that would ideally encode the information required in scoring an essay. In this network, the recurrent layer uses an LSTM unit to capture long-term dependencies.

The fourth layer is the **mean over time layer.** It is a type of layer that aggregates the hidden states of the LSTM over each time step and calculates the mean of it. The reasoning is because according to Bahdanau et al. (2015), networks show better performance if the hidden state is averaged as opposed to just use the last hidden state for each sequence as the encoding.

Figure 10. Architecture of the Neural Network

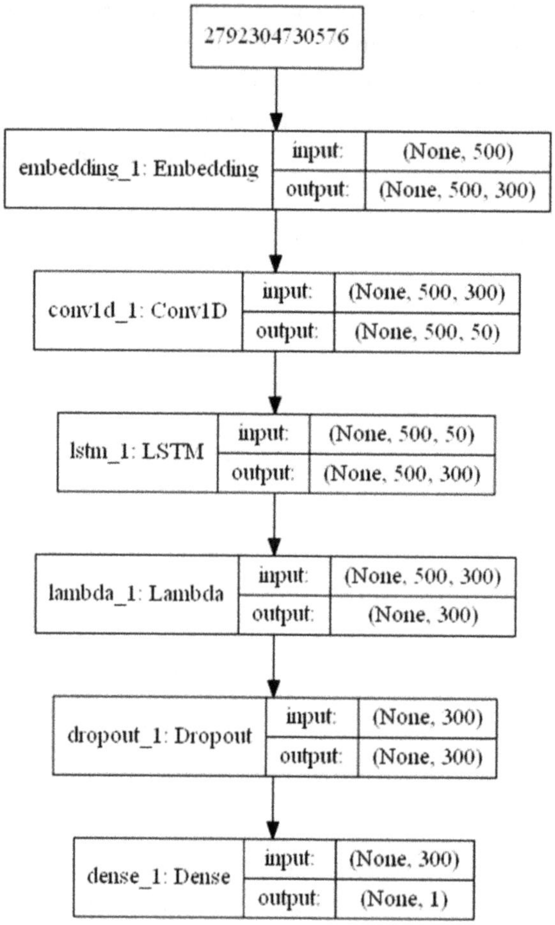

The fifth layer is the **dropout layer** which serves as a regularization method to prevent overfitting the curve. Dropout refers to the process of ignoring neurons during the training phase of the network that is chosen at random with probability p. The main reason for dropout is to prevent overfitting.

The last layer is the **dense layer** which maps the input vector from the previous layer to a scalar value. To apply a boundary to the output value of this layer, a sigmoid function is used as the activation function in this layer to bound the value to (0, 1). The mapping is done with equation 4.

$$s(x) = sigmoid(w.x+b) \quad (4)$$

where x is the input vector from the previous layer, w is the weight vector to be trained and b is the bias value. Translated to code, the architecture is shown in code snippet 4.

The hyperparameters of a model refer to the values that must be set before the training process can begin. The listed hyperparameters for the model are as below. The model will be training for 50 epochs and the network with the best results will be saved. Figure 3.15 shows a snippet of the training process.

Model hyperparameters:

i. Dropout Rate: 0.8
ii. The activation function of the final layer: Sigmoid
iii. Loss function: Mean Squared Error
iv. Optimizer: Adaptive Moment Estimation (ADAM)
v. Evaluation Metrics: Mean Squared Error, Mean Absolute Error

Figure 11. Training of the model

Figure 11 shows the code that is used to check for spelling errors in the uploaded essays and suggests some recommendations. The function accomplishes this by using the *pyspellchecker* Python module. First, an instance of the *SpellChecker* class is created that will load all the existing English words into the dictionary. The default dictionary does not include contractions such as "aren't" and others, hence they had to be manually loaded into the dictionary. Next, the Python code will iterate through the essay text and check for each word whether they exist in the dictionary or not. If the word exists in the dictionary, it is spelled correctly. Otherwise, *Pyspellchecker* will recommend a list of the closest corrected words to the misspelled word. This is one of the few improvise that has taken by the researcher in this work.

The initial plan is to create a spell checker using a neural network as ATS mentioned previously. However, the time taken for training the model is too long and not feasible for the researcher to proceed. Thus, a compromise was made by using a simpler algorithm developed by Norvig (2016) that is available in the *Pyspellchecker* module. Figure 11 and 12 shows an output screenshot of the corrected essay and recommended words for the misspelled words.

Figure 13 shows the code snippet for making RESTful API calls to the TensorFlow Serving endpoint. This is accomplished by using the Python *module* request and making a POST request to the */v1/models/aes_regressor: predict* endpoint. The payload of the request will be the converted essay texts into a sequence of integers as described in the previous section. If the response code is 200, the Flask server will de-normalize the returned score into the final actual score of the predicted essay. There are also other

Figure 12. Essay with misspelled words

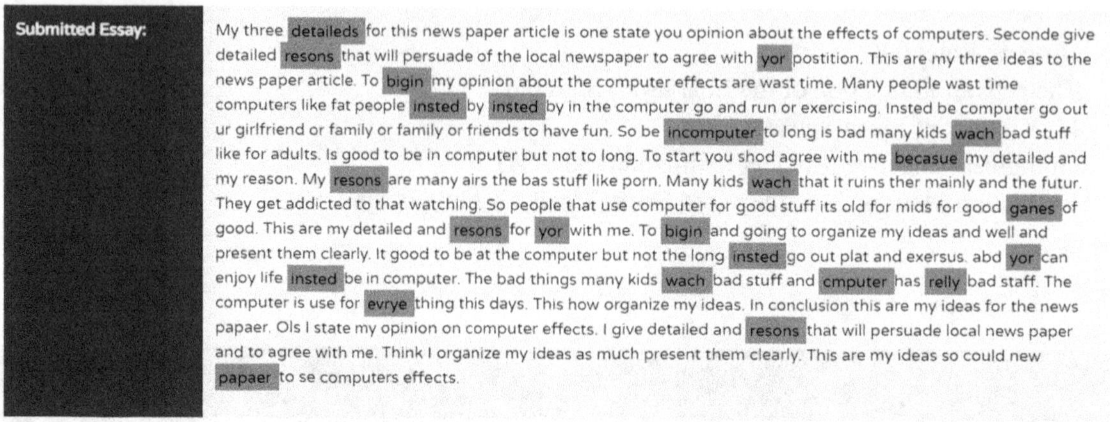

Figure 13. Recommended corrections for the misspelled words

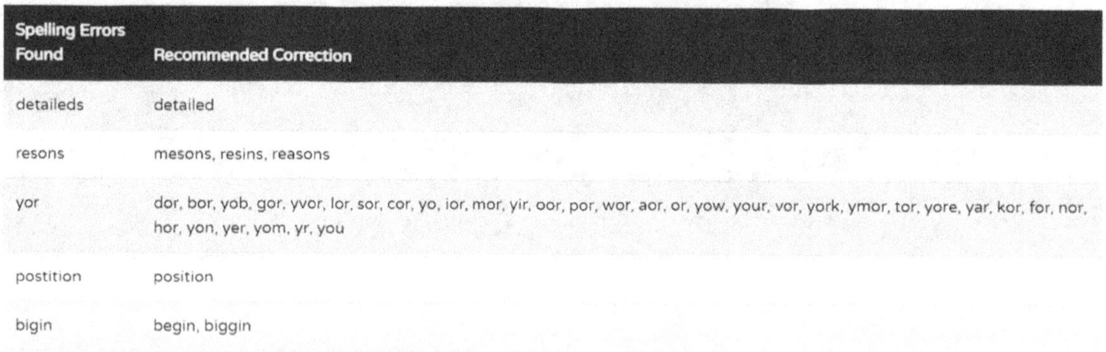

Spelling Errors Found	Recommended Correction
detaileds	detailed
resons	mesons, resins, reasons
yor	dor, bor, yob, gor, yvor, lor, sor, cor, yo, ior, mor, yir, oor, por, wor, aor, or, yow, your, vor, york, ymor, tor, yore, yar, kor, for, nor, hor, yon, yer, yom, yr, you
postition	position
bigin	begin, biggin

endpoints of the TensorFlow Serving server that can be requested by the Flask server. For example, the endpoint */v1/models/aes_regressor* can be used to check the status of the server.

Model Training Results

To critically evaluate the overall performance of the product, standardized evaluation metrics must be chosen to have a better gauge to assess the final product. As mentioned above, the product evaluation will be done on two parts: an evaluation of the front-end web application and another evaluation on the back-end model training process.

In terms of the model training, the researcher had chosen quadratic weighted kappa (QWK) to assess the overall accuracy of the model. This decision is backed by Taghipour's (2017) claim that the output of an AES system can be compared to the ratings given by human graders using various evaluation metrics such as Pearson's correlation, Spearman's correlation, Kendall's tau and quadratic weighted kappa (QWK). Since the ASAP competition which provides the AES dataset uses QWK as the official evaluation metric, this article will also be using QWK to assess the overall accuracy.

Quadratic Weighted Kappa (QWK) measures the agreement between two graders with 1 indicating complete agreement and 0 indicating random agreement between raters. The QWK score is calculated between the known scores and the predicted scores. Quadratic weighted kappa is calculated as follows with weight matrix *W* calculated using equation 5.

$$W_{i,j} = \frac{(i-j)^2}{(N-1)^2} \quad (5)$$

where *i* and *j* are the score assigned by a human grader and the score evaluated by the trained model, respectively. *N* is the number of possible ratings. Next, a matrix *O* is constructed with $O_{i,j}$ indicating the number of essays that received the score *I* by the human grader and the score *j* by the trained model. Another matrix *E* is calculated as the outer product of histogram vectors of the two (human and trained model) scores. The matrix *E* and *O* are then normalized such that the sum of elements in both matrices is the same. Finally, using matrices *O* and *E,* the QWK score is calculated using equation 6.

$$\kappa = 1 - \frac{\sum_{i,j} W_{i,j} O_{i,j}}{\sum_{i,j} W_{i,j} E_{i,j}} \quad (6)$$

Using equation 6, a QWK score is calculated for each of the essay topics that are available in the dataset. Then, all the 8 QWK scores are averaged to get a final averaged QWK score. Table 3 shows the result of the QWK scores.

Table 3. QWK Score of each essay sets

Essay Set	QWK Score
Set 1	0.73
Set 2	0.42
Set 3	0.66
Set 4	0.42
Set 5	0.77
Set 6	0.63
Set 7	0.60
Set 8	0.43
Avg. QWK Score	**0.60**

Overall, the average QWK Score for the final model is 0.60, which is slightly lower than the benchmark that is chosen which is the result of Taghipour and Ng (2016) by 0.1. This can be attributed to a few reasons. In their experiment, they trained their networks using 8 Tesla K80 GPUs which is an astronomical amount of computational power not affordable by the researcher. Given more time, money, and resources to train the network, the researcher believes that a better QWK score can be obtained. Overall, the process of training the model took 5 hours on an NVIDIA GTX 1060 GPU for 50 epochs.

Figure 14. Mean Squared Error of the training set

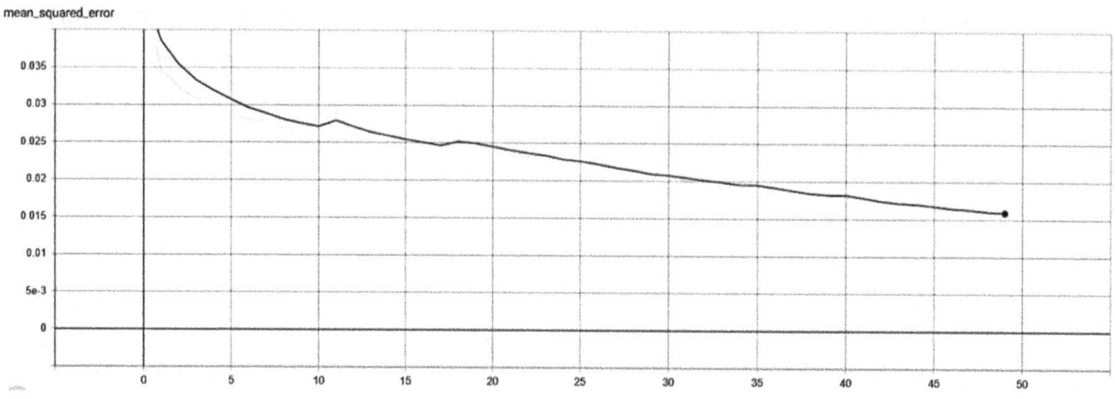

Figure 15. Mean Squared Error of the validation set

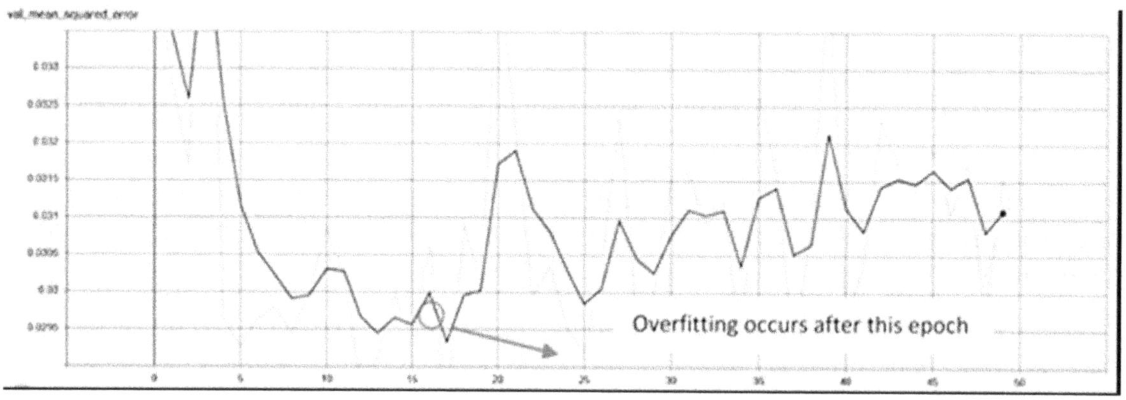

In terms of Mean Squared Error (MSE) used in training the model, the training saw a steady decline in MSE for training and validation loss. Figures 14 and 15 show the loss function for training and validation loss. The graph shows that the model starts to overfit at the 20th epoch as the validation loss starts to increase and exceeds the training loss. The researcher decided to train the model for 50 epochs and only save the model whenever there is an improvement in validation loss. As such, the model is saved before any overfitting can occur.

All in all, the performance of the deep learning model is satisfactory with a lot of room for improvement. There are several strengths and significant implementation techniques used that brought exponential returns in the long run. One of them is the use of pre-trained word embeddings like GloVe which cuts down most of the model training time. The use of Tensorboard for graph visualization also saves time for the researcher to draw graphs. However, there are also weaknesses in the trained model. One major weakness of the finalized model is its low QWK score which affects the quality of the product as a whole. Improvements can be made to the QWK score by testing out different deep learning architectures that might yield a better result. The research area of Automated Text Scoring (ATS) is quite active with better approaches built on top of works like Taghipour and Ng (2016). Taghipour (2017) in his Ph.D.

article proposed a refinement to his previous works by adding in an attention mechanism and density estimators which can help to improve the performance of the model.

Another oversight in the development of the model is also worth mentioning here. The final model aggregates all the essay topics to train in a single model. However, as discovered by the researcher late in the development phase, each of the essay topics is graded using different grading schemes with different weights on different aspects of the essay. Hence, it might be more logical to separate 8 of the essay topics and train 8 different networks to predict essays for these 8 topics. However, because of its immense work and time required to retrain 8 different models, this idea was eventually scrapped.

For the evaluation of the front-end web application, it can be said that the front-end web application yields a satisfactory result as it managed to fulfill the 3 aims identified by the researcher in Work Proposal. The system can let users login, register, view the past submission, upload essays, and get an evaluation score from the trained model. For the most part, the researcher managed to accomplish what he first set out to do for this work. However, there is one idea that is scrapped and replaced with another feature at the suggestion of the supervisor. Initially, the last feature of the web application is to detect computer-generated texts using pre-trained models such as GPT-2 or BERT. However, due to various reasons and feedback from the supervisor, the feature is eventually replaced with checking the uploaded texts for spelling errors as it is a far more fitting feature for the product. The feature of detecting computer-generated texts will be considered as a future recommendation in the last chapter.

The decision to use Flask and Jinja for developing web application also proved to be useful as the final product manages to fulfill all the functional and non-functional requirements determined in the Software Requirement Specification (SRS). Since the web application is not a full-fledge application and more to a web interface, picking Flask instead of Django is a wise decision as Django has a considerably higher learning curve.

FUTURE RESEARCH DIRECTIONS

A few improvements can be made to the product to overcome the weakness of the system, specifically by increasing the average QWK score of the model. One method of improvement is to train a different model architecture that can yield a better result such as the one that is implemented in Taghipour (2017). Next, as mentioned in the evaluation section, the QWK score can be increased by separately training the model for each essay topic as each essay has a different marking scheme. Furthermore, the experiment can be improved by replacing GloVe with state-of-the-art word embeddings such as Elmo (Peters et al., 2018) to obtain a better score.

CONCLUSION

In conclusion, EssayScore demonstrates that the task of automated text scoring (ATS) can be solved through deep learning by implementing Long Short-Term Memory Units (LSTMs). This work shows that an end-to-end approach to solving ATS is feasible with the use of deep learning for feature extraction. The system produced can grade essays based on its underlying context, grammar, and structure. Furthermore, an end-to-end approach eliminates the need for feature extraction and engineering, thereby saving a lot of time and resources. Besides, EssayScore also solves the burden for teachers in marking essays and for students to gain instant feedback on written essays for improvement. ATS systems like EssayScore aim to reduce the labor-intensive marking of essays and ensures equality in scoring.

The web application is made using Flask microframework and Bootstrap templating while the model is written and trained using TensorFlow and Keras framework. After training the model, the model is served as a RESTful API using Docker and TensorFlow Serving which will be called by the front-end via HTTP protocols. The evaluation metric used is the Quadratic Weighted Kappa (QWK) score and the model obtained a score of 0.60, which is slightly less than the baseline score of 0.76.

The web application also managed to accomplish all the functional and non-functional requirements that are proposed in Software Requirement Specification (SRS). Those included letting users upload essays to the web application either by submitting a file or copy-pasting the essay into the text area, predicting a score to an uploaded essay using the trained model and check essays for spelling errors and recommend corrections.

ACKNOWLEDGMENT

The research work has carried in UOW Malaysia KDU Penang University College. There is no funding on this work however, the author's like to thank the reviewers and collaborators contributed to this research. Special thanks to second author for his contribution and further investigate on this work.

REFERENCES

Agrawal, S. (2019). *What the heck is Word Embedding*? Retrieved 25 September 2019, from https://towardsdatascience.com/what-the-heck-is-word-embedding-b30f67f01c81

Ahire, J. (2019). *The Artificial Neural Networks Handbook: Part 1*. Retrieved 19 November 2019, from https://medium.com/coinmonks/the-artificial-neural-networks-handbook-part-1-f9ceb0e376b4

Alikaniotis, D., Yannakoudakis, H., & Rei, M. (2016). *Automatic Text Scoring Using Neural Networks*. Retrieved from https://arxiv.org/pdf/1606.04289.pdf

Arafat, A., & Raihanuzzaman, M. (2016). *Automated Essay Grading with Recommendation*. Retrieved from https://pdfs.semanticscholar.org/7bb7/22d9f01706bf3da54faae9b763fa8fd64f76.pdf

Black, B. T. V. (2019). *White Box Testing: Key Differences*. Retrieved 3 December 2019, from https://www.guru99.com/back-box-vs-white-box-testing.html

Böhm, T. (2018). *The Three Main Branches of Word Embeddings*. Retrieved 25 September 2019, from https://towardsdatascience.com/the-three-main-branches-of-word-embeddings-7b90fa36dfb9

Bornstein, A. (2018). *Beyond Word Embeddings Part 2- Word Vectors & NLP Modeling from BoW to BERT*. Retrieved 11 November 2019, from https://towardsdatascience.com/beyond-word-embeddings-part-2-word-vectors-nlp-modeling-from-bow-to-bert-4ebd4711d0ec

Chollet, F. (2018). *Deep learning with Python* (1st ed.). Manning Publications Co.

Chowdhury, G. (2003). Natural language processing. *Annual Review Of Information Science And Technology, 37*, 51-89.

Dikli, S. (2006). Automated Essay Scoring. *Turkish Online Journal of Distance Education, 7*(1). Retrieved from https://www.researchgate.net/publication/26415982_Automated_Essay_Scoring

Dong, F., Zhang, Y., & Yang, J. (2017). Attention-based Recurrent Convolutional Neural Network for Automatic Essay Scoring. *Proceedings of the 21st Conference on Computational Natural Language Learning*, 153-162. Retrieved from http://aclweb.org/anthology/K17-1017

ETS. (2013). *Practice Book for the Paper-delivered GRE General Test.* Retrieved 3 March 2019, from https://www.ets.org/s/gre/pdf/practice_book_GRE_pb_revised_general_test.pdf

Farahmand, M. (2019). *Pre-trained Word Embeddings or Embedding Layer? A Dilemma.* Retrieved 25 September 2019, from https://towardsdatascience.com/pre-trained-word-embeddings-or-embedding-layer-a-dilemma-8406959fd76c

Heidmann, L. (2019). *AI vs. Machine Learning vs. Deep Learning.* Retrieved 11 November 2019, from https://blog.dataiku.com/ai-vs.-machine-learning-vs.-deep-learning

Hochreiter, S., & Schmidhuber, J. (1997). Long Short-Term Memory. *Neural Computation, 9*(8), 1735–1780. doi:10.1162/neco.1997.9.8.1735 PMID:9377276

Hubert. (2017). *AI In Education — Automatic Essay Scoring.* Accessed 5 December 2019, from https://medium.com/hubert-ai/ai-in-education-automatic-essay-scoring-6eb38bb2e70

Jiao, W. (2018). *What are the advantages and disadvantages of Word2vec and GloVe?* Retrieved 18 November 2019, from https://www.quora.com/What-are-the-advantages-and-disadvantages-of-Word2vec-and-GloVe

Karani, D. (2018). *Introduction to Word Embedding and Word2Vec.* Retrieved 25 September 2019, from https://towardsdatascience.com/introduction-to-word-embedding-and-word2vec-652d0c2060fa

Karpathy, A. (2015). *The Unreasonable Effectiveness of Recurrent Neural Networks.* Retrieved 17 March 2019, from http://karpathy.github.io/2015/05/21/rnn-effectiveness/

LeCun, Y., Bengio, Y., & Hinton, G. (2015). Deep Learning. *Nature, 521*(7553), 436–444. doi:10.1038/nature14539 PMID:26017442

Mikolov, T., Chen, K., Corrado, G., & Dean, J. (2013). *Efficient Estimation of Word Representations in Vector Space.* Retrieved 25 September 2019, from https://arxiv.org/pdf/1301.3781.pdf

Nielsen, J. (2019). *10 Heuristics for User Interface Design: Article by Jakob Nielsen.* Retrieved 22 November 2019, from https://www.nngroup.com/articles/ten-usability-heuristics/

NLTK 3.4.5 documentation. (2019). Retrieved 21 September 2019, from https://www.nltk.org/api/nltk.tokenize.html

Norvig, P. (2019). *How to Write a Spelling Corrector.* Retrieved 1 December 2019, from https://norvig.com/spell-correct

Olah, C. (2015). *Understanding LSTM Networks -- colah's blog*. Retrieved 22 March 2019, from http://colah.github.io/posts/2015-08-Understanding-LSTMs/

Page, E. (1966). The imminence of grading essays by computer. *Phi Delta Kappan, 47*(5), 238–243. https://www.jstor.org/stable/20371545?seq=1#page_scan_tab_contents

Paruchuri, V. (2019). *Vik's Blog - Writings on machine learning, data science, and other cool stuff.* Retrieved 10 November 2019, from https://www.vikparuchuri.com/

Peters, M., Neumann, M., Iyyer, M., Gardner, M., Clark, C., Zettlemoyer, L., and Lee, K. (2018). Deep contextualized word representations. *NAACL 2018.*

Phandi, P., Chai, K., & Ng, H. (2015). Flexible Domain Adaptation for Automated Essay Scoring Using Correlated Linear Regression. *Proceedings of the 2015 Conference on Empirical Methods in Natural Language Processing*, 431-439. Retrieved 3 March 2019, from https://www.aclweb.org/anthology/D15-1049

Powell-Morse, A. (2016). *Waterfall Model: What Is It and When Should You Use It?* Retrieved 30 March 2019, from https://airbrake.io/blog/sdlc/waterfall-model

Rudner, L., & Gagne, P. (2019). *An Overview of Three Approaches to Scoring Written Essays by Computer*. Retrieved 3 March 2019, from https://pareonline.net/htm/v7n26.htm

Rudner, L., & Liang, T. (2002). Automated Essay Scoring Using Bayes' Theorem. *The Journal of Technology, Learning, and Assessment, 1*(2). Retrieved March 3, 2019, from https://ejournals.bc.edu/ojs/index.php/jtla/article/download/1668/1512

Taghipour, K. (2017). *Robust Trait-Specific Essay Scoring Using Neural Networks and Density Estimators. The*. National University of Singapore.

Taghipour, K., & Ng, H. (2016). A Neural Approach to Automated Essay Scoring. *Proceedings of the 2016 Conference on Empirical Methods in Natural Language Processing*, 1882-1890. Retrieved 29 January 2019, from https://www.aclweb.org/anthology/D16-1193

TensorFlow. (2019). *RESTful API | TensorFlow Serving | TensorFlow*. Retrieved 30 January 2019, from https://www.tensorflow.org/serving/api_rest

Walia, A. (2017). The Vanishing Gradient Problem. *Medium*. Retrieved 29 January 2019, from https://medium.com/@anishsingh20/the-vanishing-gradient-problem-48ae7f501257

Waterfall Model, S. D. L. C. (2019). Retrieved 30 March 2019, from https://www.tutorialspoint.com/sdlc/sdlc_waterfall_model.htm

Zhao, S., Zhang, Y., Xiong, X., Botelho, A., & Heffernan, N. (2017). A Memory-Augmented Neural Model for Automated Grading. *Association for Computing Machinery*. Retrieved 2 March 2019, from https://www.researchgate.net/publication/316940225_A_Memory-Augmented_Neural_Model_for_Automated_Grading

Chapter 7
Application of Machine Learning Algorithms in Stock Market Prediction:
A Comparative Analysis

Sumit Kumar
Indian Institute of Management, Indore, India

Sanlap Acharya
Indian Institute of Management, Indore, India

ABSTRACT

The prediction of stock prices has always been a very challenging problem for investors. Using machine learning techniques to predict stock prices is also one of the favourite topics for academics working in this domain. This chapter discusses five supervised learning techniques and two unsupervised learning techniques to solve the problem of stock price prediction and has compared the performances of all the algorithms. Among the supervised learning techniques, Long Short-Term Memory (LSTM) algorithm performed better than the others whereas, among the unsupervised learning techniques, Restricted Boltzmann Machine (RBM) performed better. RBM is found to be performing even better than LSTM.

INTRODUCTION

In the stock market arena, markets show all the stock information that is currently available. This makes the task of stock price prediction almost impossible for investors, as the Efficient Market Hypothesis (EMH) would. Furthermore, the Random Walk Hypothesis (RWH) states that stock prices change in a stochastic manner. Over a long period of time, x has been used by the stock experts time-series prediction models to forecast stock prices. However, the reliability of all the hypotheses have been in question. There are some studies that have been successful in predicting stock prices to a certain extent. For example, Brock et al. (1992) predicted stock prices through moving averages and trading-range breaks.

DOI: 10.4018/978-1-7998-3645-2.ch007

Both of them were technical trading rules, which means that x can use technical strategies to predict stock prices. More recent studies that use machine learning methods, such as artificial neural network, genetic algorithm, or support vector machines, argue that the market is inefficient at different levels in a wide range of stock markets, and therefore, it is feasible to predict the stock prices.

The authors' objective is to predict the behavior of a financial time-series based on the present behavior with the help of various variables. Generally, x can classify such forecasting on the basis of explanatory variables into two categories: fundamental analysis and technical analysis. Sometimes, forecasters use techniques from both of these categories simultaneously, whereas there are cases in which x has used statistical models variables that fall under both categories. Along with statistical techniques, the authors observe the emergence of machine learning techniques to solve this problem. Sometimes, x uses a machine learning technique purely, and sometimes, x combines it with another statistical technique. One of the most popular machine learning techniques that has been widely used to predict financial time-series behavior is artificial neural network (ANN).

When x uses ANN for such predictions, selection of input or independent variables is as important an aspect as the structure of ANN. Two identical ANN models can show different outputs if their input variables are different. In this book chapter, the authors focus on the prediction accuracy and error of the models. In other words, a model having more accuracy and less error is more likely to be called an efficient model. However, an accurate model or a model that has fewer errors in the form of mean average deviation (MAD), root mean square error (RMSE), mean average error (MAE), mean square error (MSE) does not guarantee a profit in stock market (Leung et al., 2000). However, it may seem more likely that a model that has more accuracy and less error will be helpful for investors. Therefore, in this book chapter, the authors will be comparing two machine learning models that will predict future stock price.

This book chapter henceforth includes a brief literature review on prediction of stock prices using machine learning algorithms. The authors then briefly describe the various machine learning techniques. Next, the authors explain the dataset and discuss implementations of all the machine learning techniques. Then, the authors present the results, the discussion, the practical implications, and finally, the future scope in this domain.

LITERATURE REVIEW

This section of the book chapter discusses studies in which researchers have used ANN along with other machine learning techniques to predict stock market. For example, Niaki and Hoseinzade (2013) used Standard and Poor's 500 (S&P 500) data to compare the performance of ANN and logistic regression. ANN performed better than logistic regression. In another study, Kara et al. (2011) used data from the Istanbul Stock Exchange (ISE) National 100 Index. They ran SVM and ANN through the data and found that ANN showed more accurate results than previous ones. Yao et al. (1999) used data from the KLSE Stock Market Index and used back-propagation neural network (BNN) techniques in forecasting returns. After comparing those techniques with auto regressive integrated moving average (ARIMA) technique by performing the same operation, they found that BNN performed better than ARIMA. De Faria et al., (2009) used data from the Brazilian stock market and compared the forecasting results of ANN with adaptive exponential smoothing (AES) technique. They did not find any significant contribution of AES in gauging the return. Interestingly, both techniques had almost the same RMSE. Furthermore, Huang et al. (2005) compared five machine learning techniques to predict the behavior of the NIKKIEI 225 index. The techniques they compared were Elman back-propagation neural networks (EBNN), random walk (RW), linear discriminant analysis (LDA), quadratic discriminant analysis (QDA), and support-vector machine (SVM). SVM was the best performer in terms of accuracy. Finally, Zhong and Enke (2017) used data from S&P 500 Index and compared the performances of ANN aligned with three types of Principal Component Analysis (PCA). They found ANN with fuzzy robust PCA to be best by a very small margin from the other two. In recent years, we have seen many articles and research papers covering these algorithms such as random forest (Khaidem, Saha, & Dey, 2016; Patel et al., 2015), Naïve Bayes (Mahajan et al., 2016), SVM (Di, 2014), and Neural Network (Kim, & Lee, 2004) with some variations. Thus, the authors derive from this fact that the performance of algorithms differ across cases. Therefore, a comparative analysis of all these algorithms would present a better picture in front of us.

MACHINE LEARNING ALGORITHM

Michie et al., (1994) stated that it will be useful for machines to learn from their experiences and increase their efficiency. Algorithmic and mathematical developments happened in the late 1970s and 1980s due to lack of computational power in the computer systems. Machine learning algorithm approaches were not implementable because all these algorithms need a lot of manipulations in the data, which require high computational power. After the 1990s, computational power of computer started becoming compatible with operation on machine learning algorithms. Finally, in the early to late 2000s, researchers and computer scientists started using machine learning techniques in different applications.

Long back, Cowles (1933) used probability to measure future movement of stock market. It is difficult to understand future price of stock market because the market is influenced by different conditions, and it is difficult to connect these conditions. In order to understand market correctly, one has to understand the relationship among all financial assets, drivers of the price of each asset, and the link between financial assets and the real economy (Schumaker, & Chen, 2009; Hsu., 2011; Wolfers, & Zitzewitz, 2006). Considering the literature cited above, it is clear that stock markets are affected by many highly interrelated economic, social, political, and even psychological factors, and these factors interact with each other in a very complicated manner. Malkiel (1991) developed the efficient market hypothesis (EMH); in this hypothesis, he indicated that it is unrealistic to forecast the share market because price change could be unpredictable. Many studies that support EMH have been published in different journals (Laffont, & Maskin, 1990; Metcalf, & Malkiel, 1994). In contrast, Yaser and Atiya (1996) claim that financial markets can be predicted to a certain extent. Their argument was based on the concept of correlation between financial market, undiscounted serials, and their vital effect on economic events affecting future financial market. This evidence opposes the theory proposed by Malkiel (1989) (i.e., EMH).

Researchers have attempted to use different statistical approaches, such as regression and maximum likelihood to predict the nature of stocks a priory (Nadaraya, 1964; Rauch et al., 1965). Researchers have observed that conventional statistical techniques for prediction have reached their limitation in applications with nonlinearities in the dataset. Even before the proposal of EMH, Felsen (1975) used learning-based pattern recognition technique for forecasting the stock market. He found that PR techniques obtained above average investment performance by using iterative probabilistic learning algorithms (based on stochastic approximation techniques). Similarly, Andreas et al. (1990) worked on limited time series data to predict behavior of non-linear dynamic systems. They found that sigmoidal functions perform better than radial basis function for high dimensional input data.

Induction-based method is popular in the early stages of machine learning development for predicting stock market price values (Salzberg, 1985; Liang, 1990). In the initial phases of machine learning development, due to complexities of data produced by the stock market, neural network seemed to be best choice. Additionally, researchers used recurrent neural networks and genetic algorithm to understand unpredictable nature of stock market (Margarita, 1991). The authors have seen that stock prices are impacted by different parameters available in the surrounding of a firm's ecosystem, and it is difficult to understand real reason behind change in stock prices of particular business firm. After understanding the structured information produced by the stock market daily, researchers started understanding unstructured data, such as the impact of news articles, economic health of nations, and political conditions of nations. In order to analyze the impact of these conditions, authors used supervised machine learning algorithms (especially neural network). In parallel, researchers were using other machine learning algorithms for understanding the mysterious patterns of stock market data. The authors used both supervised and unsupervised ways of machine learning.

Supervised Learning

Stock markets produce an abundance of data, which means researchers will never lack data to be trained. Because to this, supervised learning is always the first choice for the understanding patterns in stock market data. In supervised learning, x feeds machines with tagged labels that are the correct solutions for each row of information. Based on the rows that have the correct solutions, the machine deduces a set of rules. Later, the machine will apply the same rules on unlabeled data to estimate the correct label. Supervised learning can be categorized into two major parts: classification (when output variable is category) and regression (when output variable is real value). Supervised learning algorithms became famous in the late 1990s when storage was a big challenge. After being trained from a labelled data set, there is no need to keep labelled data in the memory because the machine has already decided its decision boundary in the form of a mathematical formula.

Due to the chaotic nature of data, researchers have used different statistical tools. Due to the large size of data application of these statistical methods, x are not efficient. Researchers have proven machine learning tools to be successful in understanding hidden rules in complex data when they are trained with proper amount of data set. The authors did a Google to understand the most famous algorithms in supervised learning for stock market prediction. They found that Neural Network seems to be the most popular algorithm. Researchers worked on all variations of Artificial Neural Network (ANN), such as RNN and Hybrid Neural Network. They even worked on combinations of Neural Network with other methods, such as Genetic Algorithm with Neural Network, which was quite popular in early 1990s. Other popular algorithms in supervised machine learning are Support Vector Machine and SVM with Genetic Algorithm.

Stability in the stock market increases or decreases unevenly with passing time. This is due to the chaotic nature of stock market. It is possible that even a small change in the beliefs of shareholders may revert directing of growth for the particular firm in the stock market. Authors have compared randomness with several natural phenomenon, such as Sand Pile Avalanche Model (in which one grain lead to collapse of whole pile) and Black Swan event (which refers to unpredictability of stock market future) (Bartolozzi, 2006; Peša, & Brajković, 2016). The cycles of rising and falling trends that occur in chaotic processes have varying time periods; quiet periods can be followed by a large jump, or vice versa. These effects make it difficult to properly predict the future in even well-trained models.

Deep Learning: LSTM

LeCun et al. (2015) said, "Deep learning allows computational models that are composed of multiple processing layers to learn representations of data with multiple levels of abstraction." One of the benefits of deep learning is in understanding stock market data due to its ability to generate new features from an available set of features. This is important in studying stock market data because knowledge of hidden information helps in explaining chaotic behavior in the market.

Decision Tree

A decision tree is basically a tree that has a structure like flowchart. The top node signifies the root node. The internal node, or the non-leaf node, signifies a test on an attribute. Each branch signifies a consequence of a test, and each leaf node signifies a class label. When the class label of a tuple is unknown, it traverses a journey from the root to leaf, and in the process, x predicts the class of that tuple. Decision tree does not require any domain knowledge or setting of parameters and is suitable for exploratory knowledge discovery.

Random Forest

There is a limitation to the decision tree algorithm. Sometimes, noise in the data may drive the tree to behave differently and give irregular patterns. The reason for such behavior is that decision trees have very low bias and high variance. The Random Forest algorithm trains decision trees on different subspace by slightly increasing the bias. As a result, none of the trees in the forest can see the entire training data. Therefore, any noise in the data can be handled with greater intelligence.

Linear Regression

Linear regression is a process in which a linear model predicts the target variable. This process is also known as Ordinary Least Square (OLS) regression. Thus, it assumes from before that the target variable will behavc linearly with the independent variables. This is much less accurate than non-linear models.

Naïve Bayes

This is a Bayesian method of classification in which x assumes that all the independent variables are statistically independent of each other. This technique is quite useful when the input variables have high dimensionality.

Unsupervised Learning

Unsupervised learning is a new machine learning technique in which the machine is not fed with prior knowledge, unlike supervised learning. Researchers use these techniques in the case of unlabeled data. In these cases, the machine learning takes place through a self-taught learning framework. In this book chapter, the authors discuss two unsupervised learning techniques: k-means algorithm and Restricted Boltzmann Machine (RBM).

K-Means Algorithm

K-means algorithm is a technique to cluster similar data used for prediction. In this technique, the probability of the most relevant function is found, and by using the probability, clusters are formed with those functions that are close to each other. Euclidean distance helps to measure the closeness.

Restricted Boltzmann Machine (RBM)

RBM is one of the most common building blocks of deep probabilistic models. It is undirected probabilistic graphical model containing a layer of unobservable variable and a single layer of latent.

DATASET

The dataset used for this work is secondary, and the authors collected it from website the nseindia.com. The authors collected the data for all NIFTY 100 companies that is available in the NIFTY. NIFTY 100 is a diversified portfolio of different business firms that have a major impact on the economy. The NIFTY 100 tracks the behavior of the combined portfolio of two indices:. NIFTY 50 and NIFTY Next 50.

The authors collected all the stock data from these 100 business firms from January 1st, 2009 to January 1st, 2019. The reason to select the NIFTY 100 is that it statistically shows the impact on all major sectors of economy. A stock index reflects the overall market sentiment and the direction of price movements of products in the financial, commodities, or any other market. Using data from before this date may cause certain problem in the results, and because of this, the authors did not consider data from before this date. The authors made predictions on the "closing price," which means the dependent variable and all other variables are independent variables, such as opening, high, low, and volume.

Abnormality in the data set is one of the major challenges. The authors have checked it so that no features are empty and no cells contain negative values. The authors found the mean, median, maximum, and minimum of the data, which are shown in Figure 1 & 2.

Figure 1. Descriptive statistics of the dataset

```
      Open              High              Low               Close
 Min.   :  87.74   Min.   :  89.29   Min.   : 86.37   Min.   :  87.58
 1st Qu.: 232.38   1st Qu.: 234.89   1st Qu.:230.40   1st Qu.: 232.44
 Median : 296.28   Median : 298.52   Median :293.64   Median : 296.05
 Mean   : 382.51   Mean   : 385.87   Mean   :378.74   Mean   : 382.35
 3rd Qu.: 544.00   3rd Qu.: 548.22   3rd Qu.:539.85   3rd Qu.: 543.65
 Max.   :1005.49   Max.   :1008.61   Max.   :996.62   Max.   :1004.28
```

After obtaining a proper understanding of the dataset, is the authors found that three variables play negligible roles in the dataset. The only significant factors in the dataset are opening price and volume. Volume of share is important as a rising market should see rising volume (i.e., increasing price and decreasing volume show lack of interest). On the other hand, share markets work on the simple funda-

Figure 2. Descriptive statistics of the dataset

```
     Volume
Min.    :  521141
1st Qu.: 1889613
Median : 2811069
Mean   : 4205708
3rd Qu.: 5232088
Max.   :41182889
```

mental rule that if at the end of the day, there arehigher closing prices than the opening prices, there issome profit. Otherwise, there is loss.

The authors use normalization to scale a value in such a manner that it will fit into particular range. The authors perform normalization because different atributes have different units, and data are avial-able in different scales. Scale by weight is a very popular method, in which operator is used to scale the pre-calculated weight. Functon is used for the scaling is min-max scalar algorithm, and it uses following formula:

$$\frac{x_i - \min(x)}{\max(x) - \min(x)}$$

Function basically shrinks down the value of the attribute between rage 0 to 1 for the all positive values, and if there are negative values, then the range varies from -1 to +1. It is specially used in cases in which distribution is not gaussian or the standard distribution is very small. It is impacted by outliers in the data. Therefore, it might not be the best choice if the data has outliers.

After normalization of data, the authors stored the data in a file named google_preprocessed. csv for future use. Both of the evaluation methods need to divide the dataset into two parts, the reason behind this is to avoid overfitting and underfitting of the model. It is standard practice to divide the dataset into two parts, approximately 70% for the training set and 30% for the testing set in the case of linear regress-sion. On the other hand, regarding LSTM, more data is needed for training. Hence, approximately 80% of data is allocated for training, and 20% is allocated for testing.

Understanding the nature of the data in this paper is important. The authors used matplotlib form the python library to generate 2D plot with the help of few lines of code. Figure 3 shows a visualization of the data using matplotlib. The figure's y- axis represents the dependent variable (i.e., closing stock price), and the x-axis represents the number of days.

Figure 3. Visualization of the data

IMPLEMENTATION

The authors have used five different machine learning algorithms to understand prediction efficiency and determine which is the best in terms of predicting the future stock price and detecting a reason for particular behaviors of algorithms. The authors discussed nature and characteristics of data in the previous section. The development of a model for predicting future prices has four major steps, and each steps has its own non-overlapping function. Figure 1 shows the name and the order of implemenatation. Before implementation of each of the algorithms, python need to be loaded with some of the foreign and pre-installed packages. The authors have described step-by-step implementation of each method and packages needed in each of their corresponding sections.

Long Short-Term Memory: LSTM

The authors will provide a brief introduction to LSTM. It is a state of art recurrent neural network (RNN) for sequence data. Sequence data are data that are in sequence. Chat box communication is one example of this. RNN is the first algorithm. It remembers its input due to its internal memory, which makes it suitable for machine learning algorithms in stock market prediction. It is a major reason for the incredible achievements of deep learning in recent years. The basic structure of RNN is shown in Figure 4.

Figure 4. Basic structure of RNN

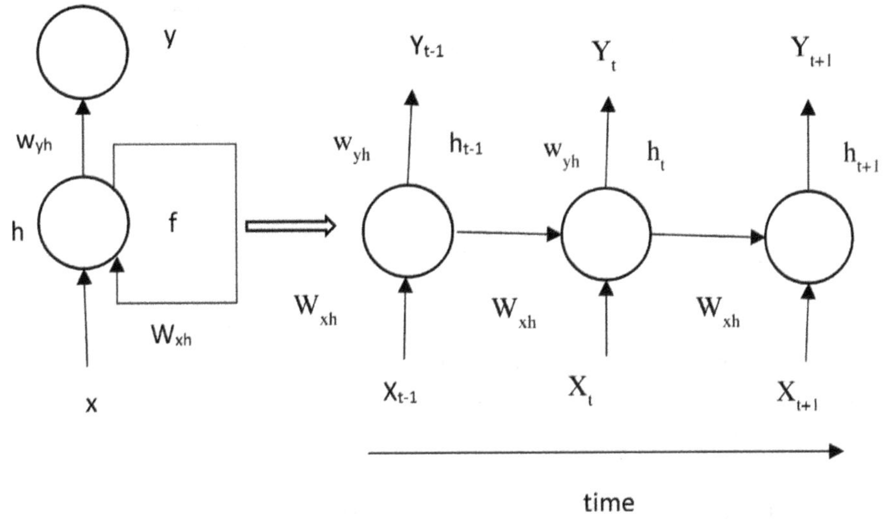

With respect to the time stamp, the authors provide the time axis and obtain output for each time stamp. Output of the previous stage remains with input of current stage (i.e., for this reason, this particular model is capable of handling sequence updates). The authors have features like open, close, high, low, and volume.

Libraries that are applicable to LSTM are Numpy, Matplotlib.Pyplot, and Pandas. Implementation of LSTM is a complex process, and it needs a lot of sub stages in order to understand data and extract patterns to perform prediction on future stock price.

1. **Reading and Analyzing Data (Panda):** This dataset is available in the form of a CSV file. The authors will upload data in this stage. They will predict the value of the opening price of stock values. For this, the authors will segregate out the opening column from rest of the dataset.
2. **Normalizing the Data:** This is the scaled down value in a particular range. In the dataset, currently dataset has different values varying in large range. The purpose of the feature is to scale it down using MinMaxScalar. All the values are between 0 and 1.
3. **Converting Data to Time-Series and Supervised Learning Problem:** The authors need to convert the dataset into a time series pattern. After normalization of dataset, the authors convert it into a special kind of data structure. This data structure is available in the form of a timestamp. Time stamp is fixed values of opening stock price.

Table 1.

t_1	t_2	t_3	t_4	t_5	t_6	t_7	t_8	....	t_{60}

As readers can see, the collection of the first 60th record it will act as an input record for the RNN model, and t_{61} will act as output for the model. Now, the next timestamp will start from t_2 and t_{61} will be the last record in the next timestamp. Output of next timestamp will be $t_{62,}$ and the diagram below shows the next time stamp:

Table 2.

t_2	t_3	t_4	t_5	t_6	t_7	t_8	t_9	....	t_{61}

Here, in the example above, 60 is an arbitrary value, and the number of records depends on the dataset and other optimization parameters. Records in the timestamp act as input for the recurrent neural network (RNN) based on the time stamp. The authors will be able to estimate output.

4. **Creating Model (Keras):** Keras is a deep learning library capable of running Tensorflow, CNTK, and Theano. Before creating recurrent neural network, researchers have to reshape their data based on input shape of RNN. In keras the authors note input shape in the following ways:
   (batch_size, timestamps, input_size)
   batch_size = total number of records in the dataset
   Timestamps = input for RNN (segregated segment)
   input_dimension = number of variables in the dataset

Creating RNN involves importing four different libraries: sequential, Dense, LSTM, and Dropout.

Sequential: Sequential wrappers for python functions allow researchers to easily define the order of execution independent of content of the function

Dense: this function is used to create last and output layers of RNN. This layer is not the LSTM layer.

Dropout: Neurons produce zero output in some of the epochs. This function will eliminate those neurons form the model to optimize it.

LSTM: This function is used to create different type of LSTM model; in this case, the authors are building stacked LSTM with four layers. In each layer, there are three different parameter units, which give the number of neurons in the LSTM. return_sequence is used to return sequences and return the hidden state output for each input time step. Additionally, input_shape has three subfields: batch size, timestamp, input dimension. The authors discussed the input shape of RNN model previously in this section.

5. **Optimization and Model Fitting:** The authors used adam as an optimizer, and the model mean square error for understanding loss. Model fitting has two main parameters: epochs and batch size. Epochs decide the number of times the dataset runs on the model to fit in it. On the other hand, batch size is the number of records given to the model in each iteration. The authors have seen that with increase in the number of epochs, loss is decreased.

Decision Tree

In order to implement the decision tree, the authors used NIFTY data from NIFTY companies from 2012 to 2019. In the previous sections, the authors described how this was evaluated using daily dataset as measuring parameter. The major aim of this work is to help daily and intraday traders because they are considering the daily stock data of NIFTY 50 stored in the CSV format. They have divided implementation into different steps to better understand implementation:

1. **Importing Data:** For importing data, the authors used pd.read.csv, which is available in the panda library, to bring the dataset from the system into the python environment. Pd stands for panda, which is the library defined in the panda package to read all kinds of data available in csv format. Pd.to_datatime helps to convert string date time into python data time object, which will be helpful in the time difference.
2. **Visualization of Data:** In machine learning, visualization of data helps in understanding nature of data and the presence of outliers in the data. When the authors analyze data, the data seems to be continuous and correct.
3. **Predictor and Target Variable:** Predictor is input to the model and understanding how input affects the output. The decision tree model takes the input and tries to understand how it results in the output. Assumption input is related to the output. It means there is cause and effect, sequence or direct dependence, or some kind of dependence between input and output. In trading, getting the predictor variable is difficult. In this implementation, the authors have taken RSI. This is basically an oscillator, and it measures the magnitude of the recent price changes to evaluate overbought or oversold conditions in the price of stock or other assets. The authors use talib to fetch the RSI value. RSI reduces the opening value in the range of 0 to 100. When this RSI value id is above 70, the authors can say that market is overbought (i.e., there is huge Bull Run and onecan expect a reversal). Similarly, if the RSI value is less than 30, one can expect bounce back. The momentum indicator tells how fast and far the momentum indicator has gone in the market. Thus, the authors use the RSI and the difference between the opening price of yesterday and opening price of today. In this way, the authors can calculate five different indicators: Diff1 to Diff4 and RSI.
   Where Diff1 = difference between opening price of today and yesterday
   Diff2= difference between opening price of yesterday and day before yesterday. Similarly, the authors can also calculate all the values of Diff 3 and Diff4.

In order to understand trends in market, the authors have created a trend column in the dataset. In the trend column, the value of each field will be either 1 or 0. If yesterday's opening value is greater than today's opening value, then the authors will label it to 0. Otherwise, the authors will allocate 1 to the respective field of trend.

Now, the authors have obtained the data for both the predictor and the target variable. As a data preprocessing step after the calculation of RSI, Diff1 to Diff4 have many NAN values. The authors have to remove all of them from the dataset, using the 'dropna' function and then create a separate predictor list and target list.

4. **Split the Data into Train and Test Dataset:** The authors use the train_test_split function in the Sklearn package. The authors will split the data into two parts (i.e., train and test). In this library,

the authors have to specify the test and training data size. The general rule is to split the majority of data in the training part and the minority of data in test set. The authors have allocated 70% of dataset to training and 30% to testing. In the hyper parameter stratify, the authors have allocated the dataset value equal to testing data set means training and test data set to have the same proportion of 0 and 1 as original dataset.

5. **Classification Tree Model:** The authors have used the Sklearn tree package to create a classification tree model. There are two ways to go for the decision tree; one is the regressor model (predicting a real value) and the other is the classification model (predicting classes from the label set). There are some hyperparameters in the decisiontreeclassifier function. The authors have used the 'gini index' as a criterion measure. Gini index measures the amount of impurity at every split of the decision tree. Gini index varies from 0 to 1. If the value is 0, it has no impurity,y and if the value is 1, it indicates a poor split. Max_depth is the amount of iteration or split that continues, and min_sample_leafs acts as an anchor condition for the tree (i.e., if the number of leaves is less than the given value, stop producing the next level of tree). Fitting the given model has been done with the help of the fit function on the training and testing set.
6. **Visualizing the Model:** This is to extract out the rule or to understand the rules of the decision tree. The authors used the graphviz package.
7. **Prediction:** After the authors train and fit the model, they will make prediction on the test data. The authors will use clf.predict on the test data set and obtain the prediction.
8. **Evaluate Model Performance:** Researchers should use the classification report function to report the performance of the model. This will provide information about precision, f1 score, and recall values.

Random Forest

Random Forest is collection of a lot of decision trees, which helps in tackling the problem of overfitting of a model. RF is an ensemble of different type of decision tree that is built using random features available in the dataset due to this accuracy of the RF is higher in comparison to single decision tree. The process of randomly selecting features with the replacement from the available dataset is called bootstrapping, aggregating, or bagging. This means single features will appear to build different DTs more than once. Once the features are selected, the trees are constructed based on the best split. Each tree gives an output that is considered as 'vote' from that tree for the given output. The output that receives the maximum 'votes' is chosen by the random forest as the final output/result or, in the case of continuous variables, the average of all the outputs is considered as the final output.

Unlike all other machine learning algorithms, this method needs numerous features to be selected in decision trees. Because of this, the authors cannot take just one variable to predict the stock market, as in the case of LSTM Model. This section shows step-by-step implementation details of the Random Forest algorithm.

1. **Importing Libraries:** The authors need two major libraries for the implementation of random forest in the stock market prediction
2. **Numpy:** This is useful for data manipulation, computing input features, and computing output.

Figure 5. Random Forest (adopted from Liaw, & Wiener, 2002)

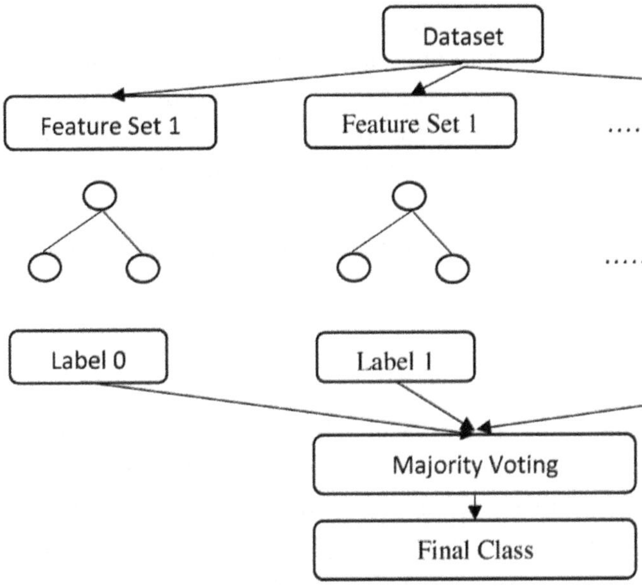

3. **Sklearn:** SkLearn is an all-in-one package for a lot of machine learning models. The RandomForestClassifier function is one of the functions defined in the Sklearn library for implementation of the Random forest model.
4. **Fetching Dataset:** The authors have already seen that daily data has been collected from NIFTY for all Nifty 50 companies from January, 2012 to November, 2019.
5. **Creating Input and Output Dataset:** In order to predict the nature of stock prices, the authors have to calculate input features and output or target variables.
6. **Input Feature:** In order to prepare four different input features that can capture information from the stock price, all of these four variables prepared from dataset are '(Open - Close)/Open', '(High - Low)/Low', standard deviation of last 5 days returns (std_5), and average of last 5 days returns (ret_5).
7. **Output Feature:** Now, after creating all input variables for the model, the authors will segregate them out in a different dataset. In order to create output variables that predict the inclination of the stock market. If tomorrow's opening price is greater than today's open price, then the output variable is set to 1, and otherwise it is set to 0. 1 indicates buying the stock, and 0 indicates selling the stock.
8. **Train and Test Split:** According to the norms of machine learning, the majority of data must be allocated for the training of model, and rest of them are dedicated to the testing purpose. The authors have also divided their dataset by allocating 70% for the training of the model and 30% for testing of the model.

Linear Regression

Step 1: In step1, one needs to split train and test the dataset for performing the 'stock_data.py' function. It helps in splitting the data for linear regression. It is a pre-developed function for splitting of data.

Step 2: In this step of the model, basic linear regression is used to fit the data by using commands defined in the 'LinearRegressionModel.py'. Codes of the function are mentioned below in the text below:

Step 3: Next, the model predicts the closing price based on the trained linear regression model. The code for this is shown below:

Step 4: In the last stage of the model, the authors plot the outpupt. Both training and testing score plotted using mean_squared_error.

Naïve Bayes

The Naïve Bayes algorithm is one of the most popular machine learning algorithms. It is based on the concept of probabilities and conditional probabilities. A basic advantage of Naive Bayes is that if one has a large number of data points with few numbers of variables in the dataset. This particular algorithm will be faster in comparison with any of the machine learning algorithms in predicting the class of any of the unknown dataset. Naïve Bayes had one assumption about independence of variables or, simply stated, a Naive Bayes classifier assumes that the presence of a particular feature in a class is unrelated to the presence of any other feature.

One can build a Naïve Bayes model in python with the help of scikit library. There are three kinds of model under it:

1. **Gaussian:** This is considered in classification kind of problems and assumes features following normal distribution.
2. **Multinomial:** This is used for discrete counts.
3. **Bernoulli:** This is used if feature vectors are binary: 0 or 1.

The authors will explain the implementation of the Naïve Bayes algorithm in a step-by-step manner. It is easy to understand the meaning of different steps involved in the implementation of Naïve Bayes.

1. **Importing Data:** The authors have already discussed the machine learning algorithm in the previous section.
2. **Visualization of Data:** In machine learning, visualization of data helps in understanding the nature of data and the presence of outliers in the data. When the authors analyze data, the data seems to be continues and correct.
3. **Predictor and Target Variable:** To create independence in the input variable of Naïve Bayes, the authors have created our independent variables in such a manner that they become un affected by each other.
4. **Momentum:** This is the market trend in last few days. The authors have considered this using the last 5 days.

$Momentum = Close\ Price(Current\ Day) - Closing\ Price(5^{th}\ Day)$

5. **Return:** This is the percentage of change from the last day's closing price.
6. **Return of Investment (ROI):** The authors have calculated ROI on 10th, 20th, and 30th day period. The authors calculate ROI on the closing price of stock value.

$$ROI = \frac{\left(current\ day\ closing\ price - n^{th}\ day\ closing\ price\right)}{n^{th}\ day\ closing\ price}$$

7. **Relative Strength Index:** This is also a momentum indicator that evaluates recent price change to analyses overbought conditions. The authors have calculated RSI for 14, 20, and 30 days.
8. **EMA:** Exponential moving average (EMA) is a type of moving average (MA) that places a greater weight and significance on the most recent data points.

$$EMA = \left(\frac{2}{(n+1)}\right) * Close\ Price + \left(1 - \left(\frac{2}{n+1}\right)\right) * Previous\ EMA$$

where Initial EMA = first day closing price

9. **Class:** One should allocate class 1 to those days when the closing price is higher than the opening price; otherwise, one should allocate class 0 to that day.
10. **Normalization of Data:** Before normalization, since the authors are using Naïve Bayes, they can take only positive points because they are running multinomial Naïve Bayes. Because of this, the authors have removed negative values from the dataset. In the end, the authors will use the 'normalize' function for normalization of all features in the dataset to achieve better output.
11. **Split the Data into Train and Test Dataset:** The authors are going to use train_test_split function Sklearn package. The authors will split the data into two parts (i.e., train and test). In this library, the authors have to specify test and training data size. The general rule is to split the majority of data in the training part and the minority of data in test set.
12. **Build Model:** In the first stage of this two-stage process, the authors will build the model using the sklearn.naive_bayes package and invoke the GaussianNB function defined in it. In the next stage, the authors fit the model with the GNB.fit function on both the training dataset of predictor and target variables.
13. **Prediction:** Once the authors train and fit the model, the will make prediction on the test data. They will use GNB.predict on the test dataset to obtain the prediction.
14. **Evaluate Model Performance:** The authors have imported metrics function defined in the sklearn library to obtain the accuracy of the model.

K-Mean Algorithm

Libraries: The authors used different libraries to perform various operations for the implementation of the K-Mean algorithm.

PyLab: This is used to plot and show the graphical representation of clustering.

Numpy: This is used to understand, manipulate, and handle array and stack related operations.

Scipy.Cluster: The authors have defined basic K-mean function in this package.

Pandas: This is data manipulation libraries of python.

Math: The authors have used this for mathematical operations.

Matplotlib: These are also libraries for data visualization.

Importing Data: For importing data, the authors use pd.read.csv, which is available in the panda library for bringing datasets from system into the python environment. Pd stands for panda, and it is library defined in the panda package to read all kinds of data available in the csv format. Pd.to_datatime helps to convert string date time into python data time objects, which will be helpful in the time difference.

Predictor and Target Variable: Theaim of using K-mean clustering is to segregate daily stock prices into two major categories: 0 or 1, where 1 represents a scenario in which closing price of that day is greater than opening price, and 0 represents a scenario in which the closing price is less than opening price. This is the target variable, which has two major clusters, meaning that value of K is 2. Other than this, the authors used variance and return as predictor variables.

Return: This is percentage of stock value from current days to nth day. The authors already discussed this formula in the previous section of the article.

Variance: To calculate variance in the stock price, the authors calculated standard deviation in the stock price.

Building Model: The authors used the elbow Method to understand the optimal number of clusters present in the dataset and compare this with pre specified clusters (i.e., 2). The authors used sum squared error (SSE) to understand optimal number of clusters in the dataset. The authors analyzed the amount of SSE with the Elbow curve in two clusters. Next, using the K-means function, the authors can segregate the dataset into two different groups.

Evaluate Model Performance: Using the classification function, the authors can report miss classified data and correctly classify data points. It is the reported accuracy of the k-mean clustering model.

Restricted Boltzmann Machine

RBM are some of the most common building blocks of deep probabilistic models. They are undirected probabilistic graphical models containing a layer of unobservable variable and a single layer of latent.

Input Dataset: For importing data, the authors used pd.read.csv, which is available in the panda library, for bringing the dataset from system into python environment. pd stands for panda, and it is the library defined in the panda package to read all kinds of data available in the csv format.

Pd.to_datatime helps to convert string date time into python data time object, which will be helpful in the time difference.

Visualization of Data: In machine learning, visualization of data helps in understanding the nature of data and the presence of outliers in the data. When the authors analyze data, the data seems to be continuous and correct.

Predictor and Target Variable: The target variable captures the final impact of trading in a single day. It is a binary variable, in which 1 represents gain at the end of the day, and 0 represents loss at the end of the day. Specifically, the authors we have two classes of labels in the target variable 1, and closing price minus opening price is a positive value. Otherwise, it is negative, and the class label will be 0. The set of predictor variables used in the model are:

1. **MACD (Moving Average Convergence/Divergence):** This is used as a momentum indicator. It is the relationship between two moving averages of security's price. The MACD is calculated by subtracting the 26th period EMA from 12 days EMA.

$$MACD = 12 - Period\ EMA - 26 - Period\ EMA$$

2. **Williams Percent Range**: This is also a momentum indicator that ranges from 0 to -100 and measures overbought and oversold value. A reading above -20 is overbought, and a reading below -80 is oversold.

$$williams\%R = \frac{Highest\ High - Close}{Highest\ High - lowest\ Low}$$

where, Highest High=Highest price in the previous period, typically measured by 14 days; Close=The most recent closing price; and Lowest Low=The lowest price in the previous period, typically measured by 14 days.

3. **True Range:** This is developed by Welles Wilder to understand the trading range of stock:
    - The distance from today's high to today's low.
    - The distance from yesterday's close to today's high.
    - The distance from yesterday's close to today's low.

Describing above statements in mathematical form:

$$TR = MAX(high[today] - close[yesterday]) - MIN(low[today] - close[yesterday])$$

Split the Data into Train and Test Dataset: The authors used the train_test_split function Sklearn package. The authors split the data into two parts (i.e., train and test). In this library, the authors have to specify test and training data size. The general rule is to split the majority of the data in the training part and the minority of the data in the test set. The authors have allocated 70% of the dataset to training and 30% to testing. In hyper parameter stratify,,the authors have allocated dataset values equal to testing dataset means, and training and test dataset have same proportion of 0 and 1 values as the original dataset.

Building Model: The authors installed rbmtf package of python to implement the binary RBM Model. Later, by calling the RBM function from the libraries, the authors implemented it. In next stage, the authors will fit the model in the RBM classifier.

Evaluate Model Performance: Using the classification_report function, the authors have seen all reports related to RBM classification.

RESULTS

LSTM Results

LSTM is a neural network method of machine learning. Along with prediction accuracy, it means that output of both the estimators are measured in the terms of mean square error, measurement of performance of an estimator. The basic concepts on which MSE relies are precision, bias, and accuracy. Measurement of MSE needs target of estimation with given data. MSE is defined as mean of the square of the error.

Error = Estimated - Estimator

Suppose $\hat{x}_i$ is the vector value of n number of predictions. Also, x_i is a vector that represents a number of values.

$$MSE = \frac{1}{n}\sum_{i=1}^{n}\left(\hat{x}_i - x_i\right)^2$$

LSTM is based on changing tuning of LSTM model at two hyper parameters. The following values show the impact of the changing their values:

Train Score: 0.00032478 MSE (0.01802172 RMSE)
Test Score: 0.00093063 MSE (0.03050625 RMSE)

LSTM deals with one more major factor due to its neural nature (i.e., time of prediction). The authors observed that the time of the prediction in the LSTM is 1237 seconds, which is acceptable because the number of epochs is 100.

Decision Tree Results

Decision tree is a machine learning algorithm, in which every node data is split according to certain parameters. A correct understanding of data is important to make any kind of prediction. Because of this, the authors visualize the data to understand continuity. Figure 6 shows a diagram of the stock market conditions between the years 2012 and 2019.

The authors analyzed the results of two basic training and testing accuracies. It is evident from the results that training and testing accuracy of decision dree are almost equal. Training accuracy relates to observed data, and testing accuracy is a measure of performance on unobserved data. This means the model is fit to the data appropriately. Table 3 presents a detailed description of the performance of the decision tree.

From the table, readers can easily understand that the testing accuracy of the model is near 67%. This means the model can make correct predictions on the direction of stock movement 67 times in 100 attempts.

Figure 6. The y-axis represents the closing price of each day, and the x-axis is represented in years

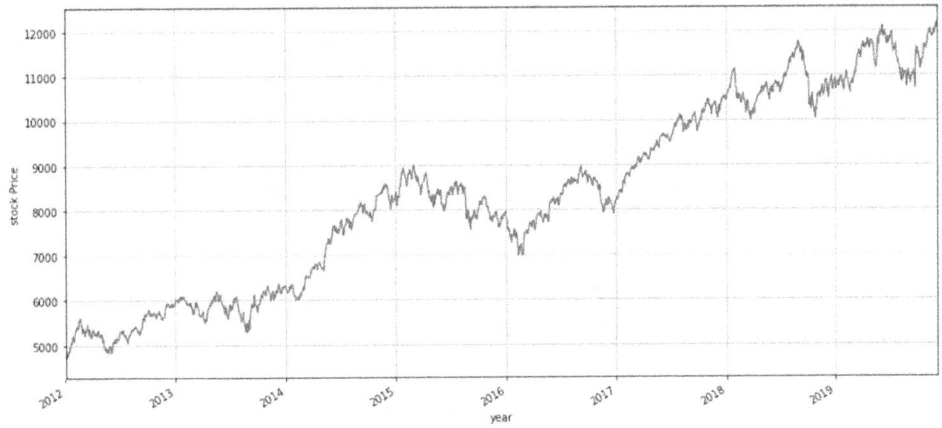

Table 3. A confusion matrix expressing correct classifications and misclassifications

	Positive (1)	Negative (0)
Positive (1)	216	88
Negative (0)	106	177

Random Forest Results

In the random forest, the authors are able to overcome the problem of over fitting by sacrificing the execution time of the algorithm. In the execution of the decision tree algorithm with a number of label three, the machine requires 217 seconds, whereas the random forest with maximum of three labels requires a much higher time – around four times more than the decision tree (i.e., 975 seconds). Still, this value is lower when one considers RBM and LSTM. The authors will explain the execution time of RBM in the coming subsection whereas they have already explained the execution of the algorithm for LSTM.

As in any other tree-based machine learning method, the authors have some standard measurement parameters to report. Regarding training and testing accuracy of the random tree, after implementation, the authors observed that accuracy improved in comparison with the decision tree. Testing accuracy of the RF model is around 72%,whereas the authors observed that the decision tree has 67% accuracy. The authors will present a detailed description of results with the help of the confusion matrix in Table 4.

Table 4. Confusion matrix of Random Forest results

	Positive (1)	Negative (0)
Positive (1)	233	90
Negative (0)	74	190

The precision of the model (also called positive predictive value) is the fraction of relevant instances among the retrieved instances. Its value for the random forest is 0.7214. The F1 score conveys the balance between the precision and the recall, and its value in this case is 0.7397.

Linear Regression Results

In linear regression, after training and testing of the model, the MSE and root mean square (RMSE) values are following:

Train Score: 0.1852 MSE (0.4303 RMSE)
Test Score: 0.08133781 MSE (0.28519784 RMSE)

In this case, the authors use LR as the classification model. Clearly, it has an output in terms of classes, just as the decision tree and the random forest have in their output. The confusion matrix in Table 5 presents the results of the classification model:

Table 5. Confusion matrix of LR

	Positive (1)	**Negative (0)**
Positive (1)	45	227
Negative (0)	278	37

Source: (IGI, 2014)

In the confusion matrix in Table 5, one can see the total number of true positive and true negative dataset classifications. It clearly shows that linear regression is poorly fit to the stock market data due to which the accuracy of the model is 13.97%. Other parameters of classification are also poor. For example, precision is 0.1654, and the F1 score is 0.1513.

Naïve Bayes Results

The naïve Bayes machine learning algorithm needs independence between its variables and uses conditional probability to predict the future stock price value. One another peculiar assumption of features is that they must be normally distributed. All restrictions on data make the model bind to perform better in almost every condition until a number of features becomes very high. Table 6 shows a confusion matrix made using the matrix score command in the Naïve Bayes.

Table 6. Confusion matrix using matrix score of Naïve Bayes

	Positive (1)	**Negative (0)**
Positive (1)	234	89
Negative (0)	72	192

In the table, one can see that testing accuracy seems to be better than some of the famous machine learning algorithms, such as decision tree and linear regression. It shows that the true power of the Naïve Bayes algorithm could be very efficient if implemented in correct ways with all constraints in tagged. The authors present other measurement parameters in Table 7.

Table 7. Results of Naïve Bayes

Parameters	Value
Accuracy	0.7257
Precision	0.7245
F1 Score	0.7440

K- Means Results

K- Mean algorithm is one of the most popular algorithms that is used as an unsupervised way of machine learning. Since xxx, due to the abundance of stock market data, unsupervised learning was ignored in stock market value prediction. The authors attempted to address this problem with the expectation that some good work will come in this field to harness the goodness of unsupervised learning algorithms.

The only problem that practitioners face in using the K-Mean clustering algorithm is selecting the optimal value of K. In order to understand this, the authors have used the elbow method, in which they see an error rate at every number of clusters. The authors have chosen two clusters, 0 and 1, for segregating their data points. The error rate at two clusters is less than 0.274, which is quite good, and the value of error continues to decrease after a further increase of clusters.

In the unsupervised learning algorithm, there is no training phase. Because of this, measuring parameters are totally different in comparison to supervised learning algorithms. The authors have two measurement criteria:

Silhouette Score: This defines the degree of separation between clusters, and its value is 0.79, which is quite good

Misclassification of Data Points: The authors observed that 72.6% of data are correctly grouped.

RBM Results

The RBM uses both unsupervised and supervised ways of learning methods. Divergence parameters, weight, and bias are fine- tuned in the unsupervised phase, leaving very little room for improvement in supervised phase. Because all parameters are already almost fine-tuned, the use of the supervised learning and the rest of tuning is complete. The authors have to select a number of layers in the RBM. They have selected three hidden layers along with one visible layer and one output layer. They have checked their output in different numbers of hidden layers, but three is provides the best result. Due to its training phase, the authors can use evaluation parameters as supervised learning algorithms.

Table 8. Confusion matrix of Restricted Boltzmann Machine

	Positive (1)	Negative (0)
Positive (1)	244	20
Negative (0)	25	298

With the help of a confusion matrix (table 6), readers can see the best performance compared to all other algorithms among supervised and unsupervised data. Table 9 shows other parameters of evaluation.

Table 9. Final result of Restricted Boltzmann Machine

Parameters	Value
Accuracy	0.9233
Precision	0.9242
F Score	0.9156

Next, the authors address the reason behind the best performance of RBM among all. is the reason for this is stealing the goodness of both supervised and unsupervised learning methods. It uses energy-based function in its back end to minimize the loss while learning and tuning parameters quickly. RBM is a deep learning algorithm. Because of this, it requires a maximum amount of time compared to all other machine learning algorithms, and it runs for 1,624 seconds.

DISCUSSION

Among the supervised learning techniques, quite expectedly, LSTM outperformed the other techniques on the basis of both accuracy and error. On the other hand, the authors found linear regression to be the most inaccurate and erroneous. Among decision tree and random forest, random forest intuitively performed better, as it took the noise into consideration. The Naïve Bayes algorithm has performed almost as well as random forest. Among the unsupervised algorithms, RBM clearly performed better than other techniques, and in fact, among all the seven machine learning techniques, RBM is the best performing.

PRACTICAL IMPLICATIONS: STOCK MARKET PREDICTION

Even after decades of work by both computer and finance experts in stock market prediction, it is still hard to predict true nature of stock prices in the future. A study in this field was conducted to understand the time effect on the stock market prices. Jegadeesh, and Titman (1993) found that stocks that have performed well during the past few months are more likely to continue their outperformance in the next month. On the other hand, a study by De Bondt and Thaler, (1985) showed that stocks that performed well in the past three to five years are more likely to underperform the market in the next three to five years, and vice versa. It is difficult to understand the true nature of stock prices in the future even after taking multiple dimensions into consideration. This article is an approach to understand the efficiency of machine learning algorithms in predicting the correct future of stock values.

1. **Finding Suitable Algorithm:** For practitioners, it is important to understand the efficiency of machine learning algorithm. This paper explains the reason the behind performance of each of the machine learning algorithms so that practitioners can find suitable algorithms for their context.
2. **Stock Market Prediction on NIFTY:** The authors have found quite a few works on stock market prediction using NIFTY data. To the best of the authors' knowledge, this study is unique in its NIFTY dataset, which evaluated the performance of different machine learning algorithms. This study will help stock predictors to understand the correct nature of stock values in the future.
3. **Breaking Myth of Supervised Learning:** The authors know that there are ample stock market data. Still, all supervised learning algorithms are not able to make correct predictions for longer periods of time. In this article, the authors have tried some unsupervised learning algorithms for understanding accuracy. K-mean and RBM are two algorithms that work differently. One is based on Euclidian Distance, and the other works on the phenomenon of an energy-based model. The authors are glad to inform the readers that the accuracy produced is after using sufficient parameters for the model development. The authors have achieved higher accuracy in comparison to many supervised machine learning algorithms, and in RBM, accuracy is better than every supervised machine learning algorithm. This approach may inspire some researchers to work in the area of unsupervised learning-based stock market predictions.
4. **Inter and Intragroup Comparison:** Unlike all other work, the authors have two kinds of comparisons: one within supervised learning and an unsupervised learning group; and later, an inter-group level between supervised and unsupervised data for stock market prediction. It will give better understanding of stock market under different scenarios.
5. **Descriptive Implementation:** The authors have described implementation details in detail with the kinds of package libraries used in it. This will give practitioners an idea to implement all of them and compare algorithms on their own to see which one performs best in their dataset.

FUTURE WORK AND CONCLUSION

Focus of our work is to compare different machine learning algorithms based on two basic classification of machine learning i.e. supervised and unsupervised learning. We know that a lot of works have been done in the area of stock market prediction. It involves many advances algorithms such as genetic algorithm, fuzzy logic, evolutionary algorithm, nature inspired algorithms. Practitioners and researchers are unaware about their competitive performance. In future, another comparison work could be done to find out best in performance between advance algorithms. Hybrid algorithms are very popular in another work in this direction could be extended to comparison of different hybrid algorithms (fusing two or more algorithm) used for stock market prediction.

Current work is based on NSE dataset, it means diversity in stock data is limited. Due to this reason results may vary for other stock exchanges. In future to get generalized and robust results on performance of algorithms. We need to train supervised algorithms on the all data collected from different stock markets.

In current work, authors have taken only few variables for prediction of future price of stocks. In future work they aim to increase variables of to achieve more accurate results. In recent literatures, it has been observed that research works on measuring impact of social media, news articles, and natural calamities on stock exchanges. The authors will try to cumulate all these factors in one study and try to understand randomness in the stock market, and predict future of stock markets across different geographical locations.

REFERENCES

Adya, M., & Collopy, F. (1998). How effective are neural networks at forecasting and rediction? A review and evaluation. *Journal of Forecasting*, *17*(5-6), 481–495. doi:10.1002/(SICI)1099-131X(1998090)17:5/6<481::AID-FOR709>3.0.CO;2-Q

An, S. C., Leung, M. T., & Daouk, H. (2003). Application of neural networks to an emerging Financial market: Forecasting and trading the Taiwan Stock Index. *Computers & Operations Research*, *30*(6), 901–923. doi:10.1016/S0305-0548(02)00037-0

Asadi, S., Hadavandi, E., Mehmanpazir, F., & Nakhostin, M. M. (2012). Hybridization of evolutionary Levenberg–Marquardt neural networks and data pre-processing for stock market prediction. *Knowledge-Based Systems*, *35*, 245–258. doi:10.1016/j.knosys.2012.05.003

Atsalakis, G. S., & Valavanis, K. P. (2009). Surveying stock market forecasting techniques –Part II: Soft computing methods. *Expert Systems with Applications*, *36*(3), 5932–5941. doi:10.1016/j.eswa.2008.07.006

Bahrammirzaee, A. (2010). A comparative survey of artificial intelligence applications in finance: Artificial neural networks, expert system and hybrid intelligent systems. *Neural Computing & Applications*, *19*(8), 1165–1195. doi:10.100700521-010-0362-z

Bartolozzi, M., Leinweber, D. B., & Thomas, A. W. (2006). Scale-free avalanche dynamics in the stock market. *Physica A*, *370*(1), 132–139. doi:10.1016/j.physa.2006.04.024

Brock, W., Lakonishok, J., & Lebaron, B. (1992). Simple technical trading rules and the stochastic properties of stock returns. *The Journal of Finance*, *47*(5), 1731–1764. doi:10.1111/j.1540-6261.1992.tb04681.x

Chao, H., Li-li, H., & Ting-ting, H. (2012). Financial time series forecasting based on wavelet kernel support vector machine. *International Conference on Natural Computation*, *8*, 79-83.

Chi-Jie, L., & Jui-Yu, W. (2011). An efficient CMAC neural network for stock index forecasting. *Expert Systems with Applications*, *38*(12), 15194–15201. doi:10.1016/j.eswa.2011.05.082

Chi-Jie, L., Tian-Shyug, L., & Chih-Chou, C. (2009). Financial time series forecasting using independent component analysis and support vector regression. *Decision Support Systems*, *47*(2), 115–125. doi:10.1016/j.dss.2009.02.001

Cowles, A. III. (1933). Can stock market forecasters forecast? *Econometrica*, *1*(3), 309–324. doi:10.2307/1907042

Dai, W., Jui-Yu, W., & Chi-Jie, L. (2012). Combining nonlinear independent component analysis and neural network for the prediction of Asian stock market indexes. *Expert Systems with Applications*, *39*(4), 4444–4452. doi:10.1016/j.eswa.2011.09.145

De Bondt, W. F., & Thaler, R. (1985). Does the stock market overreact? *The Journal of Finance*, *40*(3), 793–805. doi:10.1111/j.1540-6261.1985.tb05004.x

De Faria, E. L., Albuquerque, M. P., Gonzalez, J. L., Cavalcante, J. T. P., & Albuquerque, M. P. (2009). Predicting the Brazilian stock market through neural networks and adaptive exponential smoothing methods. *Expert Systems with Applications*, *36*(10), 12506–12509. doi:10.1016/j.eswa.2009.04.032

De Oliveira, F. A., Nobre, C. N., & Zárate, L. E. (2013). Applying artificial neural networks to prediction of stock price and improvement of the directional prediction index –Case study of PETR4, Petrobras, Brazil. *Expert Systems with Applications*, *40*(18), 7596–7606. doi:10.1016/j.eswa.2013.06.071

Di, X. (2014). *Stock trend prediction with technical indicators using SVM. Independent Work Report.* Stanford Univ.

Felsen, J. (1975). Learning pattern recognition techniques applied to stock market forecasting. *IEEE Transactions on Systems, Man, and Cybernetics*, *SMC-5*(6), 583–594. doi:10.1109/TSMC.1975.4309399

Fernandez-Rodriguez, F., Gonzalez-Martel, C., & Sosvilla-Rivero, S. (2000). On the profitability of technical trading rules based on artificial neural networks: Evidence from the Madrid stock market. *Economics Letters*, *69*(1), 89–94. doi:10.1016/S0165-1765(00)00270-6

Hsu, C. M. (2011). A hybrid procedure for stock price prediction by integrating self-organizing map and genetic programming. *Expert Systems with Applications*, *38*(11), 14026–14036. doi:10.1016/j.eswa.2011.04.210

Huang, C., Li-li, H., & Ting-ting, H. (2012). Financial time series forecasting based on wavelet kernel support vector machine. *8th International Conference on Natural Computation*. 10.1109/ICNC.2012.6234569

Huang, W., Nakamori, Y., & Shou-Yang, W. (2005). Forecasting stock market movement direction with support vector machine. *Computers & Operations Research*, *32*(10), 2513–2522. doi:10.1016/j.cor.2004.03.016

Jasic, T., & Wood, D. (2004). Theprofitability of daily stock market indices trades based on neural network predictions: Case study for the S&P 500, the DAX, the TOPIX and the FTSE in the period 1965-1999. *Applied Financial Economics*, *14*(4), 285–297. doi:10.1080/0960310042000201228

Jegadeesh, N., & Titman, S. (1993). Returns to buying winners and selling losers: Implications for stock market efficiency. *The Journal of Finance*, *48*(1), 65–91. doi:10.1111/j.1540-6261.1993.tb04702.x

Jie, W., Jian-Zhou, W., Zhe-George, Z., & Shu-Po, G. (2012). Stock index forecasting based on a hybrid model. *Omega*, *40*(6), 758–766. doi:10.1016/j.omega.2011.07.008

Kara, Y., Boyacioglu, M. A., & Baykan, Ö. K. (2011). Predicting direction of stock price index movement using artificial neural networks and support vector machines: The sample of the Istanbul Stock Exchange. *Expert Systems with Applications*, *38*(5), 5311–5319. doi:10.1016/j.eswa.2010.10.027

Khaidem, L., Saha, S., & Dey, S. R. (2016). *Predicting the direction of stock market prices using random forest.* arXiv preprint arXiv:1605.00003

Kim, K. J., & Lee, W. B. (2004). Stock market prediction using artificial neural networks with optimal feature transformation. *Neural Computing & Applications, 13*(3), 255–260. doi:10.100700521-004-0428-x

Kumar, M., & Thenmozhi, M. (2006). *Forecasting stock index movement: A comparison of support vector machines and random forest.* Indian Institute of Capital Markets 9th Capital Markets Conference Paper.

Kyoung-Jae, J. (2003). Financial time series forecasting using support vector machines. *Neurocomputing, 55*(1-2), 307–319. doi:10.1016/S0925-2312(03)00372-2

Kyoung-Jae, J., & Ingoo, H. (2000). Genetic algorithms approach to feature discretization in artificial neural networks for the prediction of stock price index. *Expert Systems with Applications, 19*(2), 125–132. doi:10.1016/S0957-4174(00)00027-0

Laffont, J. J., & Maskin, E. S. (1990). The efficient market hypothesis and insider trading on the stock market. *Journal of Political Economy, 98*(1), 70–93. doi:10.1086/261669

Leung, M. T., Daouk, H., & An-Sing, C. (2000). Forecasting stock indices: A comparison of classification and level estimation models. *International Journal of Forecasting, 16*(2), 173–190. doi:10.1016/S0169-2070(99)00048-5

Liang, T. P., Chandler, J. S., & Han, I. (1990). Integrating statistical and inductive learning methods for knowledge acquisition. *Expert Systems with Applications, 1*(4), 391–401. doi:10.1016/0957-4174(90)90048-Y

Ling-Jing, K., Chih-Chou, C., Chi-Jie, L., & Jung-Li, Y. (2013). Integration of nonlinear independent component analysis and support vector regression for stock price forecasting. *Neurocomputing, 99*, 534–542. doi:10.1016/j.neucom.2012.06.037

Mahajan Shubhrata, D., Deshmukh Kaveri, V., Thite Pranit, R., Samel Bhavana, Y., & Chate, P. J. (2016). Stock market prediction and analysis using Naïve Bayes. *Int. J. Recent Innov. Trends Comput. Commun., 4*(11), 121–124.

Malkiel, B. G. (1989). Efficient market hypothesis. In *Finance* (pp. 127–134). Palgrave Macmillan. doi:10.1007/978-1-349-20213-3_13

Margarita, S. (1991). Neural networks, genetic algorithms and stock trading. In *Artificial neural networks* (pp. 1763–1766). North Holland. doi:10.1016/B978-0-444-89178-5.50183-4

Metcalf, G. E., & Malkiel, B. G. (1994). The Wall Street Journal contests: The experts, the darts, and the efficient market hypothesis. *Applied Financial Economics, 4*(5), 371-374.

Michie, D., Spiegelhalter, D. J., & Taylor, C. C. (1994). Machine learning. *Neural and Statistical Classification, 13.*

Mingyue, Q., Cheng, L., & Yu, S. (2016). Application of the artificial neural network in predicting the direction of stock market index. *10th International Conference on Complex, Intelligent, and Software Intensive Systems.*

Nadaraya, E. A. (1964). On estimating regression. *Theory of Probability and Its Applications*, *9*(1), 141–142. doi:10.1137/1109020

Niaki, S. T. A., & Hoseinzade, S. (2013). Forecasting S&P 500 index using artificial neural networks and design of experiments. *Journal of Industrial Engineering International*, *9*(1), 1–9. doi:10.1186/2251-712X-9-1

O'Connor, N., & Madden, M. G. (2006). A neural network approach to predicting stock exchange movements using external factors. *Knowledge-Based Systems*, *19*(5), 371–378. doi:10.1016/j.knosys.2005.11.015

Paliwal, M., & Kumar, U. A. (2009). Neural networks and statistical techniques: A review of applications. *Expert Systems with Applications*, *36*(1), 2–17. doi:10.1016/j.eswa.2007.10.005

Patel, J., Shah, S., Thakkar, P., & Kotecha, K. (2015). Predicting stock market index using fusion of machine learning techniques. *Expert Systems with Applications*, *42*(4), 2162–2172. doi:10.1016/j.eswa.2014.10.031

Peša, A. R., & Brajković, A. (2016). Testing the 'black swan effect' on Croatian stock market between 2000 and 2013. *EMAJ: Emerging Markets Journal*, *6*(1), 1–16. doi:10.5195/EMAJ.2016.92

Rather, A. M., Sastry, V. N., & Agarwal, A. (2017). Stock market prediction and Portfolio selection models: A survey. *OPSEARCH*, *54*(3), 558–579. doi:10.100712597-016-0289-y

Rauch, H. E., Tung, F., & Striebel, C. T. (1965). Maximum likelihood estimates of linear dynamic systems. *AIAA Journal*, *3*(8), 1445–1450. doi:10.2514/3.3166

Salzberg, S. (1985, August). Heuristics for Inductive Learning. IJCAI, 603-609.

Sang-Hong, L., & Joon, S. L. (2011). Forecasting KOSPI based on a neural network with weighted fuzzy membership functions. *Expert Systems with Applications*, *38*(4), 4259–4263. doi:10.1016/j.eswa.2010.09.093

Schumaker, R. P., & Chen, H. (2009). Textual analysis of stock market prediction using breaking financial news: The AZFin text system. *ACM Transactions on Information Systems*, *27*(2), 12. doi:10.1145/1462198.1462204

Wolfers, J., & Zitzewitz, E. (2006). Five open questions about prediction markets (No. w12060). National Bureau of Economic Research.

Yao, J., & Tan, C. L. (2000). A case study on using neural networks to perform technical forecasting of forex. *Neurocomputing*, *34*(1-4), 78–98. doi:10.1016/S0925-2312(00)00300-3

Yao, J., Tan, C. L., & Poh, H.-L. (1999). Neural networks for technical analysis: A study on KLCI. *International Journal of Theoretical and Applied Finance*, *2*(02), 221–242. doi:10.1142/S0219024999000145

Yu, L., Wang, S., & Lai, K. L. (2009). A neural-network-based nonlinear metamodeling approach to financial time series forecasting. *Applied Soft Computing*, *9*(2), 563–574. doi:10.1016/j.asoc.2008.08.001

Zhang, G., Patuwo, B. E., & Hu, M. Y. (1998). Forecasting with artificial neural networks: The state of the art. *International Journal of Forecasting*, *14*, 35–62. doi:10.1016/S0169-2070(97)00044-7

Zhong, X., & Enke, D. (2017). Forecasting daily stock market return using dimensionality reduction. *Expert Systems with Applications*, *67*, 126–139. doi:10.1016/j.eswa.2016.09.027

Chapter 8
Wind Energy Perspectives in Myanmar

Evgenii Ignatev
Moscow Power Engineering Institute (National Research University), Russia

Galina Deryugina
Moscow Power Engineering Institute (National Research University), Russia

Htet Myat Htoon
Moscow Power Engineering Institute (National Research University), Russia

Mikhail Tyagunov
Moscow Power Engineering Institute (National Research University), Russia

ABSTRACT

One of Myanmar's problems is energy shortage. Partially, energy shortage can possibly be decreased by the construction of sizeable grid-connected offshore wind farms. Eight prospective construction sites were selected and wind turbine models chosen. This chapter describes the method for determining the optimal composition of the wind farms complex, consisting of several offshore wind farms located at a considerable distance from each other in areas with significant wind regime asynchrony. To illustrate this method, the optimal composition with an installed capacity of 47.6 MW and located off Myanmar's west coast is defined.

INTRODUCTION

Nowadays, one of the actual problems of Myanmar is an energy shortage (approximately 10% of all energy consumption). Dramatic energy consumption growth due to electrification and economic growth predicted for the next 15 years. Energy shortage can be reduced by intensifying the renewable energy development which main directions are construction of large grid-connected wind farms.

DOI: 10.4018/978-1-7998-3645-2.ch008

Earlier assessments of the wind energy potential of Myanmar territory showed that its most part (approximately 92%) belongs to areas with weak energy potential (Deryugina, Tyagunov & Zai Ir Lin, 2017; Zai Ir Mint, 2013), where the long-term average wind speed does not exceed 2 m/s at 10 m height. Thus, in such areas, the construction of large grid-connected wind farms is not feasible and perspective. The most perspective for onshore wind energy development are coastal areas in the southern and western part of Myanmar. To confirm the previously obtained conclusions, the authors assessed the Myanmar territory's wind energy potential from a long-term series of observations at onshore weather stations using the «Reliable prognosis» website's data. The chapter presents the results of this assessment.

Analysis of Myanmar's power system structure showed that it is most feasible to connect offshore wind farms to existing substations on the west coast of Myanmar. The wind energy potential of Myanmar's west coast water area assessed for the first time. The main difficulties in this assessment were the choice of an information source for these calculations since there is no actual observation data in Myanmar's water area.

The objective of the chapter is the research of the offshore wind farms usage efficiency operating in the united energy system of Myanmar using multiobjective optimization (Vasant, Zelinka, & Weber, 2019, 2020; Thomas, Karagoz, Ahamed, & Vasant, 2019).

Motivation

One of the main problems arising in the design of large grid-connected wind farms with an installed capacity of more than 30 MW is the mismatch between energy generation and consumption regimes. Because of this feature, it needs a limitation of wind farms share operating within the energy system, taking into account its value of maneuverable power reserve. Typically, such a limitation level is approximately 20% of the total installed capacity of power stations operating within the energy system.

However, there are examples of regions in which the share of wind farms in the power balance significantly exceeds this value. It is possible due to the usage of the effect of wind inflow asynchrony in different parts of a specific area on which wind farms placed and interconnected by energy, information, and infrastructure links. In this chapter considered the efficiency of such wind energy complex in Myanmar's west coast water area.

To consider this, the following problems set and solved:

- Assessment of Myanmar territory's wind energy resources;
- Assessment of wind energy resources of Myanmar's coastal water area;
- Selection of perspective sites for offshore wind farms placement according to the specific criteria;
- Wind turbine model selection for the Conditions of Myanmar's west coast water area;
- Determination of the optimal composition of wind farms complex using integral criterion method of multiobjective optimization (Vasant, Zelinka, & Weber, 2019, 2020; Thomas, Karagoz, Ahamed, & Vasant, 2019), taking into account both the efficiency of wind inflow asynchrony usage and energy efficiency.

BACKGROUND

Wind power is one of the energy sectors that grow steadily for the last years. At the beginning of 2019 world total wind farms installed capacity is 591 GW, growing by 9% during the previous year. Total newly added capacity in 2018 is 51.3 GW decreasing by 4% comparing the previous year (Global Wind Energy Council [GWEC], 2019). Although over the last four years (2015-2018), there has been a general decrease in wind farms installed capacity growth rate (from 14,7% in 2015 to 8,6% in 2018), certain wind energy industry sectors, primarily offshore wind energy, demonstrate steady growth (see Figure 1).

Figure 1. Newly added capacity (GWEC, 2019)
Source: GWEC, 2019

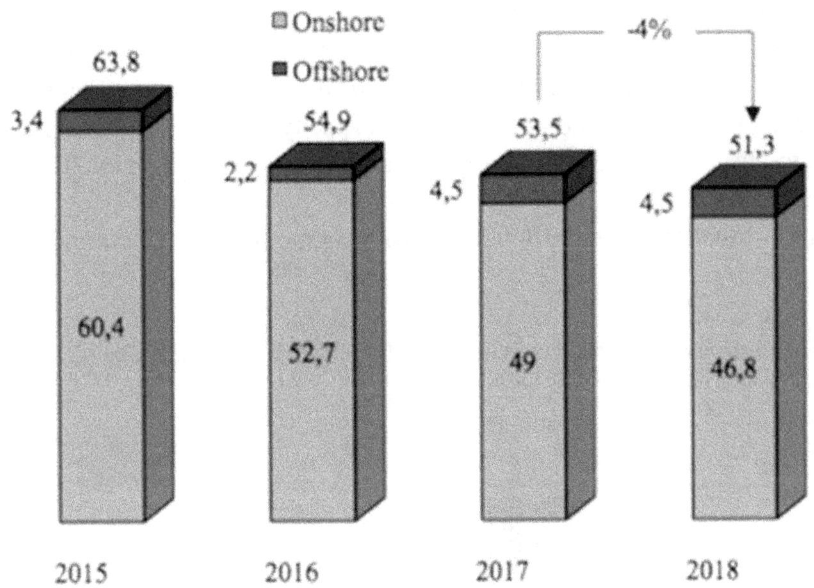

One of the fastest-growing wind energy markets is South-East Asian: Indonesia, Philippines, Thailand, Vietnam, Malaysia, Laos, Myanmar, and other smaller markets. Its rising demand for energy driven by economic growth, growth of population, and increased urbanization. At the same time, due to its technological process features, wind energy development considered in this region as one of the priority areas to combat the energy shortage.

Many of Southeast Asian governments have adopted wind energy development targets. For example, the Government of Vietnam in the National Power Development Plan VII (PDP7, released in 2011 and revised in 2016), set a target 10.7% of electricity production to be supplied by renewables by 2030, while the installed capacity of wind farms will reach 800 MW by 2020 and 6 GW by 2030 (GWEC, 2019).

The Government of the Philippines, as part of the National Renewable Energy Plan (NREP), adopted in 2011, set the goal of tripling renewable energy capacity from its 2010 level to 15,304 MW by 2030, of which 2,345 MW allocated for wind (GWEC, 2019).

The Kingdom of Thailand is a pioneer in the wind energy development in the Southeast Asia region – its development program began with the construction of a 150 kW wind farm on the Phuket Island in the early 1980s. Currently, wind farms installed capacity in the country is 648 MW. By the beginning of 2018, electricity produced from renewable sources is 15% of total energy consumption. Despite the adoption of the government's Alternative Energy Development Plan (AEDP), which sets out the target of increasing the installed capacity of wind farms to 3,002 MW, the commissioning of wind farms not produced since 2017.

The Republic of Indonesia has significant onshore economic wind energy potential, which is approximately 9.29 GW across its more than 17,500 islands. Its development carried out within the framework of the Ministry of Energy and Mineral Resources (MEMR) target to increase the share of electricity produced from renewable sources in the energy balance to 23% by 2025 and up to 31% by 2050. In 2018, the first large Sidrap wind farm with a capacity of 75 MW commissioned in the country. Supposed wind farms installed capacity by 2025 is 1.8 GW.

By the beginning of 2019, the total wind farms installed capacity in the Southeast Asian region reached 1.5 GW. According to the analytical organization Global Wind Energy Council (GWEC, 2019), one should expect an increase of wind farms installed capacity in this region to 4 GW by 2023. These indicators relate to onshore wind energy. Currently, governments of this region do not have plans for offshore wind energy development.

New wind power installations in the Southeast Asian region statistics shown in Figure 2.

Figure 2. New installations in South-East Asia (– Myanmar, Laos, Malaysia) (GWEC, 2019)*
Source: GWEC, 2019

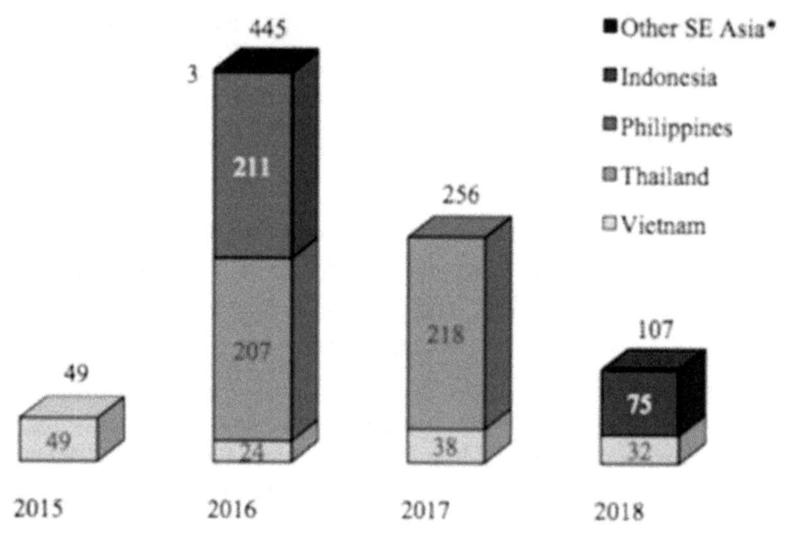

MAIN FOCUS OF THE CHAPTER

Wind Resources Information Sources

Selection of the perspective areas for wind energy usage facilitated greatly by the presence of wind resources distribution maps, the reliability of which substantially depends on the selected information source. Table 1 below shows the characteristics of specialized databases information for Myanmar wind energy resources assessment can obtain through. Different databases provide information for different periods obtained through measurements on onshore weather stations sites (as «Russia's Weather» (n.d.) or «Reliable prognosis» (n.d.)) or satellite observation of the Earth's atmosphere with its reanalysis by different computer models (as «NASA» (n.d.) database). Databases differ by type of information represented in them: such as databases in which averaged data presented for the entire observation period (average wind speed, repeatability of wind speed and wind direction, etc.) determined for different time intervals (long-term, annual, monthly or daily average) or databases containing data with different observation intervals t (from 30 minutes to several hours) (see Table 1).

Table 1. Specialized databases characteristics containing wind data

Database (developer)	Data source	Observation period	Accessibility	Data
«Russia's Weather» (IKI RAN and RFBR)	Onshore weather stations	1998 – present	http://meteo.infospace.ru (paid since 2015)	8 measures a day at 10 m height
«Reliable prognosis»	Onshore weather stations	2005 – present	https://rp5.ru (free)	8 measures per day at 10 m height
NASA	Onshore weather stations and satellite observation data assimilation	1981 – present	https://power.larc.nasa.gov (free)	Averaged series with one value a day at 10 m height
WindProMeteo Data Export ver. 5 (EMD International A/S):	Onshore weather stations and satellite observation data assimilation		http://emd.dk (paid program)	
- EmdERA		1950 – present		4 measures a day at 10, 25, 50, 75, 100, 150, 200 m height
- ERA5		1950 – present		24 measures a day at 10, 100 m height
- MERRA2		1980 – present		24 measures a day at 50 m height
- BCW		1975 – present		4 measures a day at 10, 25, 50, 75, 100, 150, 200 m height

It is possible to note the existing disadvantages of various specialized databases, such as the measurement interval in some data series at the same weather station site may vary through the observation period. Another possible disadvantage is that not all databases have free access via the Internet. Data series from some databases are available only from libraries of specialized paid programs. For example, access to the databases EmdERA, ERA5, MERRA2, BCW, and others carried out through the WindProMeteo Data Export program (n.d.).

At the early stages of wind farms designing, without wind measuring complexes data, it recommended to carry out wind energy calculations using several databases, compare them, and choose the most reliable information source for calculations. The most reliable source of information considered data of onshore weather stations. The authors of Deryugina, Tyagunov & Zai Ir Lin (2017) verified the «Reliable Prognosis» data for the territory of Myanmar, which used as the primary information source for the wind energy resources assessment. The «Reliable Prognosis» website provides long-term series of observations (from 1995 to present) of wind speed (at 10 m height), wind directions, and other climatic data at 49 sites of weather stations in Myanmar.

Wind data for Myanmar water areas is possible to obtain only from the WindProMeteo data library. The analysis of wind data from four databases (EmdERA, ERA5, MERRA2, BCW) of WindProMeteo Data Export climate data library found out that the densest coverage of Myanmar's west coastal water areas presented in the BCW database. Thus, the BCW database used as the primary data source for wind energy resources assessment: wind speed, wind direction, and other climate data series for 10 m height and observation period from 2005 to 2018 obtained from it for 43 points.

Basic Wind Energy Characteristics Calculation

Wind energy characteristics were calculated for the considered points of onshore weather stations and west coastal water areas of Myanmar for considered observation period: average wind speed - $\overline{V_0}$, m/s; wind speed coefficient of variation – C_V; wind power density – WPD_i, W/m^2 and wind energy density – WED_i, kWh, where i is a number of considered point.

Long-term average wind speed $\overline{V_0}$ calculation performed using wind speed series V_i by the formula (1):

$$\overline{V_0} = \frac{\sum_{i=1}^{n} V_i}{n}, \tag{1}$$

where n is a number of all the wind speed values V_i, m/s in the series.

Wind speed coefficient of variation calculated by the formula (2):

$$C_V = \sigma / \overline{V_0}, \tag{2}$$

where σ is a standard deviation calculated by the formula (3):

$$\sigma = \sqrt{\frac{\sum_{i=1}^{n} \left(V_i - \overline{V_0}\right)^2}{n-1}}, \tag{3}$$

The annual wind energy density *WED*, kWh is the energy flowing by the year through 1 m^2 of the cross-section is determined by the formula (4):

$$WED_i = \sum_{i=1}^{n} WPD_i \cdot \Delta t_i = \frac{1}{2} \cdot \rho \cdot \sum_{i=1}^{n} V_i^3 \cdot \Delta t_i, \quad (4)$$

where ρ, kg/m^3 is an air density; V_i, m/s is wind speed at hub height; Δt_i is a measurement interval. Air density depends on temperature and height above the sea level deviation concerning the standard conditions (t=15°C and P=760 mmHg). In the case of Myanmar average annual temperature is approximately 30°C, which leads to energy generation shrinkage, but ρ correction was not made here.

Long-term average wind power density calculated by the formula (5):

$$WPD_i = \frac{WED}{T}, \quad (5)$$

where $T = 8760$ is a number of hours in one year.

Wind power calculations carried out at all stages of wind farms designing are performed for a given hub height of considered wind turbines. Such wind turbines considered as grid-connected have a hub height of 80 m and above. Therefore, it is necessary to have maps of the wind resources distributed throughout the country's territory and its coastal water areas at heights commensurate with the hub heights of modern grid-connected wind turbines. Wind gradient modeling for onshore weather stations and coastal water areas conditions performed using a Hellman exponent:

$$V_0^h = V_0^{10} \cdot \left(\frac{h}{10}\right)^{m_0}, \quad (6)$$

where h is a height above the earth or sea surface; V_0^h, V_0^{10} is average wind speed at considered sites at h and 10 m height; m_0 is Hellman exponent of 0.2 for onshore conditions according to IEC 61400-1:2005 and 0.14 for offshore conditions according to IEC 61400-3:2009.

Assessment of Myanmar's Wind Energy Resources at 10 and 100 m Height

Basing on calculation results for Myanmar 49 onshore weather stations sites maps of long-term average wind speed and long-term distribution of wind power density at 10 and 100 m height were drawn (see Figures 3 and 4).

Table 2 below shows the variation diapason of long-term average wind energy characteristics of Myanmar territory at 10 m and 100 m height: long-term average wind speed - $\overline{V_0}$, m/s; wind power density – *WPD*, W/m^2; wind speed coefficient of variation – C_V.

Figure 3. Long-term average wind speed distribution map of Myanmar territory at the height of: a) 10 m and b) 100 m

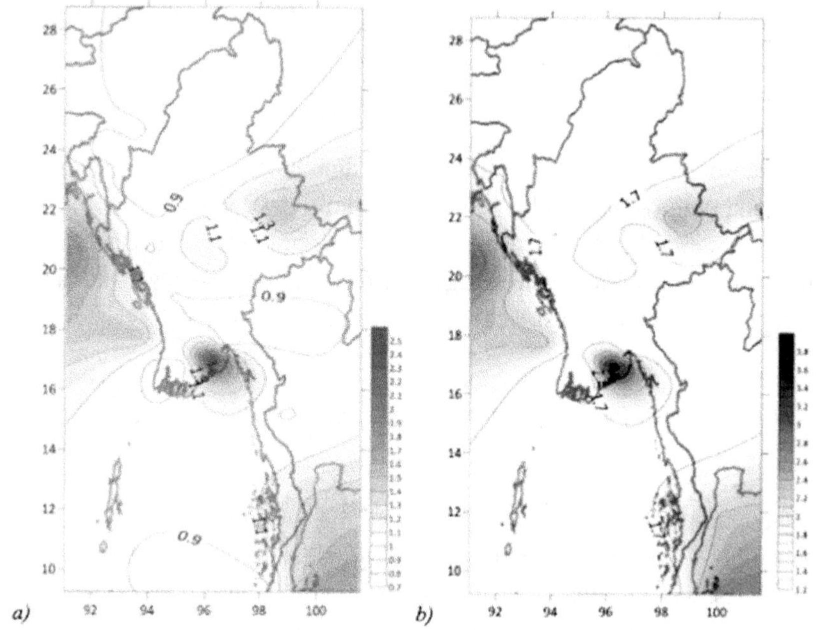

Figure 4. Long-term wind power density distribution map of Myanmar territory at the height of: a) 10 m and b) 100 m

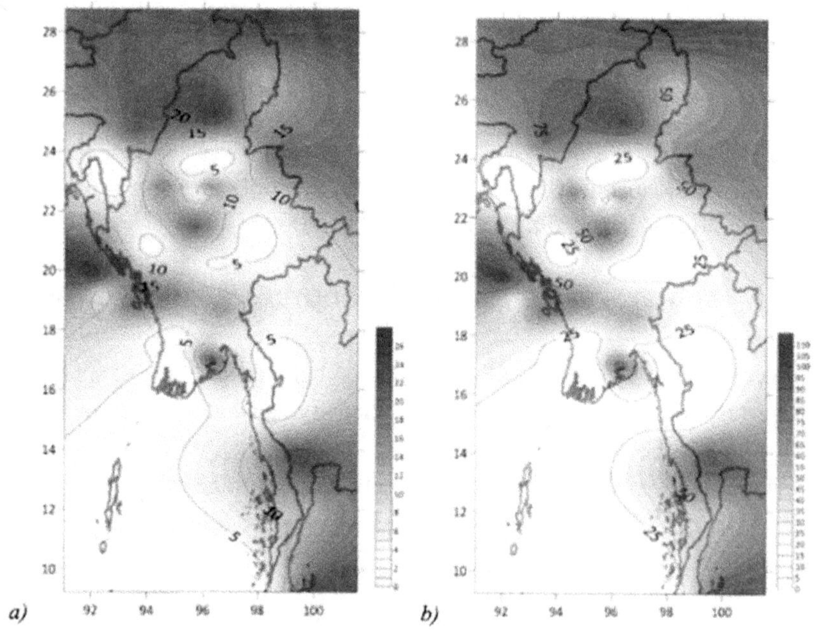

Table 2. Variation diapason of long-term average wind characteristics of Myanmar territory by «Reliable prognosis» database information

Height	$\overline{V_0}$, m/s		C_V		WPD, W/m^2	
	min	max	min	max	min	max
10 m	0.76	2.73	0.21	1.19	0.37	27.48
100 m	1.20	4.33	0.21	1.19	1.48	109.40

Assessment of Myanmar Territory's Theoretical Wind Energy Potential

The theoretical wind energy potential E^T of Myanmar territory calculation performed at 10 m and 100 m height for the time period T of one year for conditions of the typical year by formula (7):

$$E^T = \sum_{i=1}^{n} E_i^T = \sum_{i=1}^{n} \frac{\overline{WPD_i} \cdot F_i \cdot T}{20}, \quad (7)$$

where n is the number of zones Myanmar territory divided by; $\overline{WPD_i}$, kW/m^2 is an average wind power density of the i-th zone; F_i, km^2 is an area of the i-th zone (total area of all zones is the total area of Myanmar – 676,684.8 km^2).

All the territory of Myanmar divided into 49 zones with the same physical and geographical features (geographical location, landscape conditions, the presence and type of water surfaces, etc.). Corresponding zones areas as a percentage of the entire territory of Myanmar also determined. The calculation results summarized for 14 regions of the country and presented in Table 3 below. The average wind power density values for each region presented as the mean value of all values for zones related to each region.

As a result of the calculations, it found that the theoretical wind potential of the entire territory of Myanmar at 10 m height is 5,607.49 TWh per year; at 100 m height is 12,948.71 TWh per year.

Calculations to determine theoretical wind energy potential at 10 m height performed earlier in Deryugina, Tyagunov & Zai Ir Lin (2017) and Zai Ir Mint (2013). In Zai Ir Mint (2013), the wind energy potential calculations of Myanmar territory performed by the Meteonorm program, in which there are no actual observations at onshore weather stations and data series generates using computer models. The developers of the Meteonorm program (Meteotest AG, 2020) do not recommend using this data to estimate wind energy resources. In Deryugina, Tyagunov & Zai Ir Lin (2017), the theoretical wind energy potential of Myanmar territory was calculated using the same data from the «Reliable prognosis» database as in this chapter, but when dividing the territory of Myanmar into zones, as in Zai Ir Mint (2013), the country's territory was divided into zones on a grid with a step of $1^0 \times 1^0$, as is customary in the NASA methodology (NASA POWER, n.d.). That is, when dividing the territory into zones in Deryugina, Tyagunov & Zai Ir Lin (2017) and Zai Ir Mint (2013), the complex and diverse area topography within the zone was not taken into account (with a step of $1^0 \times 1^0$). The theoretical wind energy potential of Myanmar territory defined in this chapter at 10 m height is 5,607.49 TWh per year, which is 1.1 times bigger than 5230 TWh per year, determined in Deryugina, Tyagunov & Zai Ir Lin (2017) and more than 3 times bigger than defined in Zai Ir Mint (2013), which is 1820 TWh per year.

Table 3. Theoretical wind energy potential of Myanmar territory

N°	Region	$\overline{WPD_i}$, kW/m²		F_i, km²		E_i^T , TWh per year	
		10 m	100 m	km²	%	10 m	100 m
1	Bago	9.97	39.70	39,402.30	5.82	172.11	685.18
2	Ayeyarwady	2.61	10.38	35,140.00	5.19	40.12	159.72
3	Kayin	1.96	7.78	30,382.80	4.49	26.02	103.58
4	Kachin	15.11	60.17	89,041.20	13.16	589.41	2346.47
5	Kayah	9.27	36.91	11,731.50	1.73	47.64	189.66
6	Magway	3.07	12.22	44,820.60	6.62	60.26	239.91
7	Mandalay	9.87	39.30	37,945.60	5.61	164.08	653.22
8	Mon	2.81	11.19	12,296.60	1.82	15.14	60.28
9	Rakhine	16.24	64.67	36,778.00	5.44	2,616.58	1041.68
10	Sagaing	14.71	58.55	93,704.50	13.85	603.63	2403.08
11	Yangon	24.25	96.56	10,276.70	1.52	109.17	434.61
12	Chin	7.39	29.44	36,018.80	5.32	116.66	464.45
13	Shan	10.16	40.46	155,801.30	23.02	693.54	2761.02
14	Tanintharyi	18.60	74.05	433,44.90	6.41	353.13	1405.84
Total		146.04	581.38	676,684.80	100.00	5,607.49	12,948.71

According to the results of the calculations, it is possible to note that most of Myanmar territory (approximately 93%) belongs to areas with weak wind energy potential, where at 100 m height the long-term average wind speed is less than 3.5 m/s (Figure 3b), and the wind power density is less than 100 W/m² (Figure 4b). In these areas, wind energy usage is not prospective for grid-connected wind turbines. The highest wind activity observed on the west and south coast of Myanmar. In connection with the previous, it decided to analyze the feasibility of offshore wind farms construction in Myanmar's western coast water area.

Assessment of Myanmar's West Coast Water Area Theoretical Wind Energy Potential

Assessment of the available area for offshore wind farms placement in the water area of the Andaman sea near the west coast of Myanmar carried out for waters with a distance to shore from 1.5 km to 50 km and a sea depth of less than 50 m. The length of the considered coastline is 2224 km. To assess the depths of the designated water area, they reviewed using the Google Earth Pro. The necessary data on the seabed topography obtained, and a bathymetric map drawn, combined with a topographic map of the Andaman Sea (Figure 5). On the map, there is a contour line with a sea depth of 50 meters. At sea depths above this value, the offshore wind farms construction currently is not feasible for technical and economic reasons. The available assessed area for offshore wind farms placement in the Andaman Sea is 80,400.9 km² (highlighted in Figure 5). As a result of the analysis, no environmental and other restrictions found for offshore wind farms placement in the considered area; for example, there is no economic, military, or environmental activity.

Figure 5. Area suitable for offshore wind farms placement in the Andaman Sea

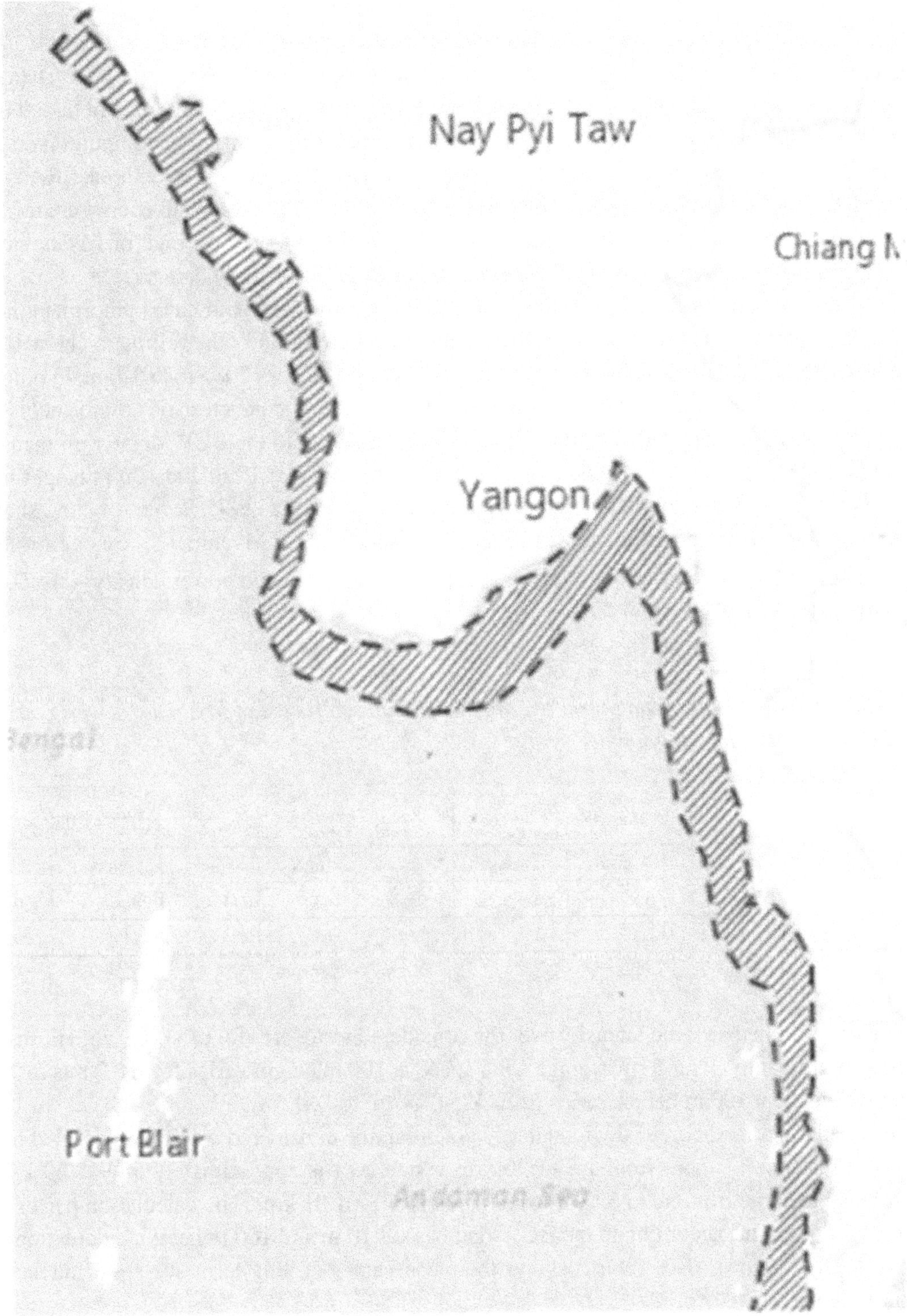

Myanmar's West Coast Water Area Wind Energy Resources

Currently, there are no observations of climatic characteristics at the Myanmar's west coast water area. It is possible to obtain climatic data for the Myanmar water area from the WindProMeteo Data Export program. Four databases (EmdERA, ERA5, MERRA2, BCW) of WindPro with data obtained through the satellite observation of the Earth's atmosphere with its reanalysis by different computer models considered. EmdERA database has data series for 19 points located at considered water area, ERA5 – for 21 points, MERRA2 – for 21 points, BCW – for 43 points. Comparison of the wind energy characteristics did not found out their significant discrepancy. Because of the densest coverage of Myanmar's west coastal water areas presented in the BCW database, it chose as the primary data source.

In the BCW database for the considered water area for the offshore wind farms placement there is a long-term data series (wind speed at 10 m height, wind directions, and other climatic characteristics) with the measurement interval of 3 hours (measurement period is from 2005 to 2018) at 43 points. Using these data calculations of basic long-term average wind energy characteristics by formulas (1)-(6) was made. Based on the calculation results (at 43 points according to the BCW database), maps of the long-term average wind speed and wind power density distribution at 10 m and 100 m height for considered offshore wind farms placement area were drawn (see Figures 6-7). Table 4 below shows the variation diapason of long-term average wind energy characteristics of Myanmar's west coast water area at 10 m and 100 m height: long-term average wind speed - $\overline{V_0}$, m/s; wind power density – *WPD*, W/m²; wind speed coefficient of variation – C_V.

Table 4. Variation diapason of long-term average wind characteristics of Myanmar's west coast water area by BCW database information

Heigh	$\overline{V_0}$, m/s			C_V		WPD, W/m²		
	min	max	average	min	max	min	max	average
10 m	4.8	6.8	5.6	0.37	0.63	161.91	283.06	230.00
100 m	6.7	8.6	7.8	0.37	0.63	425.86	744.53	600.00

It was found out that the wind intensity over the considered water area for offshore wind farms placement varies significantly: long-term average wind speed at 100 m height varies from 6.7 m/s to 9.4 m/s; wind power density at 100 m height varies from 425.0 W/m² to 744,5 W/m².

When calculating the theoretical wind energy potential, the considered water area divided into two parts: with a distance to shore from 1.5 to 30 km, which area is approximately 58,999.89 km²; and with a distance to shore from 30 to 50 km, which area is 21,401.01 km². The calculation performed by formula (7) for 43 points according to the BCW database at 10 m and 100 m height, calculation results showed in Table 5 below. Also, Table 5 shows the mean values of long-term average wind speed and wind power density.

Figure 6. Long-term average wind speed distribution map of Myanmar's west coast water area at the height of: a) 10 m and b) 100 m

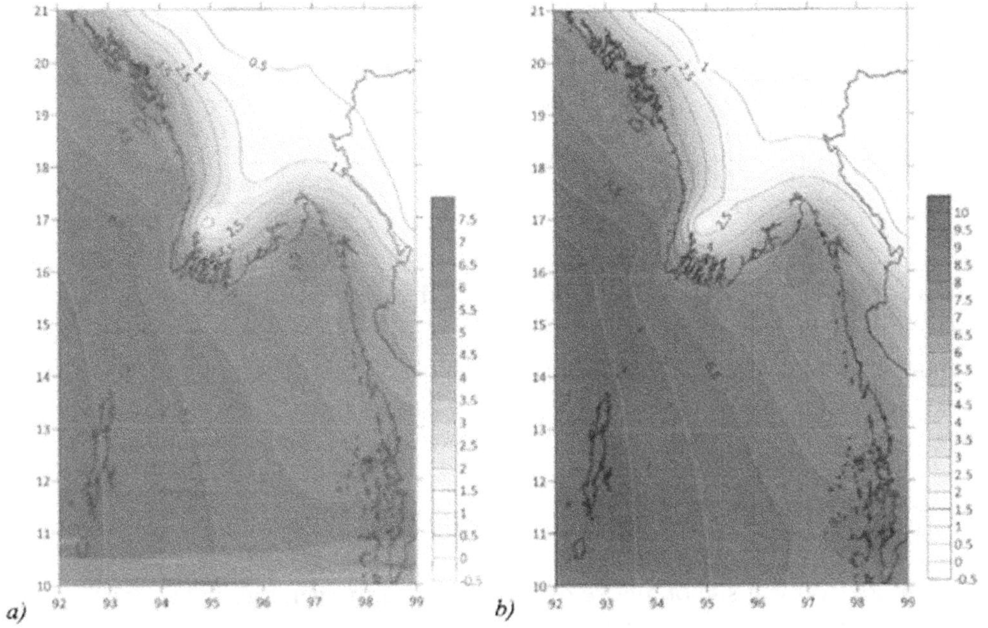

Figure 7. Long-term wind power density distribution map of Myanmar's west coast water area at the height of: a) 10 m and b) 100 m

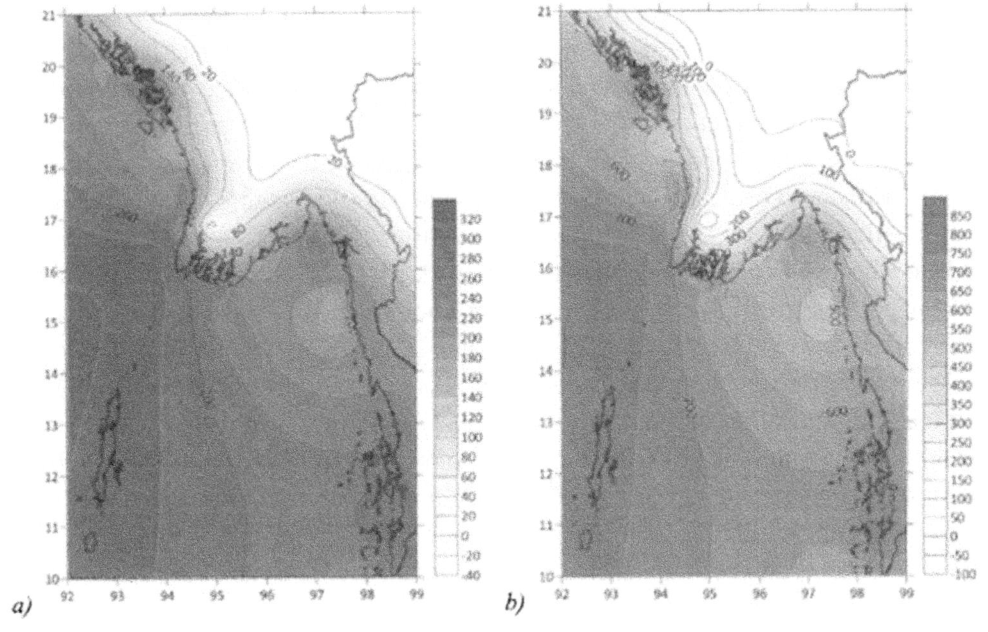

Table 5. Theoretical wind energy potential of considered Myanmar's west coast water area for offshore wind farms placement

Distance to shore diapason, km	$\overline{V_0}$, m/s		$\overline{WPD_i}$, kW/m^2		F_i, km^2		E_i^T, TWh per year	
	10 m	100 m	10 m	100 m	km^2	%	10 m	100 m
1.5-30	4,8	6,7	160	420	58,999.89	73.38	5,918.13	15,566.28
30-50	6,6	8,2	260	720	21,401.01	26.61	2,146.68	5,646.35
1.5-50	5,6	7,8	230	600	80,400.9	100	8,064.82	21,212.63

An analysis of the calculation results found out that the wind energy potential at 10 m and 100 m height of water area at a distance to shore from 1.5 to 30.0 km (this area is 8% of Myanmar territory area) is comparable to the wind energy potential of whole Myanmar territory. Thus, it is possible to conclude that the construction of offshore wind farms in the Myanmar's west coast water area is feasible and perspective.

Selection of the Offshore Wind Farms Placement Sites

The selection of a placement site of a large offshore wind farm operating in the united power system should be carried out in accordance with the following criteria: wind speed intensity; sea depth; accessibility of transport infrastructure; technical interconnection possibility of the wind farm to the power system; distance from the offshore wind farm site to the electrical substation; lack of environmental and other restrictions (for example, economic, military or environmental activities). Offshore wind farms site selection made, taking into account the absence of environmental and other restrictions. It also found out that almost the entire considered water area for offshore wind farms placement is perspective in terms of wind intensity because long-term average wind speed over the considered area at 100 m height is above 6.7 m/s. A further selection of offshore wind farms sites on a considered water area carried out according to the criterion of the technical interconnection possibility of the wind farm to the power system.

Analysis of existing substations with a voltage of 230 kV on the west coast of Myanmar allows selecting eight prospective offshore wind farms placement sites (see Figure 8) taking into account the sea depth of less than 30 m, i.e. the possibility of wind turbines installation on fixed foundations, and technical interconnection to 230 kV electrical substation. Analysis of the offshore wind farm's technical interconnection possibility to the nearest electrical substations found out that installed capacity diapason of considered offshore wind farms is from 30 MW (sites 1, 3, 5, 6) to 80 MW (site 2). Characteristics of electrical substations located near the considered offshore wind farms sites showed in Table 6 below: average distance from the offshore wind farm to shore – r_1, km; diapason of sea depth variation at the offshore wind farm site, m; average distance from the offshore wind farm to electrical substation – r_2, km; electrical substation transformer power, MVA; load level of electrical substation, %; free electrical substation power, MBA; the approximate area available for the placement of wind farm – F, km^2.

Figure 8 shows the location of the selected offshore wind farms sites and the nearest electrical substations to them.

Table 6. Characteristics of electrical substations located near the considered offshore wind farms sites

N° of site	Nearest electrical substation	r_1, km	Sea depth diapason, m	r_2, km	Transformer power, MVA		Load level, %	Free power, MVA	F_{WF}, km²
					Units by power	Total			
1	Ponnagyun	17.5	8-17.1	49.3	1×100	100	70	30	17.56
2	Kyauk Phyu	12.8	25.0-32.0	23.1	1×100, 1×50	150	43	85	17.56
3	Toungup	32.8	20.0-30.0	77.8	1×100	100	64	36	17.56
4	Ah Thoke	5.0-9.0	6.0-43.0	74.0	1×100	100	55	45	17.56
5	Pathein	11.0-16.0	17.0-36.0	116.0	2×50	100	65	35	17.56
6	Pathein	20.0-30.0	13.0-27.0	165.0	2×50	100	65	35	17.56
7	Maw La Myaing	16.0-20.0	9.0-12.0	52.0	3×50	150	60	60	17.56
8	Maw La Myaing	26.0-30.0	22.0	130.0	3×50	150	60	60	17.56

Thus, taking into account criteria of the minimum distance from the wind farm site to the electrical substation and the maximum free power of the electrical substation, site N°2, accessible for construction of an offshore wind farm with an installed capacity of 50 MW, chosen as the most possible with its technical interconnection to the Kyauk Phyu electrical substation. The sea depth at the selected site varies insignificantly (25-32 m). The site has dimensions of 10.50 x 1.67 km; thus, its area is 17.56 km^2.

Figure 8. Location of the selected offshore wind farms sites and the nearest electrical substations to them

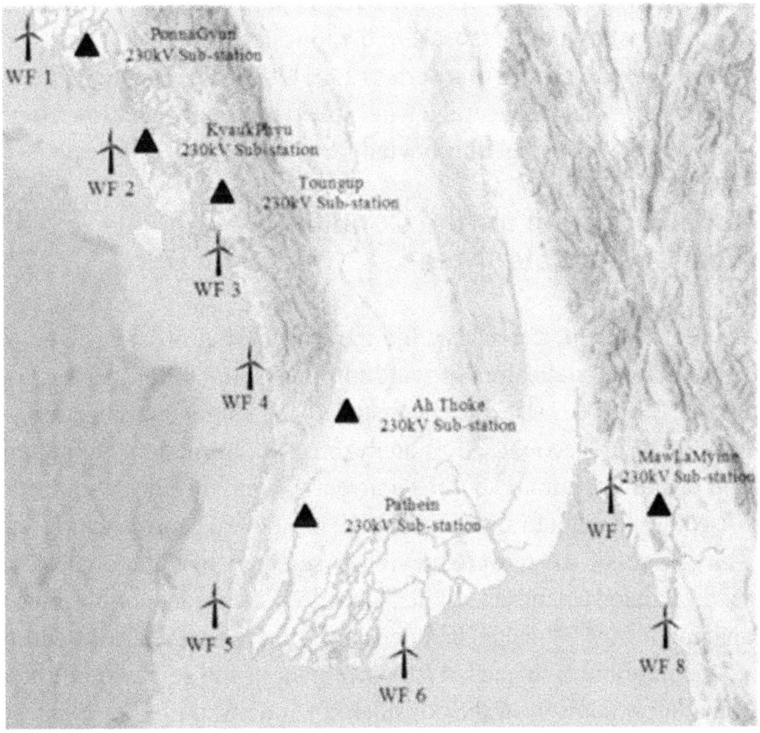

At the selected site N°2 basing on the long-term climatic data series (2005-2018) obtained for the point A (19,75 N 92,75 E) basic long-term wind energy characteristics at 100 m height were calculated: long-term average wind speed – 7.5 m/s; wind power density – 540 W/m^2; the most repeatable wind speed – 6.5 m/s. Figure 9 shows: a) long-term average wind speed variation; b) annual variation of the long-term average monthly average wind speed. Analysis of annual wind speed variation at the selected offshore wind farm site shows that prevailing wind speed observed during the summertime (June to August). Analysis of long-term wind speed variation does not found out such tendency; moreover, during the long-term period, the average annual wind speed does not change significantly. Their deviations from the long-term average do not exceed 6%.

Figure 9. Average wind speed variation at offshore wind farm site N°2 for point A (19.75 N 92.75 E) located in the water area of Andaman sea: a) annual average b) long-term average monthly average

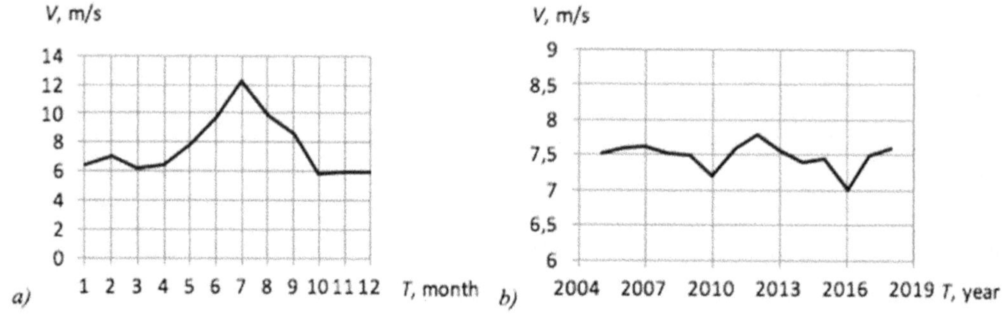

Further calculations for wind turbines and wind farms power generation assessment made for the typical year, which characteristics are the closest to the long-term average year conditions.

The typical year was selected from 13-year series (2005-2018) for years with the completeness of no less than 95% by the criterion of annual average wind speed minimal deviation from long-term average wind speed. As the typical year for the offshore wind farm site N°2 2018 selected.

Wind Turbine Model Selection for the Conditions of Myanmar's West Coast Water Area

Wind turbine model selection for placement within the considering offshore wind farm carried out by comparing the energy efficiency indicators of wind turbines suitable for safety class and climate performance, according to IEC 61400-3. During the installed capacity value selection, transportation, and installation issues not taken into account. After analyzing the international experience in the offshore wind farms design and operation, it found out that currently, wind turbines with installed capacity from 3.0 MW to 8,0 MW used. Used wind turbines must have offshore performance. 11 variants of wind turbines models with a hub height of 100 m were previously selected. Installation possibility on considered offshore wind farm site validated for these variants. S, km^2 – area needed for placement of the wind farm with an installed capacity of 50 MW calculates depending on the number of wind turbines within the wind farm – n_{WT}, and the minimal area needed for placement of the sole wind turbine – S_1, km^2, which is, in turn, depends on a rotor diameter of the considered wind turbine – D_{rotor}, m.

Calculation results and wind turbines characteristics are in Table 7 below.

Checking the considered wind turbines models variants by its climate performance and safety class in accordance with IEC 61400-3 showed that all of them can be placed on the considered offshore wind farm site.

Table 7. Wind turbines characteristics and area needed for its placement at the wind farm site

N° of site	Country manufacturer	Model	Nameplate capacity, kW	Safety class	D_{rotor}, m	n_{Wt}	S_1, km² (k = 8)	S, km²
1	Germany	Aerodyn scd 8/168	8000	Ia	168	6	1.56	10.84
2	Denmark	Bonus b82/2300	2300	IIs	82	21	0.37	9.47
3	Germany	Multibrid M5000	5000	Ib	112	10	0.70	8.03
4	Denmark	Vestas V112 Offshore	3000	IIb	112	16	0.70	12.85
5	Denmark	Vestas V164-8.0	8000	IIb	164	6	1.49	10.33
6	Denmark	Siemens SWT-3.6-130	3600	IIb	130	13	0.94	15.14
7	Denmark	Siemens SWT-3.6-120	3600	IIb	120	13	0.80	12.90
8	Germany	Senvion 3.4M140	3400	IIIb	140	14	1.09	17.56
9	Germany	Senvion 3.2M122	3200	IIIb	122	15	0.82	15.24
10	Germany	Aerodyn scd 5/139	5000	IIb	139	10	1.07	12.37
11	Germany	Aerodyn scd 5/130	5000	IIb	130	10	0.94	10.82

The final wind turbine model selection for the proposed site is carried out as a result of a feasibility study. Preliminarily reduce the number of considered variants according to energy efficiency criterion, which is a maximum of the wind turbine's capacity factor k_{WT}^{CF} and the energy density factor of the wind turbine rotor area E_{WT}^{ED}.

Comparison by energy efficiency criteria made only for wind turbines variants with the same hub height of 100 m. Annual energy generation of the sole wind turbine calculation made using wind data of typical year (2018) for point A (19,75 N 92,75 E) and wind turbine power curve. Table 8 shows the main energy efficiency indicators calculation results of considered wind turbines variants with a hub height of 100 m.

The most energy effective wind turbine model for usage within considering wind farm is a Senvion 3.4M140 with a nameplate capacity of 3.4 MW. Figure 10 shows its power curve.

Annual energy generation of the wind farm consisting of 14 wind turbines Senvion 3.4M140 is 201,180 MWh per year.

The Wind Turbines Placement Layout at the Wind Farm Site and Adjustment of Wind Farm Energy Generation

During the wind turbines placement layout at the wind farm site, various factors take into account: wind directions repeatability, seabed topography, point of wind farm grid interconnection, etc. In accordance with this, it decided to place all the wind turbines at a distance of $8 \cdot D_{rotor}$ in one row at a sea depth of 20 m. Figure 11 shows 14 wind turbines Senvion 3.4M140 placement layout at wind farm site N°2.

Table 8. Energy efficiency indicators for considered wind turbines variants

N° of site	Country manu-facturer	Model	Nameplate capacity, kW	Cut-in wind speed, m/s	Rated wind speed, m/s	Cut-out wind speed, m/s	E_{WT}^{annual}, GWh	k_{WT}^{CF}
1	Germany	Aerodyn scd 8/168	8000	3.5	13.0	25.0	23.65	0.34
2	Denmark	Bonus b82/2300	2300	3.5	15.0	25.0	6.53	0.32
3	Germany	Multibrid M5000	5000	3.5	12.0	25.0	13.89	0.32
4	Denmark	Vestas V112 Offshore	3000	3.0	12.0	25.0	10.69	0.41
5	Denmark	Vestas V164-8.0	8000	4.0	13.0	25.0	24.72	0.35
6	Denmark	Siemens SWT-3.6-130	3600	4.0	12.0	25.0	13.40	0.42
7	Denmark	Siemens SWT-3.6-120	3600	3.0	12.5	25.0	12.27	0.39
8	Germany	Senvion 3.4M140	3400	3.0	11.0	22.0	14.37	0.48
9	Germany	Senvion 3.2M122	3200	3.0	12.0	20.0	11.43	0.41
10	Germany	Aerodyn scd 5/139	5000	3.5	10.5	25.0	17.11	0.39
11	Germany	Aerodyn scd 5/130	5000	3.5	12.0	25.0	15.40	0.35

Figure 10. Senvion 3.4M140 wind turbine's power curve

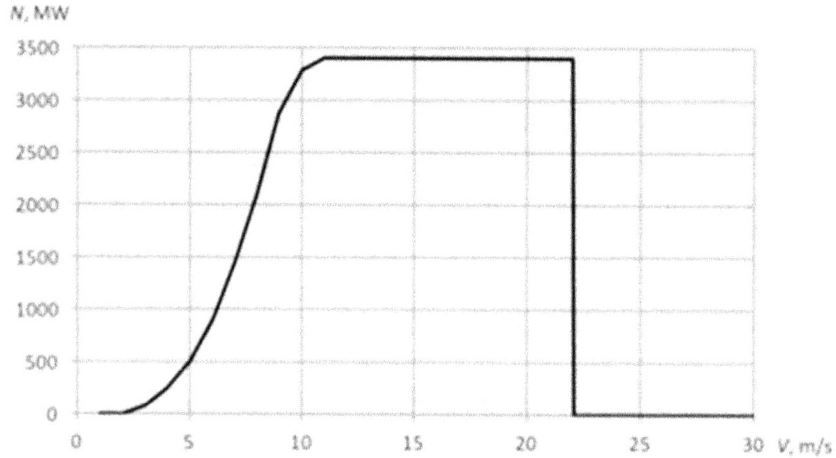

Factors, Characterizing Wind Farm Operation Regime

Factors, characterizing wind farm operation regime is possible to divide into two categories:

- gross factors characterizing the amount of wind farm electricity generation for the considering period;
- factors, which characterizing the heterogeneity of wind farm electricity generation over time.

Factors belonged to the second category determined based on an array of values of power generated by the wind farms. These values determined with a given discreteness within the considering period, which is usually taken to be one year. It should also note that it is advisable to consider such factors in relative form relative to the wind farm installed capacity value.

These factors include:

- C_{VN} – coefficient of variation of the power generated by the wind farm over the considering period – this factor shows the intensity of fluctuations of a random variable around the average value (the value of the power generated by the wind farm, determined with a given discreteness within the considering period, as a function, the argument of which is a random variable – wind speed) (Obrezkov, 1981);
- N_p/N_{IC} – the relative value of power generated by wind farm with high wind probability – this factor shows the wind farm installed capacity share, which is guaranteed to be generated with a probability above a certain accepted value (generally taken in the diapason of 85-95%).

In this chapter, the following factors are considered as factors characterizing the wind farm operation regime: wind farm electric energy generation for the considering period – E_{WF}, GWh per year; the relative value of the power generated by the wind farm with a wind probability of 90% for the considering period – $N_{P=90\%}/N_{IC}$.

The value of the power generated by a wind farm with a wind probability of 90% determined by the empirical probability curve, by analogy with the calculation of the power output probability, when regulating river flow by a hydroelectric power station reservoir (Obrezkov, 1981).

To draw an empirical probability curve, the array of values of power generated by the wind farm for the considering period is arranging in decreasing order. Moreover, each of these values assigned a certain probability of excess (in percent) by formula (8):

$$p(m) = \frac{m}{n+1} \cdot 100\%, \tag{8}$$

where m is the number of the element of the array of values of power generated by wind farm for the considering period, in decreasing order; n is the total number of values in the array.

After linear interpolation of the ordered array of values of power generated by the wind farm over the considering period, it is possible to determine its value corresponding to the probability value of 90% (see Figure 12).

Subsequently, a specific value of the power generated by the wind farm for the considering period with a probability of 90% presented in percent as wind farm installed capacity share.

Wind Inflow Asynchrony Influence on Wind Farm Operation Regime

Wind inflow asynchrony can understand as the difference in wind conditions, characterized by the wind speed and direction, in different parts of a specific area on which one or several wind farms placed and interconnected by energy, information, and infrastructure links.

Figure 11. Wind turbines placement layout at the wind farm site

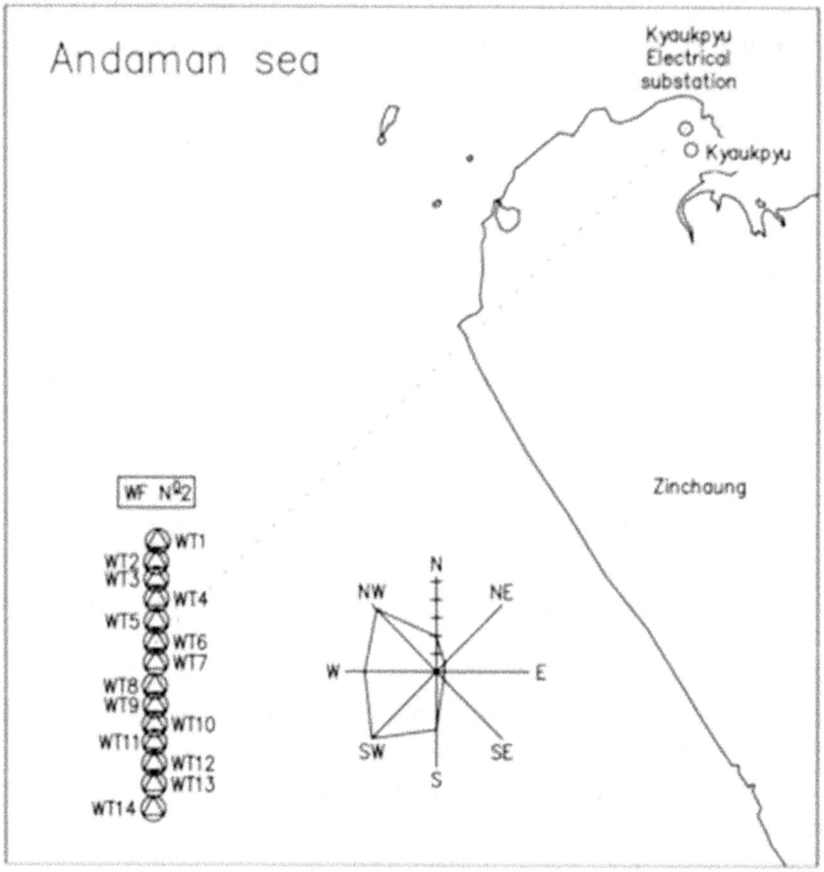

Figure 12. Drawing of an empirical probability curve and determination of the value of power generated by wind farms for the considering period with a probability of 90%

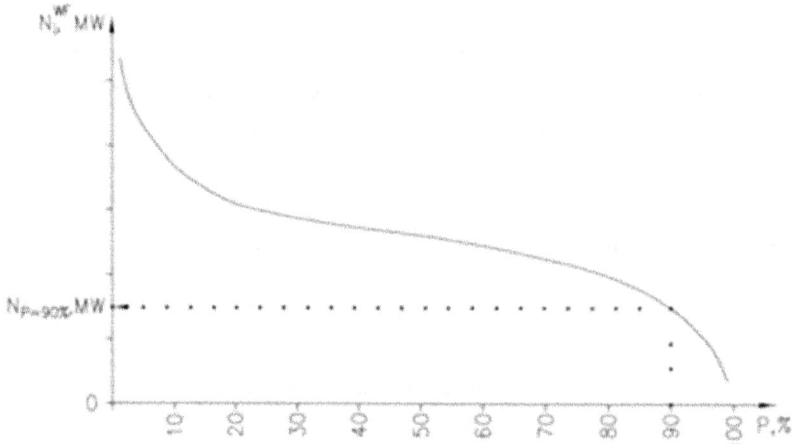

Wind inflow asynchrony has a positive effect on the management of power systems with wind farms and electricity its quality indicators. Wind inflow asynchrony usage effect can be achieved in two cases:

- an increase in the wind inflow asynchrony due to an increase in the number of wind turbines installed at the wind farm, and, as a result, an increase in the area of the wind farm with different wind regime in its different parts (see Figure 13);
- distribution of wind farms, which are interconnected by energy, information, and infrastructure links, over a wide geographical area with different wind regimes in its different parts.

In the first case, an increased number of wind turbines reduces the impact of the turbulent peak (Figure 14) as gusts do not hit all the wind turbines at the same time.

In the second case, a wider geographical distribution reduces the impact of the diurnal and synoptic peak significantly (Figure 14), as changing weather patterns do not affect all wind turbines at the same time (Söder, & Ackermann, 2005). The exact effect of reducing the uneven generation of electric power over time from the geographical distribution of interconnected wind farms depends on local wind conditions and the size of the geographical area on which the wind farms are distributed.

Basing on international experience in the design and operation of wind farms operating in the united power system, the distribution of wind farms over a wide geographical area with different wind regimes in its different parts has the greatest positive effect on operational dispatch planning.

Figure 13. The effect of increasing the number of wind turbines on wind farm. The timeline on the chart graduated in seconds. Based on simulations by Rosas (2003)
Source: Rosas, 2003

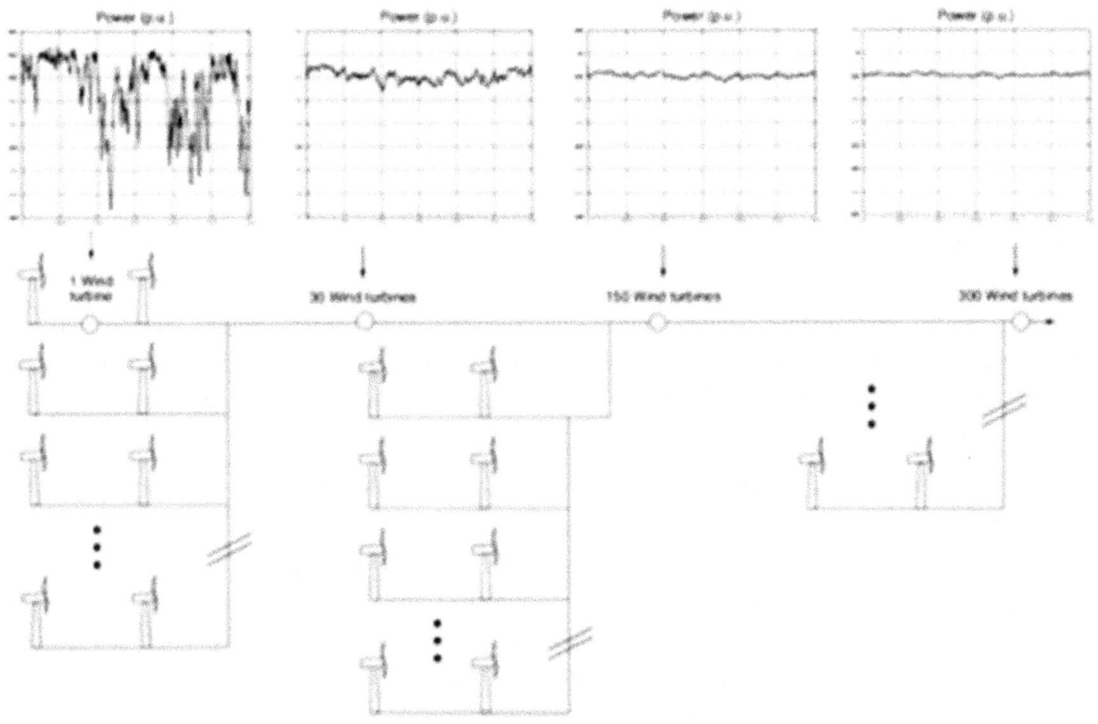

Figure 14. Wind spectrum based on work by van Hoven (Burton, Sharpe, Jenkins, & Bossanyi, 2001)
Source: Burton, Sharpe, Jenkins, & Bossanyi, 2001

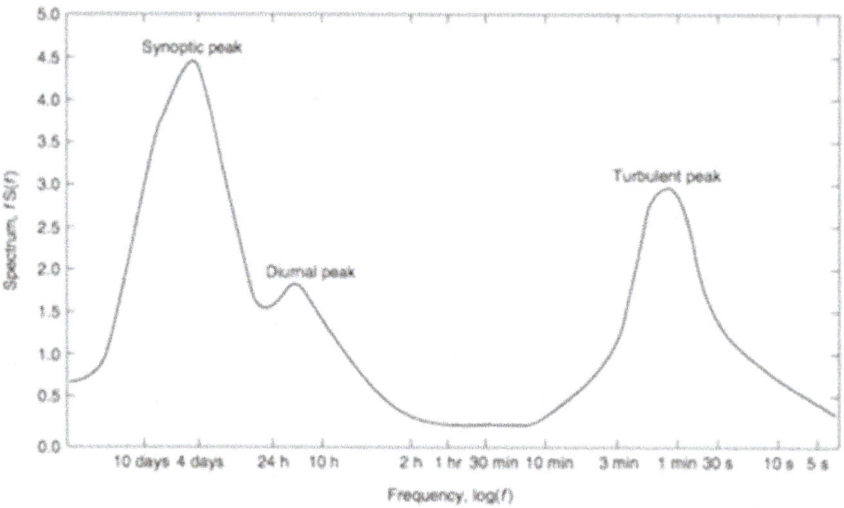

Description of the Wind Farm Complex Structure

In this chapter, a complex consisting of several offshore wind farms located at a considerable distance from each other in areas with wind regime significant asynchrony considered.

The offshore wind farms in the complex operations within a united power system according to the wind inflow regime. At the same time, due to the wind inflow asynchrony to the wind farms of the complex, power generation of wind farms with low wind activity at the sites complemented by the power generation of wind farms with high wind activity at the sites.

Wind farms complex interconnected with the united power system through the electrical substations closest to the offshore wind farms, which have a voltage class of 230 kV and a load level of less than 70%.

The wind farm's complex structural diagram shown in Figure 15.

The Methodology for Determining the Wind Farms Complex Optimal Composition

As noted above, by wind inflow asynchrony usage within the wind farms complex, which has the described structure, it is possible to increase the values of the power generated with high probability values and reduce the uneven generation of electric power over time. The efficiency of wind inflow asynchrony usage within the wind farms complex, in turn, depends on N_{WF_i}, MW – installed capacity of offshore wind farms and $n_{WT}^{WF_i}$ - number of wind turbines within wind farms operating on asynchronous wind flows (where i is an index of the considered wind farm).

At the same time, the regime with the most efficient wind inflow asynchrony usage characterized by a simultaneous decrease in the uneven generation of power by the wind farms complex and a decrease in the electric energy generation. Based on this, an integral criterion developed to determine the optimal composition of the wind farms complex, taking into account both the efficiency of wind inflow asynchrony usage within the complex and the energy efficiency.

Figure 15. The structural diagram of the wind farms complex functioning (where OWF is offshore wind farm) (Ignatev, 2019)
Source: Ignatev, 2019

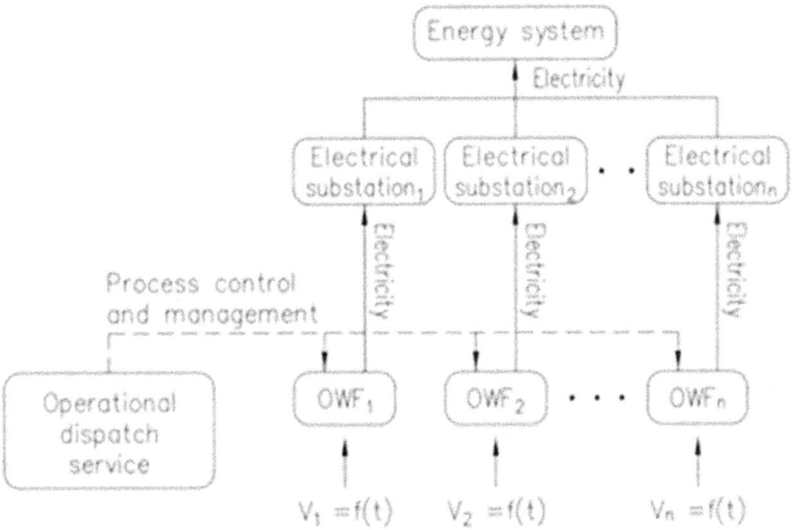

This integral criterion based on the ranking of factors characterizing the regime of wind farms joined operation within the complex with different variants for its composition. A variant of the complex composition at which the lowest average rank by the considered criteria takes as optimal (9):

$$R_{opt} = min\left\{R_{COMP_i}^{A@}\right\}, \tag{9}$$

$$R_{COMP_i}^{average} = \frac{1}{n} \cdot \sum_{j=1}^{n} R_j, \tag{10}$$

where R_{opt} – the optimal rank of the variant of the complex composition by the considered criterion which has the lowest rank;

$R_{COMP_i}^{average}$ – the average rank value of the *i*-th variant of the complex composition according to *n* criteria;
R_j – the rank value of the *i*-th variant of the complex composition according to *j*-th criterion;
n – the number of considered criteria.

The first criterion is a maximum of wind farms complex electric energy generation for the considering period. The rank R_1 increases with decreasing the wind farm's complex electric energy generation with *i*-variants of its composition. The second criterion is a maximum of the relative value of the power generated by the wind farms complex with a wind probability of 90% for the considering period. The rank R_2 increases with decreasing the relative value of the power generated by the wind farms complex with a wind probability of 90% with *i*-variants of its composition.When the $R_{COMP_i}^{average}$ values coincide, we assume that:

1. The most significant criterion is the second for which a benefit $k=\{1{,}5\}$ is specified;
2. The first criterion is considered less significant, so its value remains $k=\{1\}$.

The variant of the complex composition is determined by varying the number of wind turbines installed on each of the wind farms of the complex. Moreover, the total number of wind turbines within the complex is constant. This means that in each of the considered variants for the complex composition, the WF_i consists of a variable number of wind turbines:

$$n_{WT}^{WF_i} = \sum n_{WT_k}^{WF_i} = var \tag{11}$$

$$K = \sum n_{WT_{ik}}^{WF_i} = const \tag{12}$$

where $n_{WT}^{WF_i}$ is a number of wind turbines within the i-th wind farm; i is an index of the considered wind farm; and k is a wind turbine index within the i-th wind farm.

The calculation is performed iteratively by sequentially changing the number of wind turbines at the WF_i at $k=\{0;K\}$, where K is a total number of wind turbines installed within the wind farms complex, by condition (12) (Ignatev, 2019).

Assessment of the Offshore Wind Farms Operating Regime Characteristics Located in the Water Area of the Andaman Sea Near the West Coast of Myanmar

According to data obtained from the BCW database for eight selected prospective offshore wind farms sites (basic long-term wind energy characteristics at the midpoints on the sites – see Table 9), in accordance with the methodology above, were calculated the electric energy generation values and the relative value of the power generated with a wind probability of 90% of wind farm with installed capacity of 47,6 MW and consist of 14 wind turbines Senvion 3.4M140 (see Table 10 and Figure 16).

Table 9. Basic wind energy characteristics for eight selected prospective offshore wind farms sites

N° of site	Coordinates		Long-term wind speed characteristics at 100 m height	
	latitude	longitude	$\overline{V_0}$, m/s	WPD, W/m²
1	20.50	92.25	7.7	550.29
2	19.75	92.75	7.5	556.28
3	18.25	94.00	8.2	525.73
4	17.50	94.25	8.5	744.53
5	16.00	94.00	8.7	729.68
6	15.50	95.50	8.0	637.75
7	16.50	97.25	8.3	425.86
8	15.50	97.75	7.9	463.12

Table 10. Offshore wind farms operating regime characteristics with specified parameters and located on the eight selected prospective sites

N° of site	1	2	3	4	5	6	7	8
$N_{P=90\%}/N_{IC}$, %	4,49	4,43	5,71	8,05	8,57	6,12	4,83	4,76
1 WT power generation, million kWh	14,71	14,37	14,49	15,59	16,68	15,60	15,34	13,95
Capacity factor	0,48	0,48	0,49	0,52	0,56	0,52	0,52	0,47
14 WT power generation, million kWh	205,94	201,18	202,86	218,26	233,52	218,40	214,76	195,30

Figure 16. Empirical probability curves of the relative values of the power generated by the wind farms located on the selected prospective sites

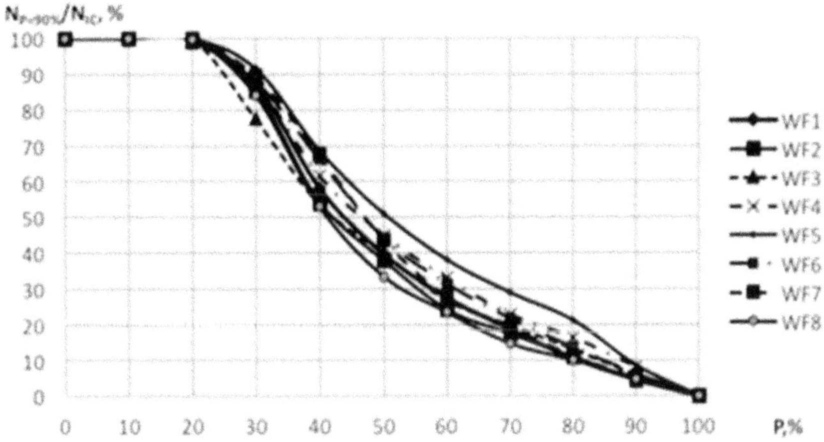

Assessment of the Offshore Wind Farms Complex Operating Regime Characteristics Located in the Water Area of the Andaman Sea Near the West Coast of Myanmar

In order to reduce the electric power generation heterogeneity over time compared to the variant of all the wind turbines concentration at the sole wind farm site was made an assessment of the creating possibility of a wind farms complex with the structure described above, within which it is possible to use the wind inflow asynchrony.

The previously selected offshore wind farm site N°2, located near the Kyauk Phyu electrical substation, is considered as the leading of the offshore wind farms complex. Optimization considered when placing wind turbines on sites N°2, 5, and 6. In this case, the maximum number of wind turbines at sites 5 and 6 is 7 following the load value of the electrical substations closest to them. Table 11 shows the characteristics of the operating regime of the three most preferred composition variants of the wind farms complex and the variant of all wind turbines concentration at the sole wind farm site. Results of the average rank value determining by integral criterion depending on the variant of wind farms complex composition see in Table 12.

Table 11. Characteristics of the operating regime of the wind farms complex composition variants

N° of complex composition variant	1		2			3		Just leading WF
N° of site	N°2	N°5	N°2	N°5	N°6	N°2	N°6	N°2
Number of WT on WF site	7	7	4	6	4	7	7	14
Power generation by WF, million kWh	100.59	116.76	57.48	100.08	62.40	100.59	109.20	201.18
Power generation by WFs compex, million kWh	217.35		219.96			209.79		201.18
$N_{P=90\%}/N_{IC}$, %	6.50		6.37			5.27		4.43

Table 12. Results of the average rank value determining by the integral criterion

N° of complex composition variant	1	2	3	Just leading WF
R_1	2	1	3	4
R_2	1	2	3	4
R_2 with benefit	1.5	3	4.5	6
$R_{COMP_i}^{average}$	1.75	2	3.75	5

Taking the benefit into account, the variant of the wind farms complex composition has the lowest rank, at which the highest value of the power generated by the wind farms complex with a wind probability of 90% for the considering period observed. In this variant, 7 wind turbines installed at the site N°2 and 7 wind turbines installed at the site N°5. This composition variant of the offshore wind farms complex allows to increase electricity generation by 8% and to increase the relative value of the power generated by the wind farms complex with a wind probability of 90% for the considering period from 4.43% to 6.50% compared with the variant of concentration of all the wind turbines at the site of the sole leading offshore wind farm.

LIMITATIONS AND ASSUMPTIONS

When resolving the issues discussed in the chapter, several assumptions and simplifications made.

One of the main problems is the lack of actual observational data at points of Myanmar's west coast water area. To select the primary information source considered four WindProMeteo Data Export program's integrated libraries of meteorological data obtained through the assimilation of atmospheric data obtained from different sources by hybrid multi-level models. The number of points of Myanmar's west coast water area suitable for consideration from these libraries: EmdERA – 19 points, ERA5 – 21 points, MERRA2 – 21 points, BCW – 43 points. The accuracy of these data can only be estimated by comparing it with actual observational data, which the authors did not have. Therefore, the assumption made to consider the data reliable if their deviations in various information sources are insignificant.

For wind gradient modeling used power function (see formula (6)), which Hellman exponent is 0.14 for offshore conditions according to IEC 61400-3:2009. The modeling error for hourly average wind speed values, in this case, can be up to 100%, and for long-term average values (monthly average, annual average), as a rule, do not exceed the data rounding accuracy (10-20%). That is when modeling the wind gradient, the effect of atmospheric stratification not taken into account, and this assumption can significantly affect the determination of wind turbines energy efficiency indicators, both overestimation, and understatement. This, in turn, significantly distorts such indicator as the value of power generated by wind farms with high wind probability.

SOLUTIONS AND RECOMMENDATIONS

The calculations for this chapter have confirmed the feasibility of offshore wind farms complex construction in Myanmar's west coast water area. The methodology developed by the authors allows expanding the number of criteria used for determining the wind farms complex optimal composition.

In this chapter, data used for calculations obtained from open sources, which do not have the accurate completeness, especially by the grid map resolution and distribution by the height. Thus, to obtain more reliable estimates of the offshore wind farms effectiveness, it is necessary to have data obtained through actual observations at the different height at points of Myanmar's west coast water area have high grid map resolution.

FUTURE RESEARCH DIRECTIONS

Most of the electricity in Myanmar (approximately 75%) generated by hydropower plants. In this regard, it is advisable to consider the possibility of the offshore wind farms complex combining into an operational dispatch complex with hydropower stations. Offshore wind farms complex unloads hydraulic units of hydropower stations during the period of high wind activity, which allows saving water in their reservoirs, creating additional, so-called doubling capacity. Using the storage capacity of hydropower stations reservoirs, varying their parameters, it is possible also to ensure its electric power generation during the period of low wind activity under the power system load pattern. The hydropower stations compensation of wind farms electric power generation shortage requires specific storage capacity of their reservoirs, which is possible to determine during future researches.

Since Myanmar has abundant solar resources, it is also advisable to consider the feasibility of large energy complexes construction combining offshore wind farms and PV solar farms. The problem for determining such energy complexes optimal composition is also possible to solve using the integral criterion based on the ranking of factors characterizing the regime of wind farms and PV solar farms joined operation within the complex (Vasant, Zelinka, & Weber, 2019, 2020; Thomas, Karagoz, Ahamed, & Vasant, 2019).

CONCLUSION

1. As a result of the calculations, the theoretical wind energy potential of Myanmar was clarified, which is 5,607.49 TWh per year at 10 m height and 12,948.71 TWh per year at 100 m height. Previous conclusions about the unfeasibility of onshore wind farms construction for the most part (93%) of Myanmar territory confirmed. The most prospective areas for grid-connected wind farms construction found, the total area of which is about 7% of the entire country's territory.
2. The available area for offshore wind farms placement in the water area of the Andaman sea near the west coast of Myanmar assessed as 80.4 thousand km^2 and its theoretical wind energy potential at 10 m and 100 m height calculated as 8,064.82 TWh per year and 21,212.63 TWh per year, correspondingly. The calculation of the wind energy characteristics at 41 points of the water area showed that offshore wind farm construction is feasible and perspective in any part of it.
3. Selected eight perspective placement sites of offshore wind farms with capacities of from 30 to 80 MW using the criteria of technical interconnection possibility to the power system. Among the selected sites, the placement site of the leading offshore wind farm chosen.
4. The most effective wind turbine model for usage within considering wind farms (Senvion 3.4M140, Germany, nameplate capacity – 3.4 MW, capacity factor, k_{WT}^{CF} =0,42-0,48 depending on wind conditions of offshore wind farms sites) was selected. Annual energy generation of 14 wind turbines of selected model (with a cumulative installed capacity of 47,6 MW) calculated. It determined that the leading offshore wind farm's energy generation value is the lowest.
5. Using the methodology developed by the authors, determined the optimal composition of the complex of two offshore wind farms located at the sites with the asynchronous wind flow. Combining wind farms into an operational dispatch complex allows to increase the value of power generation of these stations by 8% and increase the wind farms complex capacity factor.

REFERENCES

Burton, T., Sharpe, D., Jenkins, N., & Bossanyi, E. (2001). *Wind Energy Handbook*. John Wiley & Sons, Ltd. doi:10.1002/0470846062

Deryugina, G. V., & Tyagunov, M. G., & Lin, Z. I. (2017). Data verification for usage in regional Geographic Information System «Renewable energy sources». *Energetik*, *5*, 36–41.

EMD International A/S. (n.d.). *WindProMeteo Data Export* (Version 5) [Software]. Available from https://www.emd.dk/windpro/downloads/

emd.dk. (2013, June 6). *Blended Coastal Winds*. Retrieved from http://www2.emd.dk/admin/helpWiki/index.php/Blended_Coastal_Winds

emd.dk. (2013, October 18). *MERRA Data*. Retrieved from http://www2.emd.dk/admin/helpWiki/index.php/MERRA_Data

Global Wind Energy Council. (2019, April). *Global wind report 2018*. Retrieved from http://jbr.org/articles.html

Ignatev, E. V. (2019). *Energy efficiency increasing assessment of the wind energy complex in the united power system* (Unpublished candidate of engineering sciences dissertation). Moscow Power Engineering Institute (National Research University), Moscow, Russia.

Meteotest, A. G. (2020). *Meteonorm* (Version 7.3.4) [Software]. Available from https://meteonorm.com/en/download

NASA POWER. (n.d.). *Project Data Sets*. Retrieved from https://power.larc.nasa.gov

Obrezkov, V. I. (Ed.). (1981). *Hydropower: textbook for students of higher educational institutions*. Energoizdat.

Reliable prognosis. (n.d.). *Weather for 243 countries of the world*. Retrieved from http://www.rp5.ru

Rosas, P. (2003). *Dynamic Influences of Wind Power on the Power System* (Unpublished doctoral dissertation). Ørsted Institute and Technical University of Denmark, Copenhagen, Denmark.

Russia's Weather. (n.d.). *Weather archive*. Retrieved from http://www.meteo.infospace.ru

Söder, L., & Ackermann, T. (2005). Wind Power in Power Systems: An Introduction. In T. Ackermann (Ed.), *Wind Power in Power Systems* (pp. 25–51). John Wiley & Sons, Ltd.

Thomas, J. J., Karagoz, P., Ahamed, B. B., & Vasant, P. (Eds.). (2019). *Deep Learning Techniques and Optimization Strategies in Big Data Analytics*. IGI Global.

P. Vasant, I. Zelinka, & G. W. Weber (Eds.). (2019). Intelligent Computing & Optimization. In *Proceedings of the 1st International Conference on Intelligent Computing and Optimization 2018 (ICO 2018)*. New York, NY: Springer International Publishing.

P. Vasant, I. Zelinka, & G. W. Weber (Eds.). (2020). Intelligent Computing & Optimization. In *Proceedings of the 2nd International Conference on Intelligent Computing and Optimization 2019 (ICO 2019)*. New York, NY: Springer International Publishing. 10.1007/978-3-030-33585-4

Zai Ir Mint. (2013). *Research of energy characteristics of regional wind energy in the Republic of the Union of Myanmar* (Unpublished candidate of engineering sciences dissertation). Moscow Power Engineering Institute (National Research University), Moscow, Russia.

ADDITIONAL READING

Elistratov, V. V. (2013). *Renewable energy: tutorial*. Nauka.

Ignatev, E. V., Deryugina, G. V., & Tyagunov, M. G. (2019). Study of the possibility of compensatory regulation of the offshore wind farms. *New in Russian electrical energy, 4*, 49-58.

Ignatev, E. V., Deryugina, G. V., & Tyagunov, M. G. (2019). Selection of the information sources for wind energy calculations in the coastal water area of the Murmansk region. *New in Russian electrical energy, 3*, 42-53.

Ignatev, E. V., & Tyagunov, M. G. (2018, December). *Definition of the optimal configuration of the offshore wind farms complex.* Paper presented at the Youth Scientific Conference Alternative energy in the regions of Russia "AER-2018", Astrakhan, Russia.

Ignatev, E. V., & Tyagunov, M. G. (2018). Definition of the optimal configuration of the offshore wind farms complex. *Magazine C.O.K. (Plumbing, Heating and Air Conditioning), VII*(199), 78-84.

Milligan, M. (2018). Sources of grid reliability services. *The Electricity Journal, 31*(9), 1–7. doi:10.1016/j.tej.2018.10.002

Vas'kov, A. G., Deryugina, G. V., Malinin, N. K., & Pugachev, R. V. (2016). *Wind energy: tutorial.* Publishing House of MPEI.

Vissarionov, V. I., Kuznecova, V. A., Malinin, N. K., Deryugina, G. V., & Shwan, D. E. (1997). *Calculation of wind energy resources.* Moscow, Russia: Publishing house of MPEI.

KEY TERMS AND DEFINITIONS

Theoretical Wind Energy Potential: The average long-term total wind energy of the air movement over a specific territory for one year, which is available for use.

Wind: Air movement relative to the earth's surface resulting from uneven distribution of atmospheric pressure and directed from the high-pressure area to the low-pressure area.

Wind Farm: Several electrically integrated wind turbines with a common control system located compactly on a specific territory.

Wind Gradient: The vertical gradient of the mean horizontal wind speed in the lower atmosphere showing the increase of wind speed with a unit increase in height above ground level.

Wind Probability: A parameter showing the share of the time during a specific period when the wind speed is equal to or higher than a specific value.

Wind Repeatability: A parameter showing the share of the time during a specific period the wind blows at a specific speed and direction.

Wind Turbine: A device converting the kinetic energy of the air movement into other types of energy (mechanical or electrical).

Chapter 9
Improved Distributed Energy Systems Based on the End-User Consumption Profile:
A Review on How Consumers Can Drive the Energy Transition

Daniel Adrian Perez-Moscote
Moscow Power Engineering Institute (National Research University), Russia

Mikhail Georgievich Tyagunov
Moscow Power Engineering Institute (National Research University), Russia

ABSTRACT

Nations are facing today the transition to cleaner, more reliable and affordable energy systems where power grids are becoming less centralized, more flexible and digitalized, and where not only power utilities, but also consumers are playing a significant role. Distributed Energy Systems (DES) constitute a key element in such transition, with decentralized renewable energy generation near the consumption points, energy storage, electric vehicles, and energy management systems, with the potential to ensure continuous supply and achieve higher efficiency, while reducing costs and adverse environmental impacts. This chapter presents a review of the recent advances in the design and development of DES, focusing on the effect of taking into consideration the consumption profile and behaviour of the end-users. The chapter also revises the limitations of DES and summarizes the future directions of DES development.

INTRODUCTION

Access to electrical energy is essential to overcome many of the complexities that the world faces today, such as food production, water supply, industrialization, and better health and education systems, particularly in developing countries and remote regions. However, according to the International Energy

DOI: 10.4018/978-1-7998-3645-2.ch009

Agency (IEA, 2018), about 1 billion people worldwide still lack access to modern electricity. Furthermore, electricity and heat generation accounts for around 40% of total global CO_2 emissions, being thus the process that contributes most significantly to climate change (IEA, 2019).

To secure access to affordable, reliable and sustainable energy for everyone, the 2030 Agenda for Sustainable Development (UNDP, 2015) proposes to increase the share of renewable energy in the energy mix, duplicate the global rate of energy efficiency improvement and upgrade technology for energy supply in developing countries. For this, it is required a complete shift in the energy paradigm, the nowadays called 'energy transition': a transformation of the power sector from the conventional, centralized and unidirectional model, based on fossil fuels, to a new system mainly based on clean energy, decentralized and more dynamic where energy, information and money flow in multidirectional ways (IRENA, 2019a), as illustrated in Figure 1.

Distributed Energy Systems (DES) constitute a fundamental element to achieve such transformation (Tyagunov, 2017) and can be defined as power systems where energy assets, such as small-scale renewable energy generation, energy storage, electric vehicles (EV), demand response (DR) and energy efficiency management, are distributed throughout the array and energy is generated near the consumption points (Siemens, 2018).

Figure 1. Shift from centralized energy systems to distributed ones [adapted] (Vezzoli, 2018)

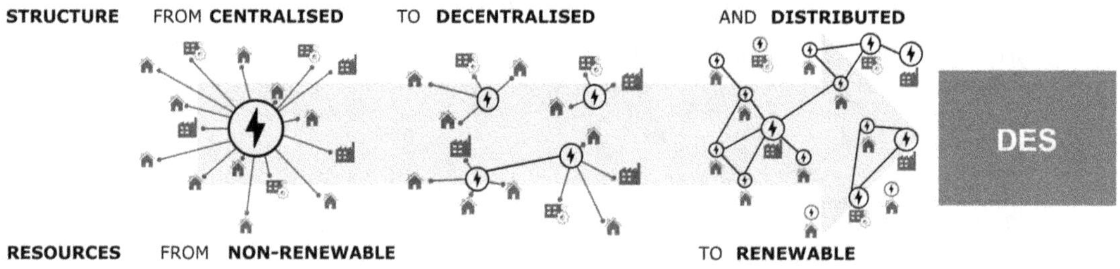

Through renewable energy generation and energy management, DES offer opportunities to ensure optimal use of energy recourses, to improve energy supply reliability, and to reduce carbon emissions. DES can also signify cost reductions for the development of the power sector, as a result of the possibility to add new installed capacity gradually, in smaller increments, depending on the location and the actual demand tendency, thus decreasing the risk of oversizing (Khokhlov et al., 2018).

Respectively, energy consumers are now playing a meaningful role in the energy transition. Active and informed energy users with the possibility to produce their own energy (prosumers), control their energy consumption and even sell energy back to the grid are transforming how DES are designed and managed. The objective of this chapter is to present a review of the recent advances in the design and development of DES, and how they can be improved by taking into consideration the end-user consumption pattern and behaviour and their influence on the electrical system. The chapter also revises the limitations of DES and summarizes the future directions of development of DES.

BACKGROUND

Historically, power systems have been developed similarly around the world. Large central power plants located near the energy resources, predominantly fossil fuels, and far from the cities, with electricity flowing in one direction from generation plants to consumers over long distance with the use of transmission and distribution lines. The larger the power plant, the more cost-effective and the structure concept remained the same for several decades.

The development of DES, where small-scale technologies are used to produce electricity closer to the energy consumers, started as a response to the 1970s oil crisis, with the use of innovative, affordable and efficient technologies for electricity generation such as combined-cycle power plants, subsequently changing the classical power grid model. Furthermore, global environmental concerns regarding climate change along with the global aim to ensure access for everyone to modern, affordable and reliable energy have driven the development of renewable energy technologies in recent decades and the world is currently undergoing a paradigm shift in the energy system: the energy transition to a low-carbon power system providing universal access.

According to the special report on global warming from the Intergovernmental Panel on Climate Change (IPCC, 2019), approximately 1°C of global warming above pre-industrial levels has been caused by human activities by increasing the amounts of greenhouse gases (GHG) in the atmosphere of the Earth, as presented in Figure 2, and it is expected to reach 1.5°C between 2030 and 2052 if the rise continues at the current rate. Moreover, many land regions and seasons are experiencing larger warming than the global average; in the Arctic, warming can be two to three times higher.

Human-caused global warming is creating long-term changes in the climate system, such as sea-level rise, increase in mean temperature in most land and ocean regions, more droughts and heatwaves and changes in precipitation patterns. The effects on biodiversity and human systems have already been detected: many land and ocean ecosystems have already changed due to global warming. The Paris Agreement, which was adopted by consensus in December 2015 within the United Nations Framework Convention on Climate Change (UNFCCC, 2015) and entered into force in November 2016, aims to limit the increase of global average temperature to 1.5 °C above pre-industrial levels. To achieve such an ambitious goal, the IPCC proposes significant investments in the energy sector to restructure power generation, transport and other energy uses, for both providers and consumers. Renewable energy technologies, together with energy efficiency tools and supervised final energy demand, offer one of the most practical and prompt ways to realise the Paris Agreement intention.

According to the Deloitte Global Millennial Survey 2019 (Deloitte, 2019), 'climate change, protecting the environment, natural disasters' was identified as the top personal concern among a list of twenty challenges facing society, even above 'income inequality, distribution of wealth'. Social views are changing, consumers and communities are requiring governments and institutions to address climate change and sustainable development issues and are taking actions to accelerate the energy transition. Additionally, the United Nations Development Programme estimates that almost 1 billion people worldwide still live without access to electricity and rely on diesel or oil to supply their energy needs, which is expensive and causes environmental pollution. The United Nations Agenda for Sustainable Development (UNDP, 2015) proposes to increase the use of renewable energy, double the global rate of energy efficiency improvement and improve technology for energy supply in developing countries in order to secure universal access to affordable, reliable and sustainable energy by 2030.

Figure 2. Global warming relative to 1850-1900 (°C) [adapted] (IPCC, 2019)

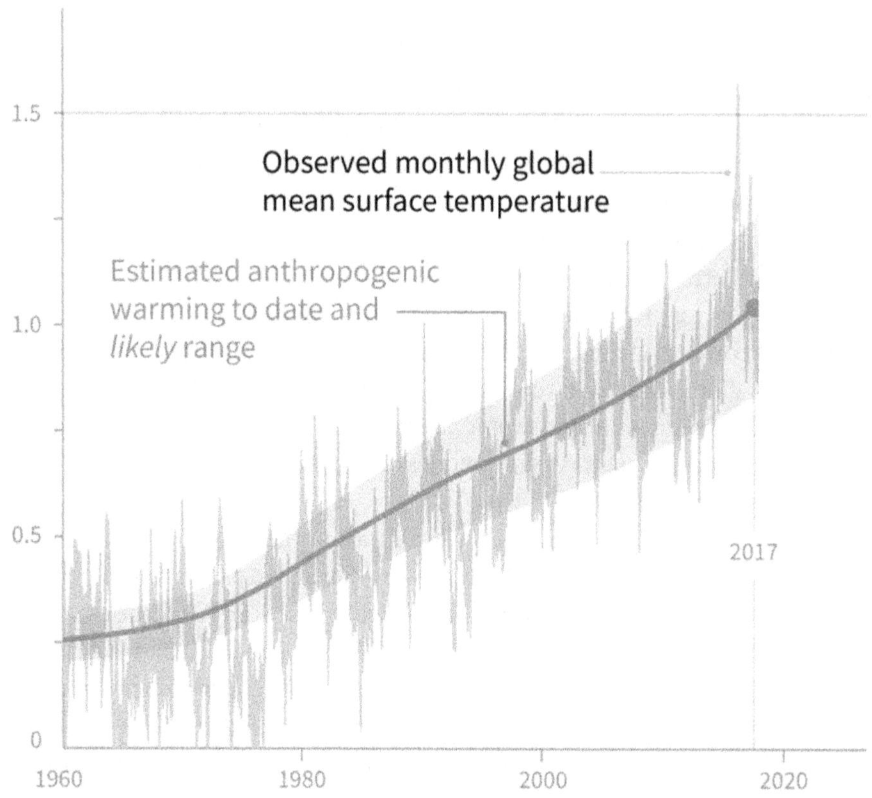

The most recent report (IRENA, 2019c) tracking the progress of the United Nation's sustainable development energy goal (SDG7), reports that, at the current rate, the world will not achieve the targets proposed by 2030. Current and planned policies to be implemented would lead to 570 million people gaining access to electricity worldwide by 2030, however, around 650 million people would still live with no access to electricity in 2030. Regarding energy efficiency, an average of 2.4% annual improvement in energy intensity is projected between 2017 and 2030 from the 3.5% energy efficiency full potential. Further systematic planning, investments in technology development and political commitment are required to transform the energy sector and achieve the SDG7 targets.

The flexibility of DES associated with their scalability and effective use of local energy resources and infrastructure makes them essential for a more sustainable future scenario (Alanne & Saari, 2006). They are mainly base on renewable energy technologies and offer reliability to existing power systems as a result of a more diverse energy mix. In contracts with centralized power systems where there is only one or a few energy sources, and a fault in a power plant fault affects the entire system or a large part of it, DES can cope better with individual energy source failures, due to the distribution of varied energy sources throughout the system, which makes them more reliable and resilient (Vezzoli et al. 2018). DES can expand electricity supply in regions without access, and it has been estimated that they have the potential to provide energy to more than 1 billion people by 2025 (Vezzoli et al. 2018). Moreover,

DES could also represent commercial opportunities to small economic entities willing to participate as prosumers, supporting financial security and inequality reduction in communities.

In the last decade there has been an important development of renewable energy technologies. Large-scale programs to support renewables, especially solar and wind energy, have been adopted by different countries around the world to promote low carbon energy and to reduce their dependence on energy imports. Wind and solar energy technologies have been deployed extensively, with a significant increase in their efficiency and reduction in costs. The average cost of electricity produced by solar and wind technologies decreased by 67-86% from 2009 to 2017 (Khokhlov et al., 2018) and according to BloombergNEF energy outlook (2019), new solar and wind energy will get cheaper than energy generation coal or gas by 2030. In the United States, more than 50% of new power generation capacity added to the grid came from renewable energy, mainly wind and solar (RMI, 2018). By 2050, consumer PV installations will make up 11% of the total installed capacity and 50 billion dollars per year over the next 32 years will be invested in small-scale PV systems (BloombergNEF, 2019).

However, the mismatch in the variable balance between renewable energy supply and electricity demand continues to be a challenge, particularly when the low price of renewable energy projects displaces other more conventional power generation projects. For instance, solar energy generation is higher around midday while peak consumption typically happens in the afternoon and early evening. This disparity subjects the grid to high stress, and it needs to be adjusted continuously.

The traditional DR tools used by utilities to manage emergency and peak load events, such as changes in the price of electricity for customers to change their consumption patterns, are not fully adequate to the dynamic impact of renewable energy on the grid. For this reason, in addition to renewable energy, new DR opportunities, energy efficiency technologies and energy-saving tools are emerging in the energy industry to shift electricity consumption to hours with high renewable availability, and further decrease energy-related emissions. Together with decentralized renewable energy generation near the consumption points, energy storage, EV and new energy management systems, DES have the potential to ensure a continuous supply and achieve higher efficiencies, while reducing GHG and adverse environmental impacts in a cost-effective way.

Among the immediate actions at the sector level needed to accomplish the energy transition, the International Renewable Energy Agency (IRENA, 2019b) advises to support the deployment of DES, encouraging energy consumers to become prosumers, supporting regulatory and pricing policies (microgeneration, tariff regulation and grid access), enabling energy aggregators to stimulate the use of distributed energy resources in communities and social collectives, accelerate the availability and use of smart meters and develop smart-home and digitalization systems to implement demand-side management.

CONSUMERS: THE KEY TO IMPROVED DISTRIBUTED ENERGY SYSTEMS

The design of DES involves a generation analysis, where the availability of renewable and conventional energy sources is studied, e.g., solar photovoltaic (PV), wind, micro-hydropower, small-scale conventional and biomass combined heat and power units, diesel generators, and correspondingly an evaluation of the electricity demand, where actual or estimated consumption data are used. However, in the design process, the end-user may be generally perceived merely as a passive consumer with energy requirements but limited participation and command over the energy system. Traditionally, the development of power systems was based on the perception that consumption profiles are rigid. Conversely, it is now argued that harnessing the real potential of DES to achieve sustainability in future energy scenarios requires the input of active end-users involved in the electricity consumption dynamics and local generation processes; consumers that are flexible and able to adopt new initiatives, technologies and technical solutions and well as modifying their energy consumption profile and behaviour.

In the classic centralised energy paradigm, end-users are not involved in the decision to build large power plants or any electric power infrastructure, and when they buy energy from the grid there is no control concerning the source of that energy. By contrast, in the new energy paradigm with DES, active customers can participate at all levels in the power system structure. As well as being able to choose more convenient energy consumption devices, such as energy-efficient home appliances or EV, or scheduling their consumption, customer can own distributed generation technologies and batteries, and could even choose the source of the energy they are buying from the grid (IEA, 2017). This not only applies to homeowners, but also to electricity aggregators, large and small enterprises, and local and state governments, which are from now participating in DES, either for their environmental commitment to address climate change, to secure continuous energy supply or for the economic benefits associated.

By formulating an adequate analysis of the end-user energy consumption patterns and procuring their participation through on-site energy generation, better DR models and energy efficiency management, it is possible to improve the design and management of DES. Active consumers that dynamically participate in their energy systems and even produce their own power, now called prosumers, can significantly contribute to realising a better integration of distributed energy generation, energy storage and energy management systems in DES, particularly in developing countries, isolated regions, and in rural and agricultural applications.

As an illustration, Figures 3 and 4 show the current state of the Russian power sector architecture and the improved target vision to enable the energy transition. In the target vision, it can be appreciated how the power system contains consumers with distributed renewable energy resources, energy storage systems and controlled demand, as well as distributed generation in distribution networks near consumers, as opposed to the current state where energy transmission is unidirectional from large generators to consumers, operational dispatch control is centralized, and consumers are passive; they are not able to generate any electricity and loads cannot be regulated. The envisioned architecture also contains controlled interfaces and service platforms to enable the integration of active end-users to the system and electricity exchange between systems, with fair access and conditions, technical and financial, to all participants in the power system.

Consumers' geographical location, socioeconomic conditions and behaviour can have a significant impact on the power system (Manjunath et al., 2014). Commonly, utilities group consumers into classes (domestic, commercial, industrial and rural classes), and differentiate tariff according to this classification, however, the particular characteristics of consumers such as location, load shape (level and times of consumption), power quality, and other factors like age group, income, education attained, behaviour and their views on environmental issues (which impact the types of loads, how and when they are used), are not considered (Manjunath et al., 2014). This classification can lead to poor approximations and incorrect power systems planning. The design and development of modern DES must consider all the aspects mentioned to accurately manage and forecast loads, increase efficiency, avoid extra investments costs, establishing adequate tariff structures and reduce energy waste.

The economic benefits of considering active end-users in DES planning have been studied by Spanish researches (Vallés et al., 2016), establishing that high involvement of final consumer could bring significant savings in postponed or avoided power system reinforcements, and could also reduce energy

Figure 3. Russian power system (current state) (Khokhlov et al., 2018)

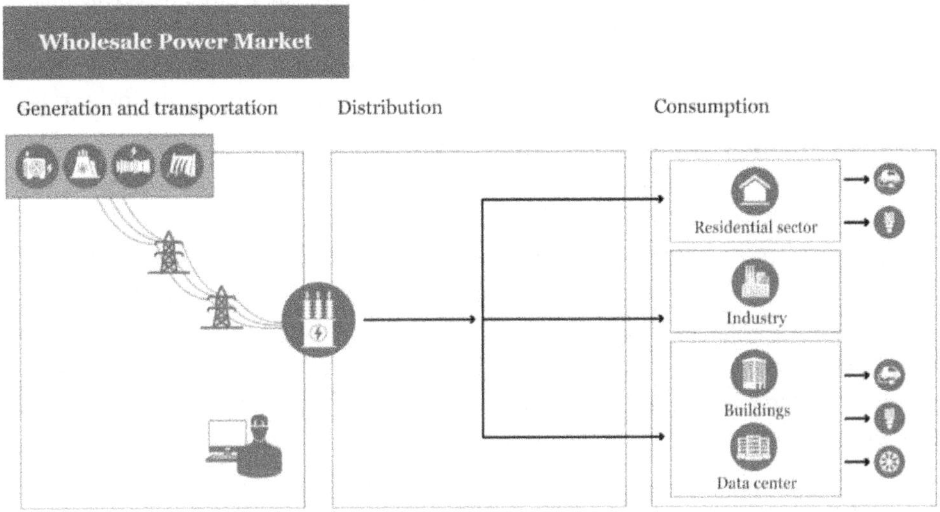

Figure 4. Russian power system – improved target vision (Khokhlov et al., 2018)

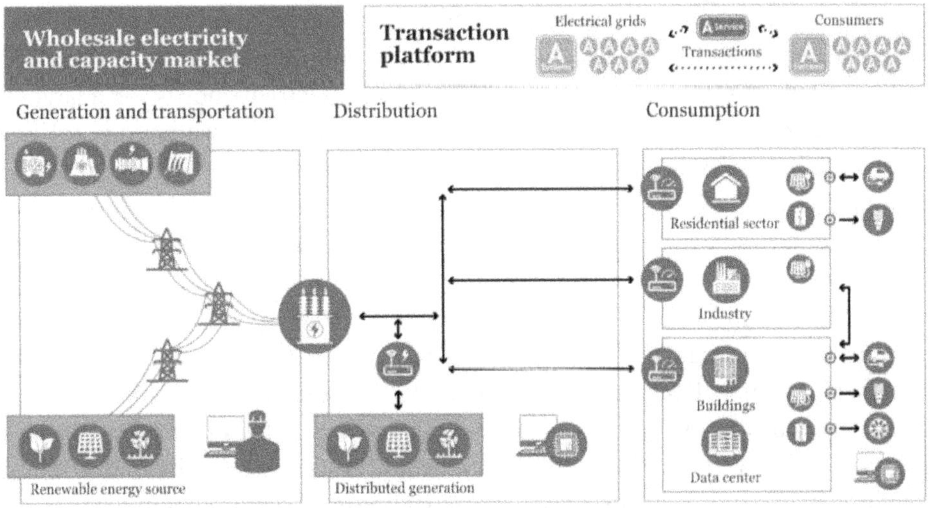

losses. Economic benefits are highly dependent on DES configuration and levels of network congestions. Correspondingly, power generation and consumption profiles of active customers can be improved with the use of smart bidirectional metering, which allows automatic measurement, storage and analysis of electricity information. Big Data technologies can help to optimize DES for customers, addressing issues such as the most cost-effective energy generation configuration, the best times to use electricity from the grid or self-produced, optimal PV dimensioning or battery sizing (IEA, 2017).

Distributed energy resources can be integrated with home appliances, EV, charging stations and other electricity-consuming devices using Internet of Things (IoT) and other wireless communication technologies. This is the concept behind the technological term *Internet of Energy (IoE)*, first employed by the economist Jeremy Rifkin in his book The Third Industrial Revolution, to refer to energy sharing systems with increased use of renewable energy resources (Kabalci & Kabalci, 2019). IoE infrastructure includes conventional power plants and transmission systems, DES, control centre, smart metering and data management system, providing data analysis and storage. Energy routers are the devices responsible for communication and energy transmission throughout the network.

Demand response, energy efficiency and changes in consumer behaviour constitute the core of Demand-Side Management (DSM) initiatives and programs, based on the principle of Demand Flexibility (DF), where the customer's level and pattern of energy consumption can be shaped to better match the electricity supply from renewable energy sources. This load shift saves energy, reduces peak demand and overgeneration, and improves the design and management of renewable energy-based DES. All this must be achieved without compromising the quality of energy services and the consumer's experience. Demand flexibility can be a powerful tool to cost-effectively balance renewable energy and facilitate the transition to a low-carbon power sector (RMI, 2018). Additionally, certain consumers, especially in areas not connected to the national gird, have specific power loads and consumption profiles for which it is not required, for instance, to ensure energy supply 24 hours a day at full load. New improved design model should be fed with intraday consumption data of the customers and projected energy demand responses and behavioural changes for better balancing supply and demand, and even reducing the energy generation and/or storage required, bringing up the efficiency and reliability of the system.

It has been determined that by focusing on the specific end-user energy services, the design and modelling of DES allows a higher satisfaction of the consumer electricity needs and a larger penetration of distributed energy sources (Mabaso, 2018). Defining the end-user consumption profile in terms of energy services enables a more adequate selection of energy supply systems in the model, and allows the integration of additional energy efficiency and energy storage technologies. Moreover, communities in developing countries have unique electricity needs and particular demand curves, consequently pertinent approaches to design and model DES specifically for these communities should be used by power system planners (Mabaso, 2018). The complete and quantitative evaluation of consumer' energy needs, and not only kilowatt-hour [kWh] usage, can lead to a more effective combination of technologies to better satisfy such needs with the use of distributed renewable energy.

Dynamic electric loads in the design of DES with active consumers can be estimated employing energy aggregators (Paoletti et al., 2011). Each aggregator has a group of individual consumers and coordinates their consumption behaviour with the power system, sending them price-volume signals (monetary reward if consumption is below, above or in-between specified values, during certain hours of the day), to which consumers respond changing the schedule of energy-consuming tasks. The aggregator works closely with the utility operator to balance the system in times of energy overload or renewable energy overproduction. Aggregators can also look for profit opportunities, operating according to the market energy price. In the model, to effectively forecast energy loads in the presence of active demand (AD) coordinated by aggregators, the load is divided into base load (seasonal patterns) and residual load (stochastic variations plus AD effects), however, the performance of the model is highly dependent on data availability and resolution.

Mabaso (2018), has explored how energy systems manage to meet energy needs in developing countries and has established a modelling approach to explore and evaluate the arrangement of energy supply solutions in DES to improve the energy service in communities within larger systems, combining a bottom-up energy system model with a top-down end-user criteria-based model. The model proposed by the study uses the "discomfort level" indicator to evaluate how community scale energy systems offer an improvement in terms of satisfying the end-user energy service needs. Discomfort level is a demand-side indicator and reflects "the energy services unmet at each hour for a selection of technology combinations" (Mabaso, 2018, p. 44), influenced not only by the type of energy supply technologies but also by the unique end-user context and behaviour. Figure 5 shows the model of a DES which integrates the various energy supply sources, the blocks of detailed end-use energy services and relevant factors for the system improvement such as energy efficiency, population, their income and access to the grid. The study highlights how the energy systems are set in a social context which cannot be ignored, and the importance of integrating a consumer-based approach with the energy supply technologies to develop better energy systems models able to overcome the current challenges.

Figure 5. Model of a DES with demand blocks [adapted] (Mabaso, 2018)

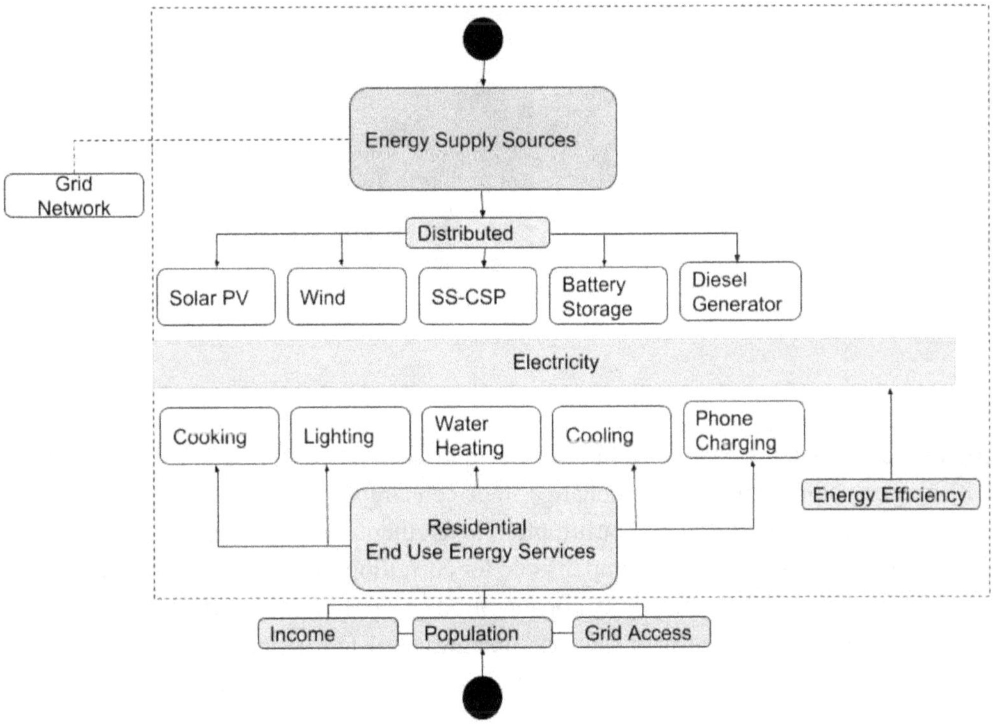

The Rocky Mountain Institute in a recent report (RMI, 2018) presents how a residential customer could possibly change electricity consumption across hours of the day to better match PV generation peak by using technical solutions, communication and control technologies, such as battery energy storage, electric vehicle charging and air conditioning controls, without any substantial impact on the quality of service or major disruption of the customer's routine; the result is illustrated in Figure 6. According to this study, demand flexibility technologies applied to power systems can increase self-balancing and energy bill savings between 10% and 40% at the end-user level.

Figure 6. Impact of demand flexibility on residential load profile (RMI, 2018)

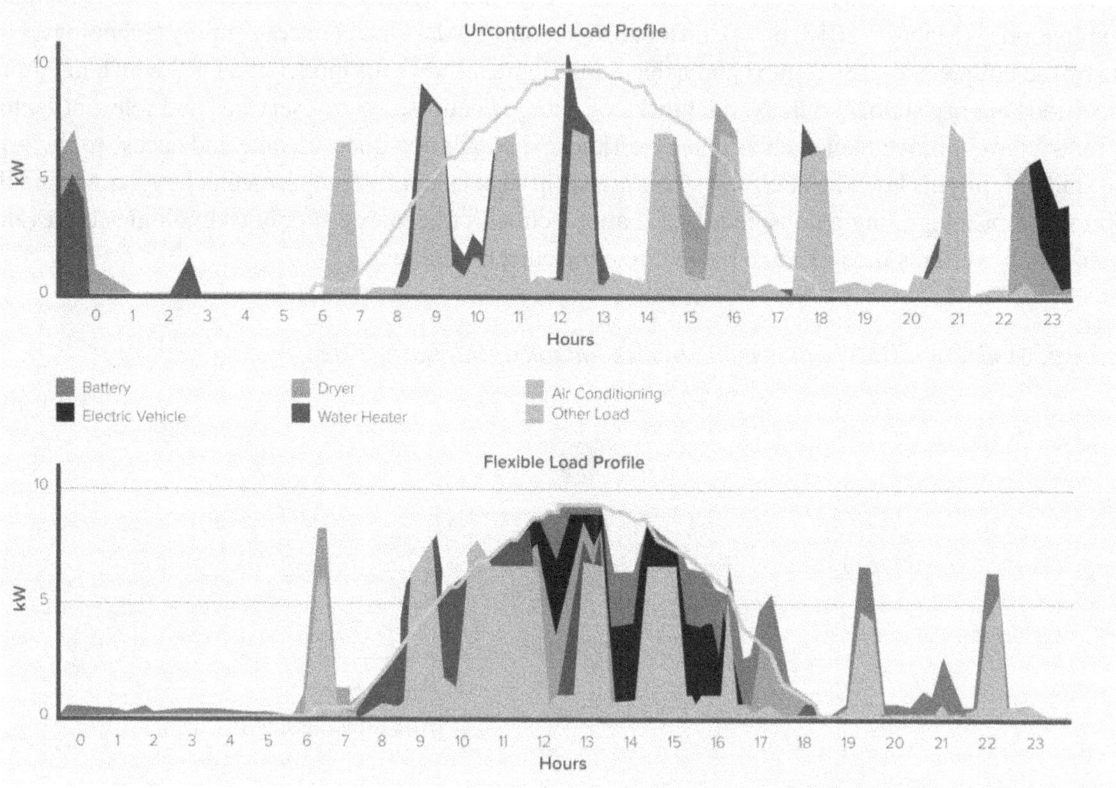

DR programs can also reduce the operational costs of grid-connected and island microgrids (Farsangi et al., 2018) by persuading consumers to change their consumption patterns. These programs can be price-based, where customers shift their consumption to another time responding to the electricity price, or incentive-based when customers are compensated for curtailing loads at times of grid stress. Figure 7 illustrates a 10% and 20% shift in electricity consumption by applying price-based DR programs. Utility pricing schemes are vital not only for the implementation of DR strategies, but also to promote the penetration of renewables at the end-user level, such as rooftop PV, and the use of local energy storage and EV as mobile batteries (Jin et al., 2018).

Research results presented at the 24th International Conference & Exhibition on Electricity Distribution (CIRED) in 2017 (Kamphuis et al., 2017), show how demand response strategies have the potential to reduce total peak demand between 5 and 15%, and the reduction can even reach 30% depending on the user willingness to participate and the technical capabilities of equipment. In this scenario, not only consumers become prosumers, but the distribution network operators also have to act as system operators to be able to deal with multidirectional and variable energy and information flow.

Figure 7. Electrical load curve before and after implementing price-based DR programs (Farsangi et al., 2018)

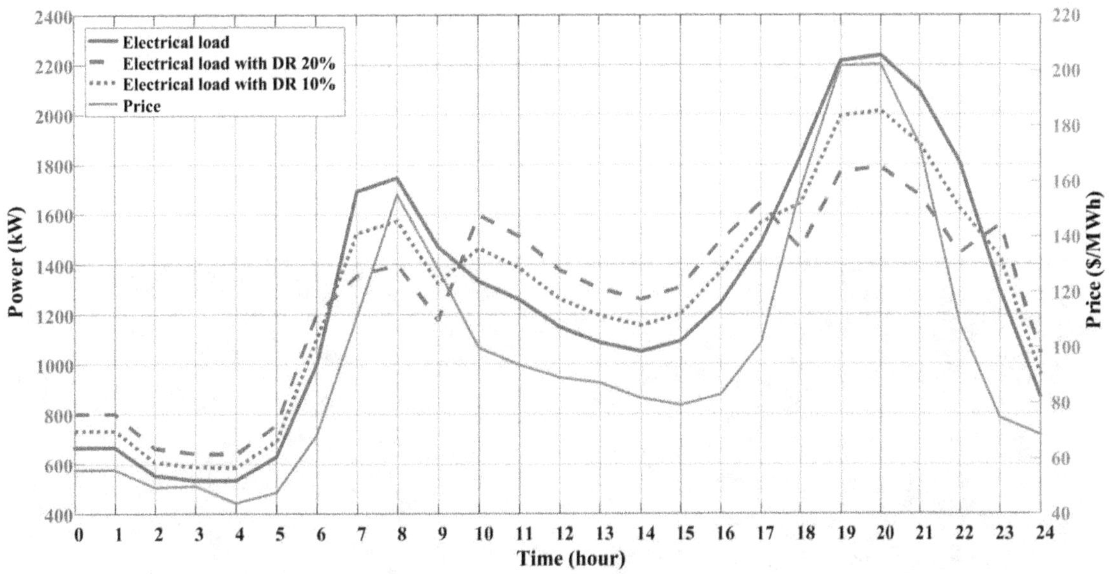

The role of consumers in the power system as participants also able to produce and store energy has been discussed for over ten years now. A prosumer-based smart grid architecture was presented in 2011 (Grijalva & Tariq, 2011), describing the implementation requirements and advantages for the energy industry. Grijalva and Tariq describe the 'flat' electricity industry, where not only utilities, but now houses, buildings, small businesses, neighbourhoods, microgrids, and such, are recognized as participants with the potential to offer services of energy supply, storage, demand response, and even transportation of energy. Information is available for all participants and they can interact with each other. In this order, consumers could choose where they get their energy from; based on economical or environmental considerations, and utilities would mainly provide the transportation services. Prosumers distributed throughout the entire electricity network should possess controls systems to operate their nano, micro and mini-grids. Moreover, the connectivity among all the prosumers participating in the energy system would allow the reduction of the supply-demand variability, since they can synchronize their consumption with schedulable loads or the use of electric vehicles and batteries.

The progress in the technological development of distributed renewable energy generation, energy storage and demand response are enabling these predictions to become a reality today. BloombergNEF forecasts that, by 2050, 40% of all battery deployment will be owned by end-users as battery systems will get more cost-effective over time. The payback period of a PV system with battery will decrease by half, from 13 years now to 6 years by 2040. More interestingly, 75% of end-user battery capacity will be used to reduce system peak demand and PV systems in households equipped with batteries will perform as virtual power plants to provide flexible capacity and bring balance to the power system, in direct competition gas peaking power plants and utility-scale batteries.

In the United States, electric utilities have witnessed a flat electricity demand over the last 10 years and, according to the U.S. Energy Information Administration (EIA, 2019), electricity demand growth will be slow and modest. Demand is stable because of improvement in energy efficiency, consumers producing their own energy and energy-intensive industry being outsourced. Electricity consumption is no longer directly related to GDP growth and the two indicators became disconnected around 1998, as seen in Figure 8.

Figure 8. U.S. GDP and U.S. electricity sales 1997-2016 (VOX, 2019)

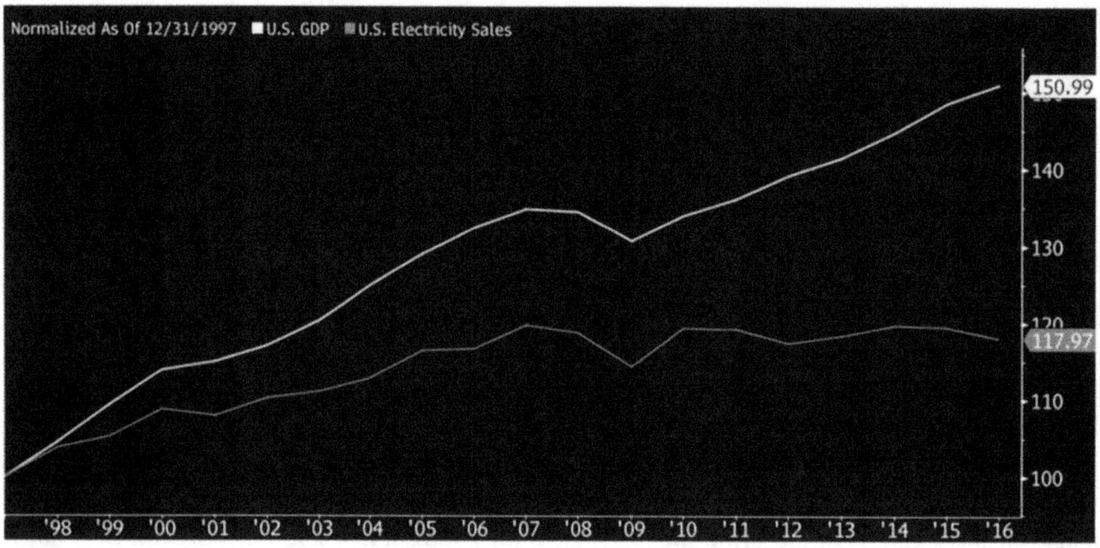

Energy regional planning agencies are having to develop and release their long-term energy plans more frequently due to the rapid changes in the power systems; future estimations for the next 20 years fall short just after a few years. As electricity demand is no longer increasing, which is good from the environmental point of view, utilities' business model is not anymore about investing in very large infrastructure like power plants and/or transmission lines, but to manage the new power systems, matching the flexible, diverse and cleaner energy generation with a steady and even decreasing controllable demand, through "sophisticated real-time sensing and software" (VOX, 2019).

As the development of DES and DSM grow, utility companies must innovate and develop new, complete and integrated solution sets (Fairbrother et al., 2017). Today, utility companies are facing an increasing number of pilot projects and demonstrations of DER, digital power systems, dispatch optimization, energy transactions, which test their adaptability and modernization capacity, and those that fail to transform their business model run the risk of falling behind the new participants in the electric power system and the energy market. Utilities, but also regulators and technology developers, who seek to adapt and innovate for the energy transition, must apply strategic planning and build a roadmap for innovation, design and explore pilot and demonstrations projects for research and learning purposes before large-scale deployment, build support and accountability from energy-sector leaders and collaborate among industry stakeholders and utilities, share best practices and lessons learned to speed up effective innovation processes (Fairbrother et al., 2017).

Blockchain: Developing Consumer-driven Cleaner Power Systems

Besides the DSM strategies to improve the design of DES, the complexity of evolving prosumers in the system requires efficient communication among all the DES participants and an effective trading scheme (Li et al., 2019). Blockchain, as a novel and rapidly evolving technology that allows to securely and instantly decentralize transactions of any kind without the participation of a central intermediary. Blockchain has the potential to actually decentralize information in energy systems, providing consumers with control over where they obtain their energy from as well as energy production information in real-time, therefore stimulating competition and promoting sustainable energy (Ellsmoor, 2019a). Blockchain-based DES will allow end-users to compare energy providers and buy directly from the ones more suitable to their needs and priorities. Blockchain can be used to provide cleaner and cheaper energy to DES already connected to the grid as well as to isolated systems in rural areas and developing countries and regions, reducing inequality and promoting global sustainable development.

Blockchain can help enhance DES design in the following ways (Li et al., 2019):

1. Blockchain is a system designed for peer-to-peer (P2P) interactions. Using blockchain in DES, end-users will be able to buy and sell clean energy between them in a safe and reliable environment. In a P2P system, every participant is able to interact with others, choosing their own behaviours discreetly. All the actions in the power system, such as information exchange, energy transmission and pricing, are now customized, recorded and trackable, without the need of a third party.
2. Blockchain can increase the participation of local enterprises in the electricity market, by using a decentralized system that allows numerous distributed small-scale energy suppliers without a single third-party intermediary, reducing the cost of daily operations. Furthermore, end-users can play an active role in energy pricing and settlement in DES.
3. Smart contracts, programming tools using blockchain to automatically execute transactions between the energy users when predetermined terms and conditions are met, will support the development and use of new concepts such as demand response, energy auction mechanisms and scheduled power demand, allowing a synchronous flow of information and value, with transparent transaction records.
4. The concept of a globally unique digital decentralized identifier (DID) in blockchain will permit the construction of a trusted network in DES. Energy behaviours can be evaluated by P2P participants through DID without compromising privacy. All transaction records are safely stored on a blockchain and authorised by a digital signature so that any suspicious information can be traced back.

In a future scenario where consumers have access to on-site renewable energy generation and storage capacity and can sell surplus energy back to the system, secure and transparent P2P transactions can be realised through the implementation of smart contracts based on blockchain technology.

Chinese researchers (Wang et al., 2020) have also recently conducted studies on how distributed P2P energy transactions can allow end-users to participate as autonomous prosumers in Urban Community Microgrid Systems (UCMS) while protecting their private data. In the study, it is proposed an innovative

P2P transaction model based on a double auction mechanism. First, prosumers in the UCMS generate information regarding their energy demand and supply, then set the price according to their profit expectations, and finally initiate P2P energy transactions among them in the double auction market without the need of transaction brokers. The application of this method allows prosumers to achieve coordination, complementary of energy sources and privacy protection in the system, promoting renewable energy consumption, energy security and economic value.

The UCMS considers two types of prosumers, residential and enterprise, with energy loads but also equipped with distributed renewable energy generation and energy storage, which enables them to dispatch electricity stimulated by financial/environmental motives. The microgrid also includes an Energy Balance Service Provider, which deals with energy transactions between the microgrid prosumers and utility grid, when distributed renewable generation is not available to supply energy demands. The model proposed includes the optimal management of each prosumer's energy resources (supply and demand), price strategy to obtain maximum benefits, and the implementation of the distributed P2P energy transaction scheme (Wang et al., 2020).

Table 1 outlines the key aspects of research projects working on DES development and improvement based on active end-user consumption pattern and behaviour.

Currently, there is a large number of leading companies and pilot projects working on blockchain for energy worldwide (Ellsmoor, 2019b), such as Energy Web Foundation (global platform), LO3 Energy (USA), Electron (Great Britain), Power Ledger (Australia), Sun Exchange (South Africa), Acciona Energy (Spain) and WePower (Estonia). According to researchers from the LO3 initiative and the Karlsruhe Institute of Technology in Germany (Mengelkamp et al., 2018), blockchain technology has the potential to encourage further penetration of renewable energy generation into DES thought efficient operation of local energy markets. In this study, the researchers also found that the BMG has successfully increased the local security of energy supply, integrating local renewable energy sources and encouraging the consumers and prosumers from the community. The microgrid is able to work in island mode as well, ensuring energy supply for key facilities like hospitals or police stations. In this case, prices vary according to demand and supply, along with other socioeconomic considerations, and participants can set their limits for renewable energy price and configure their generation mix (Mengelkamp et al., 2018).

Power Ledger, an Australian start-up, has developed an energy trading platform based on blockchain, and their technology is currently being used being used in different countries around the world, including Australia, India, Japan and the United States. They started in 2016, raising funds through the sale of their own cryptocurrency tokens – POWR, in the largest crowdfunding project in Australia, and have also received financial support from the Australian Government. In the system proposed, consumers are able to trade clean energy anywhere in the world without the need for intermediaries in a safe ecosystem. The system works with two tokens, POWR and Sparkz; POWR tokens are standard and maintain their value internationally, while Sparkz tokens are relative to the local currency and are used for energy transactions. Participants can convert POWR tokens into Sparkz to purchase or sell electricity, and they can specify where they want to buy their energy from (Power Ledger, 2019).

Similarly, Energy Web Foundation (EWF) is a global organization established in 2017 and co-founded by Rocky Mountain Institute (RMI) and blockchain developer Grid Singularity (GSy), focused on exploring the potential of blockchain to accelerate the energy transition to cleaner, decentralized and sustainable energy systems. EWF is the largest energy blockchain network with over 100 member organizations. For the last three years, it has worked on the development of the first public blockchain architecture for the energy sector – the Energy Web Chain (EW Chain), where blockchain-based decentralized applications

Table 1. Key aspects of research projects on consumer-based DES development

Institution	Country	Concept	Findings	Research gaps
University of Siena	Italy	Electricity load forecasting for distribution networks with Active Demand (AD)	Decomposition of electricity load into base load and residual load (stochastic fluctuations and AD effects) employing energy aggregators could improve DES development. (Paoletti et al., 2011).	Load predictions computing and real-time availability.
National Institute of Technology Goa	India	Impact of consumer characteristics on power systems	Consumer's annual income, location, age group, behaviour, environmental concerns and household conformation play determining role in power consumption and should not be overlooked when designing DES (Manjunath et al., 2014).	Significance of consumer behaviour and consumption routines according to particular characteristics in load forecasting and management.
Comillas Pontifical University	Spain	Economic benefits of active end-users in DES	Active demand (AD) actions could potentially substitute investments in distribution network reinforcements and lead to costs savings from 20% to 60% (Vallés et al., 2016).	Distribution system operators (DSO) remuneration schemes for network operation considering AD actions. Energy losses savings when applying AD actions.
Stellenbosch University	South Africa	DES model with demand blocks	Design and development of DES in term of consumer discomfort level (energy services not met) could reduce electricity load and increase system efficiency. For developing countries, modelling approach should correspond to particular context (Mabaso, 2018).	Elasticity of demand in developing countries. Development of software with graphical user interface to simulate/ project discomfort framework.
Rocky Mountain Institute	United States	Demand flexibility (DF)	Applying DF, consumers could shift their electricity consumption to match renewable energy supply, lowering peak demand of renewables by 24% and generating energy bill savings between 10% and 40% (RMI, 2018).	Specific results not necessarily generalizable to other cases in other regions, where possible generation and load evolution could lead to different DF strategies
University of Science and Technology	Iran	Stochastic microgrid energy management	Operational costs of microgrids could be reduced with stochastic management using Demand Response (DR). Electricity consumption can be shifted between 10% and 20% by applying price-based DR programs (Farsangi et al., 2018).	Enhancing the proposed method to consider environmental objectives in microgrid management.
Tianjin University	China	Peer-to-peer (P2P) energy transaction for Urban Community Microgrid Systems (UCMS).	P2P distributed energy transaction method based on double auction market, improving renewable energy self-consumption ratio (30%), peak-to-valley difference (49%), and overall daily cost (72.74%) in a sunny scenario (Wang et al., 2020).	Impact of differentiated energy supply and demand, and pricing strategies on P2P energy transactions.

can be built and run. More than 100 use cases have been studied and tested using the EW Chain and commercial applications have been launched. Currently, according to the last report (EWF, 2019), their technology approach is transitioning to focus on the development of tools to realise the full potential of grid flexibility propelled by end-users.

The EWF estimates that customers are to invest more in the grid than utilities by 2030 (830 billion dollars on distributed energy resources and 7 trillion dollars on electric mobility), and above 30% of global installed power capacity will be customer-owned (EWF, 2019) as illustrated in Figure 9. However, power grids and electricity markets are not ready for a future centred on end-users. Energy supply is no longer controllable due to renewables' variability, demand is no longer fixed and is becoming flexible, and grid investment is no longer a centralized function of grid operators, but rather it is projected a major customer investment in decentralized energy. The EWF plans to develop a set of decentralized open-source technologies and standards, named EW-DOS, to allow end-users, retailers, installers of renewables and distributed energy resources and other participants in the power system to execute administrative functions, which currently only grid operators can perform, among other services aimed at the democratization of the energy system.

Figure 9. Global installed capacity by resource type (EWF, 2019)

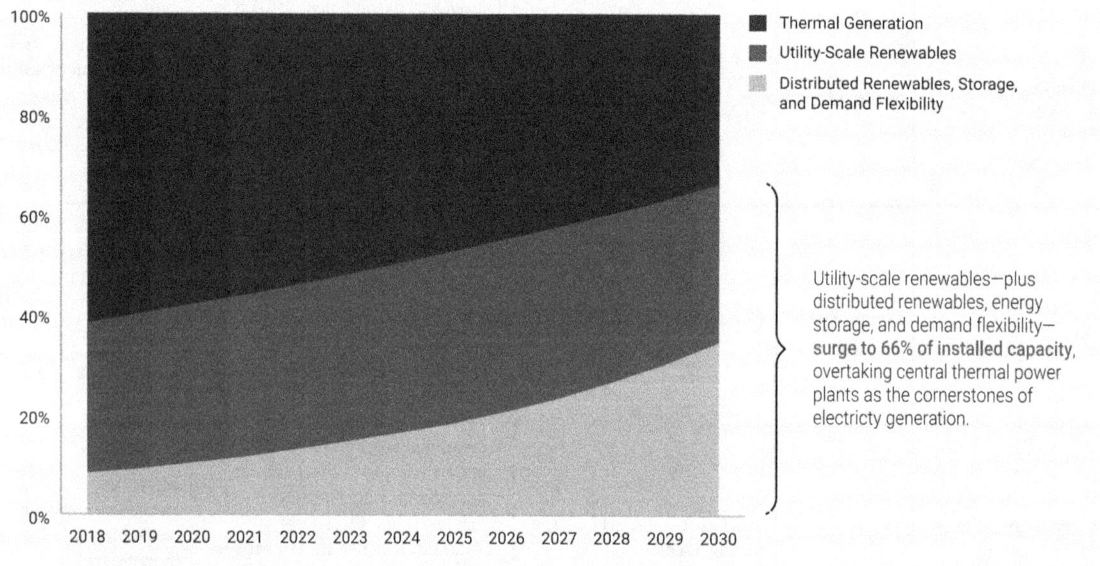

Table 2 presents an outline of main projects currently working on the creation of blockchain-based tools to develop improved energy systems driven by active consumers.

The energy transition can be shaped by active end-users not only with individual actions, but as collectives. Customers' individual decisions are significantly moulded by social structures and shared routines. These social systems that go beyond individual engagements also need to be considered in the development of DES to achieve the energy transition (Schot et al., 2016). There are other ways in which consumers can play significant roles in facilitating the energy transition, besides participating actively

Table 2. Current projects working on blockchain for DES with active consumers

Project's name	Country	Concept	Description
Energy Web Foundation	Global organization	Blockchain architecture for energy applications (EWF, 2019)	Public global blockchain architecture for the energy sector (EW Chain), where blockchain-based decentralized applications can be built and run.
Power Ledger	Australia	Blockchain energy trading platform (Power Ledger, 2019)	International energy trading platform based on blockchain using two tokens, POWR – global standard, and Sparkz – relative for each country's market.
LO3 Energy	United States	Energy pool trading system (LO3 Energy, 2020)	Pando, their technology platform, works with utilities to pool local distributed energy resources so customer can buy and sell energy, allowing flexible trading, personal energy management and grid optimization.
The Brooklyn Microgrid	United States	Local solar energy marketplace (Mengelkamp et al., 2018)	Energy network created by LO3 Energy that connects solar energy prosumers and consumers, allowing New York City residents and business owners to participate in the local energy marketplace through a mobile app.
Acciona Energy	Spain	Blockchain for energy traceability (Ellsmoor, 2019b)	First utility to use blockchain technology to certify the origin of its energy to consumers.
WePower	Estonia	Energy marketplace (Ellsmoor, 2019b)	Allow companies to buy energy directly from renewable energy generators transparently at competitive rates using their own tokens.
The Sun Exchange	South Africa	Crowdfunding platform (Ellsmoor, 2019b)	With the use of blockchain-based smart contracts and bitcoin for payments, the platform connects small and medium scale solar projects in developing countries with investors worldwide, who can then sublet them and benefit schools, businesses and local communities.

in distributed energy production, energy storage or demand management. Some of the additional forms that have been identified for consumers to drive the energy transition, framed in the social system to which they belong, are (Schot et al., 2016):

- As entrepreneurs; consumers that research, create and experiment with innovative technological/ organizational energy solutions, for instance pioneers developing local energy systems in their homes or communities using small-scale renewable technologies or P2P energy trading systems.
- As mediators; leaders, consumers that have influence over their collectives and can orient their values and perspective for the future. They analyse and interpret energy issues for their communities, guiding social commitment to support or participate in solutions to accelerate the energy transition.
- As citizens; consumers that are engaged in policy-making for the energy transition, rising awareness about needed regulatory reforms to promote the participation of their communities in the development of cleaner and more dynamic power systems. Some examples are environmental activists or individuals involved in non-governmental organizations that promote sustainable energy legislation.

Considering the different functions that active consumers can perform in the energy transition scenario, three types of P2P market structures have been identified (Sousa et al., 2019), full P2P market, community-based market and hybrid P2P market.

1. In the full market structure, consumers trade energy with other consumers directly, without the intervention of supervision of a third agent. They agree on the conditions and execute the transaction for a certain amount of energy. In this type of structure, end-users can express preference for specific energy attributes, such as origin or how green it is.
2. In community-based P2P markets, there is an administrator in charge of managing energy transactions between consumers in a group or community. The community manager is also a mediator between the community and the rest of the system. Generally, members of the group are located close to each other, e.g., in neighbourhoods, or they can form smart communities through virtual platforms, but most importantly, they share values and common interests; consumers are involved in the system as collectives with a cooperative mindset. For example, a community might be interested in buying energy from another community in a different location for the environmental or the socioeconomic effects the transactions can generate.
3. Finally, the hybrid P2P market structure, as the name indicates, is a combination of community-based and full P2P structures. There are different transaction layers, where members of the community can trade energy through the manager or between them directly. The hybrid model takes some of the advantages of the previous two; it empowers individual consumers with choice, support collective values and goals, but also ensures consistency and scalability of the structure with community managers, as controlling and negotiator agents for all the participants of the system.

Researchers from University College London (UCL) and Universidad EIA are working on a P2P pilot project in Medellin, Colombia, together with the utility company (Empresas Públicas de Medellin – EPM), to evaluate how end-users can undertake energy trading, particularly in a developing country context where not all energy users are able to produce their own electricity, but are willing to buy energy from other users, considering environmental and social effects of their consumption choices. This pilot intends to connect 14 end-users, located at different points across the city and with different income levels, through a P2P trading platform. Three low-income users and a community centre will have rooftops PV installations and will be able to trade energy with high-income users (UCL, 2019). The project will operate for a year until the end of 2020 to evaluate the impact of technical, economic, environmental and social aspects on energy consumption and energy trading for diverse end-users. The project also aims to elaborate a series of recommendations to the government and regulators and a roadmap for commercialization of P2P energy trading systems.

The socio-technical revolution needed to build more sustainable energy systems and shift the energy paradigm is a collective process, profoundly shaped by the work of active participants. Continuous advances and improvements in innovative and smart technologies and applications based on information systems are central for the transformation of the energy industry. This progress offers stimulating opportunities for researchers, pioneers, technology firms, investors and active consumers willing to be part of the change and work towards the development of sustainable energy systems.

Inspiring Active Consumers Through Policy

To encourage and promote the expansion of DES with the participation of active end-users, policy and regulatory changes in the power sector are necessary. Particularly in Russia, the major players of the electric power system and market are lacking interest or concern in the legislative changes needed to ensure the energy transition. The required regulatory framework should create new opportunities for consumers, validate and legalise the appearance of new energy entities, such as active consumers, prosumers, aggregators of distributed energy resources and microgrids, establish conditions for interactions between them and utility grids, and ensure fair conditions for all, new and former, power system and electricity market participants (Khokhlov et al., 2018).

With the implementation of changes in the power sector architecture and regulation in Russia, even partial development of DES potential could cover up to half of the projected capacity shortage in the national power system during the period 2025-2035 (about 36 GW by 2035). Harnessing the full potential of DES, including co-generation and self-generation with renewable energy, and DSM strategies, could eliminate capacity shortages completely in Russia (Khokhlov et al., 2018). At the end of December 2019, the Russian government passed the law N° 471-ФЗ, for the development of micro-generation in the electric power industry (RIA Novosti, 2019). With the implementation of this law, consumers who own a microgeneration facility, such as a PV system or small wind generators, with a capacity of no more than 15 kW will be able to sell surplus energy not consumed by the household to the grid and other energy retail companies. Customers interested in becoming prosumers should connect the microgeneration station to the local grid and sign a sales contract establishing the terms and conditions of the transactions.

Around the world, policies implemented to promote energy efficiency and conservation are technical-focused, aimed at improving the efficiency of products, systems, or buildings, and stimulating end-user to invest in more energy-efficient solutions. However, innovative policies to reduce end-users consumption through behavioural energy conservation are now emerging, such as personal carbon allowances and energy-saving feed-in tariffs (Bertoldi, 2019). In personal carbon allowance policy, for example, the government establishes the amount of carbon emissions that people can produce linked to their energy consumption and allocates annual budgets of carbon allowances. Every time a person uses energy, the emissions related to that consumption are deducted from the budget. Carbon allowances can be traded in a market, where people can buy or sell them, in case they need to increase their budget or, on the contrary, have extra allowances, depending on their consumption behaviour, thus benefiting those who save more energy. Accounting of personal allowances can be carried out through smart cards, smartphones or cloud applications, and end-user should be well informed and trained.

Several policies have been adopted recently in the European Union aimed at enabling energy end-users to overcome the barriers that prevent them from investing in energy efficiency technologies and adopting practices and lifestyle changes to reduce electricity consumption and save energy, such as financing schemes for investments in energy efficiency, providing end-users with individual metering and informative billings that shows their actual electricity consumption and other instruments to promote behavioural change like fiscal incentives, access to grants or subsidies. In 2006, The EU introduced the target to achieve 20% savings or more on primary energy consumption by 2020 in the Energy Efficiency Action Plan, and in 2014 the target was updated to 27% improvement in energy efficiency compared to 2006. Additionally, the EU's Horizon 2020 research and innovation programme has proposed several calls to support or finance projects to investigate and better understand the dynamics of energy consumer behaviour from economic, social and psychological points of view (Bertoldi, 2019).

Finally, the called 'regulatory sandboxes' are mechanisms to allow new energy technologies, products, services and business models to be tested before taking them to market, by providing closed and safe experimenting environments, reducing the developing time, and costs in some cases, providing innovators with greater access to financing opportunities and allowing them to work with regulators to ensure appropriate regulatory frameworks and consumer protections (Deloitte, 2019). Regulatory sandboxes are proposed to include tools to support innovation, such as inquiry service, to orient how pioneering projects can be feasible under current laws and regulation, temporary exemption of regulatory requirements for innovators, temporary change of current regulation or introduction of new ones in order for innovator to progress without the barriers that current regulation can impose to the development of the new concept. As a result, participants can test and evaluate their innovative energy solutions faster and possibly facing fewer risks.

Limitations to Distributed Energy Systems Development

Particular technical, social and political issues that need to be further studied and overcome in order to fully take advantage of DES in future energy scenarios are associated with:

- Low diffusion of renewable energy systems at local (end-user) level, particularly PV, even though prices have become competitive compared to conventional energy sources, such as generation from diesel (Vezzoli et al. 2018).
- Technological complexities in the integration of small renewable energy generation into the distribution system, impact on reliability and extra costs associated with electrical network modernization (Khokhlov et al., 2018).
- Unavailability of local skilled manpower for the installation, operation and maintenance of renewable energy systems and DSM technologies, especially in remote regions.
- Former consumer's social norms and value system; some consumers are not ready to assume their role as prosumers or even adopt new technologies or make any changes in their consumption behaviour. There is no enough awareness about the environmental impacts of consumption decisions or the benefits/opportunities of renewable energy, and changes may be perceived as risky or impractical.
- Lack of adequate DES planning and promotion policies from governments. Particularly, in developing countries and regions, renewable-based DES development may not be recognised as a priority in the power system improvement. There is a weak institutional framework, poor management capacity and are corruption issues (Vezzoli et al. 2018).
- Particularly in the Russian Federation, the electric power industry needs to transition to a new business model which supports renewable energy producers, stimulate active consumers and recognises the flexibility of the market (Khokhlov et al., 2018).
- Correspondingly, large-scale changes in the Russian power sector legislation are needed to create new opportunities for electricity customers willing to participate in the market, authorise new entities such as active consumers, prosumers, aggregators, microgrids and DES operators, and regulate the relationship among them (Khokhlov et al., 2018).

FUTURE RESEARCH DIRECTIONS

The successful design of efficient, resilient and sustainable DES is required to also consider power grid analysis (island and/or grid-connected operation, increasing load, power flow, stability, protections), economic evaluation (pre-operating, capital and operational expenditure), and regulatory considerations (energy policy, energy prices, economic incentives, grid access and connection mechanisms, rural and isolated areas electrification programs), as well as social issues (public acceptance, commitment).

To continue research activities on improved design and management of DES and the active role of end-users in the energy transition, the directions that may need to be explored in the near future are related to the financial (e.g., bidirectional energy transactions) and social complexities involved in the design and management of systems based on the needs of the end-user and not only on the energy supply technologies.

Regarding blockchain-based energy systems, they can be expanded by further studying the scalability of such systems, since an increasing number of participants and model complexity could lead to longer processing times to solve all the interactions among energy consumers, larger than the limits for appropriate real-time operation of the system. Also, blockchain and P2P energy trading are still developing technologies and there are numerous aspects to be improved, such as high-frequency trading and system efficiency vs. system security issues, so they can achieve better technical performance and can be adapted to existing industries and adopted by traditional power grids, end-users, developers, etc.

CONCLUSION

It is fundamental to acknowledge the key role that distributed energy systems (DES) can play in the transition to a cleaner and sustainable energy sector and the importance of developing DES design and administration tools to realise their full potential and associated benefits. Improved DES also requires the continuous assessment of barriers and difficulties that hold back the development of such systems.

Consumers' values and actions are driving the energy transition, assuming themselves the processes of power generation with renewable energy systems and energy storage with batteries and electric vehicles (EV), wisely controlling their energy consumption and behaviours, and demanding governments, regulators and industry stakeholders to act against climate change, and to ensure sustainable development.

By considering demand-side management (DSM) strategies in the design and operation of DES, more renewable energy can be added into the energy mix. As a result, active end-users play an important role in the realisation of low-carbon, low-cost power systems. DSM should be used as a fundamental tool in DES to balance renewable energy variability and moderate peak demand, avoiding extra investment in conventional power plants.

The development of demand flexibility (DF) technologies and strategies depends on the decision and disposition of customers to participate. It is essential to create appropriate incentives to increase DSM programs adoption and have a tangible effect on energy consumption, such as time-varying energy rates or other incentives related to improving the customers' experience with the use of automation and control technologies.

Under the new energy paradigm, selling electricity is not the only source of value in power systems. When active end-user in DES are linked with conventional power systems, automation, control and data management systems, the information associated with energy use has also significant value, making energy sharing and trading, DSM and energy optimization innovative business opportunities.

ACKNOWLEDGMENT

The contributions to this work come from research activities at the National Research University "Moscow Power Engineering Institute", partly funded by the Russian Government Scholarship Program for international students from the Ministry of Education and Science.

REFERENCES

Alanne, K., & Saari, A. (2006). Distributed energy generation and sustainable development. *Renewable & Sustainable Energy Reviews, 10*(6), 539–558. doi:10.1016/j.rser.2004.11.004

Bertoldi, P. (2019). Overview of the European Union policies to promote more sustainable behaviours in energy end-users. In M. Lopes, C. H. Antunes, & K. B. Janda (Eds.), *Energy and Behaviour* (pp. 451–477). Academic Press.

Bloomberg NEF. (2019). *New Energy Outlook 2019*. Retrieved from https://about.bnef.com/new-energy-outlook/

Deloitte. (2019). *Beyond the energy transition. When the demand-side is demanding change*. Retrieved from https://www2.deloitte.com/au/en/pages/energy-and-resources/articles/beyond-energy-transition.html

EIA (U.S. Energy Information Administration). (2019). *Annual Energy Outlook 2019 with projections to 2050*. Retrieved from https://www.eia.gov/outlooks/aeo/

Ellsmoor, J. (2019a, April 27). *Blockchain is the next big thing for renewable energy*. Retrieved from https://www.forbes.com/sites/jamesellsmoor/2019/04/27/blockchain-is-the-next-big-thing-for-renewable-energy/

Ellsmoor, J. (2019b, April 27). *Meet 5 Companies Spearheading Blockchain For Renewable Energy*. Retrieved from https://www.forbes.com/sites/jamesellsmoor/2019/04/27/meet-5-companies-spearheading-blockchain-for-renewable-energy/#5dac8ac7f2ae

EWF (Energy Web Foundation). (2019). *EW-DOS: The energy web decentralized operating system*. Retrieved from https://www.energyweb.org/reports/ewdos-vision-purpose/

Fairbrother, C., Guccione, L., Henchen, M., & Teixeira, A. (2017). *Pathways for Innovation: The Role of Pilots and Demonstrations in Reinventing the Utility Business Model*. Retrieved from Rocky Mountain Institute website: www.rmi.org/insights/reports/pathwaysforinnovation/

Farsangi, A. S. N., Hadayeghparast, S., Mehdinejad, M., & Shayanfar, H. (2018). A novel stochastic energy management of a microgrid with various types of distributed energy resources in presence of demand response programs. *Energy, 160*, 257–274. doi:10.1016/j.energy.2018.06.136

Grijalva, S., & Tariq, M. U. (2011). Prosumer-based smart grid architecture enables a flat, sustainable electricity industry. In *Proceedings Innovative Smart Grid Technologies (ISGT)* (pp. 1–6). IEEE. doi:10.1109/ISGT.2011.5759167

IEA. (2017). *Prospects for Distributed Energy Systems in China*. Retrieved from https://webstore.iea.org/prospects-for-distributed-energy-systems-in-china

IEA. (2018). *Population without access to electricity falls below 1 billion*. Retrieved from https://www.iea.org/commentaries/population-without-access-to-electricity-falls-below-1-billion

IEA. (2019). *CO2 Emissions from Fuel Combustion*. Retrieved from https://www.iea.org/data-and-statistics

IPCC. (2019). *Global warming of 1.5 °C*. Retrieved from https://www.ipcc.ch/sr15/

IRENA. (2019a). *Energy transition*. Retrieved from https://www.irena.org/energytransition

IRENA. (2019b). *Transforming the energy system – and holding the line on the rise of global temperatures*. Author.

IRENA. (2019c). *Tracking SDG7: the energy progress report*. Retrieved from https://www.irena.org/publications/2019/May/Tracking-SDG7-The-Energy-Progress-Report-2019

Jin, M., Feng, W., Marnay, C., & Spanos, C. (2018). Microgrid to enable optimal distributed energy retail and end-user demand response. *Applied Energy*, *210*, 1321–1335. doi:10.1016/j.apenergy.2017.05.103

Kabalci, E., & Kabalci, Y. (2019). *From smart grid to internet of energy*. Academic Press.

Kamphuis, R. (2017). Integrating demand flexibility with distributed generation – renewable energy sources (DG-RES) at the residential household and commercial customer level in electricity grids. *CIRED - Open Access Proceedings Journal, 1*, 1827-1830.

Khokhlov, A., Melnikov, Y., Veselov, F., Kholkin, D., & Datsko, K. (2018). *Distributed energy resources in Russia: development potential*. SKOLKOVO Energy Centre.

LO3 Energy. (2020). *Pando*. Retrieved from https://lo3energy.com/pando/

Power Ledger. (2019). *Technical white paper*. Retrieved from https://www.powerledger.io/wp-content/uploads/2019/11/power-ledger-whitepaper.pdf

Li, Y., Yang, W., He, P., Chen, C., & Wang, X. (2019). Design and management of a distributed hybrid energy system through smart contract and blockchain. *Applied Energy*, *248*, 390–405. doi:10.1016/j.apenergy.2019.04.132

Mabaso, M. (2018). *Community distributed energy systems: an end-user energy services approach* (Unpublished master's thesis). Stellenbosch University, South Africa.

Manjunath, M., Singh, P., Mandal, A., & Parihar, G. S. (2014). Consumer behaviour towards electricity–A field study. *Energy Procedia*, *54*, 541–548. doi:10.1016/j.egypro.2014.07.295

Mengelkamp, E., Gärttner, J., Rock, K., Kessler, S., Orsini, L., & Weinhardt, C. (2018). Designing microgrid energy markets: A case study: The Brooklyn Microgrid. *Applied Energy*, *210*, 870–880. doi:10.1016/j.apenergy.2017.06.054

RIA Novosti. (2019, December 30). *Putin signed a law on the development of green micro-generation.* Retrieved from https://ria.ru/20191230/1563033599.html

Paoletti, S., Casini, M., Giannitrapani, A., Facchini, A., Garulli, A., & Vicino, A. (2011). Load forecasting for active distribution networks. In *Proceeding 2nd IEEE PES International Conference and Exhibition on Innovative Smart Grid Technologies* (pp. 1-6). Manchester, UK: IEEE.

RMI. (2018). *Demand flexibility - the key to enabling a low-cost, low-carbon grid.* Retrieved from https://info.rmi.org/demand_flexibility

Schot, J., Kanger, L., & Verbong, G. (2016). The roles of users in shaping transitions to new energy systems. *Nature Energy, 1*(16054).

Siemens. (2018). *Design of distributed energy systems.* Erlangen: Siemens AG.

Sousa, T., Soares, T., Pinson, P., Moret, F., Baroche, T., & Sorin, E. (2019). Peer-to-peer and community-based markets: A comprehensive review. *Renewable & Sustainable Energy Reviews, 104*, 367–378. doi:10.1016/j.rser.2019.01.036

Tyagunov, M. (2017). Distributed energy systems is the future of the world's power industry. In *Proceedings of the 2nd International Conference on the Applications of Information Technology in Developing Renewable Energy Processes & Systems (IT-DREPS)* (pp. 1 - 4). Amman, Jordan: IEEE.

UCL. (2019). *Could Colombia lead the peer-to-peer energy revolution?* Retrieved from https://www.ucl.ac.uk/bartlett/sustainable/sites/bartlett/files/policybriefingp2pcolombia.pdf

UNDP. (2015). *Sustainable Development Goals (SDGs).* Retrieved from https://www.undp.org/content/undp/en/home/sustainable-development-goals.html

UNFCCC. (2015). *The Paris Agreement.* Retrieved from https://unfccc.int/process-and-meetings/the-paris-agreement/the-paris-agreement

Vallés, M., Reneses, J., Frías, P., & Mateo, C. (2016). Economic benefits of integrating Active Demand in distribution network planning: A Spanish case study. *Electric Power Systems Research, 136*, 331–340. doi:10.1016/j.epsr.2016.03.017

Vezzoli, C. (2018). Distributed/decentralized renewable energy systems. In *Designing Sustainable Energy for All. Green Energy and Technology* (pp. 23–39). Springer. doi:10.1007/978-3-319-70223-0_2

VOX. (2019). *After rising for 100 years, electricity demand is flat. Utilities are freaking out.* Retrieved from https://www.vox.com/energy-and-environment/2018/2/27/17052488/electricity-demand-utilities

Wang, Z., Yu, X., Mu, Y., & Jia, H. (2020). A distributed Peer-to-Peer energy transaction method for diversified prosumers in Urban Community Microgrid System. *Applied Energy, 260*, 1–13. doi:10.1016/j.apenergy.2019.114327

ADDITIONAL READING

Kharchenko, V., & Vasant, P. (Eds.). (2018). *Handbook of Research on Renewable Energy and Electric Resources for Sustainable Rural Development*. IGI Global. doi:10.4018/978-1-5225-3867-7

Kharchenko, V., & Vasant, P. (Eds.). (2020). *Handbook of Research on Energy-Saving Technologies for Environmentally-Friendly Agricultural Development*. IGI Global. doi:10.4018/978-1-5225-9420-8

Kharchenko, V., & Vasant, P. (Eds.). (2020). *Handbook of Research on Smart Computing for Renewable Energy and Agro-Engineering*. IGI Global. doi:10.4018/978-1-7998-1216-6

Thomas, J. J., Karagoz, P., Ahamed, B. B., & Vasant, P. (Eds.). (2020). *Deep Learning Techniques and Optimization Strategies in Big Data Analytics*. IGI Global. doi:10.4018/978-1-7998-1192-3

Vasant, P., Zelinka, I., & Weber, G.-W. (Eds.). (2018). *Intelligent Systems and Computing*. Springer.

P. Vasant, I. Zelinka, & G.-W. Weber (Eds.). (2019). Intelligent Computing & Optimization*: Proceedings of the 2nd International Conference on Intelligent Computing and Optimization 2019 (ICO 2019).* Cham, Switzerland: Springer.

KEY TERMS AND DEFINITIONS

Blockchain: A system in which a digital record of transactions made in cryptocurrency (a digital currency such as bitcoin), are maintained across several computers that link the participants in a network.

Demand Flexibility (DF): Portion of electricity demand in a power system that can be reduced, increased or shifted within a specific duration by providing price signals or controlling energy loads.

Demand Side Management (DSM): Actions that reduce or limit end-use energy consumption through the management of equipment or processes: usually during peak demand and/or in times of limited supply.

Distributed Energy System (DES): Power system where electricity is generated near the consumption points and energy assets, such as small-scale renewable energy generation, energy storage, electric vehicles (EV) and consumption management tools, are distributed throughout the array.

Energy Transition: Global transformation of the energy sector from fossil-based systems of energy production and consumption to renewable energy sources and low-carbon technologies.

Peer-to-peer (P2P) trading: Direct purchase and sale of electricity between two or more parties connected to a power grid. Participants can usually decide with whom they trade energy.

Prosumer: A person or entity who both consumes and produces electricity.

Renewable energy: Energy produced using resources that are naturally replenishing: such as sunlight, wind, rain, tides, waves, biomass, and geothermal heat.

Chapter 10

Selection of an Information Source and Methodology for Calculating Solar Resources of the Kyrgyz Republic

Alisher F. Narynbaev
https://orcid.org/0000-0003-2981-7519
Moscow Power Engineering Institute (National Research University), Russia

Baatai M. Maksatov
https://orcid.org/0000-0001-5133-8958
Moscow Power Engineering Institute (National Research University), Russia

Alexey Gennad'evich Vaskov
Moscow Power Engineering Institute (National Research University), Russia

Galina V. Deryugina
Moscow Power Engineering Institute (National Research University), Russia

Roman V. Pugachev
Moscow Power Engineering Institute (National Research University), Russia

ABSTRACT

Detailed data on incoming solar radiation are needed in the design of solar energy systems of any scale: from large PV plants to small off-grid systems. However, in most cases, obtaining data on measurements of solar radiation is connected with difficulties due to financial or technical restrictions. Often, ground-based measurements of solar radiation are either not carried out at all or only the value of the global horizontal intensity of solar radiation is measured. The aim of the present study is to review and to verify some existing empirical models of the global solar radiation and its components for the climatic conditions of Kyrgyzstan as well as to estimate the applicability of Meteonorm database model for the available solar radiation in the territory of Kyrgyzstan. The necessity to select the most suitable models of the solar radiation is called by the lack of similar studies on this direction for the conditions of the country.

DOI: 10.4018/978-1-7998-3645-2.ch010

INTRODUCTION

Solar radiation is the most significant type of renewable energy. It impacts on various physical, chemical, biological processes on Earth. Solar energy has been using by humans in wide range of technical applications. However solar energetics grows rapidly, solar radiation data is not completely available by several reasons: absence of meteorological stations or ground measurements. The data plays important role for designing and exploiting solar power systems, such as photovoltaic and thermal systems. Therefore, it is needed to develop modelling approaches for calculating solar radiation in a given point where ground measurements are not available. The paper gives an analytical review of empirical models that estimate amount of incoming solar radiation and are built on the basis of linear statistical methods. The areas of applicability of each of the considered approaches for estimating solar radiation are determined. The classification of the main statistical models by used calculation parameters is carried out. In case of Kyrgyzstan, there are four actinometric stations (AS), located in the North-Eastern part of the country, and for which there is no solar radiation data in open sources for the last 40 years. Databases based on mathematical modeling and results of satellite sensing of the Earth's atmosphere and surface can be used to estimate solar resources in other regions of the country. Among the developed global and regional databases in the field of solar energy in the world, the Swiss Meteonorm database is the most widely used, the adequacy of the data of which can be verified by comparison with the data of actinometric stations. The paper compares the long-term monthly global horizontal solar radiation (GHI) from Meteonorm the database with the data of seven actinometric stations of Kyrgyzstan and neighboring countries from two available databases: World Radiation Data Center (WRDC), Handbook on climate of the USSR (HC). The considered databases are briefly described. Based on the results of the analysis, it is shown that there is an excess of values in the data from Meteonorm database relatively to the data obtained from actinometric stations from WRDC and HC databases. Moreover, the excess varies not only within the year, but also depends on the latitude of the terrain. The analysis of the long-term variability of GHI according to the data of three ASs in the territory of neighboring countries with Kyrgyzstan from the WRDC database did not show a trend of its change. The methods used for a mathematical model developed by the authors to restore the average annual and monthly GHI values according to Meteonorm database in Kyrgyzstan, taking into account the latitude of the terrain are described. The paper presents a refined theoretical potential of solar energy in Kyrgyzstan and shows that average annual amount of GHI varies insignificantly: from 1656 kWh/m^2 to 1856 kWh/m^2. The average GHI value is 1756 kWh/m^2.

BACKGROUND

The evaluation of solar resources is one of the first and, obviously, one of the most important stages in the development of projects for solar energy systems of any type. Indeed, only a reliable evaluation of the available resource allows you to make the right technical or economic decision. Currently, developers and researchers in the field of solar energy have become available databases containing the necessary information. These sources include such global databases as the World Radiation Data Center (WRDC) (WRDC, 2019), Swiss database Meteonorm (Meteonorm, 2019), USA database NASA POWER (NASA, 2019) and others. There are also regional databases, which is considered in this research, e.g. Handbook on the Climate of the USSR database (Handbook on the Climate of the USSR, 1989) .

MAIN FOCUS OF THE CHAPTER

Classification of Solar Radiation Estimating and Predicting Methods

The problem of estimating and predicting solar radiation on the Earth's surface is usually solved using methods that are usually divided into physical and statistical ones, but in practice the boundaries between these approaches are blurred. Depending on the specific problem, hybrid models can also be used that combine both of the above methods. Physical forecasting techniques model atmospheric phenomena using numerical weather prediction (NWP) models or satellite imagery of the sky.

The output of the NWP model can also be used in statistical models to improve the forecast, but only if this kind of data is available (Letendre, Makhyoun, & Taylor, 2014). Due to the high complexity of implementation and the need for large computing power, which in turn requires enormous financial investments, NWP models are developed and operate only in developed countries of the world - the USA, Canada, Germany, Great Britain, Japan, and Russia. National meteorological services of other states use finished products of model calculations of leading countries. Empirical models predict the level of insolation based on values obtained statistically. This kind of model can be divided into two categories: 1) traditional empirical models (linear and non-linear regression models), 2) models based on data mining.

In modern solar radiation estimating and predicting systems the companies that supply solar radiation forecasts in practice use different combinations of forecasting methods (Letendre, Makhyoun, & Taylor, 2014). The best way to determine the amount of insolation at a given geographical point is to conduct and analyze long-term ground-based measurements and observations of incoming solar radiation using appropriate measuring equipment (actinometers, pyranometers, pyrheliometers, etc.). Since the implementation of such measurements is to a certain extent labor-consuming and expensive, and access to the database is often limited, in order to estimate the incidence of solar radiation in many developing countries of the world, statistical methods based on available local factors and measurement data from meteorological stations. Ground-based observations and measurements of the main meteorological parameters are carried out everywhere and in most cases are available. Empirical models are usually based on the following parameters (Besharat, Dehghan, & Faghih, 2013): 1) astronomical parameters (solar constant, distance between the Earth and the Sun, the Sun declination angle and hour angles), 2) geographical factors (latitude and longitude of the area, altitude), 3) geometric factors (azimuthal angle of inclination of the receiving surface, angle of inclination of the receiving surface to the horizon, the height of the sun above the horizon, azimuthal angle of the position of the sun), 4) physical parameters (scattering of air molecules, water vapor content, scattering of dust and other atmospheric components), 5) meteorological parameters (extraterrestrial solar radiation, sunshine duration, temperature, precipitation, relative humidity, cloud cover, soil temperature, etc.).

In view of the relatively large number of published works and studies devoted to the search for the most accurate empirical models for different geographical locations, the choice of the optimal ratio for a specific geographical point seems rather complicated and depends on the input parameters used and the resulting accuracy (Besharat, Dehghan, & Faghih, 2013).

a. Traditional Empirical Models for Estimating Solar Radiation

Empirical models, obtained in the course of a regression analysis of data on various types of accessible parameters that affect the insolation value, are conventionally considered traditional. At meteorological stations, global solar radiation is measured at a horizontal receiving plane using a pyranometer, which records the total incoming solar radiation from the entire sky. To estimate global solar radiation at an inclined receiving surface, there are corresponding conversion models.

i. Global Solar Radiation

Global solar radiation on a horizontal receiving plane (GHI) is represented as the sum of the components of the form:

$$H = H_d + H_b, \tag{1}$$

where H is global solar radiation on a horizontal surface, H_d and H_b are the diffuse and direct components of solar radiation, respectively. Based on the measurement data for the horizontal receiving plane during modeling for an inclined plane, it is necessary to clarify the values of each of the components. If only the H quantity is known, it is necessary to enter such a parameter as clearness index K_T (Iqbal, An Introduction to Solar Radiation, 1983):

$$K_T = \frac{H}{H_0}, \tag{2}$$

where H_0 is the amount of extraterrestrial solar radiation on a horizontal surface.

The first model for estimating the monthly average daily amount of GHI was presented by Angstrom (Angstrom, 1924). This well-known expression estimates monthly average daily amount of global solar radiation $\overline{H}$ as a function of sunshine duration and in the modification of Prescott (Ma & Iqbal, 1982) is expressed as the following formula:

$$\frac{\overline{H}}{\overline{H_0}} = a + b\left(\frac{\overline{T_{sd}^m}}{\overline{T_{sd}^c}}\right), \tag{3}$$

where $\overline{H_0}$ is monthly average daily amount of extraterrestrial solar radiation on a horizontal surface, a and b are empirical coefficients, $\overline{T_{sd}^m}$ is the actual measured average monthly sunshine duration in hours, $\overline{T_{sd}^c}$ is the average monthly sunshine duration, calculated for a given month as follows:

$$\overline{T_{sd}^c} = \frac{\left\{\sum_{n=n_1}^{n=n_2}\left[\frac{2}{15}\cos^{-1}\left(-\mathrm{tg}\varphi \cdot \mathrm{tg}\delta\right)\right]\right\}}{n_2 - n_1}, \tag{4}$$

where n_2,n_1 are the ordinal (from the beginning of the year) numbers of the last and first days of the month, φ is the geographical latitude of the locality, δ is the declination angle of the Sun.

Daily amount of extraterrestrial solar radiation on a horizontal surface at a given latitude can be calculated on the basis of the formula proposed by Duffy and Beckman, reduced to the calculation of daily values (Duffie, Beckman, & Worek, 2003):

$$H_0 = I_{SC} \cdot \frac{24}{\pi}\left(1 + 0.033\cos\frac{360n}{365}\right)\left(\frac{\pi\omega_7}{180}\sin\varphi\sin\delta + \cos\varphi\cos\delta\sin\omega_7\right), \quad (5)$$

where I_{SC} is the solar constant, taken equal to 1367 W/m²; ω_s – hourly sunset angle, which can be calculated as $\cos\omega_s^{-1} = -\tan\varphi\tan\delta$; n is the serial number of the day in the year (Bhardwaj & Bhola, 2016).

Empirical coefficients a and b in expression (3), which vary greatly depending on the locality, were analyzed in (Lowe, Duffie, & Smith, 1965) and by many other researchers based on data on solar irradiation received at different geographical points and climatic conditions. The following are some of the many linear regression models based on the Angstrom-Prescott expression (3).

Table 1. Models based on Angstrom-Prescott expression

Authors	Model	Geographical location
(El-Metwally, 2005)	$\frac{\overline{H}}{H_0} = 0.228 + 0.527\left(\frac{\overline{T_{sd}^m}}{T_{sd}^c}\right)$	Egypt
(Jin, Yezheng, & Gang, 2005)	$\frac{\overline{H}}{H_0} = 0.1332 + 0.6471\left(\frac{\overline{T_{sd}^m}}{T_{sd}^c}\right)$	China
(Bakirci, 2009)	$\frac{\overline{H}}{H_0} = 0.2786 + 0.416\left(\frac{\overline{T_{sd}^m}}{T_{sd}^c}\right)$	Turkey

Rietveld's modification (Rietveld, 1978) is a universalization of Angstrom's method described above. Based on the analysis of data on the distribution of the coefficients a and b, he found that a and b are linearly related to

$$\left(\frac{\overline{T_{sd}^m}}{T_{sd}^c}\right): a = 0.1 + 0.24\left(\frac{\overline{T_{sd}^m}}{T_{sd}^c}\right), \quad b = 0.38 + 0.08\left(\frac{\overline{T_{sd}^m}}{T_{sd}^c}\right).$$

Then, after appropriate substitutions, expression (3) takes the following form:

$$\frac{\overline{H}}{H_0} = 0.18 + 0.62\left(\frac{\overline{T_{sd}^m}}{T_{sd}^c}\right). \quad (6)$$

According to the author, equation (6) is applicable worldwide.

Glover and McCulloch modification (Glover & McCulloch, 1958) is based on attempt to clarify Angstrom's equation (3) by introducing the factor of the geographical latitude φ of a given point:

$$\frac{\overline{H}}{H_0} = 0.29\cos\varphi + 0.52\left(\frac{\overline{T_{sd}^m}}{T_{sd}^c}\right). \tag{7}$$

Expression (7) is valid for φ<60°.

Page model (Page, 1961) is also considered applicable for any geographic location:

$$\frac{\overline{H}}{H_0} = 0.23\cos\varphi + 0.48\left(\frac{\overline{T_{sd}^m}}{T_{sd}^c}\right). \tag{8}$$

Dognio and Lemoine (Dogniaux & Lemoine, 1983) and Soler (Soler, 1990), who processed the data from 100 European stations, received separate calculated expressions for each month of the year.

Models obtained by the method of nonlinear regression analysis are also presented in significant numbers. Bahel nonlinear model (Bahel, Bakhsh, & Srinivasan, 1987) was created in the process of data analysis with 48 stations widely scattered around the world. It is believed to be applicable at any geographical point:

$$\frac{\overline{H}}{H_0} = 0.16 + 0.87\left(\frac{\overline{T_{sd}^m}}{T_{sd}^c}\right) - 0.61\left(\frac{\overline{T_{sd}^m}}{T_{sd}^c}\right)^2 + 0.34\left(\frac{\overline{T_{sd}^m}}{T_{sd}^c}\right)^3. \tag{9}$$

The following are some of nonlinear models based on sunshine duration as an argument.

Table 2. Nonlinear models based on sunshine duration as the argument

Authors	Model	Geographical location
(El-Metwally, 2005)	$\frac{\overline{H}}{H_0} = a^{\left(\frac{\overline{T_{sd}^m}}{T_{sd}^c}\right)}$	Egypt
(Almorox & Hontoria, 2004)	$\frac{\overline{H}}{H_0} = a + b \cdot e^{\left(\frac{\overline{T_{sd}^m}}{T_{sd}^c}\right)}$	Spain
(Bakirci, 2009)	$\frac{\overline{H}}{H_0} = a\left(\frac{\overline{T_{sd}^m}}{T_{sd}^c}\right)^b$	Turkey

Black's model (Black, 1956) is based on the fact that data on the total cloud cover of the sky are much smoother than data of solar irradiation. The relationship between solar radiation and cloudiness was established by Kimball (Kimball, 1919). In this regard, the modification includes an indicator of daily fraction of sky covered by clouds:

$$\frac{\overline{H}}{\overline{H_0}} = a_1 - b_1 C , \tag{10}$$

where C is monthly average daily fraction of sky covered by clouds.

Based on the processing of various data from a wide geographical locations, Black proposed the following expression (Black, 1956):

$$\frac{\overline{H}}{\overline{H_0}} = 0.803 - 0.34C - 0.458C^2 ,\ C \leq 0.8 . \tag{11}$$

Badescu (Badescu, Correlations to estimate monthly mean daily solar global irradiation: application to Romania., 1999) offered the following relationship:

$$\frac{\overline{H}}{\overline{H_0}} = a_1 + b_1 C , \tag{12}$$

where $a = 0.8683$; $b = -0.6254$ for Romania conditions.

Since observational data on cloudiness and sunshine duration are not widely available, there is a need for models that take as a basis such a parameter as air temperature. Temperature models suggest a directly proportional relationship between the difference between the maximum and minimum air temperatures and the fraction of extraterrestrial solar radiation reaching the Earth surface (Besharat, Dehghan, & Faghih, 2013).

Hargreaves and Samani model (Hargreaves & Samani, 1982) expresses as

$$\frac{\overline{H}}{\overline{H_0}} = a\left(T_{max} - T_{min}\right)^{0.5} , \tag{13}$$

where a is empirical coefficient, for arid regions is equal to 0.17; for coastal regions $a = 0.19$; and for internal regions a is recommended to be taken equal to 0.16.

Bristow-Campbell model (Bristow & Campbell, 1984) is an exponential expression of the following form:

$$\frac{\overline{H}}{\overline{H_0}} = a\left(1 - e^{-b\Delta T^c}\right) , \tag{14}$$

where ΔT is the temperature difference; a is empirical coefficient reflecting the maximum amount of solar radiation on clear sky conditions; b and c are empirical coefficients that control the value to which a approaches as the temperature difference increases.

(Donatell & Campbell, 1998), (Hutchinson, Goodin, Vanderlip, & Knapp, 1999) considered modifications of Bristow-Campbell model. (Allen, 1997) and (Annandale, Jovanic, Benade, & Allen, 2002) presented modifications of Hargreaves-Samani model.

A lot of studies are devoted to attempts of using different meteorological parameters available in each individual case, such as precipitation, relative humidity, soil temperature, evaporation and pressure, along with the main factors for estimating GHI: sunshine duration, air temperature and cloudiness (Besharat, Dehghan, & Faghih, 2013).

The dependences between GHI and average relative air humidity *RH* and sunshine duration were derived by Swartman and Ogunlade (Swartman & Ogunlade, 1967):

$$\overline{H} = a\left(\frac{\overline{T_{sd}^{m}}}{T_{sd}^{c}}\right)^{b} RH^{C}, \tag{15}$$

$$\frac{\overline{H}}{H_0} = a + b\left(\frac{\overline{T_{sd}^{m}}}{T_{sd}^{c}}\right) + cRH, \tag{16}$$

where *a*, *b*, *c* are empirical coefficients.

Gariepy (Gariepy, 1980) proposed calculating the coefficients *a* and *b* in the Angstrom-Prescott equation (3) as follows:

$$a = 0.3791 - 0.0041T - 0.0176P, \tag{17}$$

$$b = 0.481 + 0.0043T + 0.0097P, \tag{18}$$

where *T* is the average air temperature, °C; *P* is the average precipitation, mm.

In the process of verification of 5 different models for estimating GHI based on data from 48 weather stations in China, Chen proposed his own expression (Chen, Ersi, Yang, Lu, & Zhao, 2004):

$$\frac{\overline{H}}{H_0} = a\ln(T_{max} - T_{min}) + b\left(\frac{\overline{T_{sd}^{m}}}{T_{sd}^{c}}\right)^{c} + d, \tag{19}$$

where the averaged values of the empirical coefficients *a*, *b*, *c* and *d* for Chinese conditions respectively are 0.04, 0.48, 0.83 and 0.11.

Various expressions are proposed in the following studies (Maghrabi, 2009), (Trabea & Shaltout, 2000) and (Supit & Van Kappel, 1998) connecting several meteorological parameters with the global solar radiation. Monthly average daily global solar irradiation on a tilted surface is determined by the corresponding conversion factors from a horizontal to an inclined surface (Foster, Ghassemi, & Cota, 2010). One need to carry out calculations for a specific day of the month. Klein (Klein, 1977) proposed the most suitable characteristic days for each month of the year.

ii. Empirical Models for Estimating the Diffuse Component of GHI

Liu and Jordan model (Liu & Jordan, 1960) is widely used expression, despite the fact that the calculations are based on daily rather than hourly measurements and performed in the climatic conditions of Blue Hill, Massachusetts, USA. Liu and Jordan model can be described as follows:

$$\frac{\overline{H_d}}{\overline{H}} = 1.39 - 4.027\overline{K_T} + 5.531\overline{K_T}^2 - 3.108\overline{K_T}^3;\ 0.3 < \overline{K_T} < 0.7\,. \tag{20}$$

Page model (Page, 1961) was obtained on the basis of data analysis from ten widely located stations between 40°N. and 40°S:

$$\frac{\overline{H_d}}{\overline{H}} = 1 - 1.13\overline{K_T}\,. \tag{21}$$

Iqbal model (Iqbal, Correlation of average diffuse and beam radiation with hours of bright sunshine, 1979) uses only the number of sunshine hours as measured initial data and for the conditions of Canada:

$$\frac{\overline{H_d}}{\overline{H_0}} = 0.163 + 0.478\left(\frac{\overline{T_{sd}^m}}{T_{sd}^c}\right) - 0.655\left(\frac{\overline{T_{sd}^m}}{T_{sd}^c}\right)^2. \tag{22}$$

Hourly extraterrestrial solar radiation on a horizontal surface is calculated according to the following equation (Iqbal, An Introduction to Solar Radiation, 1983):

$$I_0 = \frac{12\cdot 3.6}{\pi} I_{SC} E_0 \left[(\sin\varphi\cos\delta)\cdot(\sin\omega_2 - \sin\omega_1) + \frac{\pi(\omega_2 - \omega_1)}{180}(\sin\varphi\sin\delta) \right], \tag{23}$$

where E_0 is the eccentricity correction factor, which can be calculated using Spencer equation (Spencer, 1971).

For hourly values of GHI, expression () takes the following form:

$$I = I_d + I_b. \tag{24}$$

Methods for calculating hourly values of diffuse solar irradiation (DHI) are divided into two categories (Maleki, Hizam, & Gomes, 2017): parametric and decomposition models. Parametric models include those that require initial data on environmental conditions, such as atmospheric transparency, cloudiness, humidity, etc. (Gueymard, 1993).

ASHRAE (American Society and Heating, Refrigerating and Airconditioning Engineers (ASHRAE), 1972) proposed the following expressions of the hourly GHI on a cloudless day:

$$I = I_d + I_{bN}\cos\theta_z, \tag{25}$$

$$I_{bN} = Ae^{\left[\frac{-B}{\cos\theta_z}\right]}, \tag{26}$$

$$I_d = CI_{bN}, \tag{27}$$

where I_{bN} is direct normal component (DNI) of global solar irradiation; A, B, C – empirical constants obtained on the basis of processing measurement data for a specific region.

In ASHRAE model, the coefficients are calculated on the basis of measurements in the USA, in Machler and Iqbal model (Machler & Iqbal, 1967) they based on data from Canada, in Parishwad model (Parishwad, Bhardwaj, & Nema, 1997) they were obtained according to the territory of India, and in Nizhegorodov model (Nijegorodov, 1996) coefficients were calculated on the basis of measurements in Botswana, Zimbabwe and Namibia.

Decomposition models are based on hourly clearness index:

$$M_t = \frac{I}{I_0}. \tag{28}$$

M_t characterizes atmospheric effect on the distribution of solar radiation and depends on climatic parameters, geographical location and season (Ayodele & Ogunjuyigbe, 2015). Decomposition models are formed in a general equation:

$$I_d = r_d I, \tag{29}$$

where r_d is the coefficient of diffuse component fraction of GHI.

Table 3. Hourly DHI decomposition models

Authors	M_t range	r_d	Geographical location
(Liu & Jordan, 1960)	$0.75 < M_t \leq 1$	$0.384 - 0.416M_t$	USA, Canada
(Orgill & Hollands, 1977)	$0 < M_t \leq 0.35$, $0.35 \leq M_t \leq 0.75$, $0.75 < M_t \leq 1$	$1 - 0.249M_t$, $1.577 - 1.84M_t$, 0,177	Canada
(Boland, Scott, & Luther, 2001)	any M_t values	$\frac{1}{1+e^{7.997(M_t-0.586)}}$	Australia
(Lam & Li, 1996)	$0 < M_t \leq 0.17$, $0.15 < M_t \leq 0.17$, $0.17 < M_t \leq 1$	0.977 $1.237 - 1.361M_t$ 0.273	China

Hourly global solar radiation on a tilted surface can be defined as follows (Iqbal, An Introduction to Solar Radiation, 1983):

$$I_\beta = I_{b\beta} + I_{d\beta} + I_r, \tag{30}$$

where $I_{b\beta}$ is hourly sum of direct component of solar radiation on a tilted surface, $I_{d\beta}$ is the hourly sum of diffuse component of solar radiation on the tilted surface, and I_r is the hourly sum of reflected component of solar radiation.

Hourly models for calculating diffuse component of solar radiation on a tilted surface are divided into isotropic (uniform distribution of the diffuse component over the sky) and anisotropic ones, which include calculations of the Sun position in the sky.

One of the most well-known anisotropic models for distributing diffuse component of solar radiation on an inclined surface is Perez model (Perez, Stewart, Seals, & Guertin, 1988), which divides the diffuse component into three elements: diffuse radiation from a near-solar part of the sky, isotropic background, and horizon zones. Perez model is expressed in the following equation:

$$I_{d\beta} = I_d \left[\frac{1+\cos\beta}{2}(1-F_1) + F_1 \frac{a_1}{a_2} + F_2 \sin\beta \right], \quad (31)$$

in which a_1 and a_2 are solid angles surrounding the circumsolar part of the sky during insolation respectively on an inclined and horizontal surface; F_1 is brightness coefficient of the near-solar section; F_2 is horizon brightness coefficient.

Table 4. Isotropic models

Authors	$I_{d\beta}$ expression	Geographical location
(Liu & Jordan, 1961)	$I_{d\beta} = \left(\frac{1+\cos\beta}{2}\right) I_d$	USA, Canada
(Badescu, 3D isotropic approximation for solar diffuse irradiance on tilted surfaces, 2002)	$I_{d\beta} = \left(\frac{1+\cos(2\beta)}{2}\right) I_d$	Romania
(Tian, Davies-Colley, Gong, & Thorrold, 2001)	$I_{d\beta} = \left(1-\frac{\beta}{180}\right) I_d$	New Zealand

For climatic conditions of different countries the most preferable results are demonstrated by the models of Perez, Hay-Davis (Maleki, Hizam, & Gomes, 2017) and Steven-Unsworth (Steven & Unsworth).

b. Data Mining Models

According to many researchers, forecasting problems are most successfully solved by data mining methods using artificial intelligence, in particular, using artificial neural networks (Abrukov, et al., 2017). Artificial Intelligence (AI) methods include such areas as artificial neural networks (ANNs), genetic algorithms, expert systems, fuzzy logic, various hybrid systems (Mellit, Artificial intelligence technique for modelling and forecasting of solar radiation data: a review, 2008).

i. Artificial Neural Networks

Artificial neural networks are more often than other AI methods used to estimate or predict the amount of solar radiation. The application of ANNs can be divided into two categories: 1) forecasting, which consists in predicting a change in the value of X at time $t + \Delta t$, that is, an estimate of $X(t + \Delta t)$ based on data at time t, $X(t)$; 2) estimation of the data value $Y(t)$ from the known i-th values of the quantity $X_i(t)$. This type of model makes it possible to determine $Y(t)$ in regions where there are no measurements of $Y(t)$, but $X_i(t)$ measurement data are fully available (Notton, et al., 2012).

Basic structure of an artificial neural network (ANN) includes an input layer with n input signals, a hidden (computational layer), an output layer consisting of i neurons, sum-function, synaptic weights, threshold and activation function. The functioning of the neural network is divided into two stages: training and generating the result. The use of ANN in calculations related to power plants based on renewable energy sources is discussed in detail in the study of Kalogirou (Kalogirou, 2001). The direct application of methods using ANN to select the optimal parameters of photovoltaic systems is described by Mellit and Kalogirou (Mellit & Kalogirou, Artificial intelligence techniques for sizing photovoltaic systems: A review, 2009).

A significant amount of studies has been presented using ANNs with different input arguments and degrees of complexity of the neural network. Compared with other methods, ANN models solve the estimation and forecasting problems with the least error, which is a key advantage along with the ability to use several available parameters at once as input features.

Table 5. ANN models for estimating monthly average daily GHI

Authors	Input parameters	Output parameters	Estimation accuracy	Geographical location
(Lazzús, Pérez Ponce, & Marín, 2011)	$\overline{V_{wind}}, \overline{t_{amb}}, \overline{\varphi_h}, \overline{t_{soil}}$ (soil temperature)	$\overline{H}$	94%	Chile
(Ouammi, Zejli, Dagdougui, & Benchrifa, 2012)	ϕ, ψ, altitude, $\overline{H}$	$\overline{H}$	98.7%	Morocco

ANN applications for estimating solar radiation incidence on a tilted surface in comparison with empirical anisotropic and isotropic models show much more accurate results and are also used to determine the optimal inclination angle of a surface for conditions of a region under consideration.

Table 6. ANN models for estimating solar radiation on a tilted surface

Authors	Input parameters	Output parameters	Estimation accuracy	Geographical location
(Celik & Muneer, 2013)	I, δ, ω	I_β	98.7%	Turkey
(Notton, et al., 2012)	I, I_0, θ_z, *hour*	I_β (β=45° and β=60°)	99.82%	France

(Mellit & Pavan, A 24-h forecast of solar irradiance using artificial neural network: application for performance prediction of a grid-connected PV plant at Trieste, Italy, 2010), (Wanga, Wanga, & Sub, 2011) and (Capizzi G, Napoli C, & F., 2012) describe the application of ANN technology to solve the problems of short-term solar radiation forecasting. In (López G, Battles, & Tovas-Pescador, 2005) the authors consider DNI forecasting model.

ii. Fuzzy Logic

Solar radiation estimation is also can be conducted by fuzzy logic modeling. In (Rizwan, Jamil, Kirmani, & Kothari, 2014), a model with fuzzy logic was proposed for estimating the monthly average daily GHI values. Monthly average daily sunshine duration, average monthly temperature, latitude, longitude, altitude and number of month were used as the input parameters of the model. RMSE for the 4 geographical points of India was 5.37%.

Şen (Şen, 1998) developed an algorithm with fuzzy logic for predicting the average daily GHI, processing data on the actual and measured sunshine duration in Turkey. (Gómez & & Casanovas, 2003) research is devoted to a model using fuzzy logic for predicting global solar radiation on an inclined surface. In (Suganthi, Iniyan, & Samuel, 2015) authors presented a detailed overview of fuzzy logic-based models use developed in the context of power plant systems based on renewable energy sources.

iii. Support Vector Machine

Support Vector Machine method is another effective data mining method used to predict solar radiation and is generally used for both classification and regression problems. Mohammadi (Mohammadi, Shamshirband, Anisi, Alam, & Petković, 2015) applied SVM to predict average daily GHI, setting the measured and theoretical sunshine duration and value of extraterrestrial solar radiation. The error was approximately 4%. In (Zeng & Qiao, 2013) a short-term SVM forecasting model is described. In (Ekici, 2014) the author presents a SVM model for predicting solar radiation a day ahead.

Mellit (Mellit & Kalogirou, Artificial intelligence techniques for sizing photovoltaic systems: A review, 2009) provides various examples of hybrid solar radiation prediction models that combine ANN and fuzzy logic, as well as combinations of SVM and ANN.

c. Comparison of Traditional Empirical Models for Estimating Solar Radiation

After classifying the models, the results of modeling solar radiation using the previously mentioned empirical relationships in comparison with the long-term averaged data of ground measurements from 4 actinometric stations in Kyrgyzstan were compared.

Due to the fact that an analysis of the degree of influence of various factors on the intensity of solar radiation in the country has not been previously conducted, the results that provided by universal empirical models for estimating monthly average daily amounts of GHI were checked firstly. Simulation results were also compared with data from NASA open database (NASA, 2019).

The comparison results shown in Table 7, Fig.1, Fig.2, Fig.3 indicate that most universal empirical models estimate the monthly average daily amounts of solar radiation with low accuracy, taking into account the averaging of indicators from 4 actinometric stations. To reduce the modeling error, it is necessary to adjust the empirical coefficients for each of the models by conducting a regression analysis based on measurements with each of 4 actinometric stations separately. In subsequent studies, it is

Table 7. Results of comparing simulation results with long-term averaged measurement data

Model	*MAPE*, %	*RMSE*, kWh/m²•day	*MBE*, kWh/m²•day	*r*
NASA POWER	9.97	0.56	0.44	0.99
GHI models Based on Sunshine Duration				
Bahel model	8.47	0.45	0.11	0.98
Rietveld model	10.58	0.53	0.33	0.98
GHI models based on other meteorological parameters				
Gariepy model	6.97	0.38	-0.2	0.98
Hargreaves model	11.67	0.64	0.26	0.98
Chen model	12.99	0.65	0.57	0.98
Badescu model	15.86	0.88	0.68	0.96
Diffuse solar radiation models				
Liu and Jordan model	30.16	0.58	0.67	0.97
Page model	24.74	0.47	0.59	0.95
Iqbal model	18.14	0.34	0.42	0.86

Figure 1. Comparison of ground-based measurements, NASA POWER database and the calculated values of monthly average daily amounts of GHI by universal empirical models that use sunshine duration as the argument

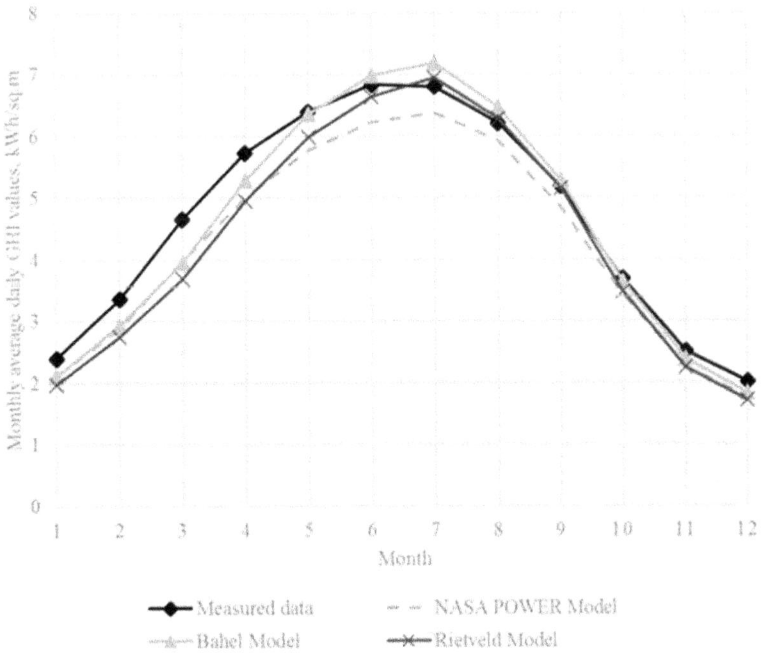

Figure 2. Ground-based measurements, NASA POWER database data, and the calculated values of monthly average daily amounts of global solar radiation by empirical models using other meteorological parameters

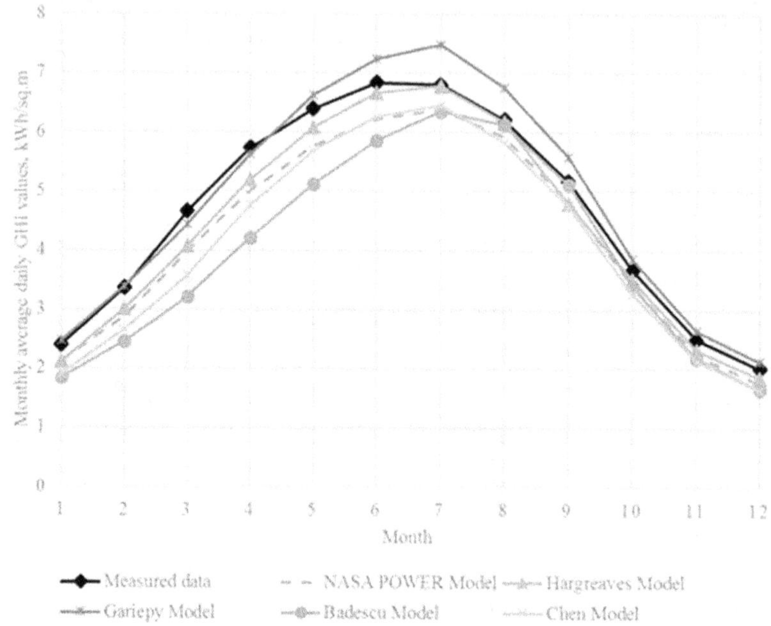

Figure 3. Ground-based measurements, NASA POWER database data, and the calculated values of monthly average daily amounts of diffuse solar radiation by empirical models

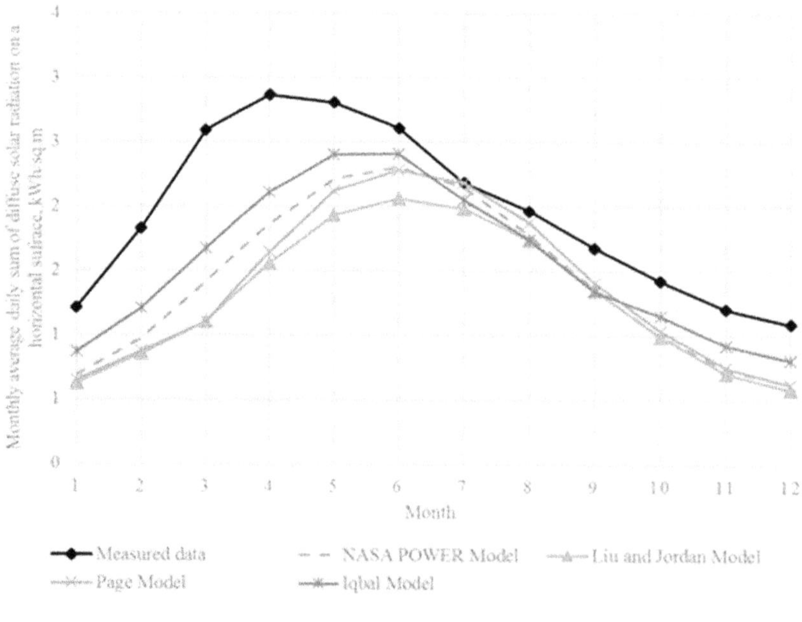

planned to carry out a detailed analysis of the applicability of statistical models to estimate the intensity of solar radiation for the climatic conditions of Kyrgyzstan, to calculate the coefficients for empirical models and to verify with actual measured data.

Features of Using Meteonorm Software for Evaluating Solar Resources of Kyrgyzstan

The main information used in the research of databases on solar resources is presented in table 8.

Table 8. Basic information of solar radiation databases used in this research

Databases	Region	Source	Observation period, years	Availability	Data
Meteonorm 7.0	Worldwide	8325 points: ASs, 5 satellite models	1980-1991 1991-2010	Paid software (commercial and educational licenses) (https://meteonorm.com/)	Long-term average monthly values, generating average hourly data.
World Radiation Data Center	Worldwide	1195 data points (29 points on the territories of Russia)	Since 1964	Free (http://wrdc.mgo.rssi.ru)	Daily and monthly observation data
HC	Former USSR territory	166 data points	1964-1980	Property of the Department of Hydropower and Renewable Energy Sources at NRU MPEI	Long-term average monthly and hourly data
NASA POWER	Worldwide	1195 data points	Since 1983	Free (https://power.larc.nasa.gov/)	Long term average monthly and hourly data

a. Databases

i. HC

HC database was developed at the Department of Hydropower and Renewable Energy Sources at the National Research University «Moscow Power Engineering Institute» (NRU MPEI). HC database includes information from 166 actinometric stations for solar radiation, obtained from Handbook on the Climate of the USSR (Handbook on the Climate of the USSR, 1989). The data is averaged over the period from 1964 to 1980. The database also includes the incidence of horizontal solar radiation (global, direct and diffuse) (in MJ and kWh per 1 m^2). These values are presented in the form of hourly incidence of solar radiation for monthly average days of each month in the average long-term year; in the form of monthly average daily sum of solar radiation; in the form of monthly amounts of solar radiation. In addition, HC database contains monthly average surface reflectance coefficients (albedo) and daily extraterrestrial horizontal solar radiation. The selection of 166 actinometric stations is carried out according to any of three features: by the serial number of the actinometric station in the database; by the name of the actinometric station; by the geographical coordinates of the actinometric station. When setting the calculated point A (φ, ψ), which does not coincide with the data of the actinometric station, the nearest actinometric station will be selected.

ii. Meteonorm

Meteonorm offers unique access to the Global Energy Balance Archive Data (GEBA). The GEBA data is provided from national weather services and fulfils the quality criteria of the World Meteorological Organization (WMO). It also includes high quality measurement networks such as the Baseline Surface Radiation Network (BSRN) or the networks of MeteoSwiss and the German Weather Service. Meteonorm delivers accurate weather data for any place on Earth: radiation, temperature, humidity, wind, precipitation and other climate parameters. Based on data from 8325 weather stations, 5 geostationary satellites and 30 years of experience, Meteonorm offers interpolation models, delivering global weather data with good accuracy. Meteonorm provides data in several ways: monthly GHI values as ASCII files (named xxxxMON.DAT), which may be directly read by PVsyst for creating *.SIT file and hourly GHI values as ASCII files (named xxxxHOUR.DAT). A stochastic model builds the hourly data. Moreover, Meteonorm proposes other benefits e.g. calculation solar irradiance for tilted planes.

iii. NASA SSE

NASA-SSE (Surface Meteorology and Solar Energy programmer) are monthly GHI data, averaged of satellite measurements from 1983-2005 period, provided for any cell in a grid of 1°x1° worldwide (1° of latitude is 111 km). In contrast to ground measurements, the SSE dataset is continuous and consistent 22-year global climatology and meteorology data. It is derived from several databases, including Goddard Earth Observing Systems (GEOS-1), International Satellite Cloud Climatology Project (ISCCP D-1) by data from the Geostationary and Polar satellite for Environmental Observation (GOES and POES), the European Geostationary Satellite Meteosat, and Japanese ones. etc.

Although the SSE data within a particular grid cell are not necessarily representative of a particular microclimate, or point, within the cell, the data are considered to be the average over the entire area of the cell. For this reason, the SSE data set is not intended to replace quality ground measurement data. Its purpose is to fill the gap where ground measurements are missing, and to augment areas where ground measurements are available.

Accuracy of satellite measurements has been evaluated using numerous ground-based measurements. Although the reliability of these ground measurements themselves is not always well assessed, error in NASA estimates of monthly values is around 13-16%, and Mean Bias Error (MBE) lies between -2% and +0.7%.

Solar irradiation data is generated using Pinker-Laszlo shortwave algorithm. Cloud data is taken from International Satellite Cloud Climatology Project DX dataset (ISCCP). ISCCP DX data is on an equal area grid with an effective 30x30 km pixel size. The output data is generated on a nested grid containing 44,016 regions. The nested grid has a resolution of 1° latitude globally, and longitudinal resolution ranging from 1° in the tropics and subtropics to 120° at the poles. This, in turn, is mapped to 1° equal-angle grid (360 longitudes by 180 latitudes). The remapping method is by replication, wherein any grid region that is larger than 1°x1° is subdivided into 1°x1° degree regions, each with the same value as the original.

iv. World Radiation Data Centre

The World Radiation Data Centre (WRDC) (St. Petersburg, Russian Federation), sponsored by WMO, provides data on solar radiation, radiation balance and sunshine duration in convenient and readily accessible form. From 1964 up to 1992, solar radiation data had been published once a month. Since

1993, the publication has been issued four times a year. Observational data received by WRDC with certain delay are published in the last issue for the year, i.e. in the 4th quarterly issue. The issues are distributed among National Meteorological Services which send their solar radiation data to the WRDC to be archived and published.

In 1981 the World Radiometric Network changed over to the use of a new pyrheliometric scale - the World Radiometric Reference (WRR). Data on solar radiation and radiation balance are expressed in J/cm^2, data on sunshine duration are given in hours. Values can be converted to W/m^2 by dividing by 0.36. Observational data in each table of the publication are arranged according to the regions adopted by WMO. Countries within regions are listed in the order of English alphabet, stations of each country are placed by their coordinates from north to south and, if their latitudes are equal, from west to east. Hourly intervals are given in True Solar Time (TST). For convenience each issue is provided with a list of stations with their synoptic WMO index (if available), full name, latitude, longitude and elevation above sea level, as well as types (symbols) of the instruments which were used for measurements.

Of the available ground-based observation data in HC database, there are annual averaged (over the period 1964-1980) monthly values of GHI and its components (direct, diffuse) at 4 actinometric stations in Kyrgyzstan (Frunze, Cholpon-Ata, Suusamyr, Tjan-San). Moreover, all 4 actinometric stations are located in the north-east region of the country (Fig. 4). Therefore, for some idea of how the incidence of solar radiation in Kyrgyzstan is changing, actinometric stations of two neighboring states were considered: Kazakhstan – Aral Sea actinometric station; Uzbekistan – Tashkent actinometric station (Fig. 4).

Figure 4. Location of actinometric stations in the territory of Kyrgyzstan and neighboring countries

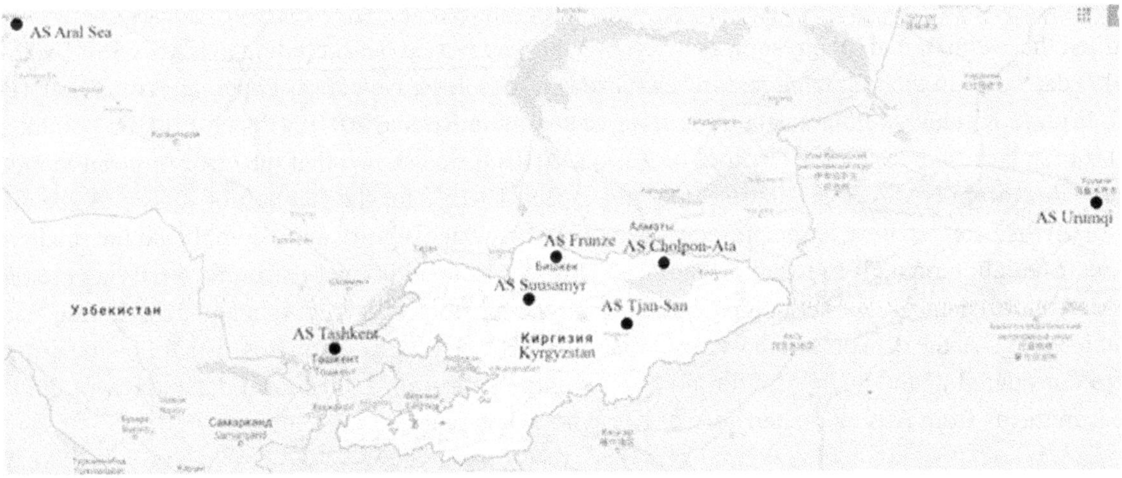

WRDC database does not contain ground-based observation data on the territory of Kyrgyzstan, but there are some of solar radiation observations in period from 1964 to 1991 on the territory of neighboring states: Uzbekistan - AS Tashkent; Kazakhstan - AS Aral Sea, China - AS Urumqi (Table 9), allowing to obtain averaged solar radiation data for the total period with the data of HC database (1964-1980) and the Meteonorm database (1980-1991 years).

Table 9. Availability of information from actinometric stations in Kyrgyzstan and neighboring countries in the databases

AS name	Geographical coordinates			Availability of information in the databases			
	Country	φ⁰, nl.	ψ⁰, el.	HC	WRDC	Meteonorm	NASA SSE
Frunze	Kyrgyzstan	42.8	74.5	+	-	+	+
Cholpon-Ata	Kyrgyzstan	42.7	77.1	+	-	+	+
Tjan-San	Kyrgyzstan	41.4	76	+	-	+	+
Suusamyr	Kyrgyzstan	41.7	78.2	+	-	+	+
Tashkent	Uzbekistan	41.2	69.2	+	+	+	+
Aral Sea	Kazakhstan	46.7	61.6	+	+	+	+
Urumqi	China	43.8	83.6	-	+	+	+

Meteonorm database was chosen as a source of data on solar radiation at any place in Kyrgyzstan. In Meteonorm database, there are no ground surveillance data on the territory of Kyrgyzstan and neighboring states, but there are computer models that generate data at any geographical point in Kyrgyzstan and neighboring states. Like any simulation results, the data provided by the Meteonorm database requires verification with ground-based actinometric measurements.

b. Evaluation of the Applicability of Meteonorm Database Data for Kyrgyzstan

Earlier, the evaluation of solar resources in Kyrgyzstan was carried out according to NASA SSE (NASA, 2019) database. In the following researches (Gridasov, Kiseleva, Nefedova, Popel, & Frid, 2010), (Popel, Frid, & Kiseleva, Climate data for renewable energy in Russia, 2010), (Popel, Frid, & Kolomiets, Atlas of solar resources on the territory of Russia, 2010) it was shown that the error in mean monthly GHI values from NASA SSE database does not exceed 10-15% for the flat part of Russia. A specificity of Kyrgyzstan territory is the presence of mountains, which significantly complicate the modeling of solar radiation on a given receiving surface. This fact is an additional reason for verifying the data of Meteonorm database for actinometric stations on the territory of Kyrgyzstan and neighboring states (Table 9, Fig. 4), the data on which were obtained from HC or WRDC databases. The following criteria were considered as the main criterion for verifying the data from Meteonorm database with ground measurements from ASs on the territory of Kyrgyzstan and neighboring states:

$$\delta_{Met-HC(WRDC)} \leq \delta_{add}, \tag{32}$$

where $\delta_{Met-SARC(WRDC)}$ is relative deviation between the compared GHI values in two databases, determined by the formula:

$$\delta_{Met-HC(WRDC)} = \frac{H^{Met} - H^{HC(WRDC)}}{H^{HC(WRDC)}} \cdot 100\%, \tag{33}$$

where H^{Met} и $H^{HC(WRDC)}$ are long-term monthly and annual GHI values from two databases for the total observation period; δ_{add} = 10% is permissible error threshold accepted in the research (Vissarionov, Deryugina, Kuznetsova, & Malinin, 2008).

The compared data should be averaged over the same periods. The data in Meteonorm database were averaged over the observation period from 1980 to 1991, therefore, they were compared with the averaged data for the same period with the data from WRDC database by the following actinometric stations: AS Tashkent, AS Aral Sea, AS Urumqi (Table 10). As can be seen from the table 10 and Fig. 5, the data of Meteonorm database are overstated by 16% – 46% (on average by 31.6%) relatively to the data from WRDC database, the dynamics of excess within the year is not revealed. Criterion (32) is not satisfied for all compared values, that is, Meteonorm database computer models are not applicable in the conditions of the considered ASs.

Table 10. Comparison of monthly GHI values of three ASs on the territory of neighboring countries according to WRDC and Meteonorm databases

Month	AS Tashkent			AS Aral Sea			AS Urumqi		
	WRDC	Meteonorm	δ, %	WRDC	Meteonorm	δ, %	WRDC	Meteonorm	δ, %
1	59	76	28.8	58	64	16.4	46	62	34.8
2	77	111	44.2	87	109	28.2	63	92	46.0
3	110	150	36.4	127	157	19.8	108	145	34.3
4	151	213	41.1	170	216	26.3	151	209	38.4
5	209	281	34.4	227	265	16.2	188	253	34.6
6	234	323	38.0	234	278	17.8	195	271	39.0
7	239	322	34.7	232	272	17.7	179	251	40.2
8	218	289	32.6	208	242	18.6	175	235	34.3
9	164	227	38.4	158	195	25.8	138	192	39.1
10	110	145	31.8	101	117	23.2	97	130	34.0
11	65	89	36.9	57	76	46.2	52	72	38.5
12	44	61	38.6	40	51	30.8	39	53	35.9
Year	1680	2287	36.1	1698	2042	21.3	1431	1965	37.3

The rationale for the assumption of comparing the data from Meteonorm database with the data from HC database averaged over different observation periods was carried out as follows. A comparison was made of the averaged data from HC database with WRDC database dataset averaged over two observation periods: for the total observation period by HC dataset (1964-1980); for the observation period that does not coincide with the averaging period of HC database, but is common with the averaging period of Meteonorm database data (1980-1991) (Table 11). As WRDC database does not include any information on actinometric stations in Kyrgyzstan, the comparison was made according to two actinometric stations in neighboring countries: Uzbekistan – AS Tashkent, Kazakhstan – AS Aral Sea (Table 11).

Figure 5. Deviations between the data of three AS on the territory of neighboring countries according to WRDC and Meteonorm databases

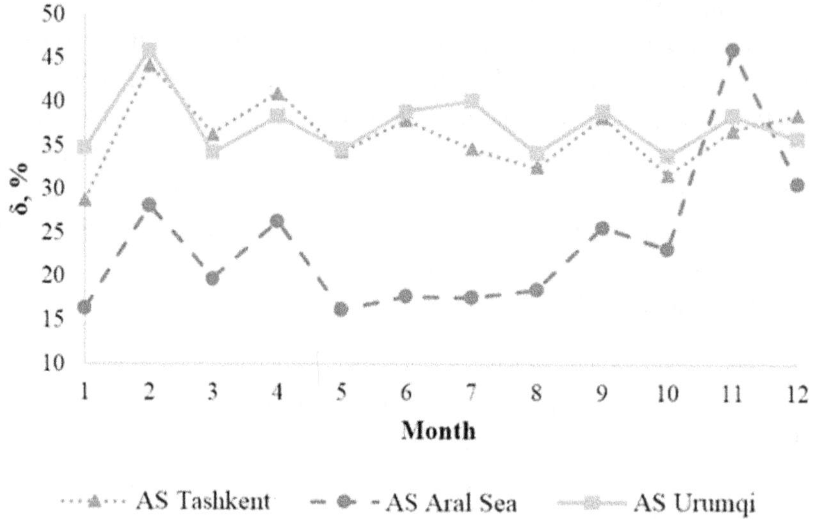

Table 11. Comparison of long-term monthly GHI values (kWh/m^2) according to two databases for total and non-coincident observation periods

Month	Comparison of WRDC database and HC database datasets for the total observation period (1964-1980)						Comparison of WRDC database and HC database datasets for inconsistent observation periods					
	AS Tashkent			AS Aral Sea			AS Tashkent			AS Aral Sea		
	WRDC	HC	δ, %	WRDC	HC	δ, %	WRDC 1980-1991	HC	δ, %	WRDC 1980-1991	HC	δ, %
1	59	64	-7.2	58	55	4.6	61	59	4.2	56	58	-1.8
2	77	79	-2.8	87	85	2.2	76	77	4.8	85	87	0.1
3	110	117	-5.9	127	131	-3.4	108	110	7.9	128	127	2.3
4	151	159	-4.7	170	171	-0.4	158	151	0.3	163	170	4.9
5	209	215	-2.7	227	228	-0.4	203	209	5.6	230	227	-0.9
6	234	240	-2.6	234	236	-0.8	229	234	4.9	231	234	2.2
7	239	241	-0.7	232	231	0.5	227	239	6.0	227	232	1.8
8	218	221	-1.1	208	204	1.9	203	218	8.6	202	208	1.0
9	164	169	-2.8	158	155	1.7	155	164	8.9	154	158	0.6
10	110	120	-8.0	101	95	6.3	111	110	7.8	97	101	-2.1
11	65	72	-9.7	57	52	9.6	70	65	2.9	56	57	-7.1
12	44	49	-9.7	40	39	3.1	47	44	3.7	41	40	-4.9
Year	1680	1743	-4.8	1698	1683	2.1	1648	1680	5.5	1670	1698	-0.3

As can be seen from table 11, the relative deviations between the compared data for two databases that was averaged over the total observation period (from 1964 - 1980) are from 0% to 9.7%, i.e. do not exceed ± 10% in accordance with the criterion (32). It should be also noted, that the relative deviations between the data of HC and WRDC databases that was averaged over different observation periods are from 0.1% to 8.9%, i.e. they also do not exceed ± 10% in accordance with the criterion (32). Analysis of changes in the annual GHI amount for the period from 1964 to 1991 was carried out according to the data of the ASs (Tashkent and Aral Sea) from the WRDC database (Fig.6, Table 12). The annual income of GHI in a long-term context did not demonstrate a tendency to change. It changes insignificantly: coefficient of variation of long-term GHI values on the AS Tashkent is 0.033 and on the AS Aral Sea is equal to 0.026 (Table 12).

Table 12. Ranges of long-term GHI values for the period of 1964-1991 according to WRDC database

AS name	Ranges of annual GHI values for the period of 1964-1991, kWh/m²		C_v	Long-term GHI values for the period of 1964-1991, kWh/m²
	Minimum	Maximum		
Tashkent	1604	1743	0.033	1678
Aral Sea	1535	1737	0.026	1683

Figure 6. Long-term GHI values on AS Tashkent and AS Aral Sea

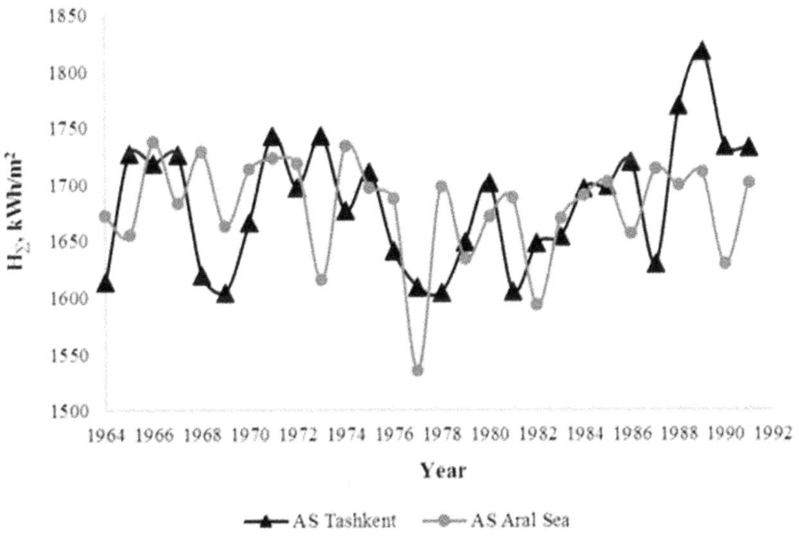

The conducted studies allowed to make the assumption described above, according to which a comparison was made of the data of 4 actinometric stations of Kyrgyzstan from HC and Meteonorm database averaged over inconsistent observation periods (Table 13, Fig.7). As can be seen from Table 13 and figure 7, the relative deviations between the compared data of the 4 AS of Kyrgyzstan are from -2.4% to 62.7%, i.e. in most cases, criterion (32) is not fulfilled. It was also noted that the intra-annual change in relative deviation depends on the latitude of the area. To use the data from Meteonorm database on the territory of Kyrgyzstan, it is necessary to develop a mathematical model or correction factors.

Table 13. Comparison of long-term monthly GHI values (kWh/m²) at 4 AS in Kyrgyzstan from HC and Meteonorm databases averaged over non-coincident observation periods

Month	AS Cholpon-Ata			AS Frunze			AS Suusamyr			AS Tjan-San		
	Meteonorm	HC	δ, %	Meteonorm	HC	δ, %	Meteonorm	HC	δ, %	Meteonorm	HC	δ, %
1	91	62	46.8	83	62	33.9	80	82	-2.4	96	91	5.5
2	126	85	48.2	112	76	47.4	115	103	11.7	126	113	11.5
3	176	141	24.8	164	105	56.2	163	156	4.5	195	176	10.8
4	213	173	23.1	218	143	52.4	199	166	19.9	227	205	10.7
5	248	207	19.8	247	185	33.5	240	186	29.0	253	215	17.7
6	268	209	28.2	263	206	27.7	282	202	39.6	282	204	38.2
7	268	212	26.4	271	218	24.3	312	208	50.0	287	196	46.4
8	240	192	25.0	245	193	26.9	260	198	31.3	245	187	31.0
9	196	160	22.5	208	148	40.5	231	156	48.1	209	158	32.3
10	133	111	19.8	135	98	37.8	152	116	31.0	145	131	10.7
11	86	69	24.6	96	59	62.7	89	79	12.7	106	93	14.0
12	74	53	39.6	61	47	29.8	71	69	2.9	74	80	-7.5
Year	2119	1674	29.1	2103	1541	39.4	2181	1719	23.2	2245	1850	18.4

Figure 7. Relative deviations of monthly GHI values between Meteonorm and HC database at ASs

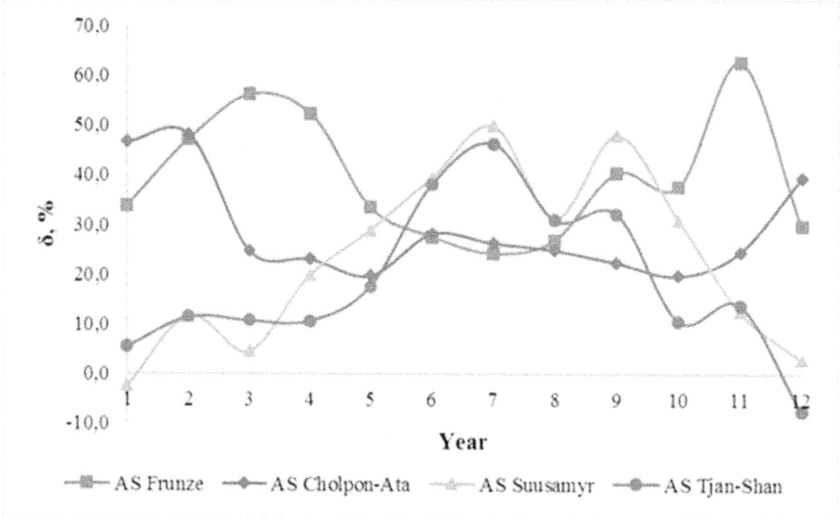

c. Developing of a Mathematical Model for Mean Long-Term Monthly GHI Restoration According to Meteonorm Database in Kyrgyzstan

Considering the method of restoring monthly GHI data in Kyrgyzstan according to Meteonorm database, the possibility of constructing a linear model of the relationship was analyzed:

$$H_{\Sigma}^{mod} = a \cdot H_{\Sigma}^{Met} + b, \quad (34)$$

where a, b are coefficients of model; H_{Σ}^{Met} – the average long-term monthly GHI at the selected place according to Meteonorm database; H_{Σ}^{mod} – reconstructed (modeled) GHI value at the selected place. As a criterion for the reliability of the model, the coefficient of determination R^2 is used. R^2 characterizes the presence (or absence) of a correlation between the generated data of Meteonorm database H_{Σ}^{Met} and the averaged data from HC database (or WRDC database) $H_{\Sigma}^{HC(WRDC)}$. Figure 8 shows the empirical relationship between the data of 5 AS (Frunze, Cholpon-Ata, Suusamyr, Tjan–San, Tashkent) from HC database (or WRDC database) and Meteonorm database and its approximating dependence in the following form:

$$H_{\Sigma}^{mod} = 0.7212 \cdot H_{\Sigma}^{Met} + 10.531, \quad (35)$$

with the coefficient of determination is $R^2 = 0.9499$.

Figure 8. Empirical and approximating relationships between average long-term monthly GHI values according to the data from HC and Meteonorm databases

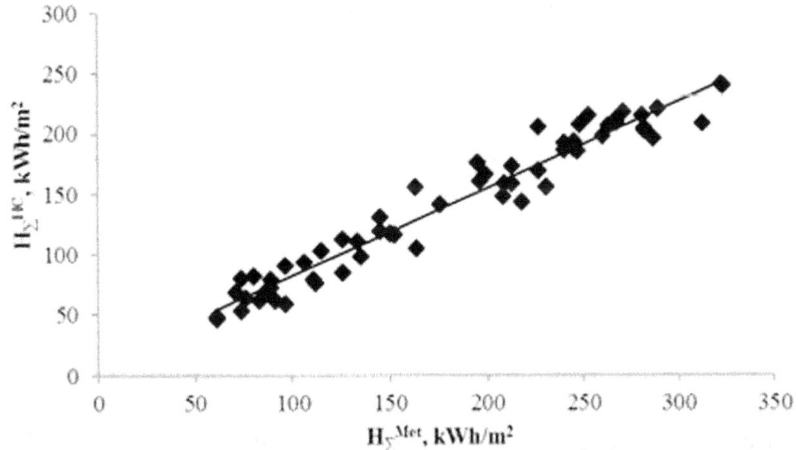

The assessment of the developed model (35) was evaluated in accordance with criterion (32). It was found that the average relative deviation between the values that was calculated according by the model (35) and according to the data of HC database is 8.9%. To reduce the error of model (35), similar models were obtained for each AS (Table 14). Figure 9 shows the relationship between the data from Meteonorm and HC database for 5 AS (Frunze, Cholpon-Ata, Suusamyr, Tjan-San, Tashkent).

Analysis of the data given in Table 14 revealed approximating dependences of the coefficients a and b of model (35) on the latitude of the terrain:

$$a(\varphi) = 0.1157\varphi - 4.129; \quad (36)$$

$$b(\varphi) = -23.86\varphi - 1010.4. \tag{37}$$

The addition of model (35) by determining the coefficients *a* and *b* from dependences (36) and (37) allowed one to reduce the average value of the relative deviation between the model values that was calculated from the Meteonorm database and from the HC database to 7.5%.

Table 14. Coefficients of the model (34), showing the relationship between the data of Meteonorm and HC database for 5 AS

Name of AS	φ, n.l.	*a*	*b*	R^2
Cholpon-Ata	42.7	0.84	-8.45	0.99
Frunze	42.5	0.8	-12.55	0.97
Suusamyr	41.7	0.6	33.48	0.95
Tjan-San	41.4	0.64	34.27	0.92
Tashkent	42.7	0.73	5.5	0.99

d. Development of a Mathematical Model for Long-Term Monthly Incomes of Diffuse Solar Radiation (DHI) Restoration According to the Data From Meteonorm Database in Kyrgyzstan

The development of a linear model of the relationship between the average long-term monthly values of DHI according to the data from HC and Meteonorm databases was carried out for 4 ASs (Frunze, Cholpon-Ata, Suusamyr, Tjan-San), because DHI data at the AS Tashkent is absent in WRDC and HC databases. As a result of comparing the above data (Fig. 10), an analytical dependence was obtained with the coefficient of determination $R^2 = 0{,}6788$ in the following form:

$$H_{diff}^{mod} = 0.5885 \bullet H_{diff}^{Met} + 15.481, \tag{38}$$

where H_{diff}^{Met} – the average long-term monthly DHI value.

The assessment of the developed model (38) was evaluated in accordance with criterion (32). It was found that the average relative deviation between the values that was calculated by the model (38) and according to the data from HC database is 14.51%. To reduce the error of the model (38), similar models were obtained for each AS (Table 15). Figure 11 shows the relationship between the data from Meteonorm and HC databases for 4 ASs (Frunze, Cholpon-Ata, Suusamyr, Tjan-San).

The analysis of the data in Table 15 showed the analytical dependences of the coefficients *a* and *b* of model (38) on the latitude of the terrain:

$$a(\varphi) = -0.1464\varphi - 6.7376, \tag{39}$$

$$b(\varphi) = 0.7801\varphi - 16.87. \tag{40}$$

Figure 9. Empirical and approximating relationships between average long-term monthly values of the global solar radiation according to the data of the HC database and the Meteonorm database on the AS: Cholpon-Ata, Frunze, Suusamyr, Tjan-San and Tashkent

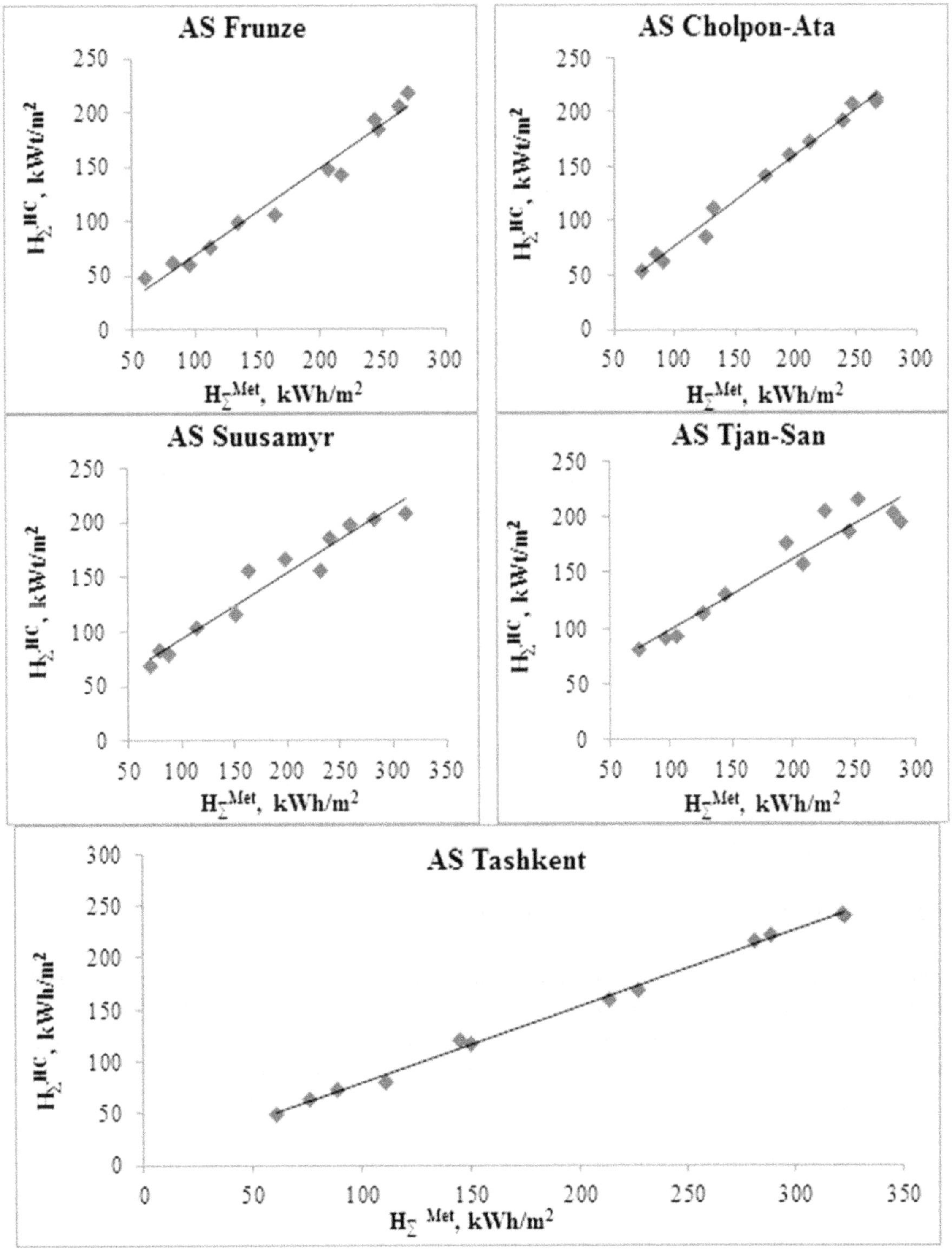

Figure 10. Empirical and approximating dependences between average long-term monthly DHI values according to the data from HC and Meteonorm databases

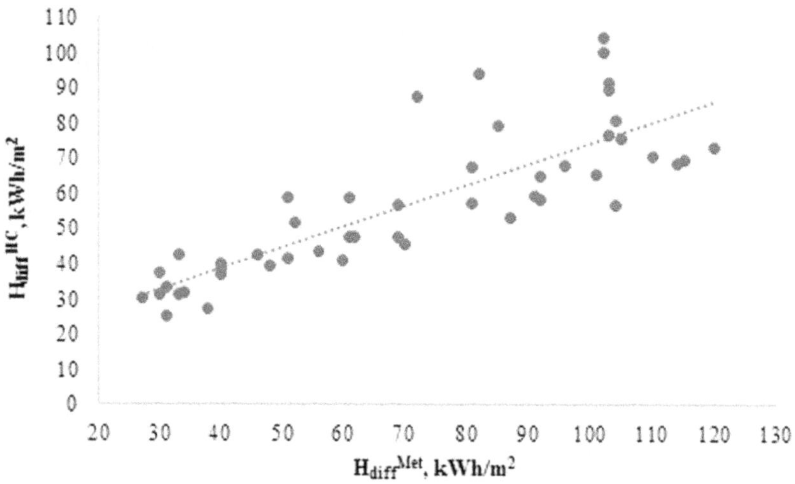

Table 15. Model coefficients (37), showing the relationship between the data from Meteonorm and HC databases for 4 ASs

Name of AS	φ, n.l.	a	b	R^2
Cholpon-Ata	42.7	0.5629	11.28	0.94
Frunze	42.5	0.4735	18.861	0.85
Suusamyr	41.7	0.5192	23.531	0.55
Tjan-Shan	41.4	0.7576	10.144	0.74

The addition of model (37) by determining the coefficients *a* and *b* using formulas (38) and (39) allowed one to reduce the average value of the relative deviation between the modelled DHI values by Meteonorm database and HC database values to 12.65%.

e. The Theoretical Potential of Solar Energy

Comparison of the average long-term GHI according to the data of 5 ASs from Meteonorm and HC databases showed that the overstatement of GHI values by Meteonorm database is from 21.4% to 36.5% (Table 13), i.e. it is need to be corrected to determinate annual global solar radiation on the territory of Kyrgyzstan by the data of the Meteonorm database. Correction coefficients were determined using the data of 5 actinometric stations by the following expression:

$$K_{\Sigma l} = H_{\Sigma l}^{HC} / H_{l}^{Met}, \tag{41}$$

where l – the number of actinometric station; H_l^{Met} – average long-term income of GHI at the selected place l by the data from Meteonorm database; $H_{\Sigma l}^{HC}$ – average long-term income of GHI at the selected place l by the data from HC database (or in case of their absence according to WRDC database) (Table 16).

An analysis of the results showed an analytical dependence of the average annual coefficient on the latitude of the area (φ):

$$K_o(\varphi) = -0.0177\varphi + 1.5209. \tag{42}$$

Figure 11. Correlation between averages the long-term monthly values of diffuse solar radiation according to the data from HC and Meteonorm databases on the following ASs: Cholpon - Ata, Frunze, Suusamyr, Tjan-San

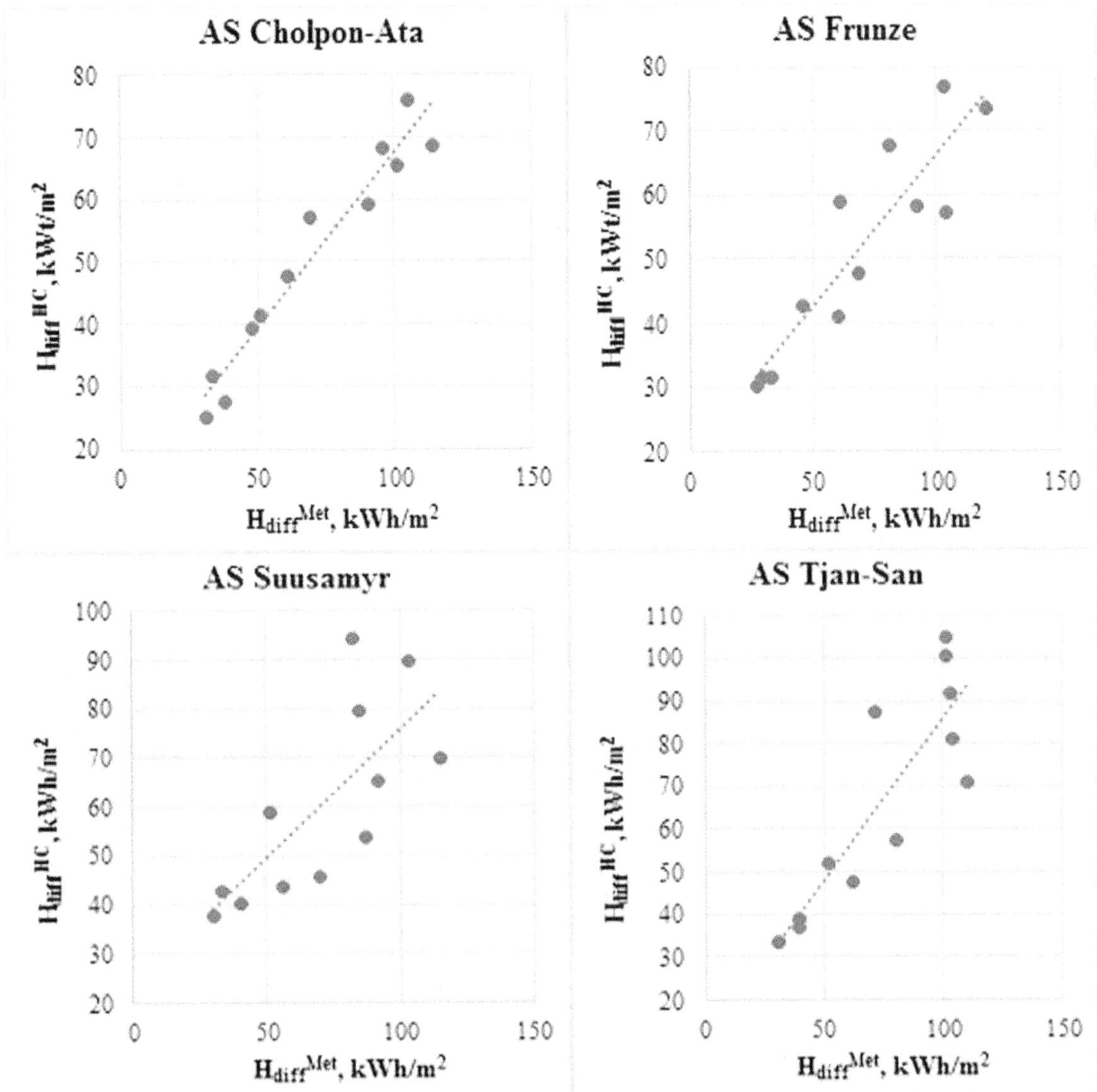

Table 16. Average long-term and month correction coefficients

Month	Tashkent	Cholpon-Ata	Frunze	Suusamyr	Tjan-San
1	0.84	0.68	0.75	1.03	0.95
2	0.71	0.67	0.68	0.90	0.90
3	0.78	0.80	0.64	0.96	0.90
4	0.74	0.81	0.66	0.83	0.90
5	0.76	0.83	0.69	0.78	0.85
6	0.74	0.78	0.73	0.72	0.72
7	0.75	0.79	0.72	0.67	0.68
8	0.76	0.80	0.76	0.76	0.76
9	0.74	0.82	0.68	0.68	0.76
10	0.82	0.83	0.68	0.76	0.90
11	0.81	0.80	0.61	0.89	0.88
12	0.80	0.72	0.77	0.97	1.08
Year	0.77	0.78	0.70	0.83	0.86

The verification of the obtained model was carried out in accordance with criterion (32) by comparing the data on the global solar radiation at the selected place of 5 actinometric stations from HC database with model data that was calculated by the following expression:

$$H_{\Sigma l}^{\mathrm{mod}} = K_{>}(\varphi)\cdot H_l^{Met}, \tag{43}$$

where $K_o(\varphi)$– average long-term corrections coefficients determined by the formula (7); H_l^{Met} – average long-term GHI in l place by the data of Meteonorm database; $H_{\Sigma l}^{mod}$ – model value of average long-term GHI in place l.

As can be seen from table 17, the deviations between the compared values do not exceed 10%, which confirms the applicability of model (43) for the territory of Kyrgyzstan.

Table 17. Comparison average long-term GHI values by the data of HC and Meteonorm databases

Parameters / AS	Cholpon-Ata	Frunze	Suusamyr	Tjan-San	Tashkent
H_l^{Met}, kWh/m²	2119	2103	2181	2245	2287
$H_{\Sigma l}^{HC}$, kWh/m²	1674	1541	1719	1850	1743
$\delta_{Met\text{-}HC}$, %	26.6	36.5	26.9	21.4	31.2
K_{ol}	0.79	0.73	0.78	0.82	0.76
$K_o(\varphi)$	0.77	0.77	0.78	0.79	0.79
$H_{\Sigma l}^{mod}$, kWh/m²	1631.63	1619.31	1701.18	1773.55	1806.73
$\delta_{mod\text{-}HC}$, %	2.5	5.1	1.03	4.1	-3.6

The calculation of the theoretical potential of solar energy in the country is carried out according to the formula:

$$E_{pot} = T \cdot \sum_{i=1}^{n} H_{\Sigma i}^{mod} \bullet F_i, \tag{44}$$

where n is quantity of zones into which the territory of Kyrgyzstan is divided by grid in 1°x1°; $H_{\Sigma i}^{mod}$ – average long-term GHI in i zone, kWh/m^2, that was calculated according to expressions (37) and (38) according to the database Meteonorm H_i^{Met} (Table 17); F_i is part of the country in i zone. The area of Kyrgyzstan is 199800 km^2.

The theoretical potential of solar energy of Kyrgyzstan, clarified by the authors, is 350847 billion kWh/m^2 or 43.05 billion tons of fuel equivalent, that 1.3 times more than, as determined by research (Kiseleva, Popel, & Kolomiets, 2015), which is 35 billion tons of fuel equivalent.

Calculating $H_{\Sigma i}^{mod}$ in Surfer 8 software (Surfer, 2019) distribution map of average long-term GHI by the territory of Kyrgyzstan (Fig. 12) was built.

As can be seen from Figure 12, there are favorable conditions for the use of solar energy throughout Kyrgyzstan: average long-term GHI by the territory of country changes insignificantly: from 1656 kWh/m^2 to 1856 kWh/m^2; average value is equal to 1756 kWh/m^2.

Figure 12. Distribution map of long-term averaged GHI on the territory of Kyrgyzstan

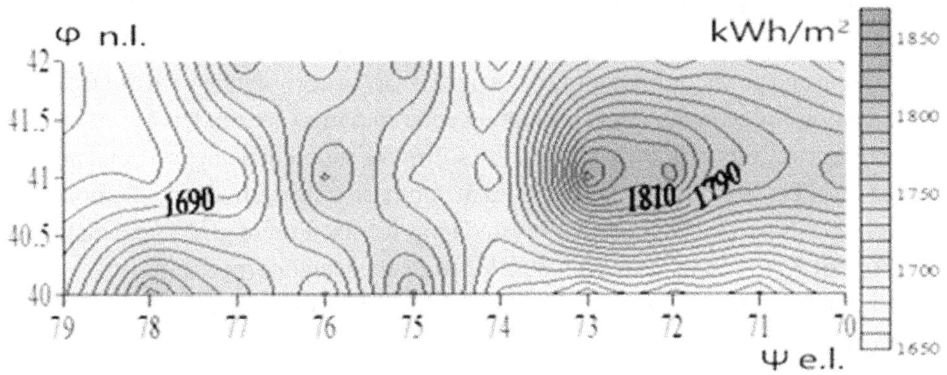

FUTURE RESEARCH DIRECTIONS

The analysis of solar radiation models from the point of view of their use on the territory of Kyrgyzstan conducted in the study showed that it is necessary to develop special models, taking into account local climatic conditions and country characteristics. The most significant errors in modelling are observed in highlands, therefore, it is need to study the factors affecting the distribution of solar radiation in such regions. In subsequent works, it is planned to carry out a more detailed verification of empirical models, conduct an analysis using the ground-measurement data, propose refinement empirical coefficients for the most suitable models and build solar radiation estimation and prediction models using data mining approaches.

Table 18. Average long-term GHI values in different geographical zones of Kyrgyzstan (kWh/m²)

Ψ e.l.	φ=40 n.l.		φ=41 n.l.		φ=42 n.l.	
	H_i^{Met}.	$H_{\Sigma i}^{mod}$	Meteonorm	$H_j^{(\varphi,\Psi)}$	Meteonorm	$H_j^{(\varphi,\Psi)}$
70	2174	1696	2294	1789	2213	1726
71	2125	1658	2294	1789	2240	1747
72	2135	1665	2353	1835	2234	1743
73	2131	1662	2380	1856	2229	1739
74	2164	1688	2191	1709	2158	1683
75	2273	1773	2203	1718	2253	1757
76	2177	1698	2259	1762	2210	1724
77	2226	1736	2154	1680	2240	1747
78	2278	1777	2135	1665	2155	1681
79	2176	1697	2136	1666	2123	1656

CONCLUSION

A general classification of methods for estimating and predicting solar radiation is given. Various types of statistical models were considered. Traditional empirical models were categorized according to the parameters used and evaluated. As a first approximation, comparison of universal empirical models is carried out using the example of long-term average data for Kyrgyzstan. The main methods of data mining used for assessing and predicting solar radiation are listed. In accordance with the final accuracy, the most preferable models for estimating and predicting solar radiation are models based on NWP data, in the case of their inaccessibility it is recommended to solve this problem using artificial neural networks, due to a small error in the results they demonstrate in comparison with traditional empirical models.

Mathematical models for calculating monthly and long-term values of global and diffuse solar radiation in the territory of Kyrgyzstan have been developed using Meteonorm database.

Inapplicability of Meteonorm database computer models for generating monthly and annual global and diffuse solar radiation in the conditions of Kyrgyzstan has been proved.

The theoretical potential of solar energy in Kyrgyzstan was clarified, which amounted to 350 847 billion kWh/m² or 43.05 billion tons of fuel equivalent.

REFERENCES

Abrukov, V. S., Kochakov, V. D., Abrukov, S. V., Anufrieva, D. A., Vasilev, A. I., & Smirnov, A. V. (2017). Intelligent Forecasting System for Solar Power Plants. *ISJAEE*, *16*, 30–42. doi:10.15518/isjaee.2017.16-18.030-042

Allen, R. (1997). Self-calibrating method for estimating solar radiation from air temperature. *Journal of Hydrologic Engineering*, *2*(2), 56–67. doi:10.1061/(ASCE)1084-0699(1997)2:2(56)

Almorox, J., & Hontoria, C. (2004). Global solar radiation estimation using sunshine duration in Spain. *Energy Conversion and Management*, *45*(9-10), 1529–1535. doi:10.1016/j.enconman.2003.08.022

American Society and Heating, Refrigerating and Airconditioning Engineers (ASHRAE). (1972). Handbook of Fundamentals. ASHRAE.

Angstrom, A. (1924). Solar and terrestrial radiation. *Quarterly Journal of the Royal Meteorological Society*, *50*(210), 121–126. doi:10.1002/qj.49705021008

Annandale, J. G., Jovanic, N. Z., Benade, N., & Allen, R. G. (2002). Software for missing data error analysis of Penman–Monteith reference evapotranspiration. *Irrigation Science*, *21*(2), 57–67. doi:10.1007002710100047

Ayodele, T. R., & Ogunjuyigbe, A. S. (2015). Prediction of monthly average global solar radiation based on statistical distribution of clearness index. *Energy*, *90*, 1733–1742. doi:10.1016/j.energy.2015.06.137

Badescu, V. (1999). Correlations to estimate monthly mean daily solar global irradiation: Application to Romania. *Energy*, *24*(10), 883–893. doi:10.1016/S0360-5442(99)00027-4

Badescu, V. (2002). 3D isotropic approximation for solar diffuse irradiance on tilted surfaces. *Renewable Energy*, *26*(2), 221–233. doi:10.1016/S0960-1481(01)00123-9

Bahel, V., Bakhsh, H., & Srinivasan, R. (1987). A correlation for estimation of global solar radiation. *Energy*, *12*(2), 131–135. doi:10.1016/0360-5442(87)90117-4

Bakirci, K. (2009). Correlations for estimation of daily global solar radiation with hours of bright sunshine in Turkey. *Energy*, *34*(4), 485–501. doi:10.1016/j.energy.2009.02.005

Besharat, F., Dehghan, A. A., & Faghih, A. R. (2013). Empirical models for estimating global solar radiation: A review and case study. *Renewable & Sustainable Energy Reviews*, *21*, 798–821. doi:10.1016/j.rser.2012.12.043

Bhardwaj, S., & Bhola, P. (2016). Solar Energy Estimation Techniques: A Review. *Conference Paper*.

Black, J. N. (1956). The distribution of solar radiation over the earth's surface. *Arch. Met. Geoph. Biokl,* (7), 165-189.

Boland, J., Scott, L., & Luther, M. (2001). Modelling the diffuse fraction of global solar radiation on a horizontal surface. *Environmetrics*, *12*(12), 103–116. doi:10.1002/1099-095X(200103)12:2<103::AID-ENV447>3.0.CO;2-2

Bristow, C. L., & Campbell, G. S. (1984). On the relationship between incoming solar radiation and daily maximum and minimum temperature. *Agricultural and Forest Meteorology*, *31*(2), 159–166. doi:10.1016/0168-1923(84)90017-0

Capizzi, G., Napoli, C., & Bonanno, F. (2012). Innovative second-generation wavelets construction with recurrent neural networks for solar radiation forecasting. *IEEE Transactions on Neural Networks and Learning Systems*, *23*(11), 1805–1815. doi:10.1109/TNNLS.2012.2216546 PMID:24808074

Celik, A. N., & Muneer, T. (2013). Neural network based method for conversion of solar radiation data. *Energy Conversion and Management*, *67*, 117–124. doi:10.1016/j.enconman.2012.11.010

Chen, R., Ersi, K., Yang, J., Lu, S., & Zhao, W. (2004). Validation of five global radiation models with measured daily data in China. *Energy Conversion and Management*, *45*(11-12), 1759–1769. doi:10.1016/j.enconman.2003.09.019

Dogniaux, R., & Lemoine, M. (1983). Classification of radiation sites in terms of different indices of atmospheric transparency. *Solar energy research and development in the European Community, series F, 2*.

Donatell, M., & Campbell, G. S. (1998). A simple model to estimate global solar radiation. *Proceedings of the fifth European society of agronomy congress*, 133-134.

Duffie, J., Beckman, W., & Worek, W. (2003). *Solar Engineering of Thermal Processes.* Wiley.

Ekici, B. (2014). A least squares support vector machine model for prediction of the next day solar insolation for effective use of PV systems. *Measurement*, *50*, 255–262. doi:10.1016/j.measurement.2014.01.010

El-Metwally, M. (2005). Sunshine and global solar radiation estimation at different sites in Egypt. *Journal of Atmospheric and Solar-Terrestrial Physics*, *67*(14), 1331–1342. doi:10.1016/j.jastp.2005.04.004

Foster, R., Ghassemi, M., & Cota, A. (2010). *Solar Energy*. CRC Press.

Gariepy, J. (1980). *Estimation of global solar radiation. International report*. Service of Meteorology.

Glover, J., & McCulloch, S. (1958). The empirical relation between solar radiation and hours of sunshine. *Quarterly Journal of the Royal Meteorological Society*, *84*(360), 172–175. doi:10.1002/qj.49708436011

Gómez, V., & Casanovas, A. (2003). Fuzzy modeling of solar irradiance on inclined surfaces. *Solar Energy*, *75*(4), 307–315. doi:10.1016/j.solener.2003.08.006

Gridasov, M. V., Kiseleva, S. V., Nefedova, L. V., Popel, O. C., & Frid, S. E. (2010). *"Renewable energy sources in Russia" information system developing*. Retrieved October 15, 2019, from www.gis-vie.ru

Gueymard, C. (1993). Critical analysis and performance assessment of clear sky solar irradiance models using theoretical and measured data. *Solar Energy*, (51), 385–397. doi:10.1016/0038-092X(93)90059-W

Handbook on the Climate of the USSR (Vol. 32). (1989). Leningrad.

Hargreaves, G. H., & Samani, Z. A. (1982). Estimating potential evapotranspiration. *Journal of Irrigation and Drainage Engineering*, *108*(IR3), 223–230.

Hutchinson, J. M., Goodin, D. G., Vanderlip, R. L., & Knapp, M. C. (1999). Estimating solar irradiance for crop modeling using daily air temperature data. *Agronomy Journal*, *91*(5), 845–851. doi:10.2134/agronj1999.915845x

Iqbal, M. (1979). Correlation of average diffuse and beam radiation with hours of bright sunshine. *Solar Energy*, *23*(23(2)), 169–173. doi:10.1016/0038-092X(79)90118-X

Iqbal, M. (1983). *An Introduction to Solar Radiation*. Academic Press.

Jin, Z., Yezheng, W., & Gang, Y. (2005). General formula for estimation of monthly average daily global solar radiation in China. *Energy Conversion and Management*, *46*(2), 257–268. doi:10.1016/j.enconman.2004.02.020

Kalogirou, S. A. (2001). Artificial neural networks in renewable energy systems applications: A review. 2001;5(4). *Renewable & Sustainable Energy Reviews*, *5*(4). Advance online publication. doi:10.1016/S1364-0321(01)00006-5

Kimball, H. H. (1919). Variations in the total and luminious solar radiation with geographical position in the United States. *Monthly Weather Review*, *47*(47(11)), 769–793. doi:10.1175/1520-0493(1919)47<769:VITTAL>2.0.CO;2

Kiseleva, S. V., Popel, O. S., & Kolomiets, Y. G. (2015). Evaluation of solar resources in the Central Asia. *Geliotechnics*, 69-74.

Klein, S. A. (1977). Calculation of monthly average insolation on tilted surfaces. *Solar Energy*, *19*(4), 325–329. doi:10.1016/0038-092X(77)90001-9

Lam, J. C., & Li, D. (1996). Correlation between global solar radiation and its direct and diffuse components. *Building and Environment*, *31*(6), 527–535. doi:10.1016/0360-1323(96)00026-1

Lazzús, J. A., Pérez Ponce, A. A., & Marín, J. (2011). Estimation of global solar radiation over the city of La Serena (Chile) using a neural network. *Applied Solar Energy*, *47*(1), 66–73. doi:10.3103/S0003701X11010099

Letendre, S., Makhyoun, M., & Taylor, M. (2014). *Predicting Solar Power Production: Irradiance Forecasting Models*. Applications and Future Prospects.

Liu, B., & Jordan, R. (1961). Daily insolation on surfaces tilted towards equator. *ASHRAE*, (10), 53-59.

Liu, B. Y., & Jordan, R. C. (1960). The interrelationship and characteristic distribution of direct, diffuse and total solar radiation. *Solar Energy*, *4*(4(3)), 1–19. doi:10.1016/0038-092X(60)90062-1

López, G., Battles, F. J., & Tovas-Pescador, J. (2005). Selection of input parameters to model direct solar irradiance by using artificial neural networks. *Energy*, *30*(9), 30. doi:10.1016/j.energy.2004.04.035

Lowe, G., Duffie, J., & Smith, C. (1965). *World Distribution of Solar Radiation*. Academic Press.

Ma, C., & Iqbal, M. (1982). Statistical Comparison Of Solar Radiation. *Solar Energy*, *33*(2).

Machler, M. A., & Iqbal, M. A. (1967). A modification of the ASHRAE clear sky irradiation model. *ASHRAE Transactions*, (91), 106–115.

Maghrabi, A. H. (2009). Parameterization of a simple model to estimate monthly global solar radiation based on meteorological variables, and evaluation of existing solar radiation models for Tabouk, Saudi Arabia. *Energy Conversion and Management*, *50*(11), 2754–2760. doi:10.1016/j.enconman.2009.06.024

Maleki, S. A., Hizam, H., & Gomes, C. (2017). Estimation of Hourly, Daily and Monthly Global Solar Radiation on Inclined Surfaces: Models Re-Visited. *Energies*, 10.

Mellit, A. (2008). Artificial intelligence technique for modelling and forecasting of solar radiation data: A review. *International Journal of Artificial Intelligence and Soft Computing*, *1*(1), 52–76. doi:10.1504/IJAISC.2008.021264

Mellit, A., Kalogirou, S. A., Hontoria, L., & Shaari, S. (2009). Artificial intelligence techniques for sizing photovoltaic systems: A review. *Sustainable Energy Reviews, 13*(2), 406–419. doi:10.1016/j.rser.2008.01.006

Mellit, A., & Pavan, A. (2010). A 24-h forecast of solar irradiance using artificial neural network: Application for performance prediction of a grid-connected PV plant at Trieste, Italy. *Solar Energy, 84*(5), 84. doi:10.1016/j.solener.2010.02.006

Meteonorm. (2019). Retrieved October 7, 2019, from http://meteonorm.ru

Mohammadi, K., Shamshirband, S., Anisi, M. H., Alam, K. A., & Petković, D. (2015). Support vector regression based prediction of global solar radiation on a horizontal surface. *Energy Conversion and Management, 91*, 433–441. doi:10.1016/j.enconman.2014.12.015

NASA. (2019). Retrieved October 7, 2019, from The NASA Surface Meteorology and Solar Energy Data Set: https://eosweb.larc.nasa.gov/sse/

Nijegorodov, N. (1996). Improved ashrae model to predict hourly and daily solar radiation components in Botswana, Namibia, and Zimbabwe. *WREC*, (9), 1270-1273.

Notton, G., Paoli, C., Vasileva, S., Nivet, M. L., Canaletti, J., & Cristofari, C. (2012). Estimation of hourly global solar irradiation on tilted planes from horizontal one using artificial neural networks. *Energy, 39*(1), 166–179. doi:10.1016/j.energy.2012.01.038

Orgill, J. F., & Hollands, K. G. (1977). Correlation Equation for Hourly Diffuse Radiation on a Horizontal Surface. *Solar Energy, 19*(4), 357–359. doi:10.1016/0038-092X(77)90006-8

Ouammi, A., Zejli, D., Dagdougui, H., & Benchrifa, R. (2012). Artificial neural network analysis of Moroccan solar potential. *Renewable & Sustainable Energy Reviews, 16*(7), 4876–4889. doi:10.1016/j.rser.2012.03.071

Page, J. K. (1961). The estimation of monthly mean values of daily total short wave radiation on vertical and inclined surfaces from sunshine records for latitudes 40°N-40°S. *Proc. U.N. Conf. on New Sources of Energy, 4*, 378-390.

Parishwad, G., Bhardwaj, R. K., & Nema, V. K. (1997). Estimation of hourly solar radiation for India. *Renewable Energy, 12*(12), 303–313. doi:10.1016/S0960-1481(97)00039-6

Perez, R., Stewart, R., Seals, R., & Guertin, T. (1988). *The Development and Verification of the Perez Diffuse Radiation Model*. Sandia National Labs. doi:10.2172/7024029

Popel, O. C., Frid, S. E., & Kiseleva, S. V. (2010). Climate data for renewable energy in Russia. Academic Press.

Popel, O. C., Frid, S. E., & Kolomiets, Y. G. (2010). *Atlas of solar resources on the territory of Russia*. MIPT.

Rietveld, M. (1978). *A New Method For Estimating The Regression Coefficients*. Agrtcultural Meteorology.

Rizwan, M., Jamil, M., Kirmani, S., & Kothari, D. P. (2014). (2014). Fuzzy logic based modeling and estimation of global solar energy using meteorological parameters. *Energy*, *70*, 685–691. doi:10.1016/j.energy.2014.04.057

Şen, Z. (1998). Fuzzy algorithm for estimation of solar irradiation from sunshine duration. *Solar Energy*, *63*(1), 39–49. doi:10.1016/S0038-092X(98)00043-7

Soler, A. (1990). Monthly specific Rietveld's correlations. *Solar & Wind Technology*, *7*(2-3), 305–308. doi:10.1016/0741-983X(90)90100-G

Spencer, J. W. (1971). Fourier series representation of the position of the sun. *Applied Optics*, (10), 2569–2571.

Steven, M. D., & Unsworth, M. H. (1979, July). The diffuse solar irradiance of slopes under cloudless skies. *Quarterly Journal of the Royal Meteorological Society*, *105*(445), 593–602. doi:10.1002/qj.49710544507

Suganthi, L., Iniyan, S., & Samuel, A. A. (2015). Applications of fuzzy logic in renewable energy systems – A review. *Renewable & Sustainable Energy Reviews*, *48*, 585–607. doi:10.1016/j.rser.2015.04.037

Supit, I., & Van Kappel, R. R. (1998). A simple method to estimate global radiation. *Solar Energy*, *63*(3), 147–160. doi:10.1016/S0038-092X(98)00068-1

Surfer. (2019). Retrieved October 15, 2019, from Surfer 8 Software: http://www.goldensoftware.com

Swartman, R. K., & Ogunlade, O. (1967). Solar radiation estimates from common parameters. *Solar Energy*, *11*(3-4), 170–172. doi:10.1016/0038-092X(67)90026-6

Tian, Y. Q., Davies-Colley, R. J., Gong, P., & Thorrold, B. W. (2001). Estimating solar radiation on slopes of arbitrary aspect. *Agricultural and Forest Meteorology*, *109*(1), 67–74. doi:10.1016/S0168-1923(01)00245-3

Trabea, A., & Shaltout, M. (2000). Correlation of global solar radiation with meteorological parameters over Egypt. *Renewable Energy*, *21*(2), 297–308. doi:10.1016/S0960-1481(99)00127-5

Vissarionov, V. I., Deryugina, G. V., Kuznetsova, V. A., & Malinin, N. K. (2008). *Solar energy*.

Wanga, Z., Wanga, F., & Sub, S. (2011). Solar irradiance short-term prediction model based on BP neural network. *Energy Procedia*, *12*, 12. doi:10.1016/j.egypro.2011.10.065

WRDC. (2019). Retrieved October 5, 2019, from World Radiation Data Centre: http://wrdc.mgo.rssi.ru/wwwrootnew/wrdc_ru_new.htm

Zeng, J., & Qiao, W. (2013). Short-term solar power prediction using a support vector machine. *Renewable Energy*, *52*, 118–127. doi:10.1016/j.renene.2012.10.009

KEY TERMS AND DEFINITIONS

Actinometric Station: A type of meteorological station which observes and measures solar radiation incidence on the Earth's surface.

Diffuse Solar Radiation: The total amount of solar energy received by a horizontal surface from all parts of the sky apart from the direct sun.

Direct Normal Radiation: Solar radiation arising from a narrow solid angle (5° aperture) centered around the sun's disk and impinging on a surface normal to the direction of the radiation.

Extraterrestrial Solar Radiation: Solar radiation received at the limit of the Earth's atmosphere.

Global Solar Radiation: Hemispherical solar radiation received by a horizontal plane.

Mathematical Model: An abstract mathematical representation of a process, device, or concept; it uses a number of variables to represent inputs, outputs and internal states, and sets of equations and inequalities to describe their interaction.

Meteonorm: A widely used and accepted solar radiation data source in the solar energy industry. It has been around for 30+ years (first released in 1985), and became the standard meteorological database for solar energy simulations. It is also the default meteorological database of some popular PV design software, such as PVsyst or PVSOL.

Solar Radiation: Incident energy per unit area of a surface, found by integration of solar irradiance over a specified time interval, often an hour or a day.

Sunshine Duration: The length of time that the ground surface is irradiated by direct solar radiation (i.e., sunlight reaching the Earth's surface directly from the sun).

Chapter 11
Smart Tourist Destination Management Using Demand Forecasting Techniques:
Using Big Data for Destination Demand Forecasting as Part of a Destination Management System

Leo Mrsic
Algebra University College, Croatia

Gorazd Surla
Valamar Riviera, Croatia

Mislav Balkovic
Algebra University College, Croatia

ABSTRACT

Tourism destination is the place where tourism demand and supply meet. Destination is often the main reason why people travel. In the first part of the chapter, the research is focused on the fast growth of tourism in the past decade, which does not come without problems. As one of the most discussed problems, the focus is on overtourism, which started as growth potential and moved into a sustainability issue. The scope of this chapter was to continue previous conducted research in the Croatian coastal town Šibenik and build additional scenarios using technological advancements and available data to build a cornerstone for a data-driven Destination Management System. The results of two experiments suggest that usage of simple techniques can be widely adopted in the search for sustainable management of the destination. In the second part of the research, the authors were able to combine data before outbreak of Coronavirus COVID-19 and during its early growth phase in Croatia to show what devastating impact it has on tourist arrivals in a short period of time and demonstrate in real time how fragile tourism demand is.

DOI: 10.4018/978-1-7998-3645-2.ch011

INTRODUCTION

Sustainability became the buzzword in the tourism sector somewhere in the mid-1990's mostly due to the revived involvement of the international development cooperation in the field of tourism promotion. UNWTO defines sustainable tourism as tourism that takes full account of its current and future economic, environmental and social impacts, addressing the needs of all tourism stakeholders - visitors, the environment, the industry, but also host communities.(UNWTO, no date) Looking at other definition, maybe one that best describes sustainable tourism is given by J. Swarbrooke in 1999, defining it as tourism that is "economically viable, but does not destroy the resources on which future of tourism will depend, notably the physical environment, and the social fabric of the host community."(Swarbrooke, 1999). All forms of tourism should strive to be more sustainable. Tourism is in a very special position to benefit local communities, both economically and socially. Tourism is also raising support and awareness for conservation of the environment. Within the tourism sector, economic development and environmental protection should not be opposing forces. Instead, they should both be pursued hand in hand as mutually reinforcing aspirations. Making tourism more sustainable is not just about managing and controlling the negative impacts of the industry. The aim of policies and actions must be to reduce the cost and strengthen the benefits of tourism. (UNEP-WTO, 2005)

Overtourism is quite a new term that started to be used on twitter as #overtourism in August 2012. Today the term Overtourism can be found in Collins online dictionary explained as "The phenomenon of a popular destination or sight becoming overrun with tourists in an unsustainable way." (Collins, no date) Moreover, this term can also be known as "overcrowding" or "tourismphobia." UNWTO has defined overtourism in its paper from September 2018 as "the impact of tourism on a destination, or parts thereof, that excessively influences perceived quality of life or citizens and/or quality of visitors experiences in a negative way." (UNWTO, 2018). It is a truly global issue (Thomas, 2019). Main examples are European destinations like Venice where the population of 55.000 is welcoming 20 million international overnight visitors per year(Ross, no date) or Barcelona, one of the first cities to elect a mayor who ran on a platform of managing overcrowding.(Becker, no date) Thai authorities were recently forced to act when the number of tourists visiting the beach Maya Bay, made famous by Danny Boyle's film "The Beach," led to shocking environmental damage.(Milano, Cheer and Novelli, no date) Other examples of worldwide destinations where overtourism has reached disruptive proportions include Paris, Kyoto, Berlin, Palma de Mallorca, Reykjavik, and Dubrovnik.

One of the first major analyses of that issue was produced by the WTTC and McKinsey & Company in December 2017 and was called "Coping with success: Managing overcrowding in tourism destinations." This report identified practical actions ranging from spreading visitors across sites, adjusting pricing to balance supply and demand and smoothing visitor numbers over time, or to limit access and activities and regulating accommodation supply. The main conclusion of that report was that "local tourism managers must work with all public and private stakeholders to develop a coherent plan to create and manage tourism growth that puts people and communities at its heart." (McKinsey&Company, 2017).

BACKGROUND

Smart tourism will shape the future direction of smart destinations, with huge benefits for both tourism and city development. Smart technology offers exciting opportunities for tourism attraction, with lasting impacts on smart city infrastructure, travel and planning. Every tourist destination wants to grow making it "smart way" is an opportunity not to jeopardize its authenticity and sustainability. Sustainable tourism development must be pursued at the level of destinations. Among others, Ritchie & Crouch argue that "every destination must examine its ability to maintain all dimension of sustainability if it wants to develop and preserve true competitiveness. Competitiveness without sustainability is illusionary."(Ritchie, J. R. B. Crouch, 2003) They argue that "from a destination management standpoint, it is important to appreciate that sustainable tourism does not attempt simply to control development, but that it also seeks to encourage the development and promotion of appropriate forms of tourism – many of which can actually enhance the environmental, social and cultural well-being of a destination in addition to increasing its economic prosperity."

Principles behind sustainable tourism management are given by Bramwell et al. (Bramwell, B., Henry, I., Jackson, G., Prat and Richards, G. and van der Straaten, 1996) and later on summarized by Swarbrooke (Swarbrooke, 1999). Two key points are essential for understanding sustainable tourism and are stated by Butler in his "state-of-the-art review" of sustainable tourism. (Butler, 1999) A policy makes and destination managers need to recognize that sustainable tourism acknowledges that there must be limits to growth in tourist development. Meaning that each destination needs to find its unique solution, its strategy for sustainable development. In McKinsey & Company and WTTC report for managing overcrowding, (McKinsey&Company, 2017) we can find the best practices that are foundational to the successful growth and management of tourism destinations. Their findings are showing that many destinations are experimenting with different tactics to mitigate the issues in destinations, but in many cases, these tactics are taken outside the context of a broader understanding or management plan. According to the report, four best practices to follow to improve destination's ability to alleviate overcrowding are:

1. Build a comprehensive fact base and update it regularly
2. Establish a sustainable growths strategy through rigorous, long-term planning
3. Involve all sections of society – commercial, public and social
4. Find new sources of funding

Smart technology can provide the ultimate visitor experience, boost sustainability and give destinations an improved longevity and ability to react quickly to shifts in visitor demographics or market demands. Solutions like smart apps and advanced visualization support systems can guide the visitor experience. No longer will visitors have to wait hours in a queue to score tickets to a top attraction. Smart technology apps will improve the navigation of a city's tourism attractions, helping tourists find the best times to visit, buy tickets in advance or give accurate directions on the go. This type of smart technology makes every trip as easy as possible for new visitors to interact with a city, improving their experience from the start. Tourists can also use smart technology to instantly connect with a local guide on social networks, giving them a unique and valuable tourist experience. Combined with architecture and structure in smart cities, IoT solutions will help tourists to personalize their surroundings, including light and temperature in a hotel, or use beacon technology to instantly optimize routes to a tourist location. Smart technology produces masses of data, and smart tourism is no different. While trawling through the output of smart data can be a challenge, using data effectively has game changing powers for tourist attractions and city planners alike. The direction of the conversation will change, with future tourism experience coming close towards personalization of every tourist trip.

Future technologies like, smart beacon technology or geo-tagging, will allow tourism destinations to see how visitors interact and flow with their attraction, as well as the destination as a whole. Dwell time, areas of congestion and popular hot spots can all be monitored, giving attractions and city planners the tools to effectively improve the tourist experience, every single time. City planners will also be able to utilize this data for infrastructure planning. Attractions can use smart technology to market to tourists directly in their local area. Tourism destinations can promote their site in quieter periods and highlight when busier times are, as well as reach new target markets quickly and easily. Finally, smart tourism will shape the structure of smart cities in terms of transport and travel but also city flow and infrastructure. While every destination relies on tourism revenue, smart technology for tourism has the potential to impact on smart cities in a much bigger way. Impacts from smart tourism will filter down to infrastructure developments and the allocation of resources, as city planners and developers will see how new visitors interact with a city space. For any destination and its tourism industry, efficient transportation is essential. Smart tourism and smart tech will give city planners the tools to create dynamic and responsive routes that optimize journeys around a city to beat congestion. Sustainable tourism and city development will be improved by data from smart technology too, highlighting where resources need to be deployed and utilized to be most effective. Data analytics from smart technology will allow cities to respond to every new visitor in real time, helping new visitors respond to different scenarios easily such as a delayed flight or hotel overbooking.

SMART TOURIST, SMART DESTINATION AND SMART TOURISM

In many ways, smart tourism can be seen as a logical progression from traditional tourism and more recently eTourism in that the groundwork for the innovations and the technological orientation of the industry and the consumers were laid early with the extensive adoption of ICT in tourism.(Buhalis, 2002)(Gretzel, 2011b). Different smart tourism initiatives around the world are followed by incredible institutional support, and in some instance, even pressure to realize smart tourism. In Europe, smart tourism initiatives are mostly spin-off from smart city projects. The focus is more put on innovation and competitiveness and developing smart end-user applications that support enriched tourism experiences using already existing data combined and processed in a new way. (Boes, Buhalis and Inversini, 2015) (Lamsfus et al., 2015) (Hunter et al., 2015). However, in practice, smart tourism has become very fuzzy concept often used to drive specific political agendas and to sell technological solutions and is frequently used in contact of open data initiates or for rather trivial projects as promoting free Wi-Fi or developing mobile applications (Kharchenko 2018, 2109). While these technologies and new approaches to data sharing and collection management are an important stepping stone in implementing smart tourism, they do not provide the full picture of what smart tourism should be. Instead laying on the theoretical foundations for its advancement and/or critique, academic work is largely focused on describing the phenomenon in the form of case studies (Gretzel et al., 2015). The Smart Experience component focuses on technology-mediated tourism experiences and their enhancement through personalization, context-awareness, and real-time monitoring. (Buhalis and Amaranggana, 2015).

It is important to notice that smart tourism spans three layers across these three components:

1. A smart information layer that aims at collection data
2. A smart exchange layer that supports interconnectivity

3. A smart processing layer that is responsible for the analysis, visualization, integration and intelligent use of data

In the end, on these components and layers considerations, they have identified ICT as the key to the conceptualization as well as the development of smart tourism. ICT with the ability to support tourism intelligently has been discussed, developed and envisioned for quite a long time. Within a smart tourism setting, this kind of technology is the key component of an information system that promises to supply tourism service providers and consumers with better decision support, more relevant information, greater mobility, and at the end, more enjoyable tourism experiences.(Gretzel, 2011a) (Werthner, 2003) These smart systems include a wide range of technologies in direct support of tourism such as decision support systems, context-aware systems, recommender systems, autonomous agents searching and mining Web sources, artificial and ambient intelligence as well as systems that create augmented realities. In smart tourism, technology is seen as an infrastructure, rather than as individual information system, and encompasses a variety of smart computing technologies that integrate network technologies, software and hardware to provide real-time awareness of the real world and advanced analytics to help people make more intelligent decisions about alternatives, as well as actions that will optimize business performances and business processes.(Washburn, D., Sindhu, U., Balaouras, S., Dines, R. A., Hayes, N. M., & Nelson, 2010) Visualization is considered to be an incredible opportunity for its supposed capacity to provide answers to practically any question that could be asked about people's behaviors, views, and feelings (Mayer-Shönberg and Cukier, 2013)(McAfee and Brynjolfsson, 2012). In tourism, analytics has been proposed as a new paradigm and a toolbox for tourism design, tourism marketing, and destination management.(Xiang and Fesenmaier, 2017) Tourists generate an incredible amount of information, before, during and after their trips. Most of this data could be collected and used by decision-makers in destinations to efficiently assign public funds to increase tourist attendance and satisfaction and thus making their territory attractive. However, these data are still mostly unused due to the lack of suitable tools (Soualah-Alila et al., 2016). As a toolbox, analytics is diverse regarding the nature of data, analytical operation, and business application.(Xiang et al., 2015) Compared to traditional methods of research and development, advanced analytics improve our capability to understand the consumer market at unprecedented scale, scope and depth.(Boyd and Crawford, 2012)(Xiang and Fesenmaier, 2017).

"Smart tourism" and "smart destinations" have been gaining attention as new frameworks powered by technology. They are crucial in understanding the impact the latest information and communication technologies (ICTs) have on the relationship between businesses, destinations and tourists. However, one centerpiece of the smart tourism ecosystem that has been rather neglected in research so far is: the tourists themselves. This new type of tourist is described through their attitudes and behaviors, and their role within the smart destination scenario. "To understand smart tourism, we need to understand who the 'smart tourist' is." (Fremenia-Serra et al. 2018).

How and why did the tourist behavior transform? Today, tourists are informed, empowered and wishes to receive more personalized and better-designed experiences. Technology have made it possible: service providers can co-create together with customers, who get tailored and richer experiences. But not only tourists are changing, also destinations. Technology have given rise to multiple opportunities for destinations to improve their performance and processes. For example, intelligent systems, the Internet of Things (IoT) or cloud computing are all examples of systems and applications that are currently shaping and reshaping tourism. They are forcing us to rethink traditional frameworks and consider unprecedented situations for destinations and tourists alike. Tourism, destinations, tourists and experiences can no longer be considered in the same way as before. Destination management organizations (DMOs) play a critical role in managing these shifts and ensuring a strong performance and collaboration among all the stakeholders to move towards becoming a "smart destination". Smart destinations (SDs) are a response to these changes, capturing the opportunity to provide improved and more personalized experiences through better interconnection between stakeholders in the smart service system. Is it possible to progress from eTourism to Smart Tourism? Smart tourism may be understood as an evolution, a step forward of eTourism. While eTourism refers to the usage of technology in the processes carried out by tourism organizations, smart tourism combines the physical world to the digital sphere through a range of technological systems and solutions. Smart tourism is an 'ecosystem', formed by a smart business network, smart destinations and a smart technologies infrastructure. This ecosystem causes benefits for the whole system and for each individual. One question always remains the same: How can destinations facilitate tourists with experiences that better address their individual preferences and create value? Intelligent systems can provide better tourist experiences due to a better management of information and a higher support in decision making. They offer flexibility and adaptability to the context and have the capacity to learn from experience and provide feedback. Finally, is it possible for an eDestination become a Smart Destination? While eDestinations emphasized the utilization of technology to provide information, in smart destinations, various technologies are embedded in all elements thanks to new developments, like the IoT. Technology here becomes the space, where all the interconnections between stakeholders happen. Therefore, improving tourist experiences through the use of the latest I technologies and smart services is the main objective of future destinations (Femenia-Serra, Barbara et al., 2018).

Moving from technology towards experience, we are evaluating possibilities on how experiences can be enhanced in a smart destination. Central technological platforms that interconnect all the stakeholders, integrate the data input from different sources, and allow for dynamic, real-time information sharing is just one of the multiple answers and solutions. What is needed are systems that maximize efficiency, improve decision making and enhance tourist experiences, while co-creating value for all involved stakeholders connected to the tourism ecosystem. Tourist behavior has undergone a great transformation because of tourists' use of technology during travel. Tourists have become more demanding, active, independent, informed and skilled, and have discovered new ways of searching for information, comparing, booking, interacting, sharing, complaining, reviewing and recommending. Making a digital tourist become a "smart tourist". Conceptualizing the smart tourist, model shows the development from eTourism to Smart Tourism, from eDestination to Smart Destination and from the Digital Tourist to the Smart Tourist.

Smart tourist is one that shares data, uses the range of technologies available and is ready to interact and co-create. Our research conceptualizes that smart tourists are active users of smart technologies. They not only to enhance their own experience, but perform dynamic, real-time interaction and co-creation with other stakeholders in the destination service ecosystem. Smart tourists need to trust stakeholder(s) in each situation, feeling they still have control over the experience by employing smart technologies according to their needs and preferences, but involving voluntarily other agents because they will get added value by doing so. Therefore, we see smart tourist as: "The tourist who, by being open to sharing his or her data and making use of smart technologies, interacts dynamically with other stakeholders, co-creating in this way an enhanced and personalized smart experience. This tourist is open to innovations, social and pro-active and finds his or her natural environment in the smart tourism ecosystem and the smart destination. (Femenia-Serra et al., 2018) (Harrison et al., 2010) (Vicini and Sanna, 2012). In the context of tourism, smart is used to describe a complex amalgam of all of the above (Gretzel et al., 2015).

Figure 1. Smart transformation of tourism and emergence of the smart Tourist (Femenia-Serra, Barbara et al., 2018)

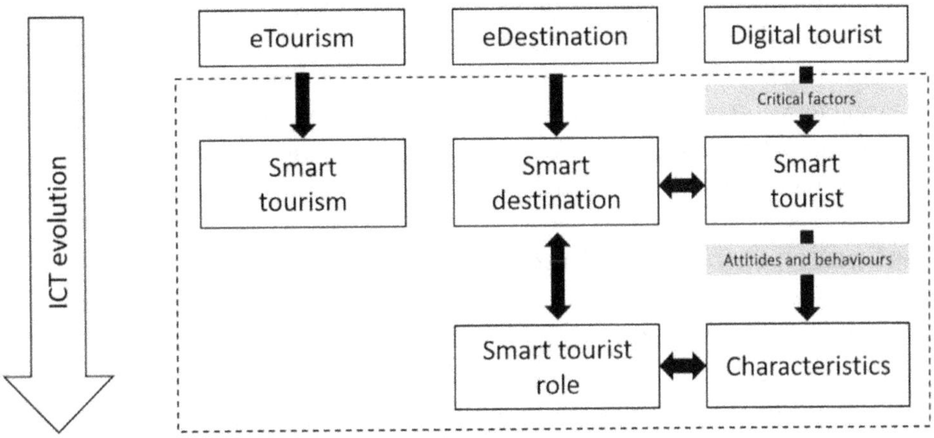

DESTINATION MANAGEMENT SYSTEMS FOR MODERN ERA

DMS is often considered as advanced DMOs web platform, since their inception somewhere in the middle of '90s the evidence clearly shows that not many destinations were able to develop and implement such systems successfully (Alford & Clarke, 2009) (Egger & Buhalis, 2008). More and more, DMOs use digital technology in order to facilitate the tourist experience before, during and after destination visit (all parts of travelling process), as well as for the coordination of all stakeholders involved in the experience and service delivery of tourism (Dimitrios Buhalis, 1997). Nowadays DMOs are attempting to provide information and accept reservations for different local enterprises and coordinate their facilities, but also utilize digital technology to promote their policies, harmonize their operational processes, increase the expenditure of tourist, increase overall experience level and boost the multiplier effects in the local economy (Dimitrios Buhalis & Spada, 2000). To do that DMS needs to administrate a wide range of request and provide efficient and appropriate information on an increasing supply of tourism products.

National and regional governments are employing DMSs to facilities DMOs management, as well as to support local ecosystem at the destination level (UNWTO, 2008) (D. Buhalis, Leung, & Law, n.d.).

In recent studies like "Åre Case" (Fuchs, Höpken, & Lexhagen, 2014) authors made research on Swedish mountain tourism destination. They have explored different customer-centric knowledge sources, like tourists' search (Web navigation), booking and feedback-behavior (review platforms, surveys), with a goal to create a business intelligence-based destination management information system based on data collected from pre-trip and post-trip experiences and facts. System in overall represents decentralized presentation and advanced visualization of data models with data that rests on the knowledge-based transactional layer (Höpken, Fuchs, Keil, & Lexhagen, 2011), generally called DMS.

Besides a sophisticated technology application, the effective use of a DMS demands development and implementation of learning processes as part of organizational behavior. In Åre Case authors find it crucial to integrate both, public and private, stakeholders and defined their user and location specific and unique/individual knowledge requirements. Also, for the post-trip phase, they were using feedback as customer perceptions and experience indicator (Vasant 2019). Following the research by so-called "Swedish collective" we have built our version of data-driven DMS that we are going to call eDestination.(Mršić, Surla and Balković, 2019) During research on Swedish mountain tourism destination, the focus was on activities before and after trip. Customer-based knowledge sources, like tourists' search (web logs), booking and feedback-behavior (surveys, reviews) were used. eDestination project was be developed for Šibenik, Croatia, and it had objective to test what data could be found and used in tourist destination during tourist stay. The authors wanted to exploit a few open and available technologies like Big Data, IoT and advanced data visualization to set up a cornerstone for future destination management system that can be built in any tourist destination.

Smart Destination Principles: Case of Šibenik Croatia

Šibenik eDestination project was be developed for Croatian sea tourist destination City of Šibenik. The author objective was to create different scenarios that will be used to test what data would be available in a real tourist destination. The main goal was to find what open data can be used and what additional data is there available for DMO – Šibenik Tourist Board (TZ Šibenik). With different scenarios, the idea was to see how different data can be put in correlation to get understanding of tourist behavior during their stay at the destination. Also, authors wanted exploit a few open and available technologies like Big Data, IoT and advanced data visualization. We set up eDestination pilot as a cornerstone for future destination management system that can be built in any Croatian tourist destination and later on transmitted to any other destination worldwide.

Authors used "Swedish collective" case as a starting point for research, but with two major differences. First one is that it is made for Šibenik and second that it uses only destination performance indicators that are happening during "On-Site" phase. eDestination Šibenik uses open data that is available to everyone (social networks), data that is available only to DMOs in Croatia (tourist statistics) and some new data that will be created for Šibenik (IoT sensors). That data from different sources are collected in Cloud-based SQL database. Figure is presenting high-level architecture of eDestination, where data collection is set up in "Backend" and under "Client" is shown what would DMO have as part of destination management. By using different scenarios, the idea was to show how different data can be put in correlation to get a needed insight into what is happening in the Šibenik.

All Croatian DMOs, have access to the eVisitor system, an information system for check in and check out of the guest. It is a mandatory system in which all accommodation providers in Croatia need to sign in all their guests. Besides signing in the guests, the system is used for accounting city tax that needs to be paid for each overnight, and for some statistical purpose. The system was introduced in 2015. Data that eVisitor can provide are about tourist and accommodation units. It can provide the length of stay, sex, age, country of origin of the guest, and type, category, location, occupancy of the accommodation unit. Ideally, the purpose of this system would be to help with destination management or to determine what marketing activity to use for the destination.

Destination Demand Forecasting: Accommodation Heat Map

The idea behind this scenario is to examine in which part of destination most tourist are staying. Looking at the statistic reports that were provided by TZ Šibenik, we can see that they are gathering information about numbers of arrival grouped for hotels, camps and private accommodation. For hotels and

Figure 2. eDestination high level architecture proposal

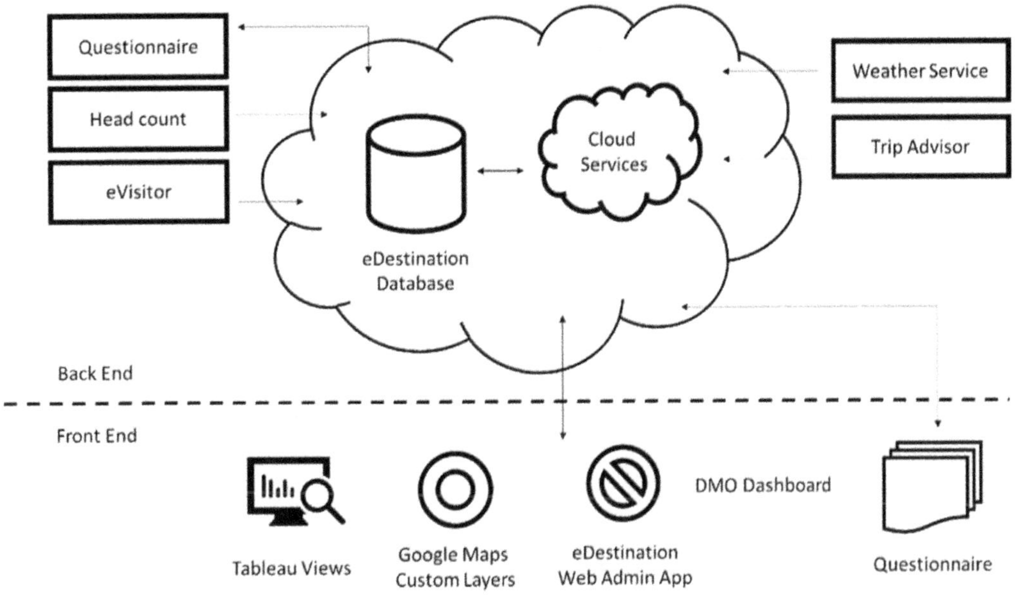

camps, they are having total numbers for each hotel and camp. And private accommodation providers are grouped into city districts, with total number of arrivals provided per district. Following the idea behind this report, we examined what data could be extracted from eVisitor. For each accommodation property registered in eVisitor till February 2019 we extracted its name, full address, accommodation type and category, number of accommodation units/rooms and maximum number beds in property. Using geocoder created in Python, we gathered geolocation for each address. And we used Tableau Software for advanced data visualization.

After importing all accommodation properties in Tableau and field mapping them using address latitude and longitude, we got map with all properties marked. But the idea behind this scenario is to compare different areas of the destination based on the number of tourists that could have overnight. For that, we would need to create a heat map of rooms or beds around Šibenik and we decided to create that with a Hexbin chart in Tableau. Hexbin chart will allow us to show aggregate data aggregated to a level that's just above a very specific granularity that will allow us to show the relative frequency of accommodation rooms or beds at an aggregate level while we will still maintain the detail of the view. If we were to try and map every single accommodation property as dot and use different size of dot for the number of beds the map could get busy very quickly and the view will not be helpful. A Hexbin chart works past this by providing a level of aggregation in the form of a hex bin and then its plotting those as hexagons on our view at which we will then color by the frequency. We have grouped all accommodation units in on a map of Šibenik using hex bin with radius of 1 kilometer. As we are using map with geolocations, to get the evenly sized hex bins that will give nicer look to our map and would get the same size area under each individual hexagon, we needed also to correct the Web Mercator projection issue, as it was stretching the sizes of hex bins showed on a map. After getting the map with hex bins we used red color scale to mark number of beds. The final result is heat map with all registered accommodation beds within area in radius of 1 kilometer marked across city of Šibenik. Final map is shown in Figure 3.

Figure 3. Heat Map of accommodation for City of Šibenik Croatia

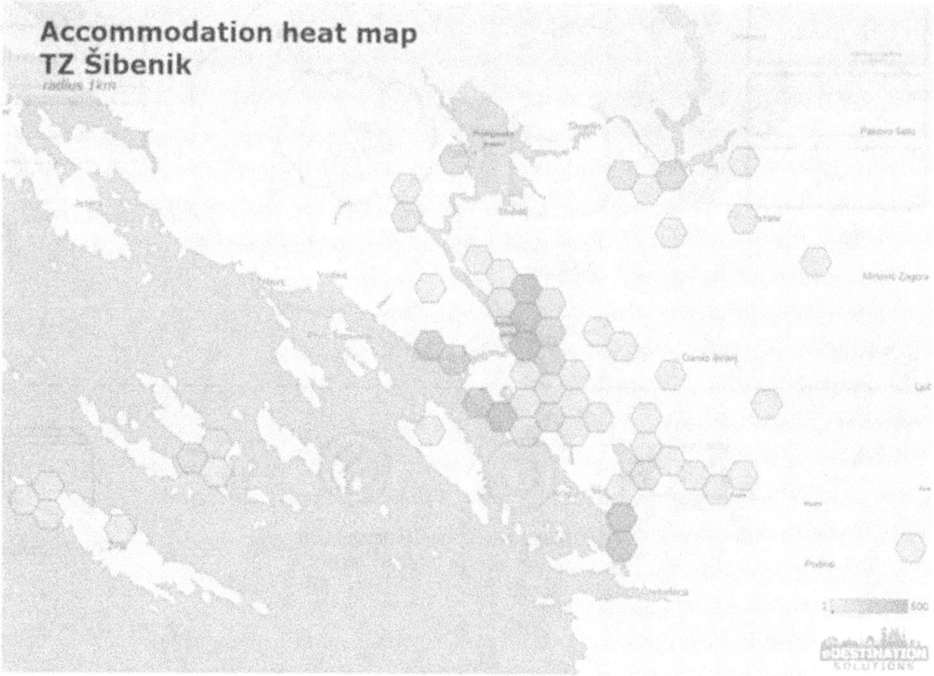

Counting total beds in area is only one possible setup of this heat map. With easy modification of this map we can mark all accommodation units, or all registered properties.

In addition, there could be shown almost real time usage of specific area. In the same way we could only mark occupied units within selected area on specific date. And as eVisitor is getting all data within 24 hours, we could setup this kind of report each day. Also, for more detailed look, radius scale can be lowered from kilometers to meters and the heat map can used to show more detailed look at smaller city areas.

Destination Demand Forecasting: Smart Tourist Demand Analysis and Forecasting

Through years, tourist demand modelling and forecasting is one of the important areas of tourism research. According to Song and Li study (Gang and Song, 2008) tourist arrivals variable is the most popular measure of tourism demand. In tourism, forecasting methods could be divided into quantitative and qualitative methods, but majority of studies uses quantitative which is divided into two subcategories; non-casual time series models and casual econometric approaches. The difference between these two subcategories is whether the model identifies any causal relationship between the tourism demand variable and its influencing factors. In their research, authors found that time series techniques are mostly used to model the demand.

Time series forecasting methods use only past patterns in a data series to extrapolate the future from the past. These implicitly assume that the course of a variable, such as tourism demand, over time is the product of a substantial number of unknown forces that give the series a momentum. This momentum can be captured in a model reflecting one of several time series methods. A time series forecasting model relates the values of a time series to previous values of that time series, its errors, or other related time series.(Makridakis, Wheelwright and Hyndman, 1998) These methods include the naïve method, the moving average method, the exponential smoothing method and autoregressive integrated moving average models. Tourist arrivals variable is the measure that is gathered in eVisitor. With this scenario we are going to create forecast of tourist arrival in Šibenik. Date collection through eVisitor started from January 2016. During first year input of data wasn't mandatory and due to that most of 2016. data is not significant. For this research we would use data from December 2016. to December 2019. And we would create monthly forecast of tourist arrivals for year 2020. From eVisitor we can also extract data about type of accommodation properties, so we will make separate forecast for different accommodation type groups - hotels, private accommodation and auto camps. As tourism demand is dealing with seasonality, something that is very pronounced in Šibenik, we used exponential smoothing as forecasting technique and we will be looking for potential seasonal patterns in every 12 months.

For forecasting we used Holt-Winters triple exponential smoothing technique. With forecast algorithms we tried to find a regular pattern in measures that can be continued into the future. We used forecasting software that automatically selects the best out of 9 models, with the best one being the one that generated the highest quality forecast. All used algorithms are simple models of a real-world data generating process. To have a high-quality forecast that simple pattern must match the pattern described by the model reasonably well.

Exponential smoothing methods are not restricted to those we have presented so far. By considering variations in the combinations of the trend and seasonal components, nine exponential smoothing methods are possible. Each method is labelled by a pair of letters (T,S) defining the type of 'Trend' and 'Seasonal' components. For example, (A,M) is the method with an additive trend and multiplicative seasonality; (Add,N) is the method with damped trend and no seasonality; and so on.

Table 1 gives the recursive formulas for applying the nine exponential smoothing methods in Figure 3. Each cell includes the forecast equation for generating hh-step-ahead forecasts, and the smoothing equations for applying the method.

Table 1. A two-way classification of exponential smoothing methods

Trend Component	**Seasonal Component**		
	N	**A**	**M**
	(None)	**(Additive)**	**(Multiplicative)**
N (None)	(N,N)	(N,A)	(N,M)
A (Additive)	(A,N)	(A,A)	(A,M)
Add (Additive damped)	(Add,N)	(Add,A)	(Add,M)

In each case, ℓtℓt denotes the series level at time tt, btbt denotes the slope at time tt, stst denotes the seasonal component of the series at time tt, and mm denotes the number of seasons in a year; αα, β∗β∗, γγ and ϕϕ are smoothing parameters, ϕh=ϕ+ϕ2+⋯+ϕhϕh=ϕ+ϕ2+⋯+ϕh, and kk is the integer part of (h−1)/m(h−1)/m.

In this scenario we made forecast for future 12 months – beginning of January 2020 to end of December 2020 – based on collected data, with seasonal pattern cycle of 12 months. Table 2 shows more exhaustive statistics and smoothing coefficient values for the Holt-Winters exponential smoothing models underlying the forecasts. For each measure that is forecasted, forecast models created is explained. In table Model section specifies whether the components Level, Trend, or Season are part of the model used to generate the forecast. The value for each component is one of the following: None (the component is not present in the model), Additive (the component is present and is added to the other components to create the overall forecast value) and Multiplicative (the component is present and is multiplied by the other components to create the overall forecast value). Quality Metrics shows set of values that are providing statistical information about the quality of the model: Root mean squared error, Mean absolute error, Mean absolute scaled error, Mean absolute percentage error and Akaike information criterion.

Table 2. Forecast statistics and smoothing coefficient values

	Model				Quality Metrics						Smoothing Coefficients		
	Level	**Trend**	**Season**		**RMSE**	**MAE**	**MASE**	**MAPE**	**AIC**		**Alpha**	**Beta**	**Gamma**
Auto Camps	**Additive**	**None**	**Additive**		**555**	**429**	**0.78**	**382,6%**	**498**		**0.165**	**0.000**	**0.000**
Private	**Multiplicative**	**Additive**	**Multiplicative**		**600**	**326**	**0.07**	**9.1%**	**507**		**0.000**	**0.040**	**0.020**
Hotels	**Additive**	**None**	**Additive**		**1,347**	**938**	**0.24**	**14.0%**	**498**		**0.025**	**0.000**	**0.000**

Last section is showing Smoothing Coefficients. Depending on the rate of evolution in the level, trend, or seasonal components of the data, smoothing coefficients are optimized to weight more recent data values over older ones, such that within-sample one-step-ahead forecast errors are minimized. Alpha is the level smoothing coefficient, beta the trend smoothing coefficient, and gamma the seasonal smoothing coefficient. The closer a smoothing coefficient is to 1.00, the less smoothing is performed, allowing for rapid component changes and heavy reliance on recent data. The closer a smoothing coefficient is to 0.00, the more smoothing is performed, allowing for gradual component changes and less reliance on recent data.

We have put the results of forecast in graphs shown in Figure 5 together with all data that is used to create forecast. On forecast, we made also 95% prediction intervals. We have also included trend line that is represented as dotted line on graph. All accommodation properties are divided into 3 main groups.

First one is Hotels. There is total of 15 hotels, categorized as 3*, 4* or non-categorized. Largest number of accommodation properties are under private accommodation. In 2019. in Šibenik were 1.443 privately rented properties. But as can been seen on Figure 5, majority of tourist are using hotel accommodation, even though there is large number of private accommodation options. Smallest percentage of arrivals is part of accommodation under auto camp group, in 2019. there were 11 auto camps. As shown in Table 3, through years there is increasing number of properties in all 3 accommodation groups, especially in private accommodation which has increase of 370 properties from 2016. to 2019.

During our research period in Šibenik were total of 896.498 arrivals, of this 494.947 were in hotels, 259.120 in private accommodation and 142.431 in camps. Table 4 shows tourist arrival in Šibenik in details, shown per year and per accommodation type during period of our research - December 2016 to December 2019.

Figure 4. Rationale used for recursive calculations and point forecasts

Trend	Seasonal N	Seasonal A	Seasonal M
N	$\hat{y}_{t+h\|t} = \ell_t$ $\ell_t = \alpha y_t + (1-\alpha)\ell_{t-1}$	$\hat{y}_{t+h\|t} = \ell_t + s_{t+h-m(k+1)}$ $\ell_t = \alpha(y_t - s_{t-m}) + (1-\alpha)\ell_{t-1}$ $s_t = \gamma(y_t - \ell_{t-1}) + (1-\gamma)s_{t-m}$	$\hat{y}_{t+h\|t} = \ell_t s_{t+h-m(k+1)}$ $\ell_t = \alpha(y_t/s_{t-m}) + (1-\alpha)\ell_{t-1}$ $s_t = \gamma(y_t/\ell_{t-1}) + (1-\gamma)s_{t-m}$
A	$\hat{y}_{t+h\|t} = \ell_t + hb_t$ $\ell_t = \alpha y_t + (1-\alpha)(\ell_{t-1} + b_{t-1})$ $b_t = \beta^*(\ell_t - \ell_{t-1}) + (1-\beta^*)b_{t-1}$	$\hat{y}_{t+h\|t} = \ell_t + hb_t + s_{t+h-m(k+1)}$ $\ell_t = \alpha(y_t - s_{t-m}) + (1-\alpha)(\ell_{t-1} + b_{t-1})$ $b_t = \beta^*(\ell_t - \ell_{t-1}) + (1-\beta^*)b_{t-1}$ $s_t = \gamma(y_t - \ell_{t-1} - b_{t-1}) + (1-\gamma)s_{t-m}$	$\hat{y}_{t+h\|t} = (\ell_t + hb_t)s_{t+h-m(k+1)}$ $\ell_t = \alpha(y_t/s_{t-m}) + (1-\alpha)(\ell_{t-1} + b_{t-1})$ $b_t = \beta^*(\ell_t - \ell_{t-1}) + (1-\beta^*)b_{t-1}$ $s_t = \gamma(y_t/(\ell_{t-1} + b_{t-1})) + (1-\gamma)s_{t-m}$
A_d	$\hat{y}_{t+h\|t} = \ell_t + \phi_h b_t$ $\ell_t = \alpha y_t + (1-\alpha)(\ell_{t-1} + \phi b_{t-1})$ $b_t = \beta^*(\ell_t - \ell_{t-1}) + (1-\beta^*)\phi b_{t-1}$	$\hat{y}_{t+h\|t} = \ell_t + \phi_h b_t + s_{t+h-m(k+1)}$ $\ell_t = \alpha(y_t - s_{t-m}) + (1-\alpha)(\ell_{t-1} + \phi b_{t-1})$ $b_t = \beta^*(\ell_t - \ell_{t-1}) + (1-\beta^*)\phi b_{t-1}$ $s_t = \gamma(y_t - \ell_{t-1} - \phi b_{t-1}) + (1-\gamma)s_{t-m}$	$\hat{y}_{t+h\|t} = (\ell_t + \phi_h b_t)s_{t+h-m(k+1)}$ $\ell_t = \alpha(y_t/s_{t-m}) + (1-\alpha)(\ell_{t-1} + \phi b_{t-1})$ $b_t = \beta^*(\ell_t - \ell_{t-1}) + (1-\beta^*)\phi b_{t-1}$ $s_t = \gamma(y_t/(\ell_{t-1} + \phi b_{t-1})) + (1-\gamma)s_{t-m}$

Figure 5. Demand forecast visualization

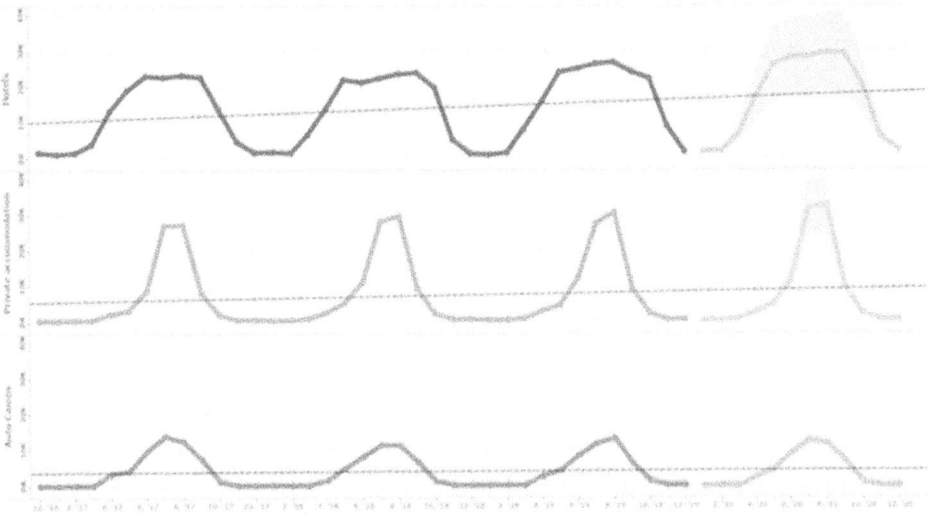

DESTINATION DEMAND FORECASTING: EFFECT OF COVID-19

In continuation of this research authors wanted to test its forecast model and see how their forecast. Beginning of year in destination Šibenik is low season and historically there not many tourism arrivals during January and February. Bigger number of tourism arrivals starts from March, with peak of high season in July and August. Comparing actual to forecasted numbers, January and February numbers are similar to each other, shown on Figure 6. Combined arrivals in these 2 months were 4037, compared to 4087 forecasted arrivals. However, March shows big difference.

Table 3. Accommodation properties divided into groups during reasearch period

	Hotels	Private accommodation	Auto Camps
2016.	13	1.073	10
2017.	13	1.241	9
2018.	15	1.442	10
2019.	15	1.443	11

Table 4. Tourist arrivals in Šibenik during research period

	Hotels	Private accommodation	Auto Camps	Grand Total
2016.	1.678	326		2.004
2017.	149.053	79.139	51.279	279.471
2018.	160.718	89.107	44.069	293.894
2019.	183.498	90.548	47.083	321.129
Grand Total	494.947	259.120	142.431	896.498

Figure 6. January to March 2020 tourist arrivals in Šibenik - comparison actual to forecasted

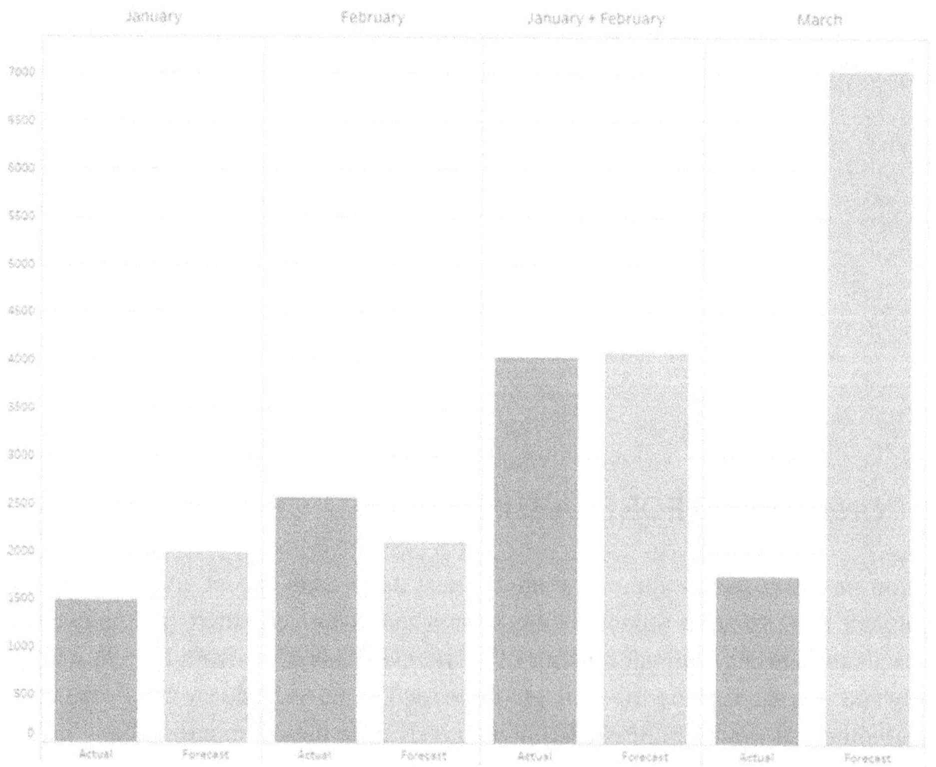

Unfortunately, this difference is not due to wrong calculation. At the end of December 2019 China reported a cluster cases of pneumonia in Wuhan, Hubei Province. On 7th January the outbreak was identified as a new coronavirus, later named as COVID-19. From there things evolved quickly, on 13th there was first confirmed case outside of China, in Thailand. On 22nd human to human transmission was recorded and on 23rd Wuhan (city with population of 11 million people) was put under quarantine with all trips to and from it suspended. In following days and weeks virus started to spread over the world, hitting Italy (Croatian neighbor country) very hard[1]. On 25th February first case of COVID-19 is recorded in Croatia[2]. World Health Organization declared that corona virus outbreak can be characterized as a pandemic on 11th March[3]. () Croatian Government first banned public gathering of more than 1.000 people, than 100 and on 19th March gathering of more than 5 persons[4]. On the same day transits through border crossing of the Republic of Croatia were banned, meaning that tourism in Croatia will be suspended[5].() This situation can be seen all over the world and according to research by World Tourism Organization, as of 6th April, 96% of all worldwide destination have introduced some kind of travel restriction due to COVID-19 pandemic[6].

Due to the pandemic, forecasting for March, but also for the rest of the 2020. became irrelevant. Forecasted number of arrivals for March was 7030. On 13th March biggest opened hotel in Šibenik during March dates, hotel Ivan, closed his doors[7].() Soon after it, same thing did all other hotels in Šibenik. On Figure 6 we have additionally analyzed number of arrivals per day in March. All hotels in Šibenik were closed after 16th March. And they are note open in April either. Their re-opening date will depend on the end of COVID-19 pandemic and opening of boarders. As majority of tourist arrivals in Šibenik are international (82,76% in 2019) revival of tourism and arrival in the rest of 2020. will not only depend on COVID-19 situation in Croatia, but to worldwide development.

Figure 7. Number of arrivals per day in Šibenik, March 2020

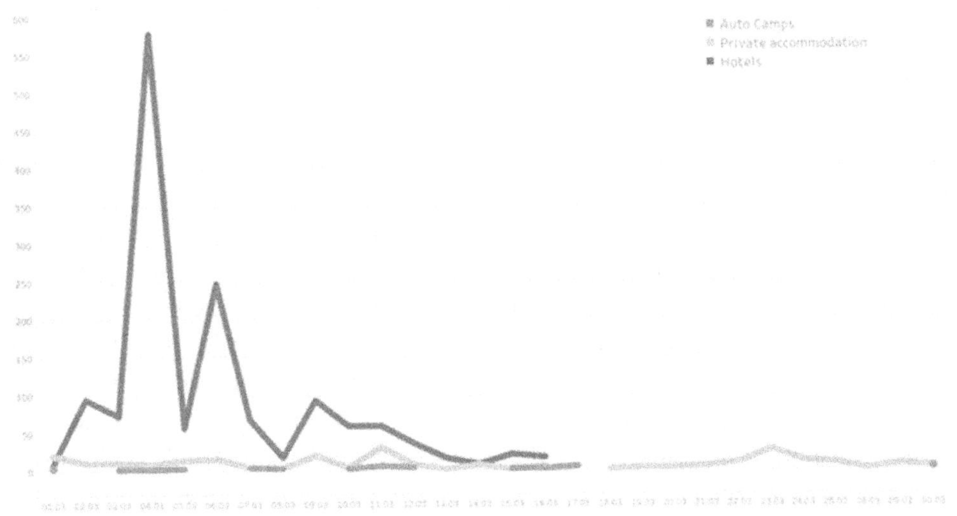

Actual number of tourist arrivals in March was 1746. After all hotels there were closed, only few private accommodation units were left opened during second part of March. UNWTO estimates that international arrivals worldwide in 2020 will be down between 20% and 30%[8]. Unfortunately for Šibenik, this might be even worse, due to the uncertainty of COVID-19 pandemic ending and high seasonality where 42,1 of total arrivals in 2019 were recorded only in 2 summer months (July and August). How big impact this is on Šibenik tourist arrivals is show on Figure 7 on which we have analyzed running total of all arrivals in 2020 till end of March. Till 11th March, year 2020 was looking to be record year for arrivals in first quarter, but unfortunately, since then everything has stopped. Or better say paused.

Figure 8. Running total of tourist arrivals in Šibenik - January to March 2020

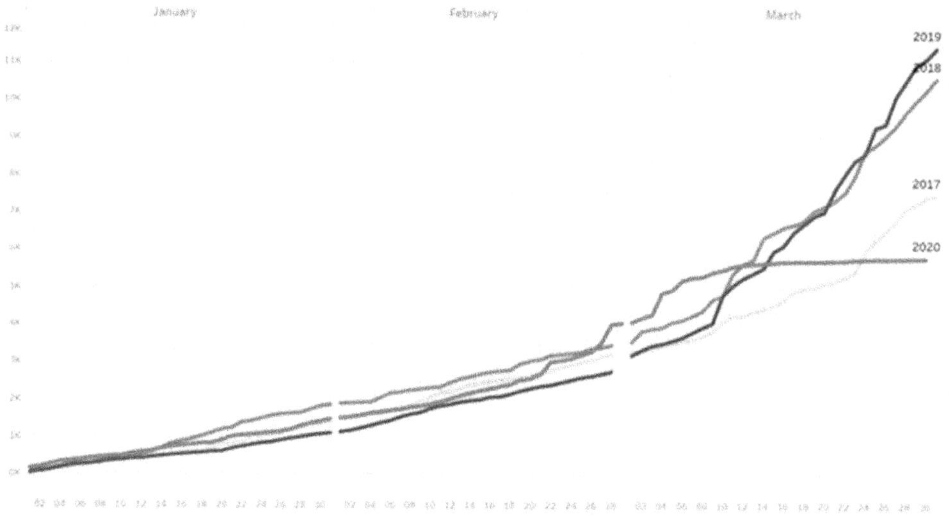

CONCLUSION

This research represents future step of DMS shown in our previous work. Here we have included additional data from longer time period and have explored what can be added to give additional value to destinations "Hawkeye". Moreover, authors have shown how demand forecasting in destination can be easily done with data that destinations already have and it can be one of the strongest part of DMS.

After exploring options that could be made from data from eVisitor future research should focus on creating a diverse mix of all presented scenarios so fare, and put them in correlation. If that future scenario would be created for attractions in Šibenik, it would contain questionnaire data from Scenario 1, data from the sensor from Scenario 2 that would be put in heat map from scenario 4 and data from eVistor from Scenario 3 with forecast from scenario 5. With these combinations, we would get information about visitor satisfaction and the number of visitors visiting each attraction and we would be able to create a heat map of Šibenik areas used by tourist on specific hour of a day. We would add a total number of tourists that had overnight in Šibenik from eVisitor and combine this date to make forecast of movement of tourist in Šibenik. From eDestination Šibenik, we could see that digital technology and tools can be easily used in tourism. Destinations already have a large amount of unused data. In addition, there are different new datasets available to be gathered. Moreover, with the use of technology, getting new data is not an issue anymore. As shown in our research, this is something that can be done with the use of simple IoT technology. ICT innovations are making life much easier in some other industries, so there is no reason that this cannot happen in the tourism industry. Even though the tourism sector is known as it does not go away with innovations well, this is something that will be changed. As the World is entering the fourth industrial revolution, the tourists are changing its behavior and with them, the tourism industry needs to change as well. Overtourism is new problem, it is problem that is not affecting all destinations and it does not have same scale in affected destinations. Nevertheless, that problem can happen to any destination if it is not managed on the right way.

Recent pandemic of COVID-19 has showed that tourism demand is indeed very fragile. Tourism industry was one of the first industries hit by this pandemic and probably it would be one of the most affected ones. By latest World Truism and Travel Council research up to 75 million jobs in tourism and travel industry are at immediate risk worldwide, among them 10 million are in Europe[9].

From history, we can see that tourism industry has shown resilience to all kind of global situation, and it will definitely bounce back this time also. It may take some time and some things might change, but human need to move and travel will keep tourism industry strong. Even though in this period of time where travel has stopped overtourism problem looks like it is in the past. However, without destination management overtourism might easily become problem again, and considering changes in perception of way to travel after the COVID-19 this problem will have bigger negative impact on destinations.

ACKNOWLEDGMENT

This research received no specific grant from any funding agency in the public, commercial, or not-for-profit sectors.

REFERENCES

Becker, E. (n.d.). *Barcelona: Taming tourism*. Available at: https://www.travelweekly.com/Europe-Travel/Barcelona-Taming-tourism

Boes, K., Buhalis, D., & Inversini, A. (2015). Conceptualising Smart Tourism Destination Dimensions. In *Information and Communication Technologies in Tourism 2015* (pp. 391–403). Springer International Publishing. doi:10.1007/978-3-319-14343-9_29

Boyd, D., & Crawford, K. (2012). Critical questions for big data: Provocations for a cultural, technological, and scholarly phenomenon. *Information Communication and Society*, *15*(5), 662–679. Advance online publication. doi:10.1080/1369118X.2012.678878

Bramwell, B., Henry, I., Jackson, G., Prat, A. G., Richards, G., & van der Straaten, J. (1996). *Sustainable tourism management: principles and practice* (B. Bramwell, Ed.). Tilburg University Press.

Buhalis, D. (2002). *eTourism: Information technology for strategic tourism management*. Financial Times/Prentice Hall.

Buhalis, D., & Amaranggana, A. (2013). Smart Tourism Destinations. In *Information and Communication Technologies in Tourism 2014* (pp. 553–564). Springer International Publishing. doi:10.1007/978-3-319-03973-2_40

Buhalis, D., & Amaranggana, A. (2015). Smart Tourism Destinations Enhancing Tourism Experience Through Personalisation of Services. In *Information and Communication Technologies in Tourism 2015* (pp. 377–389). Springer International Publishing. doi:10.1007/978-3-319-14343-9_28

Butler, R. W. (1999). Sustainable tourism: A state-of-the-art review. *Tourism Geographies*, *1*(1), 7–25. doi:10.1080/14616689908721291

Collins. (n.d.). *Collins Overtourism*. Available at: https://www.collinsdictionary.com/submission/19794/Overtourism

Crouch, G. I., & Ritchie, J. R. B. (2000). The competitive destination: A sustainability perspective. *Tourism Management*. Advance online publication. doi:10.1080/1461668042000280264

Davidson, R., & Maitland, R. (1997). *Tourism Destinations*. Hodder & Stoughton.

Ernie, H., & Wall, G. (1992). *Marketing Tourism Destinations: A Strategic Planning Approach*. Wiley.

Femenia-sierra, F., Neuhofer, B., & Ivars-Baidal, J.A. (2018). Towards a conceptualisation of smart tourists and their role within the smart destination scenario. *The Service Industries Journal, 39*(2).

Gang, L., & Song, H. (2008). Tourism Demand Modelling and Forecasting. *Tourism Management*, *29*(2), 203–220. doi:10.1016/j.tourman.2007.07.016

Gretzel, U. (2011a). Intelligent systems in tourism. A Social Science Perspective. *Annals of Tourism Research*, *38*(3), 757–779. Advance online publication. doi:10.1016/j.annals.2011.04.014

Gretzel, U. (2011b). Intelligent systems in tourism. *Annals of Tourism Research*, *38*(3), 757–779. doi:10.1016/j.annals.2011.04.014

Gretzel, U., Sigala, M., Xiang, Z., & Koo, C. (2015). Smart tourism: Foundations and developments. *Electronic Markets*, *25*(3), 179–188. doi:10.100712525-015-0196-8

Harrison, C., Eckman, B., Hamilton, R., Hartswick, P., Kalagnanam, J., Paraszczak, J., & Williams, P. (2010). Foundations for Smarter Cities. *IBM Journal of Research and Development*, *54*(4), 1–16. Advance online publication. doi:10.1147/JRD.2010.2048257

Hunter, W. C., Chung, N., Gretzel, U., & Koo, C. (2015). Constructivist Research in Smart Tourism. *Asia Pacific Journal of Information Systems*, *25*(1), 105–120. doi:10.14329/apjis.2015.25.1.105

Kharchenko, V., & Vasant, P. (2018). *Handbook of Research on Renewable Energy and Electric Resources for Sustainable Rural Development*. IGI Global. doi:10.4018/978-1-5225-3867-7

Kharchenko, V., & Vasant, P. (2019). *Handbook of Research on Energy-Saving Technologies for Environmentally-Friendly Agricultural Development*. IGI Global. doi:10.4018/978-1-5225-9420-8

Kharchenko, V., & Vasant, P. (2019). *Handbook of Research on Smart Computing for Renewable Energy and Agro-Engineering*. IGI Global, doi:10.4018/978-1-7998-1216-6

Lamsfus, C. (2015). Smart Tourism Destinations: An Extended Conception of Smart Cities Focusing on Human Mobility. In *Information and Communication Technologies in Tourism 2015* (pp. 363–375). Springer International Publishing. doi:10.1007/978-3-319-14343-9_27

Makridakis, S. G., Wheelwright, S. C., & Hyndman, R. J. (1998). *Forecasting: Methods and Applications*. Academic Press.

Mayer-Shönberg, V., & Cukier, K. (2013). *Big Data: A Revolution That Will Transform How We Live, Work, and Think*. Hodder & Stoughton.

McAfee, A., & Brynjolfsson, E. (2012). *Big Data. The management revolution*. Harvard Buiness Review. doi:10.100712599-013-0249-5

McKinsey & Company. (2017). *Coping with success: managing overcrowding in tourism destinations*. Available at: https://www.wttc.org/-/media/files/reports/policy-research/coping-with-success---managing-overcrowding-in-tourism-destinations-2017.pdf?la=en

Milano, C., Cheer, J. M., & Novelli, M. (n.d.). *Overtourism is becoming a major issue for cities across the globe*. Available at: https://www.weforum.org/agenda/2018/07/overtourism-a-growing-global-problem

Mršić, L., Surla, G., & Balković, M. (2019). Technology Driven Smart Support System for Tourist Destination Management Organizations. *International Conference on Innovative Computing and Communication (ICICC-2019)*. Available at: http://icicc-conf.com/icicc2019.pdf

Ritchie, J. R. B., & Crouch, G. I. (2003). *The competitive destination: a sustainable tourism perspective* (J. R. B. Ritchie & G. I. Crouch, Eds.). CABI. doi:10.1079/9780851996646.0000

Ross, W. (n.d.). *The death of Venice: Corrupt officials, mass tourism and soaring property prices have stifled life in the city*. Available at: https://www.independent.co.uk/news/world/europe/the-death-of-venice-corrupt-officials-mass-tourism-and-soaring-property-prices-have-stifled-life-in-10251434.html

Soualah-Alila, F. (2016). DataTourism: Designing an Architecture to Process Tourism Data. In *Information and Communication Technologies in Tourism 2016* (pp. 751–763). Springer International Publishing., doi:10.1007/978-3-319-28231-2_54

Swarbrooke, J. (1999). *Sustainable Tourism Management*. CABI.

Thomas, J., Karagoz, P., Ahmed, B., & Vasant, P. (2019). *Deep Learning Techniques and Optimization Strategies in Big Data Analytics*. IGI Global. doi:10.4018/978-1-7998-1192-3

UNEP-WTO. (2005). Making tourism more sustainable. A guide for policy makers. *Environment*. doi: 92-807-2507-6

UNWTO. (2007). *A Practical Guide to Tourism Destination Management*. Available at: https://www.e-unwto.org/doi/book/10.18111/9789284412433

UNWTO. (2018). *'Overtourism'? Understanding and Managing Urban Tourism Growth beyond Perceptions. World Tourism Organization*. UNWTO. doi:10.18111/9789284419999

UNWTO. (n.d.). *Definition of sustainable tourism*. Available at: http://sdt.unwto.org/content/about-us-5

Vasant, P., Zelinka, I., & Weber, G.-W. (2019a). Intelligent Computing & Optimization. *Proceedings of the 2nd International Conference on Intelligent Computing and Optimization 2019 (ICO 2019)*, 1. DOI 10.1007/978-3-030-00979-3

Vasant, P., Zelinka, I., & Weber, G.-W. (2019b). Intelligent Computing & Optimization. *Proceedings of the 2nd International Conference on Intelligent Computing and Optimization 2019 (ICO 2019)*, 2. DOI 10.1007/978-3-030-33585-4

Vellas, F., & Bécherel, L. (1999). *The International Marketing of Travel and Tourism: A Strategic Approach*. Macmillan Education UK. doi:10.1007/978-1-349-27486-4

Vicini, S., & Sanna, A. (2012). How to Co-Create Internet of Things-enabled Services for Smarter Cities. *The First International Conference on Smart Systems, Devices and Technologies.*

Washburn, D., Sindhu, U., Balaouras, S., Dines, R. A., Hayes, N. M., & Nelson, L. E. (2010). *Helping CIOs Understand "Smart City" Initiatives: Defining the Smart City, Its Drivers, and the Role of the CIO*. Forrester Research, Inc.

Werthner, H. (2003). Intelligent Systems in Travel and Tourism. *IJCAI 2003: 18th International Joint Conference on Artificial Intelligence.*

Xiang, Z., & Fesenmaier, D. R. (2017). *Big Data Analytics*. Tourism Design and Smart Tourism., doi:10.1007/978-3-319-44263-1_17

Xiang, Z., Wang, D., O'Leary, J. T., & Fesenmaier, D. R. (2015). 'Adapting to the Internet: Trends in Travelers' Use of the Web for Trip Planning'. *Journal of Travel Research*, *54*(4), 511–527. Advance online publication. doi:10.1177/0047287514522883

KEY TERMS AND DEFINITIONS

DMO: Organization which coordinates the many constituent elements of the tourism product; provides visitor services and the necessary information structure to market the destination in a most democratic way to enhance residents' well-being.

eVisitor: Information system called eVisitor developed by The Croatian National Tourist Board, together with the system of tourist boards and other stakeholders, with the aim of simplifying the process of tourist check-in and check-out and the control of tourist tax payments.

Forecasting: A decision-making model used by many businesses to help in budgeting, planning, and estimating future growth. In the simplest terms, forecasting is the attempt to predict future outcomes based on past events and management insight.

Overtourism: The perceived congestion or overcrowding from an excess of tourists, resulting in conflicts with locals. The term has only been used frequently since 2015 but is now the most commonly used expression to describe the negative impacts ascribed to tourism.

ENDNOTES

1 https://www.who.int/news-room/detail/08-04-2020-who-timeline---covid-19
2 https://www.hzjz.hr/priopcenja-mediji/covid-19-priopcenje-prvog-slucaja/
3 https://www.who.int/news-room/detail/08-04-2020-who-timeline---covid-19
4 https://narodne-novine.nn.hr/clanci/sluzbeni/2020_03_32_713.html
5 https://narodne-novine.nn.hr/clanci/sluzbeni/2020_03_32_714.html
6 https://www.unwto.org/news/covid-19-response-travel-restrictions
7 https://www.sibenik.in/sibenik/kljuc-u-bravu-u-solarisu-zatvoreni-svi-hoteli-uprava-se-nada-boljem-sutra/119949.html
8 https://www.unwto.org/tourism-covid-19
9 https://wttc.org/en-gb/COVID-19/Member-Hub

Chapter 12
Development of DNN Model for Predicting Surge Pressure Gradient During Tripping Operations

Shwetank Krishna
Universiti Teknologi PETRONAS, Malaysia

Syahrir Ridha
Universiti Teknologi PETRONAS, Malaysia

Pandian Vasant
Universiti Teknologi PETRONAS, Malaysia

ABSTRACT

Application of machine learning tools in drilling hydrocarbon well is still exploratory in its stage. This chapter presents a brief review of various applied research in drilling operations using machine learning (ML) tools and develop a deep neural network (DNN) model for predicting the downhole pressure surges while tripping. Tripping in or out drill-string/casing with a certain speed from the wellbore will result in downhole pressure surges. These surges could result in well integrity or well control problems, which can be avoided if pressure imbalances are predicted before this operation is engaged. Existing analytical models focus on forecasting the pressure imbalance but requires cumbersome numerical analysis. This could be solved by integrating DNN tool with the best existing analytical model predicted dataset. Consequently, the aim of this chapter is to provide an overview of various applications of machine learning tools in drilling and presenting a step-by-step process of developing a DNN model for the prediction of downhole pressure surges during tripping operation.

DOI: 10.4018/978-1-7998-3645-2.ch012

INTRODUCTION

The purpose associated with this book chapter is to explain the development of an intelligent model using deep neural network (DNN) for controlling well integrity problem during tripping operations. This model uses the historical or real-time data for predicting the surge/swab pressure associated with tripping operations. By prescribe remedial solution for lost circulation zone by integrating different data sets equip by various service companies. This model will further optimize the drilling operations, reduces non-productive time (NPT) and lowers drilling risks.

In comparison with other method, ML finds quick results from data due to its flexibility and automatic perception. Machine learning is a data analytics method that educates computers to do what comes normally to people and animals: learn from experience (Pankaj et al., 2018). Machine learning calculations utilize computational strategies to "learn" data specifically from information without depending on a foreordained condition as a model. The algorithms adaptively improve their execution as the number of tests available for learning increases (Pankaj et al., 2018). In fact, ML can deal with the lack of proper physical models or in the presence of uncertainty that usually effect the outcomes in other approaches. The thought of ML is to prepare a model on known input and yield data so that it can foresee future yields (supervised learning) or finding covered up designs or inherent structures in input data (unsupervised learning) (Raschka & Mirjalili, 2017). Then when different input arrives, the ML predict the correct output by applying what it has been taught. For this, ML will require a large set of selective data (historic and real-time data) to develop this model. The drilling business has endured a ton because of various drilling anomalies. An industry average estimate of the additional cost to the well caused by lost circulation specifically, is around 40 percent, with most of this cost due to rig down-time (Reddy, Zhang, & Ellis, 2016). This is excluding the apparatus costs because of the harm to the drill pipe and/or blowout. This could be optimized by developing predictive models.

Motivation

Currently, oil & gas business facing difficulty to exploit artificial intelligent innovations to upgrade operational decision making for drilling anomalies. According to the literature survey, there are a few restrictions in the way, well development information is as of now procured and gave by the service contractor companies or organizations in a manner, such as, constrained access to the information, numerous information sources that are not really open by means of every one of provider's frameworks and reliance on exclusive software tools and services from the providers for post-preparing/examining information to educate operational choices to forestall or relieve or mitigate or prevent any lost circulation occasions (Andia, Sant, & Whiteley, 2018). According to Unrau and Torrione, 2017, formation fluid influxes and drilling fluid losses is bound to happen during drilling operation. Detection of these phenomena during real-time operation is complicated due to varying and transient nature of fluid rheological parameters during drilling. This make the deployed detection sensor to malfunction and provide false alarm on a regular basis. Since, rate of false alarm increases with increasing complication, rig personnel start to treat these alarm as unreliable.

Novelty

The challenge is to predict surge pressure gradient during tripping operations to check most commonly occurred drilling anomaly i.e., loss circulation. The technological advancements and development of predictive tool will help in monitoring fluid losses in real time. The making of such a device requires an establishment of real-time monitoring platform that can anticipate fluid loss and progressively reduces the restrictions. This predictive model will deploy algorithms that conduct real-time computations, which utilizes static and real-time streaming data from various sources and subsequently storing the algorithm's results. At last, imagining of the raw and derived data is required, intended to determine forecast of lost dissemination.

This developed model will be based on the logic used by drillers and operators to identify fluid loss events. These logic are used to develop the predictive model whose data is used to train the DNN model, which have been never conducted according to the surveyed literature. Once DNN is trained it will check the parameters using lessons learned from training for determining surge pressure gradient. As a final piece of the finding the developed tool will benefit the operator and contractors to better manage the operations and early identify the surge pressure range to check associated problematic events.

BACKGROUND

Machine Learning

Machine learning is an application of artificial intelligence (AI) that gives frameworks the capacity to naturally learn and progress from experience without being unequivocally programmed (Raschka & Mirjalili, 2017; Vasant, Kose, & Watada, 2017; Vasant, Marmolejo, Litvinchev, & Aguilar, 2019). Machine learning focuses on the advancement of computer programs that can get to data and utilize it to learn for themselves. The method of learning starts with observations or data, such as cases, direct experience, or instruction, in order to seek for designs in data and make better decisions in the future based on the illustrations that we provide. The essential point is to permit the computers to memorize automatically without human intervention or help and alter activities accordingly. Machine learning empowers analysis of enormous amounts of data. While it generally delivers quicker, more exact results to distinguish beneficial opportunities or unsafe risks, it may also require extra time and resources to prepare it properly (Bishop, 2006).

Machine learning uses two types (Figure 1) of techniques: supervised learning and unsupervised learning. First once trains a model on identified input and output data so that it can forecast future outputs, and the latter one, discovers hidden patterns or inherent structures in input data.

Supervised machine learning builds a model that produces predictions based on evidence within the presence of uncertainty. A supervised learning calculation takes a known set of input information and known responses to the data (output) and trains a model to create sensible predictions for the response to new data (Abakouy, En-Naimi, & El Haddadi, 2017). *Unsupervised learning* finds covered up patterns or inherent structures in data. It is utilized to draw inductions from datasets comprising of input data without labelled responses (Bishop, 2006; Pandian Vasant, Ivan Zelinka, 2019).

Figure 1. Learning Style of ML

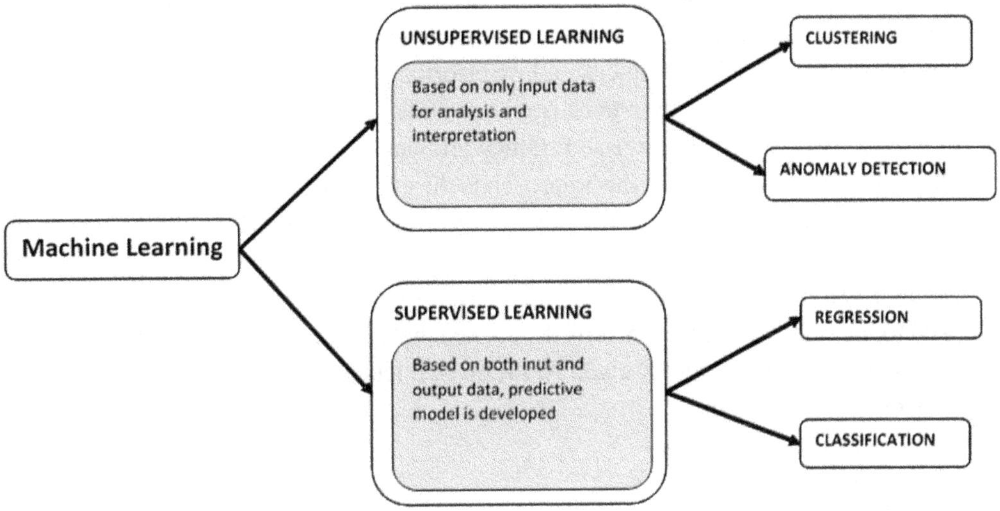

There are number of algorithms for training ML model, but the most widely used are rule-based systems, Bayesian and statistical algorithms, and neural networks. These are represented in a form of Venn's diagram (Figure 2).

Figure 2. Venn Diagram representing the location of DNN in the hierarchy of AI study

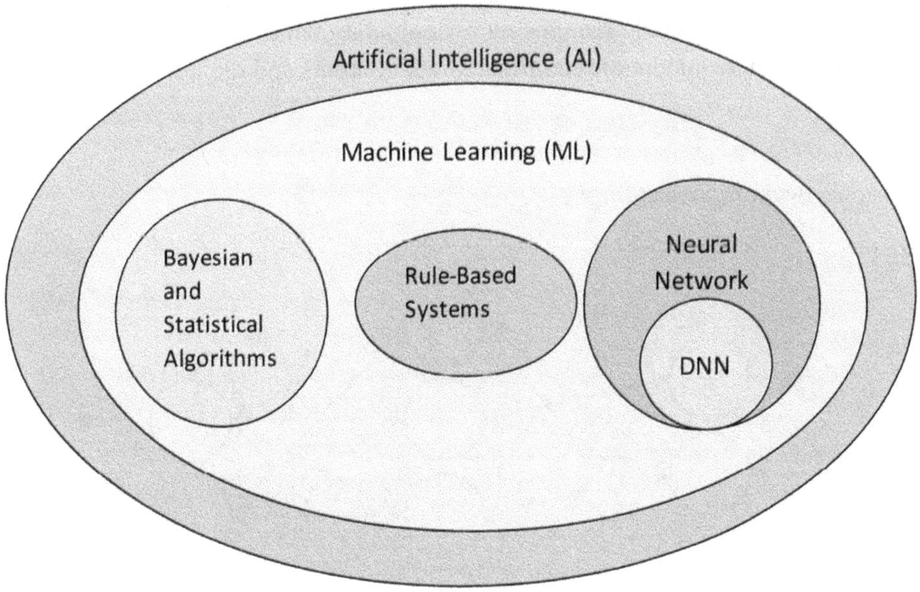

Rule-based and decision tree algorithms are the impressively basic to get it. Rule-based frameworks comprise of a collection of consistent rules or conditions based on inputs (J. Joshua Thomas, Pinar Karagoz, B. Bazeer Ahamed, 2020). A rule is activated when its conditions are met. In some cases, rules can produce clashing activities in which case a few priority components must be executed. A rule- or behaviour-based framework ordinarily moves from one state to the next, applying all the rules to each state. *Decision trees* are a structure rule-based framework where each node in a tree has conditions that permit classification. One advantage of this approach is the capacity to backtrack the logic prepare used to perform an action or reach a conclusion. Typically, exceptionally important for debugging, but it can also be valuable within the learning prepare.

Bayes' theorem describes the probability of a test result founded on preceding data of conditions that might be linked to the result. The theorem tells the chance that an event 'A' happened given the indicator X, Pr(A|X), to the probability of event X given A, Pr(X|A). It lets for alteration of measurement mistakes if the real probabilities are known.

$$P(A|X) = \frac{P(X|A) * P(A)}{P(X)}.$$

There are several Bayesian algorithms based on the theorem, including Naive Byes and Bayesian Belief Network (BBN).

A fundamental *neural network* comprises of a set of inputs and outputs with a hidden layer in between (Figure 3) (DNN has multiple hidden layers). On the basis of systems logic, the number of inputs, outputs, and hidden layers changes as well as other arrangement alternatives. Neural networks require training to create the weights utilized within the framework. Systems can learn dynamically, but these tend to be more complex as preparing regularly requires more computational drive than deduction. Training uses a feedback framework where an input is coordinated with outputs and the internal hidden layer weights

Figure 3. Neural Network representation

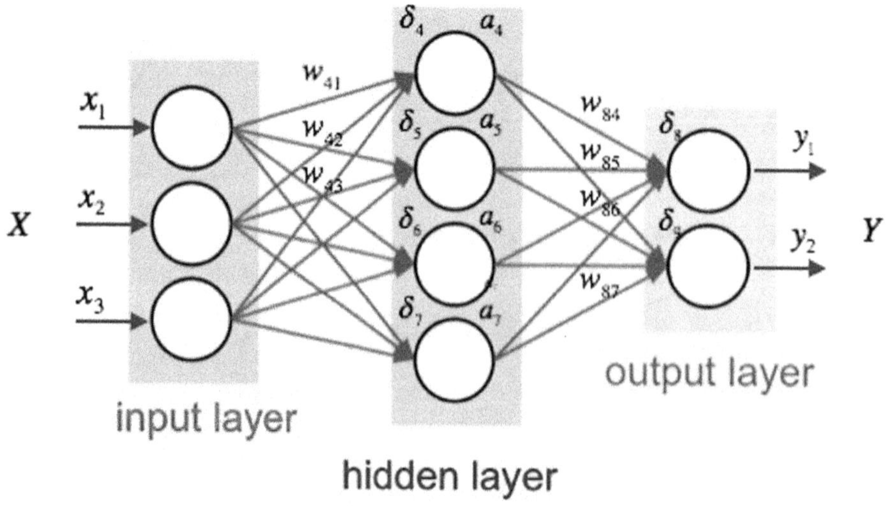

are balanced. This process requires more than just a couple of samples in fact the framework requires a huge number of inputs.

To summarize, machine learning, is an approach to AI. It is a method that permits computer frameworks to make strides with experience and data. Machine learning is the only reasonable approach to building AI frameworks that can operate in complicated, real-world situations.

Machine Learning Application in oil and gas Industry

According to the paper presented by Hoeink and Zambrano, 2017, shale discrimination and cross-referring is one of the tedious work. If done manually, cautious and obliging picking of shale interims is often exceptionally time devouring, and distinctive professionals frequently deliver different interpretations. In this paper they solved these disadvantages by utilizing a machine learning strategy. Linear logistic regression is assessed as a strategy for binary classifying of shale vs. non-shale lithology utilizing log information.

As per Bangert, 2019, research paper, condition monitoring is the method of observing a parameter of condition in apparatus (vibration, temperature etc.), in order to distinguish a noteworthy, alter which is characteristic of a developing fault. This process is known to send expansive number of wrong alarms and missing alarms. Utilizing machine learning procedures, the huge information accumulated around large equipment, or a complete plant can be examined as a single coherent whole to draw conclusions about its current state of wellbeing. This strategy is seen to be distant more effective than standard condition monitoring thus preventing wrong alarms and continuously alarming undesirable states.

Numerous features of geology include handling huge sets of information and recognizing designs. Still, geology remains a qualitative study where experts spend endless man hours pouring through reports to construct models for prescient investigation. According to the paper by Snøtun, 2018, in conventional data analytics an algorithm is connected to data which at that point yields a result, but machine learning compares expansive sets of information for analysis to make models which can then be utilized for predictive analysis. Machine Learning is a subfield of analytics and computer science which is committed to utilizing pattern recognition to construct models. In this paper they examine huge sets of curve data from the Norwegian Mainland Rack was studied and dissected using diverse Machine Learning techniques. The results of this examination classify stratigraphy for new wellbores with data and also distinguish irregularities and mistakes in classification within the database.

Directional drilling is a complex process involving the remote control of tool arrangement and force application to a really long drill string subject to variable outside strengths. Pollock et al., 2018 had created an artificial intelligence framework to memorize from the activities of expert directional drillers and the mechanics of drilling simulations. Machine learning algorithms were utilized to improve the effectiveness of directional drilling: optimized ROP, less convoluted borehole, less work force on board (POB), and consistency over operations. The system ingests historical and simulation information corresponding to the data utilized and activities taken by expert directional drillers and uses that information to create choices that result in proficient slide drilling. This approach shows great potential for creating an advisory framework for directional drillers and eventually a robotized directional drilling rig.

Design of penetrating liquids, spacers, cement slurries, and breaking liquids is frequently done by trial and error within the laboratory. Trial and error can be time devouring, very costly, and misleading. To optimize the method Shadravan et al., 2017, presented a paper on the application of machine learning-based strategies to anticipate the costs of testing, integrate available experimental information, and eliminate the requirement for staff supervision. The created intelligent prediction and design tool can extricate and relate any set of input parameters to yield factors depending on the given preparing tests, which can be both experimented and simulated.

Referring to the paper presented by Zhao et al., 2017, a method is developed to detect the precursors of drilling events based on drilling data such as surface data, wellbore geometry data, lithology (formation characteristics), and downhole measurements from various downhole tools. The method is based on various machine learning techniques to learn the changing trend of drilling parameters when the drilling events happen. This is an innovative work to apply the data mining and machine learning techniques to drilling interpretation. It provides a useful way for the remote centre to monitor the onset of abnormal drilling events and inform the drillers taking actions to optimize the drilling operation.

According to the Unrau and Torrione, 2017, paper a system is described for optimizing the false alarm rate for fluid loss or gain. The system developed utilizes machine learning algorithms to maintain an accurate estimate of what mud volumes and flow rates should be during all phases of the drilling process. Alarm thresholds are calculated and adapt in real time to the current rig activity. This results in considerable reduction in false alarms even when enforcing tight alarm bounds that enable early detection. Since the crew is left only with meaningful alarms, they are more likely to take corrective action in a timely manner.

Reservoir description and characterization is one of the main/critical engineering components which require a great understanding to guarantee the ideal reservoir development that leads to the most noteworthy recovery. Reservoir modelling uses all accessible data which incorporates at a least logs information, and liquid and rock properties. In Korjani et al., 2016 study, a deep learning neural network was created to gauge the petrophysical characteristics required building a full field earth model for a huge reservoir. This was fulfilled through a learning process whereby the model was presented with assorted and huge volumes of log information measured within the field. The study illustrates the capability of the deep learning neural network model when tried against the newly bored wells within the field. The model demonstrated to generate synthetic logs nearly indistinguishable to those recorded within the modern wells. The strategy utilized is simple, practical and can be utilized for any reservoir with a great number of wells and suite of logs. They have created the method of anticipating well logs utilizing adjacent wells data and connected effectively deep learning neural network on the unused data set to figure synthetic well logs in any area within the field.

One method valuable within the testing and advancement of drilling automation system is to utilize synthetic data. A great drilling time series simulator can enhance a dataset for testing, and empower the inference of the states of drilling in real time (Yu, Chambon, Liu, & Belaskie, 2018). However, conventional test systems don't create the "warts" of real data (noise, gaps, etc.). The proposed solution by Yu et al., 2018, is a model that learns from real data, characterizes the distinctive drilling reactions and conditions the data with a deep neural network (DNN) approach, and produce reasonable drilling time arrangement dataset. To mimic a drilling-time arrangement dataset, a DNN can demonstrate physical properties of the formation, rig, and sensors, and produces information with realistic curve patterns when it is trained on real estimations, e.g. block position, hook load, standpipe pressure, and surface torque. The neural network has different convolutional, repetitive, and fully connected layers. The model, prepared with wellsite recorded information, captures the spatiotemporal disseminations among data channels, and then uses a windowed input to anticipate the next data points, which are then fed back into the network to create the mimicked data sequence recursively. An actual sensor drilling-time arrangement dataset containing different channels are input into the DNN. The re-enacted time arrangement preserves the features of the initial training data, whereas maintaining the data dispersion of numerous channels. In conclusion, the proposed deep-neural arrange (DNN) is a low-cost, robust model that re-enacts a drilling time arrangement with complex spatio-temporal patterns. This test system models the nontrivial physical rules and properties, including formation, rig, and sensors, and creates information with reasonable curve patterns. The proposed model is the primary known simulator of drilling time arrangement utilizing DNN approach.

In spite of the large number of publications, protecting plausible geographical features when updating facies models is still one of the most challenge with ensemble-based history matching (Canchumuni, Emerick, & Pacheco, 2017). This fact propelled an intense investigation reported within the literature to create effective and strong parameterizations. The work published by Canchumuni et al., 2017 appears initial results towards the improvement of a robust parameterization based on Deep Learning (DL) for appropriate history matching of facies models with ensemble methods. The method starts with a set of earlier facies realizations, which are utilized for preparing a DL network. DL distinguishes the main features of the facies pictures, permitting the development of a reduced parameterization of the models. This parameterization is changed to follow a Gaussian distribution, which is updated to account for the dynamic observed data utilizing the strategy known as ensemble smoother with numerous data absorption (ES-MDA). After each data absorption, DL is utilized to reproduce the facies models based on the initial learning. We compare the results of the proposed strategy against the standard ES-MDA (with no parameterization) and another parameterization based on vital component analysis. The results show that the proposed parameterization has extraordinary potential. Deep learning is showing amazing results in a few zones. These procedures can be embedded in a huge number of applications and there's a large number researches working in this region. In fact, each day modern architectures are developed. In this work, they utilized deep generative models based on a autoencoder design.

According to research paper presented by Ejimuda, 2019, fit for service or erosion risk assessments of oil and gas facility systems, a key parameter required to plan and actualize an effective risk administration strategy is visual inspection. This Saudi Aromoco's paper clarifies how utilizing state of the art computer vision and deep learning methods might address such challenges. They basically utilized the python programming dialect, Tensorflow Application Programming Interface, Resnet deep learning architecture, GPU machines and cloud computing innovations to attain this. Beyond the challenges of getting adequate corrosion defects data, there final solution is a systematic method that would help field personnel's, office engineers, benefit companies and administration more precisely distinguish corrosion defect types and failure modes impartially. This leads to more cost effective and faster recommendation of preventive or remedial measures.

According to the paper published by, choosing the most beneficial wells for steam jobs is of high value given the cost and assets required for the procedure. To discover the finest candidates, they create a framework with two parts: i) An autoencoder to decrease the dimensionality of input information, ii) A kernel regression to foresee the production gain. To show the benefit of application of autoencoder-extracted features is by comparing the resulting accuracy with physically extracted features. To prepare their framework, they utilize a training data set comprising of 3200 steam jobs and assess it on a testing information comprising of 800 steam jobs. Their results showed that convolutional autoencoders essentially improve the precision of the forecast of oil gain from steam jobs.

As per the paper presented by Crnkovic-Friis and Erlandson, 2015, a geology driven deep learning Estimated Ultimate Recovery (EUR) forecast model for multi-stage hydraulically fractured horizontal wells in tight gas and oil reservoirs was created. The novel approach was made conceivable by later advancement within the field of deep learning and the use of big data. A Deep Neural Network (DNN) was prepared to memorize the relationship between geology and the normal EUR (assessed by decline analysis). The model was validated on wells from other geological locales to show its generalization capabilities. The model is appropriate within the exploration stage, because it requires only geological information (Thickness, Porosity, Water Immersion, Bulk Thickness, Vitrinite Reflectance, Add up to Natural Carbon and Brittleness). The DNN model displayed in this paper has captured the complex connections between geological parameters and the anticipated EUR. It can be straightforwardly connected to other districts that are administered by the same geological principles.

DNN MODELLING METHODOLOGY

In recent times, neural network is used extensively for developing predictive models. Here, a DNN model is developed for non-linear block box models of surge pressure gradient by tripping operations. Three rudimentary steps are used to construct the model, namely (1) data generation for DNN model training (2) parametric analysis to identify most suitable data (3) training the model for different architecture and (4) performance analysis of developed models.

Data Generation

To generate the data Schuh analytical model is used (Frank J. Schuh, 1964). It contains all the relevant dynamics of the tripping operation to be use in predictive model. F J Schuh, 1964 had developed for non-newtonian fluid which represents the rheological characteristics of drilling fluid.

Assumptions

Fluid is incompressible and its rheological characteristics does not change with temperature. Wellbore geometry is concentric in nature and does not changes with time. There is no movement of fluid between drilled formation and wellbore and vice versa. Close-ended pipe id considered for data generation. Laminar fluid flow is considered in the annulus under steady state condition.

Theoretical model formulation

In his models, power law fluid type of non-newtonian fluid is used. This model utilizes the fluid's rheological properties and wellbore geometrical configuration in combination with its effect on fluid flow velocity due to movement of drillpipe/casing in the wellbore to predict the surge pressure gradient. The calculation schemes include,

The fluid velocity between the space of drill-string and borehole is given by equation (1),

$$U_n = U_v + U_d. \quad (1)$$

Where, 'U_v' is the displaced fluid velocity in the annulus. This is calculated by using equation (2),

$$U_v = \left(\frac{\bar{d}^2}{1-\bar{d}^2} \right) .U_p . \tag{2}$$

Where, 'U_p' is the pipe tripping velocity and '$\bar{d}$. is the diameter ratio.

And, 'U_d' is the fluid velocity in the annulus due to fluid's viscous drag. This velocity depends on whether fluid flow showing laminar or turbulent flowing behaviour. If the fluid flow showing laminar behaviour, then viscous drag is calculate using equation (3),

$$U_d = \left(\frac{\bar{d}^2 - 2\bar{d}^2 ln\bar{d} - 1}{2\left(1-\bar{d}^2\right) ln\bar{d}} \right) .U_p . \tag{3}$$

If the fluid flow showing turbulent behaviour, then viscous drag is calculate using equation (4),

$$U_v = \left(\frac{1}{2} \right) .U_p . \tag{4}$$

A modified Reynolds number calculating equation is developed using Dogde and Metzner (1959) method. The equation is represented as equation (5),

$$R_m = \frac{\left(D_h - D_p\right)^n .U_n^{(2-n)} .\rho}{\frac{2}{s}.k.\left(\frac{2n+1}{2n} \right)^n .8^{(n-1)}} . \tag{5}$$

Where, 'D_h' is the wellbore diameter; 'D_p' is the inner pipe diameter; 'n' is the power law index; 'ρ' is the fluid density; and 'k' is the fluid's consistency index.

Now to calculate the surge/swab pressure gradient, equation (6) will be used

$$\frac{\Delta P}{\Delta L} = \frac{f.\rho.U_n^2}{D_h\left(1-k\right)} . \tag{6}$$

Where, '*f*. is the frictional factor. Its value changes with changing fluid flow behaviour, to calculate the friction factor following expressions are used mentioned in equation (7), (8) and (9) for laminar, turbulent and transitional fluid flow respectively.

For laminar flow

$$f_l = \frac{24}{R_m} . \tag{7}$$

For turbulent flow

$$f_t = \left(\frac{4}{n^{0.75}} \log \left(R_m \times f^{\left(\frac{1-n}{2}\right)} \right) - \frac{0.4}{n^{1.2}} \right). \tag{8}$$

For transitional flow

$$f_{tf} = \frac{16}{R_m} + \left(\frac{R_m - (3470 - 1370n)}{800} \right).(f_t - f_l). \tag{9}$$

Where, 'f_l', 'f_t' and 'f_{tf}' is the frictional factor during laminar, turbulent and transitional fluid flowing behaviour, respectively.

Now to generate the dataset for training, testing and validation of DNN model, system of equation mentioned above will be used. Table 1 provides the range of rheological and geometric parameters for generating the dataset using Schuh model. A detailed flowchart for generating the data is also provided through Figure 4 for better understanding. To categorize the parameters which are affecting the surge pressure gradient during tripping operation, a sensitivity analysis is done.

Table 1. Input Data

Parameter	Value or range	Unit
Mud Weight	8 – 12	ppg
Diameter of hole	17 – 7.875	Inches
Outer diameter of pipe	13.37 – 6.13	Inches
Tripping velocity	6 – 60	ft/min

Sensitivity Analysis of Parameters

Bottom-hole pressure differential during tripping operations depends upon various drilling parameters namely fluid rheology and wellbore geometry. A parametric study is carried out to detect the parameters affecting the surge pressure gradient. Base inputs used to carry out this analysis were: n = 0.58; K = 0.88 lbf.s/100ft^2; r_p = 0.55 inch; r_h= 1.2495 inch. As anticipated, diameter ratio has major effect on the surge pressure gradient. Figure 5 depicts with the increase in pipe tripping velocity surge pressure gradient tends to increase with increasing diameter ratio.

In addition to diameter ratio, fluid's power law index and consistency factor also has considerable effect on the surge pressure gradient. Figure 6 represents the effect of fluid consistency index on bottom-hole pressure surging which shows with the increase in fluid consistency index, surge pressure is also increases. Similar, relationship is shown by power law index on surge pressure shown in Figure 7, but effect of fluid consistency is more than power law index on pressure surging.

Figure 4. Flowchart for generating the data

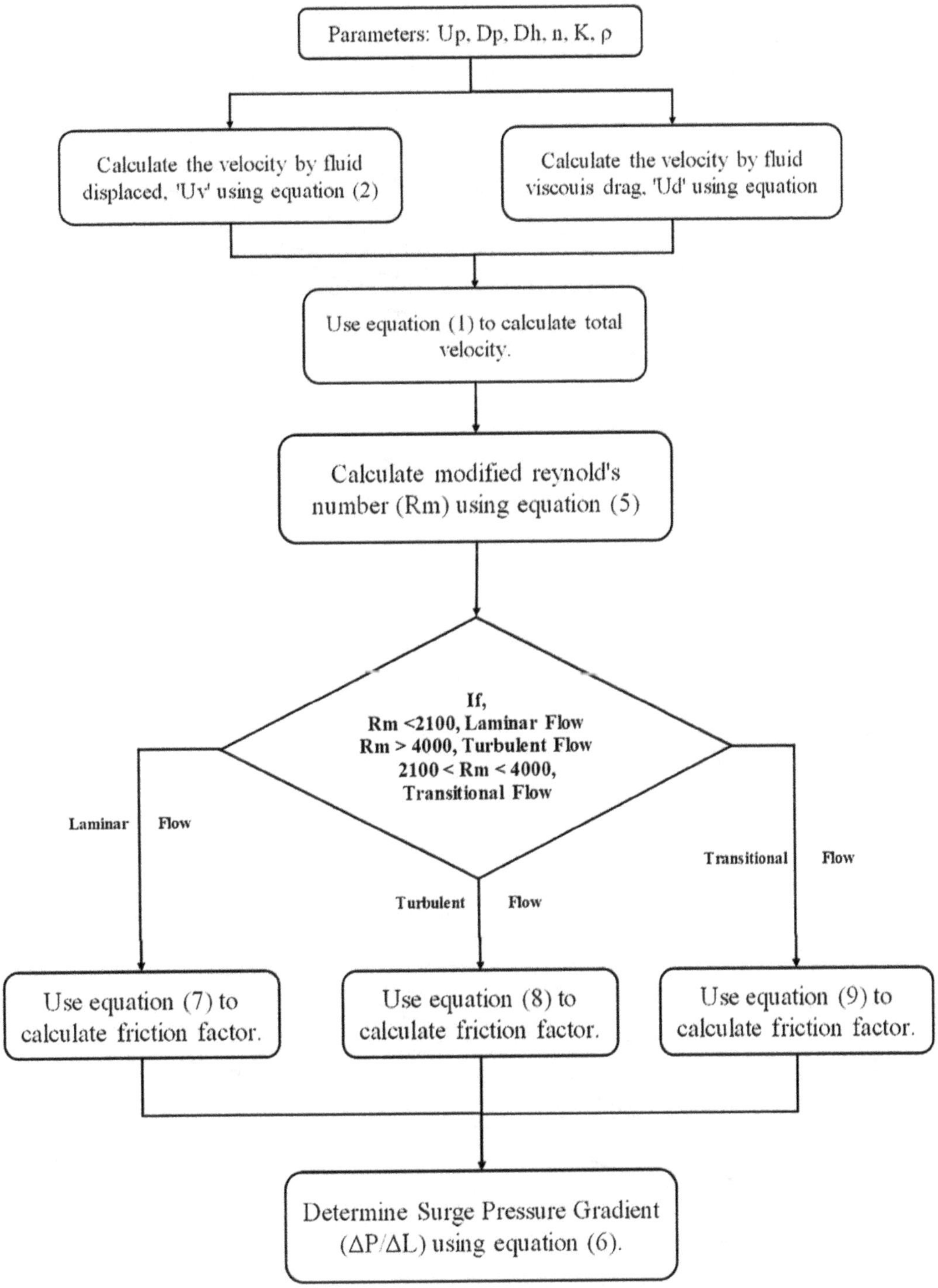

Figure 5. Effect of pipe tripping velocity on surge pressure gradient by varying diameter ratio

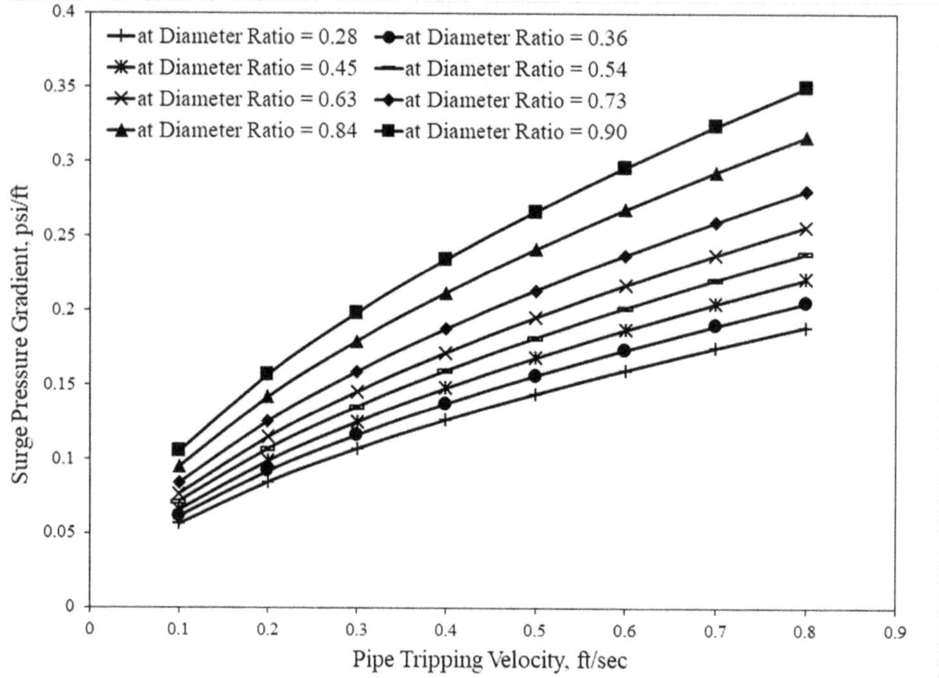

Figure 6. Effect of fluid consistency index (K) on surge pressure gradient with varying pipe tripping velocity

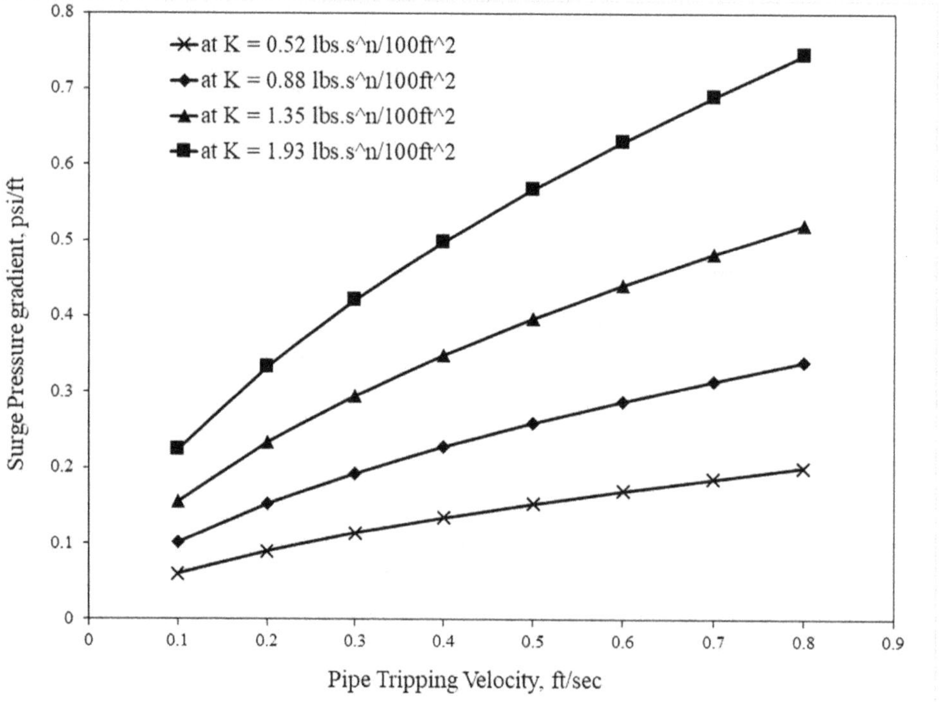

Figure 7. Effect of fluid power index (n) on surge pressure gradient under varying pipe tripping velocity

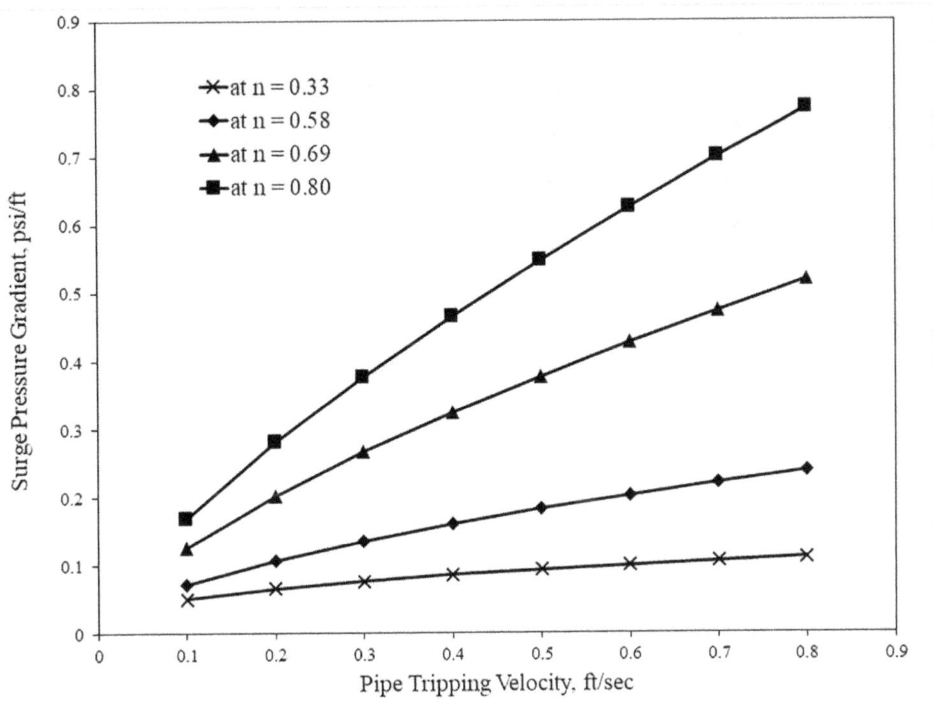

Training of DNN

Feed forward neural network with one hidden layer have the tendency to approximate any non-linear situation with certain degree of accuracy (Draeger, Engell, & Ranke, 1995). The feed forward network structure was chosen based on its applicability with different complex scenarios (Wasserman & D., 1993). For activation, sigmoidal activation function was used. Design of DNN architecture is shown in Figure 8. MATLAB is used to develop the DNN model. There is not any specified method to find the most optimal structure of DNN model thus, trial-n-error method is employed to find the best structure. This selection is based on changing the hidden layers, iteration number and neuron number. Iteration is set at 100,000; neuron in hidden layers, ranges from 2-16; and 4 and 5 number of hidden layers are used. Total of 61 networks with variations of hidden layer and number of neurons were developed and trained.

Performance of DNN Networks

To check the performance of the different networks, statistical analysis was carried out to calculate mean absolute error (MAE), root mean square error (RMSE), relative absolute error (RAE), root relative square error (RRSE) and coefficient of determination (R^2). If the value of RAE and RRSE is closer to zero, performance of the model will be excellent whereas, for R^2, value need to be closer to 1. Equation (10-14) of all performance measuring tools are mentioned below: -

Figure 8. Schematic diagram of DNN architecture

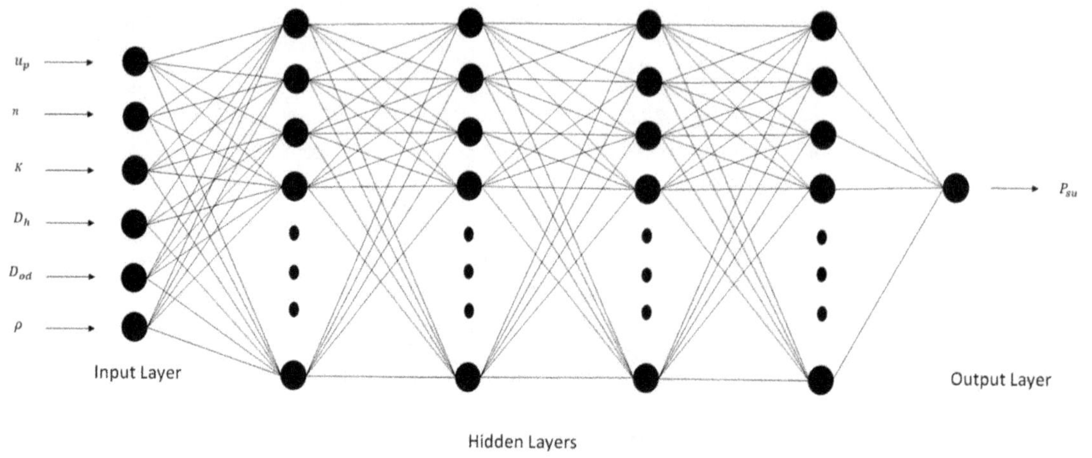

$$MAE = \frac{1}{P}\sum_{1=1}^{P}(PD - DD). \tag{10}$$

$$RMSE = \sqrt{\frac{1}{P}\sum_{1=1}^{P}(PD - DD)^2}. \tag{11}$$

$$RAE = \frac{\sum_{1=1}^{P}|PD - DD|}{\sum_{1=1}^{P}|DD - \overline{DD}|}. \tag{12}$$

$$RRSE = \sqrt{\frac{\sum_{1=1}^{P}(PD - DD)^2}{\sum_{1=1}^{P}(DD - \overline{DD})^2}}. \tag{13}$$

$$R^2 = 1 - \frac{\sum_{1=1}^{P}(DD - PD)^2}{\sum_{1=1}^{P}(DD - \overline{DD})^2}. \tag{14}$$

Where, P is the number of samples, PD represents the predicted data given by the developed model and DD represents the desired data calculated by using the above given system of equations according to flowchart given in Figure 4.

Table 2. Data of different DNN configuration and its performance analysis

Hidden Layers	Neurons					R^2	MAE	RMSE	RAE	RRSE
	1st layer	2nd layer	3rd layer	4th Layer	5th layer					
	4	2	2	6	-	0.474308	0.254499	0.309381	132	113
	14	3	8	12	-	0.970225	0.006278	0.008905	3.26	3.24
	16	8	11	15	-	0.969831	0.044502	0.060279	23.16	21.94
	3	6	16	8	-	0.572444	0.25444	0.309307	132.41	112.59
	12	10	3	6	-	0.973972	0.038925	0.043473	20.25	15.85
	8	4	13	8	-	0.268531	0.25449	0.309375	132.44	112.62
	15	14	9	10	-	0.971604	0.042305	0.058026	22.017	21.123
	16	14	5	8	-	0.96452	0.042812	0.06106	22.2812	22.2281
	12	7	12	16	-	0.969831	0.042727	0.058583	22.237	21.3263
	2	11	11	7	-	0.436657	0.254495	0.309379	132.44	112.62
	7	14	13	7	-	0.506659	0.254493	0.309377	132.44	112.62
	4	14	11	15	-	0.97338	0.043682	0.058241	22.733	21.201
	8	5	2	6	-	0.752036	0.170043	0.197965	88.49	72.06
	12	9	5	13	-	0.96924	0.042628	0.058801	22.185	21.4055
	3	16	13	6	-	0.600625	0.254482	0.309368	132.44	112.62
FOUR	12	6	11	2	-	0.139502	0.254492	0.309377	132.44	112.62
	4	10	3	9	-	0.626156	0.254519	0.309398	132.4606	112.631
	2	10	12	7	-	0.979506	0.02907	0.039783	15.12	14.48
	8	5	3	6	-	0.980298	0.0637	0.0748	8.1	10.17
	14	11	14	7	-	0.35952	0.254493	0.309378	132.44	112.62
	16	9	10	16	-	0.959616	0.009356	0.011117	4.8692	4.047
	11	7	8	15	-	0.95942	0.005894	0.008547	3.06	3.11
	15	14	15	9	-	0.971802	0.043367	0.058097	22.57	21.14
	6	5	4	10	-	0.978912	0.064069	0.069789	33.34	25.4
	11	9	2	9	-	0.559205	0.254529	0.309401	132.4654	112.63
	10	8	16	2	-	0.564752	0.254484	0.30937	132.44	112.62
	4	9	12	6	-	0.971999	0.045427	0.060328	23.64	21.96
	4	7	14	11	-	0.5616	0.254393	0.309233	132.39	112.57
	10	5	11	5	-	0.575019	0.254494	0.309378	132.44	112.62
	3	11	12	7	-	0.967076	0.050558	0.066224	26.31	24.1
	10	7	15	9	-	0.615754	0.254495	0.303977	132.4475	112.6236

continues on following page

Table 2. Continued

Hidden Layers	Neurons					R^2	MAE	RMSE	RAE	RRSE
	1st layer	2nd layer	3rd layer	4th Layer	5th layer					
FIVE	16	12	2	9	4	0.952381	0.286378	0.367367	14.07	15.17
	16	10	9	10	9	0.976342	0.161487	0.243268	7.93	10.04
	10	5	16	16	3	0.975156	0.152492	0.235148	7.496	9.71
	13	10	14	16	12	0.978517	0.149691	0.222619	7.358	9.19
	8	3	4	15	16	0.975156	0.156244	0.230353	7.68	9.51
	15	6	4	3	10	0.963342	0.124004	0.197954	6.09	8.17
	13	10	12	9	10	0.965306	0.114028	0.189298	5.6	7.82
	14	16	3	9	6	0.973182	0.152956	0.225541	7.51	9.31
	3	4	12	7	16	0.950235	0.295423	0.386836	14.52	15.98
	6	13	12	16	12	0.985255	0.0772	0.0882	7.46	8.23
	8	11	13	7	9	0.947313	0.292313	0.384167	14.36	15.87
	10	13	5	10	15	0.948871	0.290387	0.379189	14.27	15.66
	16	13	4	10	12	0.94926	0.286655	0.374509	14.09	15.47
	10	9	2	12	4	0.949845	0.284897	0.37172	14	15.35
	4	8	2	9	8	0.945367	0.293108	0.389659	14.4	16.09
	3	4	10	6	8	0.945756	0.291928	0.388019	14.35	16.02
	10	4	3	11	15	0.961772	0.233319	0.309737	11.46	12.79
	8	5	16	6	14	0.959616	0.20263	0.297063	9.96	12.27
	9	6	12	8	8	0.961772	0.231675	0.309917	11.388	12.8
	7	16	12	2	2	0.960596	0.232344	0.313426	11.42	12.94
	4	12	5	3	16	0.945562	0.268264	0.363558	13.18	15.01
	4	5	14	12	15	0.958637	0.237486	0.317431	11.67	13.11
	15	5	4	7	2	0.946729	0.263435	0.357708	12.94	14.77
	3	4	7	4	3	0.963735	0.216	0.290744	10.61	12.01
	14	8	3	13	12	0.967862	0.224598	0.316031	11.04	13.05
	9	8	7	8	3	0.969437	0.191559	0.265519	9.41	10.96
	11	16	15	12	14	0.968453	0.192528	0.277813	9.464	11.47
	10	10	2	11	16	0.967862	0.224617	0.318545	11.04	13.15
	10	16	12	4	3	0.968846	0.212002	0.303652	10.421	12.5442
	5	7	3	11	15	0.968059	0.196227	0.28754	9.64	11.87
	2	13	2	8	6	0.970028	0.178757	0.260903	8.787	10.7782

RESULTS AND DISCUSSION

Using the Schuh model, dataset of total 981 were determined for varying rheological and geometrical configuration. This dataset was further divided into training and testing dataset. Random seventy percent of original data set are used for training the model and left thirty percent are used for testing the trained model. After each selected DNN architecture are trained, its performance is analysed using the equation mentioned in the last section. Coefficient of determination (R^2) is considered as the primary factor which defines the performance of the model. If the R^2 is nearer to 1, it considered as the well-trained model. Performance analysis of all 61 architecture are mentioned in Table 2. According to table 2, DNN model with five hidden layer with 6, 13, 12, 16, 12 neurons respectively and four hidden layers with 8, 5, 3, 6 neurons respectively, shows the excellent result with R^2 value of 0.98525 and 0.98029. These optimal DNN architecture are tested using testing data. Figure 9 and Figure 10 represents the actual versus predicted data plot of testing data for both architecture, it shows R^2 of 0.983 for architecture with five hidden layers and 0.977 for architecture with four hidden layers. The new developed model showed excellent performance in prediction of surge/swab pressure. However, its shows disadvantages in combination to advantages. These merits and drawbacks are mentioned in Table 3.

Table 3. Advantages and disadvantages/limitation of new developed model

S. No	Advantages	Drawbacks
1	This model can predict pressure differential under varying geometrical characteristics of concentric annulus.	This model doesn't perform for eccentric and elliptical annulus.
2.	This model is developed for Power law fluid.	This model is not feasible for other non-Newtonian fluid models.
3.	This model can predict surge pressure in any conventional oil/gas well drilled for exploration or production of hydrocarbon.	For geothermal wells, this model does not perform.
4.	This model can integrate with existing real-time monitoring drilling anomaly console.	–
5.	This models performs exceptionally for both open-ended and close-ended pipe.	–

CONCLUSION AND FUTURE DIRECTION

Surge/swab pressure during tripping operations can result in hazardous scenario. To predict these surge pressure gradient, DNN model is developed for power law fluid. Developed model has the capability to forecast surge pressure gradient under varying fluid rheology and borehole geometry. To find the best architecture for this model, tri-n-error technique is utilized. Two DNN configuration gave the best predictive result; (a) 5 hidden layers with 6, 13, 12, 16, 12 nodes (b) 4 hidden layers with 8, 5, 3, 6 nodes. First DNN configuration showed coefficient of determination of (R2) of 0.983 with testing data and second DNN configuration showed coefficient of determination of (R2) of 0.977 with testing data. Both models are predicting the surge pressure satisfactorily without using the cumbersome time-consuming numerical analysis. This intelligent model may give vital advantages when tripping operation is carried out in

Figure 9. Comparative plot of actual surge pressure gradient and predicted surge pressure gradient by most optimal DNN architecture

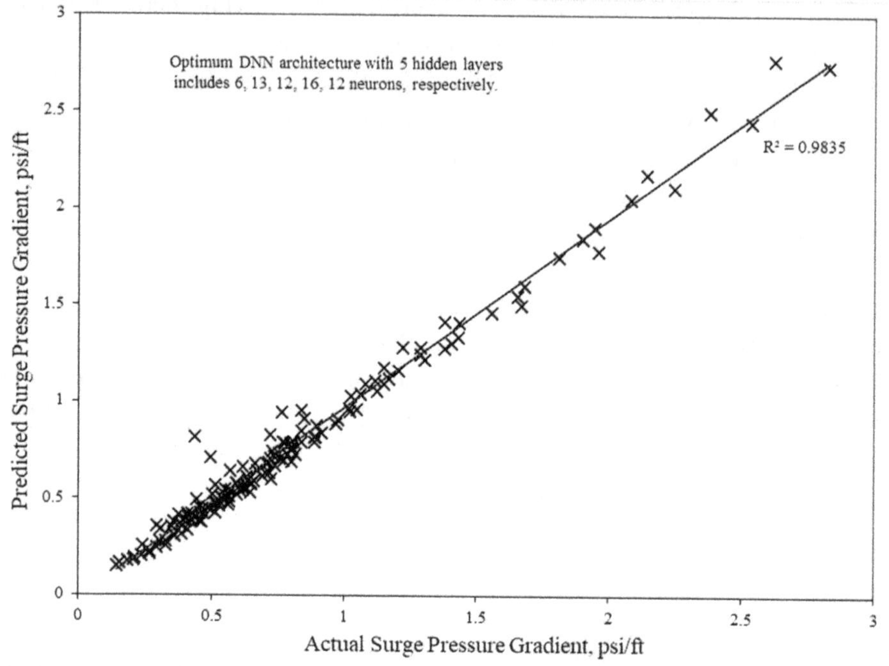

Figure 10. Comparative plot of actual surge pressure gradient and predicted surge pressure gradient by another optimal DNN architecture

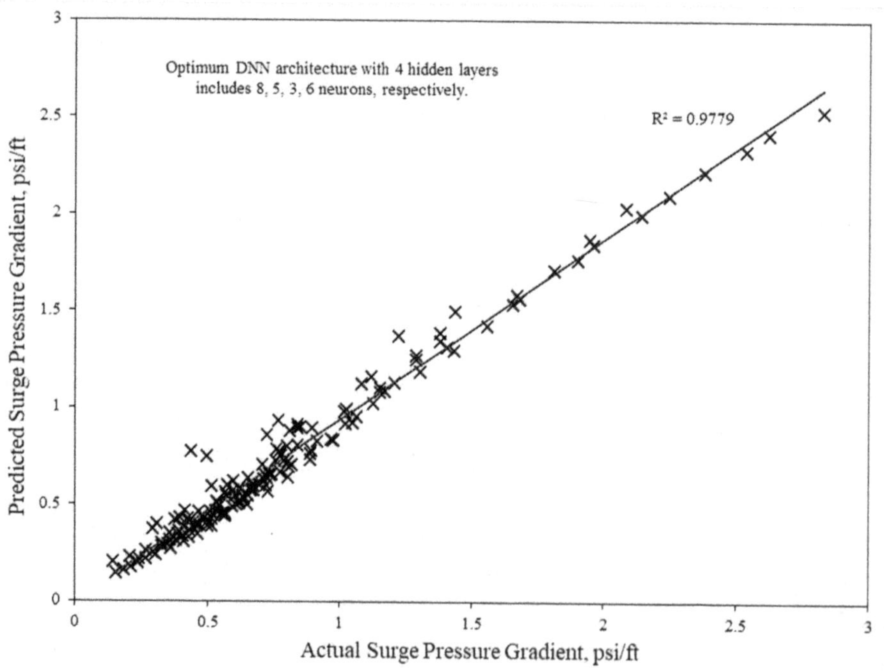

low pressure margin wells where repetitive calculation of surge/swab pressure is required. Also, this model can be extremely useful during real-time tripping operations because it will help well engineers to set the appropriate pipe tripping speed and rheological parameters so that surge/swab pressure due to centroidal action of drilling hydraulics pipe axial movement will be in between the range of drilled formation fracture and pore pressure. This model is developed to predict surge/swap pressure in concentric annulus. For future research, similar methodology can be used to develop predictive model in eccentric and elliptical annulus. Furthermore, different machine learning algorithms or genetic algorithms can be used to conduct comparative analysis with new developed model to identify its accuracy and precision.

ACKNOWLEDGMENT

This work is supported by Petroleum Engineering Department and Institute of Hydrocarbon Recovery at Universiti Teknologi PETRONAS.

REFERENCES

Abakouy, R., En-Naimi, E. M., & El Haddadi, A. (2017). Classification and Prediction Based Data Mining algorithms to Predict Email Marketing Campaigns. *Proceedings of the 2nd International Conference on Computing and Wireless Communication Systems - ICCWCS'17*, 1–5. 10.1145/3167486.3167520

Andia, P., Sant, R. V., & Whiteley, N. (2018). A Comprehensive Real-Time Data Analysis Tool for Fluid Gains and Losses. *IADC/SPE Drilling Conference and Exhibition, 2018*(March), 6–8. doi:10.2118/189576-MS

Bangert, P. (2019). Predicting and Detecting Equipment Malfunctions using Machine Learning. *SPE Middle East Oil and Gas Show and Conference, MEOS, Proceedings, 2019-March*. Retrieved from https://www.onepetro.org/conference-paper/SPE-195149-MS

Bishop, C. M. (2006). Pattern Recognition and Machine Learning. *Information Science and Statistics*. Retrieved from https://www.springer.com/gp/book/9780387310732

Canchumuni, S. A., Emerick, A. A., & Pacheco, M. A. (2017). Integration of ensemble data assimilation and deep learning for history matching facies models. *OTC Brasil, 2017*, 385–397. doi:10.4043/28015-MS

Crnkovic-Friis, L., & Erlandson, M. (2015). Geology driven EUR prediction using deep learning. *Proceedings - SPE Annual Technical Conference and Exhibition*, 1062–1071. 10.2118/174799-MS

Draeger, A., Engell, S., & Ranke, H. (1995). Model Predictive Control Using Neural Networks. *IEEE Control Systems*, *15*(5), 61–66. doi:10.1109/37.466261

Ejimuda, C. (2019). Using deep learning and computer vision techniques to improve facility corrosion risk management systems 2.0. *Society of Petroleum Engineers - SPE Nigeria Annual International Conference and Exhibition 2019, NAIC 2019*. 10.2118/198863-MS

Hoeink, T., & Zambrano, C. (2017). Shale discrimination with machine learning methods. *51st US Rock Mechanics / Geomechanics Symposium 2017, 4*, 2840–2845. Retrieved from https://www.onepetro.org/conference-paper/ARMA-2017-0769

Joshua Thomas, J., Pinar Karagoz, B., & Bazeer Ahamed, P. V. (Eds.). (2020). Deep Learning Techniques and Optimization Strategies in Big Data Analytics. doi:10.4018/978-1-7998-1192-3

Korjani, M., Popa, A., Grijalva, E., Cassidy, S., & Ershaghi, I. (2016). A new approach to reservoir characterization using deep learning neural networks. *Society of Petroleum Engineers - SPE Western Regional Meeting*. doi:10.2118/180359-ms

Pankaj, P., Geetan, S., MacDonald, R., Shukla, P., Sharma, A., & Menasria, S., … Judd, T. (2018). Application of data science and machine learning for well completion optimization. *Proceedings of the Annual Offshore Technology Conference*, *6*, 4186–4201. 10.4043/28632-MS

Pollock, J., Stoecker-Sylvia, Z., Veedu, V., Panchal, N., & Elshahawi, H. (2018). Machine learning for improved directional drilling. *Proceedings of the Annual Offshore Technology Conference*, *4*, 2496–2504. 10.4043/28633-ms

Raschka, S., & Mirjalili, V. (2017). *Python machine learning*. Packt Publishing Ltd.

Reddy, B. R., Zhang, J., & Ellis, M. (2016). Cement Strength retrogression issues in offshore deep water applications - Do we know enough for safe cementing? *Proceedings of the Annual Offshore Technology Conference*, *2*, 1612–1632. 10.4043/27012-MS

Schuh, F. J. (1964). Computer makes surge-pressure calculations useful. *Oil & Gas Journal*, *31*, 96.

Shadravan, A., Tarrahi, M., & Amani, M. (2017). Intelligent tool to design drilling, spacer, cement slurry, and fracturing fluids by use of machine-learning algorithms. *SPE Drilling & Completion*, *32*(2), 131–140. doi:10.2118/175238-PA

Snøtun, H. (2018). Using machine learning to create data. *Offshore Technology Conference Asia 2018. OTCA, 2018*. Advance online publication. doi:10.4043/28587-ms

Unrau, S., & Torrione, P. (2017). Adaptive Real-Time Machine Learning-Based Alarm System for Influx and Loss Detection. *SPE Annual Technical Conference and Exhibition*, 11. 10.2118/187155-MS

Vasant, P., Kose, U., & Watada, J. (2017). Metaheuristic techniques in enhancing the efficiency and performance of thermo-electric cooling devices. *Energies*, *10*(11), 1703. doi:10.3390/en10111703

Vasant, P., Marmolejo, J. A., Litvinchev, I., & Aguilar, R. R. (2019). Nature-inspired meta-heuristics approaches for charging plug-in hybrid electric vehicle. *Wireless Networks*. Advance online publication. doi:10.100711276-019-01993-w

Vasant, P., & Zelinka, I. G.-W. W. (2019). *Intelligent Computing & Optimization: Advances in Intelligent Systems and Computing*. Academic Press.

Wasserman, P. D., & D., P. (1993). Advanced methods in neural computing. In *Advanced methods in neural computing* (pp. 1–255). Retrieved from https://dl.acm.org/citation.cfm?id=562821

Yu, Y., Chambon, S., Liu, Q., & Belaskie, J. P. (2018). Recorded well data enriches the testing of automation systems by using a deep neural network approach. *Society of Petroleum Engineers - IADC/SPE Drilling Conference and Exhibition*. doi:10.2118/189591-ms

Zhao, J., Shen, Y., Chen, W., Zhang, Z., & Johnston, S. (2017). Machine Learning–Based Trigger Detection of Drilling Events Based on Drilling Data. *SPE Eastern Regional Meeting*. 10.2118/187512-MS

Chapter 13
A Cloud Computing-Based Model for Wildlife Conservation and Health Care Improvement in Endangered Wild Life Animals

Shahbaz Afzal
https://orcid.org/0000-0002-1217-9357
B.S. Abdur Rahman Crescent Institute of Science and Technology, India

G. Kavitha
B.S. Abdur Rahman Crescent Institute of Science and Technology, India

Amir Ahmad Dar
B.S. Abdur Rahman Crescent Institute of Science and Technology, India

ABSTRACT

The era of the Information Technology revolution has endeavored mankind with enormous technological comforts. Apart from human health care, cloud computing integrated with other technologies can be exploited for the welfare of endangered wildlife animals that are currently on the verge of facing extinction in the near future. International Union for Conservation of Nature (IUCN) has declared a huge number of wildlife species into red list falling into 'critically endangered', 'endangered' and 'vulnerable' species because of pathogenic diseases, poaching, habitat degradation, and climate change. The aim of the chapter is to propose a cloud computing based wildlife health care model to preserve, sustain and protect the global wildlife heritage and wildlife conservation by identifying and providing treatment to diseased animals, monitoring their health conditions periodically and protecting against poaching. The proposed health care model makes use of various wearable sensors implanted on and within the body of an animal. The sensors perform data gathering about different body-related parameters to be transmitted to a cloud system for analysis and reporting a required timely action. When a cloud system senses emergency with respect to the health-related or threat-related incident, the information is immediately signaled to wildlife physician and wildlife ranger respectively to act accordingly.

DOI: 10.4018/978-1-7998-3645-2.ch013

INTRODUCTION

The 18th century emerged as the age of the industrial revolution and renaissance. The 19th century was dominated by classical physics, mathematics and astronomy. The first half of the 20th century was the age of displaying military might and power, armed race which showed the world two great wars ever fought. The second half of the 20th century especially the last two decades became the era of computers and electronics on a massive scale. The 21 century appeared to be the age of information and telecommunication technology especially the second decade of information technology enabled the integration of different technologies to merge to form smart computing. The integration of different computing, communication, sensing and internet-based technologies paved the way for developing smart and intelligent systems much enough automatic and robotic to relieve human intervention to a great extent. Some of the technologies embedded in smart computing are networked devices, Internet of things (IoT), big data analytics and processing, web services, cloud computing and high generation cellular networks. Smart technologies are expected to make a lot of benefits on socio-economic and environmental aspects in traditional and non-traditional fields [1] [2][3].

Wildlife forms an essential part of the biosphere and biodiversity that regulates various ecosystems and environmental factors like food chain, food web, ecological balance, pollution breakdown, climate stability, water bodies' protection and soil formation. Apart from this, the wildlife plays an important role in social, economical, religious, cultural and educational activities like medicinal values, natural aesthetics, fun and entertainment, culture and heritage, recreation and tourism development, food security, wildlife research and development, economic development, job creation, sustainability, agriculture and farming. Figure 1 shows Panthera Tigris locally known as a tiger from the cat family in a wildlife sanctuary.

Figure 1. Photo by Pawan Kotiyal [7]

Figure 2 shows the extraction of python venom which is used for the treatment of various diseases like cancer, Alzheimer's syndrome and Parkinson's syndrome.

Figure 3 depicts how wildlife makes our planet colorful and adds aesthetics to nature.

Figure 2. Snake Venom used for medicinal properties [7]

Figure 3. Wildlife adds aesthetics to nature picture by wildlife Orrisa [8]

Unfortunately, most of the wildlife animal species are getting endangered or face extinction at the hands of humans because of the destruction of natural habitats for land development, smuggling, poaching and hunting for leather, skin, fur, jewelry, ornaments and meat that are the contributing factors towards wildlife animal extinction. Hence, it becomes of utmost importance to preserve and conserve wildlife animals for the welfare of ecology and biodiversity. According to the International Union for Conservation of Nature (IUCN) red list release 2012, out of total 63,837 species assessed the report revealed that 19,817 are threatened with extinction, 3,947 were specified as "critically endangered", 5,766 were described as "endangered" and more than 10,000 species were listed as "vulnerable" [9][10][11][12]. According to the world wildlife fund report, the global wildlife population declined by 52% between 1970 and 2014 [6].

A lot of national and international wildlife organizations have set up initiatives, wildlife sanctuaries, national parks, zoos, projects to protect these endangered animals from becoming extinct. Some of the eminent global wildlife organizations are World Wide Fund for Nature (WWF), Global Wildlife Conservation, Wildlife Conservation Society, The National Wildlife Federation, and International Fund for Animal Welfare, World Society for the Protection of Animal, African Wildlife Foundation, People for the Ethical Treatment of Animals (Peta), and Defenders of Wildlife. With the advancement

Figure 4. Some of the prominent critically endangered wildlife animals

in communication and information technology and other allied technologies, the wildlife conservation organizations are trying to effectively harness the power of technology for the welfare of endangered animals both in terms of preventing poaching and improving the healthcare of the wildlife animals. For example, wireless sensor networks have been extensively used for monitoring the physical characteristics and behavior of wildlife animals. The safety of the wildlife animals has been further enhanced by the advanced technology of IoT and wireless sensor networks for tracking animals as well [13]. Technology plays a crucial role in wildlife protection and conservation. Different kinds of evolved technologies like GSM, GPS, GPRS and RFID technologies using cellular networks have been used for domestic and wildlife animal protection like animal tracking. Because of the shortcomings of these technologies they could not prove very much effective in wildlife conservation due to geographical and other technological constraints. With the huge progress in information and communication technologies (ICT) and advances in internet and web technologies, Internet of things, big data, cloud computing and allied branches, and high-performance computing (HPC), the notion of smart computing came into existence which gave rise to high-end computing and real-time data transmission. Implementing smart computing from the perspective of wildlife conservation could be a breakthrough achievement to tackle poachers, hunters, habitat destruction, and wildlife epidemics and diseases.

Cloud computing along with its child technologies (fog and mist computing) finds a promising future in smart computing specifically to smart wildlife conservation when integrated with different associated technologies [4]. The RFID and GSM based systems dominated for some time but due to many short-comings and limitations placed a need to be replaced with much-sophisticated technology [5].

This chapter presents a cloud computing model with the integration of sensor-based IoT to improve the healthcare and safety of endangered wildlife animals and the timely effort of rescuing wildlife animals from an emergency situation to help to preserve wildlife conservation. The rest of the paper is structured as follows: Section 2 discusses the motivation behind the idea, section 3 presents the related literature in accordance with the wildlife conservation using different technologies, section 4 gives a detailed view on the proposed system model and also explains its working principle along with different system components used in the model. Section 5 highlights some of the limitations of the model while section 6 concludes the chapter and points out the future directions.

Motivation

In view of the importance of wildlife, wildlife conservation and its role in sustaining the biodiversity of the earth by checking the illegal trade, trafficking, smuggling and poaching of wildlife animals is the primary motivation that motivated the authors to come up with the idea of wildlife protection and healthcare improvement by harnessing the power of cloud computing and sensor-based Internet of Things (IoT). This will be the work of its first kind in wildlife conservation where much advanced smart computing model has been proposed. Despite wildlife acts, wildlife laws, summits, conferences, seminars and other legislative procedures being implemented and conducted, the wildlife is disappearing at an alarming rate.

Some of the technologies like GSM, GPRS, GPS, and RFID that were used in the existing systems had some shortcomings as follows:

I The use of these technologies mostly relies on cellular networks that may not be available everywhere like in forests, mountains, and other uneven terrains.
II The existing systems were mostly used for animal tracking and monitoring purposes.
III Although the systems were automatic they were not enough smart or intelligent to observe the behavior of an animal or to identify an external agent that can pose risk or danger to the animal.
IV These systems were not used to monitor the proper health conditions or to detect the presence of disease in an animal.
V The systems were mostly suitable for domestic and pet animals.

In response to the above limitations, the authors were motivated to come out with a smart, pervasive and intelligent model to deal with wildlife conservation and wildlife healthcare of endangered wildlife animals.

Literature Survey

In terms of animal conservation, the Internet of Things (IoT) provides numerous opportunities. The number of endangered animal species increased globally from about 10,000 in 2000 to over 25,000 in 2018, according to the International Union for the Conservation of Nature (IUCN) [9]. With increasingly organized, combative and heavily equipped poaching networks, it is critical to shift conservation efforts from reactive measures towards predictive approaches. This transformation is visible in many wildlife conservation projects which initiated technological solutions to encounter the alarming rate of poaching incidents. As a matter of concern, African rhinos are on the top of the list with the statistics showing that poaching occurs at a rapid rate of three per day from the last six years and are most likely to get extinct by 2025. This is because rhino horns are sold and traded in the international black markets to use as a status symbol to display success and wealth.

Welgevonden one of the largest reserve for rhinos in South Africa which covers an area of around 90,000 acres is inhabited by a large population of rhinos. To curb rhino poaching, a joint mission carried by Welgevonden, Wageningen University and IBM developed an innovative IoT solution that focuses on monitoring the activities and behavior of herd animals such as zebras and impalas instead of tracking rhinos. Researchers at Wageningen University conducted a study to model movement pattern recognition of animals in response to the type of threat they face. Like, animals run for safety as a group when advanced by a natural predator such as lion or leopard; however they scatter randomly when faced with a human poacher.

Taking advantage of behavioral changes, modeling, analyzing and detecting those changes by equipping animals with sensor fitted collars to gather information about the geospatial location and average velocity. These smart collars are fabricated on Long-range (LoRa) technology based low-power wide-area network (LPWAN) that operates on spread spectrum modulation techniques derived from the chirp spread spectrum (CSS) technology with coverage of 10-15 Kms within the radius. The sensors fitted within the collar transmit data to IBM's IoT system for analysis to detect distressed animals and alert the reserve team.

Welgevonden on the other side makes use of collars in herd animals rather than fitting them directly in rhinos and this unique strategical approach analyses the behavior pattern of herd animals (like zebras, wildebeests, impalas, and elands) that are usually not the prime target for poachers. So, practically these herds together with IoT enabled collars form up an early warning system that signal park rangers identifying poachers before an encounter takes place with a rhino. This approach acts proactively which makes it more effective than conventional reactive based IoT approaches that monitor endangered animals directly [14].

Songtao et al. proposed the application of IoT to study the behavior of animals and to aid researchers in the field of animal ecology [15]. Xareni p. Pacheco presented the technological aspects in wildlife conservation and reviewed technological benefits to ecology and animal conservation. As per the study, some of the available technologies that find a promising solution are bio-telemetry, GPS, camera traps, accelerometers and sensor-based technologies [16]. However, the author does not include the list of cutting edge potential technologies that can prove much advanced in the future of wildlife conservation like Zigbee, WSN, IoT, GSM, GPRS, LTE (A), internet technologies and cloud computing. Table 1 below discusses some of the existing methods for wildlife conservation mostly for animal tracking and animal monitoring.

A lot of approaches have been devised using the internet of things, artificial intelligence, big data analytics and machine learning to tackle the problem of wildlife animal poaching so as to protect them from getting extinct. However, the growth of cloud computing can enable a seamless integration of these technologies to come out with many fruitful tactics. As of now, there is no such cloud-based model to monitor the wildlife animals for their welfare. Further, the existing approaches do not take into account the smart disease detection and healthcare monitoring of these species. The ill and diseased animals are checked manually by vets without any automatic process.

The aim of the chapter is to present a wildlife care model using cloud computing and the internet of things to improve healthcare in endangered wildlife animals from different pathogenic agents, environmental conditions and anthropogenic agents that can pose a threat for their extinction. Also, the model can help in detecting various issues during the breeding period of these species.

Table 1. Literature Survey of Existing Works

Reference and Author (s)	Year	Underlying Technology	WildLife Application							
			Wildlife Monitoring	Animal and Location Tracking	Animal Behavior Identification and Recognition	Habitat Observation	Anti-poaching and Poacher Behavioral Recognition	Disease Detection	Threat Detection	Early Warning System
[17] Xiahon et al.	2015	IoT, GSM, 3G, LTE, Internet, GPS, RFID and WSN	YES	YES	YES	YES	NO	NO	NO	NO
[18] Shanmugasundaram et al.	2017	IoT, GPS, WSN and GSM	YES	YES	NO	NO	NO	NO	NO	NO
[19] Ayele et al.	2018	IoT, BLE and LORA	YES	NO	NO	NO	NO	NO	NO	NO
[20] Fang et al.	2017	Game Theory Approach and Artificial Intelligence	NO	NO	NO	NO	YES	NO	YES	NO
[21] Kays et al.	2011	Radio Telemetry System (RTS)	NO	YES	YES	YES	NO	NO	NO	NO
[22] Ayele et al.	2018	IoT, LORA and BLE	YES	NO	NO	NO	NO	NO	NO	NO
[23] Elias et al.	2017	IoT, Cloud and Edge Computing, Neural Networks	YES	YES	NO	NO	NO	NO	NO	NO

System Model

The proposed system model for cloud computing based wildlife health care system and early warning system in endangered wildlife animals is shown in figure 5. The proposed system is composed of different components operating under the joint integration of two primary technologies: internet of things (IoT) and cloud computing and big data analytics. The system components comprise the wildlife object as endangered wildlife animal (rhino, tiger, elephant etc), wearable sensors fitted on or within the body of wildlife object, a base station transmitting antenna, cloud computing system with big data analytics and processing system, automatic alarm and defense system, wildlife health care system, wildlife ranger and wildlife veterinary physician. The wearable sensors consist of biosensors, image capturing devices usually a camera recorder, voice capturing device usually a microphone, temperature sensor, blood pressure sensor, a biochip, accelerometer, pulse sensor, ECG sensor. The biosensors detect the presence and concentration of pathogens and toxicants in the object's body, camera recorder continuously records the object's surroundings for poachers, the microphone picks the audio signals for detecting gunshots, temperature sensor continuously or periodically monitors the object's body temperature, blood pressure sensor records the objects blood pressure, biochip sensor transmits the object's live location and other physiological information, accelerometer provides the objects motion information and the ECG sensors transmit the object's heart beat details. Fear detection levels through adrenaline and cortisone hormones can also be analyzed for identification of the threat.

Figure 5. Cloud Computing based Wildlife Conservation and healthcare System

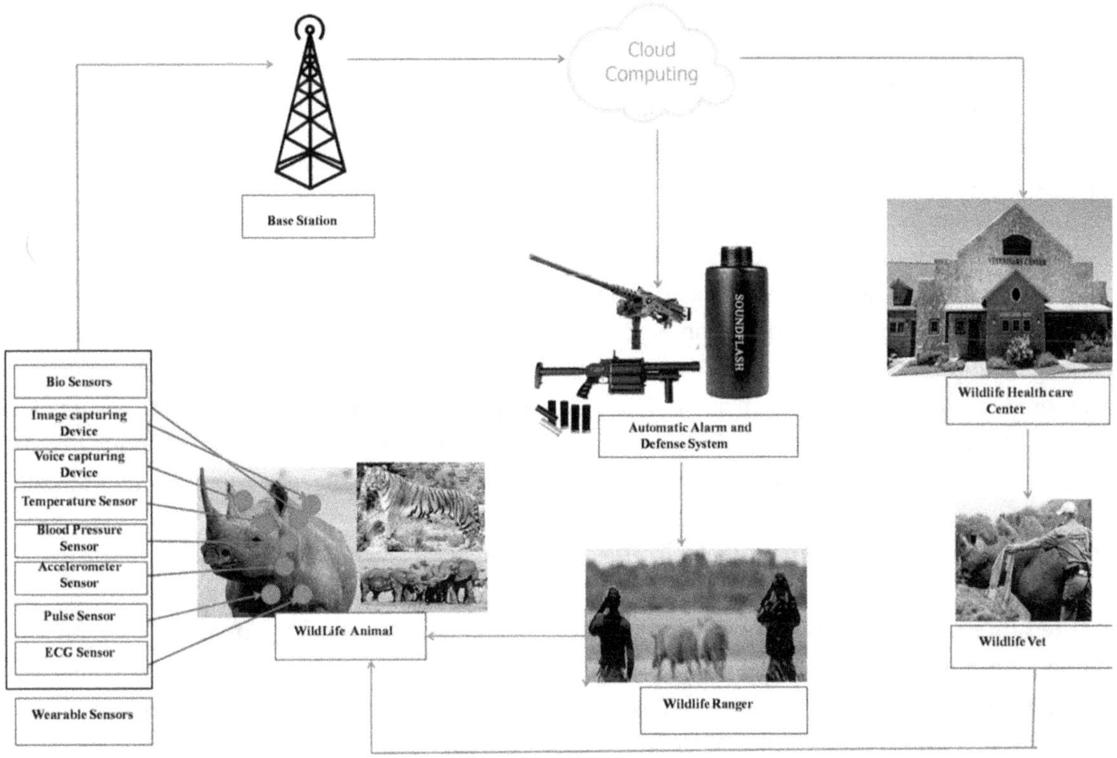

Working Principle

The wearable sensors implanted on wildlife object form a wireless sensor network (WSN) which gather object's data and transmits it to cloud system through a base station or radio beacons. The WSN continuously generating the voluminous amount of data is stored in the cloud system. The cloud system equipped with big data analytics and processing feature analysis processes and extracts certain information from this data. The information is bifurcated into two groups: health-related information and threat-related information. Health-related information is transmitted to wildlife health care center as and when required. The health-related information at the cloud system is compared with a predefined set of information related to animal health and disease. As an example, for specific animal disease, the health care center specifies the characteristics of the disease and pathogenic agent of that disease along with the physiological condition of the animal when affected by that disease. Based on this information from the health care center, a set of rules is specified at the cloud system running specific software as a service (SaaS) application. Based on the input data from an object when an animal's body conditions (like body temperature, heartbeat rate, blood pressure, intoxicant level, infection from pathogenic agent etc.) match with some predefined data it can guess the type of disease the animal is suffering from. The cloud system can send this information to the healthcare center or directly pass to a wildlife Vet alerting the vet the animal suffering about a particular disease. In a more smart way, the cloud system can also prescribe the medication directly for treatment if it has been defined in the cloud system. In case of an injury by a fellow animal or some other physical action or by an attack due to some human agent can also be monitored and both the vet and ranger can come to rescue. The pain levels of an animal can also be measured and can be treated through this system. The healthcare about pregnant wildlife animals can also be monitored on a regular basis and updates about their health conditions can be observed constantly for normal delivery which can save both animal and its baby. The wildlife health care center can also interact with the cloud system manually to monitor the condition of a wildlife animal or to check the progress of a diseased animal. In health-related information, the object's physiological condition is monitored for normal well being of the object. For an emergency situation, if an animal requires immediate attention or special care, the health care center passes the information to a veterinary physician to act upon the situation. The veterinary physician equipped with a smartphone running a special application gets updates regarding medicines or drugs to be required. In threat-related information, the objects the psychological condition is monitored to detect the object's fear level or to identify human intervention trying to cause harm to object. This critical information is sent to an automatic alarm and defense system which activates it. The automatic alarm and defense system alerts a wildlife ranger for action on his smartphone and before that the automatic alarm and defense system proactively fires some rounds in the air or tear gas canisters or sound grenades to create a fear psycho in poachers. This has also one more advantage that the animal can run for self-defense and the tear gas smoke can reduce the visibility to a poacher.

Purpose of Different Components in the System

Wildlife Object It represents the endangered wildlife animal under consideration and is an important component in the healthcare model for its welfare against diseases, physical injury and human intervention. Some of the world's wildlife endangered animals on the hit list are Javan rhinoceros, gorillas, tigers, snow leopards, Asian elephants, polar bears, jaguars, lions, deer and much more. On the basis of the pace at which their number is declining, the IUCN classified them into three groups namely critically endangered, endangered and vulnerable animals. The aim of the chapter is to propose a wildlife healthcare model to protect the wildlife animals from becoming extinct.

Wearable Sensors The wearable sensors are implanted within or on the body of wildlife animal for measuring different physiological parameters. Wearable sensors are usually fitted in a collar and placed on the neck of an animal while implanted sensors are injected within the body like biochip sensors. The sensors act as basic data collection agents which send data to be processed by a cloud system.

Base Station The base station which is usually a simplex data transmitting antenna transmits the sensor data to a cloud system located somewhere on the Internet. The sensor data cannot be sent directly to cloud system in the form it is generated. The data is coded into a proper format using any of the available data coding techniques. Multiplexing techniques can also be employed to transmit data over a single channel. In the case of a vast geographical area like in Amazon rain forests or Himalayan forests, radio beacons can be installed at different locations which can transmit data to the cloud system.

Cloud Computing cloud computing forms the backbone of the proposed model and is hosted on the Internet. Software as a service (SaaS) model based utility is hosted on the cloud which performs on user-defined rule-based system. A set of rules are framed within the software which compares sensor-based input data with some preset threshold values. If the cloud system detects some unusual behavior compared with preset values it triggers an event that needs to be addressed. Since the sensors generate data continuously with time, the cloud system is equipped with a big data analytical system. The advantage of the cloud system is that it has global access anytime, anywhere and by anyone. The cloud system contains all the information regarding animal diseases and the type of medication or treatment required if an animal is infected by a specific disease. Similarly, it has also the information when an animal gets a physical injury or when it is likely to be attacked by some external agent like poachers. Based on the type of information retrieved which is classified either as health-related information or threat-related information and sent to wildlife healthcare center or automatic alarm and defense system respectively. The cloud system can directly communicate with wildlife veterinary physician and wildlife ranger enabled with a smartphone app also in case of critical emergency.

Wildlife Health Care Center

The wildlife healthcare center periodically updates the information regarding the wildlife diseases and factors causing the disease. It also updates information regarding new drugs and medicines to cloud system. Upon receiving information from the cloud system, the health care center can signal the vet for some attention regarding the animal. The animal can be taken for treatment in the healthcare center.

Limitation of the Model

The model proposed in this chapter aims to provide better healthcare and an anti-poaching defense system for endangered wildlife animals of terrestrial origin specifically mammals. The proposed model is not considered for aquatic endangered animals, reptiles and Aves. The other limitation of the proposed system is that it does not focus on poacher behavior.

CONCLUSION AND FUTURE WORK

The chapter discusses cloud computing based model for Wildlife conservation and health care improvement in endangered wildlife animals as a means of smart computing for protection of wildlife animal resources from being getting extinct. The model harnesses the power of the latest technologies like cloud computing, big data analytics, internet of things, wireless sensor networks, GPS, LTEA, satellite imagery and RFID for smart computing wildlife conservation. The most important advantage of this model which makes it distinct from the existing models is that it not only considers animal location tracking, animal behavior, and animal monitoring but also focuses on the anti-poaching mechanism, wildlife defense system, early warning system, threat detection, disease detection and treatment.

The authors suggested a cloud computing based model for wildlife conservation and healthcare improvement as a theoretical aspect and have not been implemented in the real-time scenario. As a part of future work, the authors would like to implement the proposed model in a real-time situation. The authors also believe that implementing artificial intelligence and prediction algorithms with machine learning and deep learning algorithms [24] can emerge as a potential weapon to tackle poaching activities, poacher behavior and wildlife healthcare improvement which also forms the future work in wildlife conservation and biodiversity preservation.

ACKNOWLEDGMENT

The researchers are very thankful to the editor "J. Joshua Thomas" for providing an opportunity to consider this chapter in the handbook titled "Handbook of Research on Smart Technology Models for Business and Industry". The researchers would also like to thank anonymous reviewers for their valuable suggestions and comments. This research received no specific grant from any funding agency in the public, commercial, or not-for-profit sectors.

REFERENCES

Abdelwahab, S., Hamdaoui, B., Guizani, M., & Rayes, A. (2014). Enabling smart cloud services through remote sensing: An internet of everything enabler. *IEEE Internet of Things Journal*, *1*(3), 276–288. doi:10.1109/JIOT.2014.2325071

Ayele, E. D., Das, K., Meratnia, N., & Havinga, P. J. (2018, February). Leveraging BLE and LoRa in IoT network for wildlife monitoring system (WMS). In *2018 IEEE 4th World Forum on Internet of Things (WF-IoT)* (pp. 342-348). IEEE.

Ayele, E. D., Meratnia, N., & Havinga, P. J. (2018, February). Towards a new opportunistic iot network architecture for wildlife monitoring system. In *2018 9th IFIP International Conference on New Technologies, Mobility and Security (NTMS)* (pp. 1-5). IEEE. 10.1109/NTMS.2018.8328721

Bartels, A. H., Daley, E., Parker, A., Evelson, B., & Muteba, C. (2009). *Smart computing drives the new era of IT growth*. Forrester Inc.

Dos Santos, G. A. M., Barnes, Z., Lo, E., Ritoper, B., Nishizaki, L., Tejeda, X., . . . Kastner, R. (2014, October). Small unmanned aerial vehicle system for wildlife radio collar tracking. In *2014 IEEE 11th International Conference on Mobile Ad Hoc and Sensor Systems* (pp. 761-766). IEEE. 10.1109/MASS.2014.48

Elias, A. R., Golubovic, N., Krintz, C., & Wolski, R. (2017, April). Where's the bear?-automating wildlife image processing using iot and edge cloud systems. In *2017 IEEE/ACM Second International Conference on Internet-of-Things Design and Implementation (IoTDI)* (pp. 247-258). IEEE.

Fang, F., Nguyen, T. H., Sinha, A., Gholami, S., Plumptre, A., Joppa, L., ... Critchlow, R. (2017). Predicting poaching for wildlife protection. *IBM Journal of Research and Development*, *61*(6), 3–1. doi:10.1147/JRD.2017.2713584

Guo, S., Qiang, M., Luan, X., Xu, P., He, G., Yin, X., ... Fang, D. (2015). The application of the Internet of Things to animal ecology. *Integrative Zoology*, *10*(6), 572–578. doi:10.1111/1749-4877.12162 PMID:26338071

James Ayre. (2012). *The Red List Of Threatened Species, Annual Report Released*. planetsave.com

Jawalkar, M. S., Desale, N., Suratwala, F., Lamkhade, A., Mahalle, P. N., & Railkar, P. N. (2017). Intelligent wildlife tracking using ubiquitous technological suite. *International Journal of Synthetic Emotions*, *8*(1), 44–59. doi:10.4018/IJSE.2017010104

Jukan, A., Masip-Bruin, X., & Amla, N. (2017). Smart computing and sensing technologies for animal welfare: A systematic review. *ACM Computing Surveys*, *50*(1), 1–27. doi:10.1145/3041960

Kays, R., Tilak, S., Crofoot, M., Fountain, T., Obando, D., Ortega, A., ... Hirsch, B. (2011). Tracking animal location and activity with an automated radio telemetry system in a tropical rainforest. *The Computer Journal*, *54*(12), 1931–1948. doi:10.1093/comjnl/bxr072

Liu, X., Yang, T., & Yan, B. (2015, November). Internet of Things for wildlife monitoring. In *2015 IEEE/CIC International Conference on Communications in China-Workshops (CIC/ICCC)* (pp. 62-66). IEEE.

Mallick, P. K. (Ed.). (2015). *Research Advances in the Integration of Big Data and Smart Computing*. IGI Global.

Matthew Knight. (2012). *Extinction threat 'a call to world leaders' at Rio Earth Summit*. edition.cnn.com

Naik, G. (2014, Sept. 30). Wildlife Numbers Drop by Half Since 1970, Report Says. *The Wall Street Journal*.

Pacheco, X. P. (2018). How Technology Can Transform Wildlife Conservation. In Green Technologies to Improve the Environment on Earth. IntechOpen.

Pawan Kotiyal. (2019). *Reasons Why We Need to Save Wildlife*. Retrieved from https://www.tourmyindia.com/blog/reasons-to-save-wildlife/

Shanmugasundaram, R., Pavithra, S., Sangeetha, V., Tamilselvan, S., & Ahmed, A. T. (2017). Iot based animal tracking and monitoring system in zoo. *South Asian Journal of Engineering and Technology*, *3*(2), 162–168.

APPENDIX: ABBREVIATIONS

IUCN: International Union for Conservation of Nature
IoT: Internet of Things
WWF: World Wide Fund for Nature
PeTa: People for the ethical Treatment of animals
GSM: Global System for Mobile Communication
GPS: Global Positioning System
GPRS: General Packet Radio Service
RFID: Radio Frequency Identification
ICT: Information and Communication Technology
HPC: High Performance Computing
LoRa: Long Range
LPWAN: low-power wide-area network
CSS: Chirp spread spectrum
LTE: Long Term Evolution
LTEA: Long Term Evolution Advanced
BLE: Bluetooth Low Energy
RTS: Radio Telemetry System

Chapter 14

Modelling and Forecasting Portfolio Inflows:

A Comparative Study of Support Vector Regression, Artificial Neural Networks, and Structural VAR Models

Mogari I. Rapoo
https://orcid.org/0000-0002-3602-7016
North-West University, South Africa

Elias Munapo
North-West University, South Africa

Martin M. Chanza
North-West University, South Africa

Olusegun Sunday Ewemooje
https://orcid.org/0000-0003-3236-6018
Federal University of Technology, Akure, Nigeria

ABSTRACT

This chapter analyses efficiency of support vector regression (SVR), artificial neural networks (ANNs), and structural vector autoregressive (SVAR) models in terms of in-sample forecasting of portfolio inflows (PIs). Time series daily data sourced from Rand Merchant Bank (RMB) covering the period of 1st March 2004 to 1st February 2016 were used. Mean squared error, root mean squared error, mean absolute error, mean absolute squared error, and root mean scaled log error were used to evaluate model performance. The results showed that SVR has the best modelling performance when compared to others. In determining factors that affect allocation of PIs into South Africa based on SVAR, 69% of the variation was explained by pull factors while 9% was explained by push factor. Hence, SVR model is more accurate than ANNs. This chapter therefore recommends that banking sector particularly RMB should use machine learning technique in modelling PIs for a better financial solution.

DOI: 10.4018/978-1-7998-3645-2.ch014

INTRODUCTION

Conventional econometric models have been used in modelling portfolio inflows for decades. These econometric models are significant in analysing the data since they are linear in nature while time series data are not as they are nonlinear in nature. This is because the dynamics and patterns of the series are nonlinear whereas the linear models assume a linear structure of the series. The discovery of the financial data being nonlinear has led its centre stage taken in analysing financial data. Nonlinear models have utmost accuracy in analysis series for their best modelling properties, thus, making them the most reliable models in predicting financial time series. Recent development in nonlinear models analyses has proved machine learning models to be better and powerful approximates. These machine learning models have been utilized in analysing time series data in different disciplines. In instances where linear models cannot address the fundamentals of time series data, nonlinear models are used as they capture those fundamentals. Therefore, this chapter investigates whether machine-learning models are effective in modelling portfolio inflows or not using econometric model of Structural VAR to identify the key drivers of portfolio inflows into South Africa and furthermore assess the efficiency and performance of machine learning models namely support vector regression (SVR) and artificial neural networks (ANNs) models in modelling and forecasting portfolio inflows, respectively.

BACKGROUND

In the literature, the effect(s) of strong wave of portfolio inflows are highlighted; under ordinary conditions capital flows have valuable impacts for developing economies. In a few events, floods of strong portfolio flows have gone before scenes of money related instability, for instance, the Mexican emergency of 1994 and the Asian emergency 1997 (Lo Duca, 2012). As this is the case, the negative effect of portfolio inflows to receiving economies calls for appropriate policies to be put in place, in which case the drivers of these flows may be used in developing these policies.

There are several applications of Support Vector Regression in solving forecasting problems in many fields where the model was successfully applied such as atmospheric science forecasting (Hong, 2009) and financial time series (stock index and exchange rate) forecasting (Cao, 2003). Chen and Wang (2007) employed Support Vector Regression, back-propagation neural networks (BPNN) and Autoregressive Integrated Moving Average (ARIMA) to forecast tourism demand and genetic algorithm was employed to select the optimal parameters of the Support Vector Regression model and show that Support Vector Regression outperforms other selected models. Hong (2009) also employed chaotic particle swarm optimization (CPSO) for choosing parameters for the Support Vector Regression model and showed that CPSO outperforms both the genetic algorithm (GA) and simulated annealing algorithm. Kazem et al. (2013) forecasted stock market prices employing a model based on Support Vector Regression, chaotic mapping and firefly algorithm using a time series data of stock prices, bank shares and intel. They compared their proposed model with Genetic Algorithm based Support Vector Regression (SVR-GA), Chaotic Genetic Algorithm based Support Vector Regression (SVR-CGA), Firefly based Support Vector Regression (SVR-FA), Artificial Neural Networks (ANNs) and Adaptive-Network-based Fuzzy Inference Systems (ANFIS), and revealed that the proposed model outperformed other models. Also, Adebiyi et al., (2014) compared artificial neural networks (ANNs) and Autoregressive Integrated Moving Average (ARIMA) models as far as anticipating precision of the stock market data sourced from New York Stock

Exchange and disclosed that the Artificial neural networks (ANNs) model outperformed Autoregressive Integrated Moving Average (ARIMA) model.

Ramani and Murarka (2013) used multilayer feed forward artificial neural networks (ANNs) to predict stock price using algorithm of back propagation to the model by training the network through historical stock prices (closing) and showed that a feed forward network using back propagation is quite reasonable for stock price prediction. Bing et al., (2012) employed BPNN model to forecast Shanghai Stock Exchange Composite Index and their results revealed that the model was utilized successfully to highest, lowest and closing values of the Index. Ajide and Raheem (2015) examined the drivers of capital flows (portfolio) into Nigerian economy and the dynamic effect shocks of the drivers using structural vector auto-regression (SVAR), variance decomposition and impulse response analysis. Their study revealed that there is need to introduce policies that can be used to deal with shocks in economic activity variables both internal and external. Korap (2010) employed structural vector auto-regression (SVAR) model in analysing the key determinants of capital flows (portfolio) in the framework of Turkish economy and found that the Turkish economy push factors have a superior role in elucidating the behaviour of portfolio flows and that there is a negative correlation between real interest rate and portfolio flows. Further study by Raghavan et al., (2014) examined the correlation between portfolio capital flows to domestic credit and their impacts on the Australian economy using the structural vector auto-regression (SVAR) model and it was revealed that debt flows are driving the impact of net portfolio flows in Australia, while equity flows have no real effect on the domestic macroeconomic variables.

There are various techniques which have been utilized in literature to model portfolio inflows and its drivers to emerging markets. Commonly used conventional econometric methods are; structural vector auto-regression (SVAR) model (used by Korap 2010; Alquist 2006; Çulha 2006), VAR model (used by Egly et al., 2010) and Finite distributed lag model and VECM model (used by Ekeocha et al., 2012; Hong and Pai 2007). Consequently, this chapter examines whether machine-learning models are effective in modelling portfolio inflows or not using econometric model of Structural VAR to identify the key drivers of portfolio inflows into South Africa and furthermore assess the efficiency and performance of machine learning models.

MAIN FOCUS OF THE CHAPTER

In achieving the research objectives, this chapter used the historic time series data of portfolio inflows obtained from the Rand Merchant Bank (RMB), South Africa. The time series data were chosen based on the notion that strong portfolio inflows can be of a negative impact to the recipient economy. Different models in the literature have been utilized in modelling and forecasting portfolio inflows, and furthermore utilized to explore which models give accurate predictions of portfolio inflows. Support Vector Regression (SVR), artificial neural networks (ANNs) and structural vector auto-regression (SVAR) are the fundamental statistical techniques utilized for the analysis in this current study. Schwarz Bayesian Criterion (SBC) developed by Schwarz (1978) used as a model selection method while forecasting efficiency of the model(s) is evaluated based on the accompanying error metrics: Mean Squared Error (MSE), Root Mean Squared Error (RMSE), Mean Absolute Error (MAE), Mean Absolute Squared Error (MASE) and Root Mean Scaled Log Error (RMSLE). All analyses are carried out using R, E-views and EasyReg International statistical softwares. Appropriate preliminary statistical analyses were undertaken and the data was assessed in terms of normality assumption by applying skewness and kurtosis, and shown to be nonlinear in nature.

Data Description and Sources

In address the research objectives, the chapter uses historic time series data of portfolio inflows which were sourced from the Rand Merchant Bank (RMB), South Africa. The chapter used portfolio inflows as the dependent variable and exchange rate, inflation linked bonds, real GDP as the independent variables covering the period of 1st March, 2004 to 1st February, 2016 consisting of 3111 observations.

Figure 1. Methodology Diagram

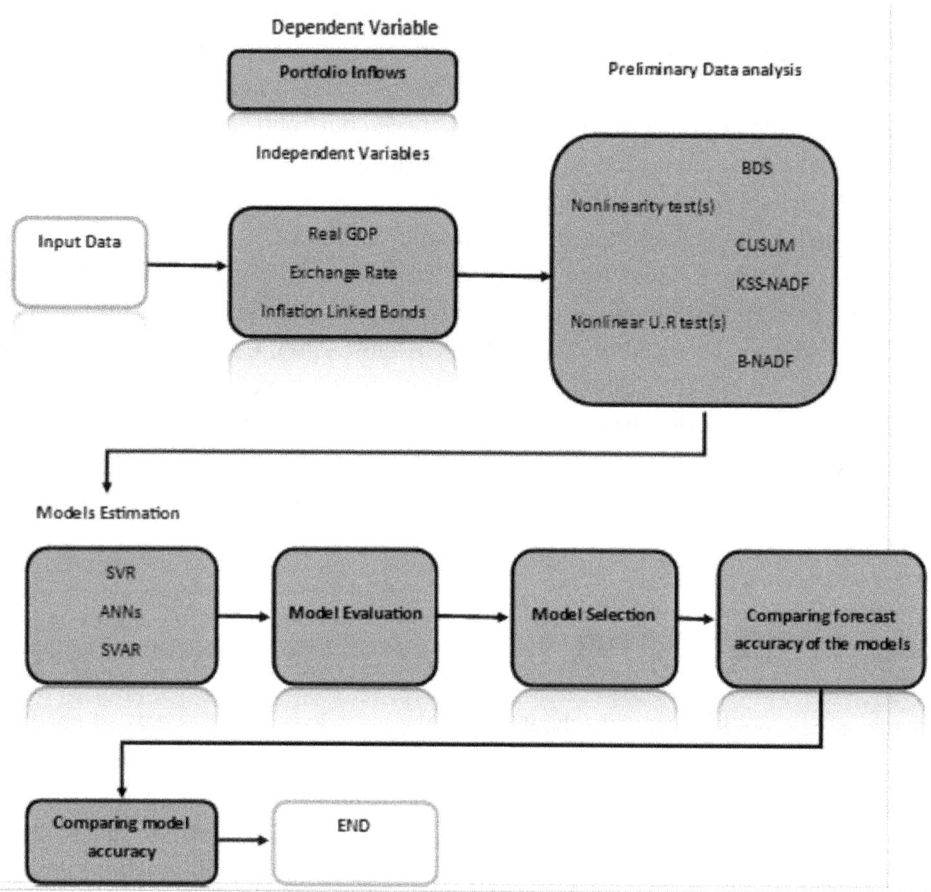

Modelling and Forecasting Methods

Support vector regression: Support vector machine is a developed machine learning classification paradigm while support vector regression (SVR) is the adaptation of support vector machine. In the SVR concept, a typical regression problem is formulated. Let's reflect on a set of data $G = \{(x_i, q_i)\}_{i=1}^{n}$, given that x_i is a model input in a vector form, q_i is actual value and n, total number of data patterns. In Support Vector Regression (SVR), the inputs are firstly nonlinear mapped into a high dimensional element space (F), wherein they are put into a linear relationship with the outputs in order to take care of nonlinear regression problem. According to Lu et al. (2009) the SVR have the following linear estimation function:

$$f(z) = (v \bullet \varphi(z)) + b \tag{1}$$

Given that $v, b, \varphi(z)$ and $(v \cdot \varphi(z))$ are the weight vector, constant, mapping function from the feature space and the dot product in the feature space F, respectively. Here, there is a transformation from a nonlinear regression problem to a linear regression problem from a lower dimension input space (z) to a higher dimension feature space (F). To formulate Support Vector Regression (SVR), there are a number of cost functions that can be used. These functions include Laplacian, Huber's Gaussian and ε-insensitive loss function ($l\varepsilon_)$. The most commonly adopted one is given in equation (2) by Lu et al. (2009).

$$l_\varepsilon\left(f(z),q\right) = \begin{cases} \left\{ \left|f(z)-q\right| - \varepsilon \quad if \left|f(z)-q\right| \geq \varepsilon \right\} \\ 0 \qquad\qquad\qquad otherwise \end{cases} \tag{2}$$

where ε denotes a radius of the tube situated in the regression function $f(z)$ and it is a precision parameter. Figure 2 presents Support Vector Regression (SVR) with ε -insensitive loss function. In figure 2 the locale encased by the tube is known as "ε -insensitive" since the loss function assumes a zero figure in this locale and does not punish the forecast errors with degrees smaller than ε as given in Lu et al. (2009). Hence, the weight (v) and constant (b) in equation (1) and the subsequent risk function can be minimized when estimating the parameters (weight and constant):

$$R(C) = C\frac{1}{n}\sum_{i=1}^{n} L_e\left(f(x_i),q\right) + \frac{1}{2}|w|^2 \tag{3}$$

Given that $L_e(f(x),q)$ is *ε-insensitive* loss function in equation (2), $\frac{1}{2}|w|^2$ is the regularization term which controls the exchange between the complexity and the estimation exactness of the regression model to guarantee that the model has an improved generalization performance. *C is* the regularization constant used to indicate the exchange between empirical risk and regularization term. The parameters *C* and ε can be determined by the user.

Two positive slack variables, ξ_i *and* ξ_i^*, $i=1,2,\ldots,n$, are used to measure deviation $(q_i - f(x_i))$ from the borders of the *ε-insensitive* zone. These denote the distance from the actual values to the conforming border value of *ε-insensitive* zone as shown in figure 2. Therefore, equation (3) is transformed by using the slack variables into the subsequent constrained method:

$$\text{Minimize: } R_{reg}(f) = \frac{1}{2}|w|^2 + C\sum_{i=1}^{n}\left(\xi_i + \xi_i^*\right) \tag{4}$$

Figure 2. A schematic representation of the SVR e-insensitive loss function

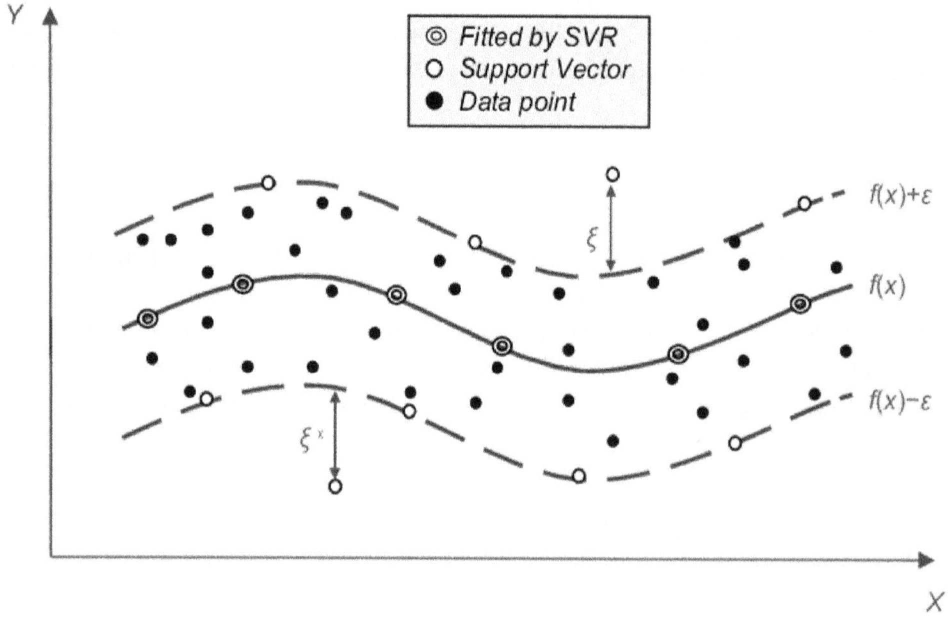

Subject to:

$$\begin{Bmatrix} q_i - \left(w \cdot \vartheta\left(x_i\right)\right) - b \le \varepsilon + \xi_i \\ \left(w \cdot \vartheta\left(x_i\right)\right) + b - q_i \le \varepsilon + \xi_i^* \\ \xi_i, \xi_i^* \ge 0, \ for \ i = 1, \ldots, n \end{Bmatrix} \tag{5}$$

By using Lagrangian multipliers and Karush-Kuhn-Tucker conditions in equation (3), this yields the following dual Lagrangian form given in equation (6).

Maximize

$$L_d\left(\alpha, \alpha^*\right) = -\varepsilon \sum_{i=1}^{n} \left(\alpha_i^* + \alpha_i\right) + \sum_{i=1}^{n} \left(\alpha_i^* - \alpha_i\right) q_i - \frac{1}{2}_i \sum_{j=1}^{n} \left(\alpha_i^* - \alpha_i\right)\left(\alpha_j^* - \alpha_j\right) k\left(x_i, x_j\right) \tag{6}$$

Subject to the constraints:

$$\begin{Bmatrix} \sum_{i=1}^{n}\left(\alpha_i^* - \alpha_i\right) = 0 \\ 0 \le \alpha_i \le C, \ i = 1,\ldots,n \\ 0 \le \alpha_i^* \le C, \ i = 1,\ldots,n \end{Bmatrix} \tag{7}$$

The equality of $\alpha_i\alpha_i^* = 0$ is satisfied by the Lagrangian multipliers in equation (6). The Lagrangian multipliers, α_i *and* α_i^* are calculated and an optimal desired weight vector of the regression hyperplan is $v^* = \sum_{i=1}^{n}\left(\alpha_i - \alpha_i^*\right)k\left(x, x_j\right)$. Support Vector Regression (SVR) function can be written according to Çulha (2006) and Amisano & Giannini (1997) as:

$$f\left(x, v\right) = f\left(x, \alpha, \alpha^*\right) = \sum_{i=1}^{n}\left(\alpha_i - \alpha_i^*\right)k\left(x, x_i\right) + b \tag{8}$$

where $k(x,x_i)$ is called kernel function. The product of the two vectors, x_i and x_j, in the feature space $\varphi(x_i)$ and $\varphi(x_j)$ will be equal to the kernel; that is, $k(x,x_i) = \varphi(x_i)\varphi(x_j)$. Any function that can be used to satisfy the Mercer's condition can also be employed as a kernel function (Chen and Wang, 2007). The broadly utilized kernel function is the radial basis function, although several functions are available. The radial basis function (RBF) is computed as follows:

$$K\left(x_i, x_j\right) = exp\left(\frac{-x_i - x_j^2}{2\sigma^2}\right) \tag{9}$$

where σ is the width of the RBF. In this manner, the current chapter utilizes the RBF as it was utilized in the study of Lu et al. (2009).

Artificial Neural Networks: Computational intelligence systems, in particular, the artificial neural networks (ANNs) are free of dynamics and have been used in approximation function and forecasting (Khashei and Bijari, 2010). The ANN models have significant advantage over other nonlinear models; ANNs are widespread approaches that can approximate a substantial class of functions with a high level of accuracy. Their capacity originates from the parallel processing of information from the data. Also, there are no prior assumptions required in the model building process. Single hidden layer feed forward network has been widely applied in time series modelling and forecasting (Khashei and Bijari, 2010; Adebiyi et al. 2012). This model is made up of a network of three layers of simple processing units connected by a cyclic links (see figure 3). The output (y_t) and the inputs ($y_{t-1},\ldots,y_{t-p}$) have a relationship that can be denoted as follows in a mathematical representation:

$$y_t = w_0 + \sum_{j=1}^{q} w_j \cdot g\left(w_{0j} + \sum_{i=1}^{p} w_{ij} \cdot y_{t-i}\right) + \varepsilon_t \tag{10}$$

where $w_{ij}(i=0,1,2,\ldots,p; j=1,2,\ldots,q)$ and $w_j(j=0,1,2,\ldots,q)$ are model parameters often called connection weights. Parameter p is the number of input nodes and q is the number of hidden nodes. There are several forms activation function can take and the reason for this activation function is to prohibit outputs from achieving huge values which can 'incapacitate' neural networks and along these lines inhibit training (Kaastra and Boyd, 1996). The logistic and tangent hyperbolic functions are used as the hidden layer activation function which is shown in equations (11) and (12), respectively;

$$sig(y) = \frac{1}{1+exp(-y)}. \tag{11}$$

$$Tanh(y) = \frac{1-exp(-2y)}{1+exp(-2y)}. \tag{12}$$

such that the artificial neural networks (ANN) model in equation (10), a nonlinear function mapping will be performed from the past observations to the future value y_t, i.e.

$$y_t = f(y_{t-1},\ldots,y_{t-p},w)+\varepsilon_t \tag{13}$$

where w is a vector of all parameters and $f(\bullet)$ is a function determined by the network structure and connection weights.

Figure 3. Artificial Neural Network Structure $N^{(p-q-1)}$

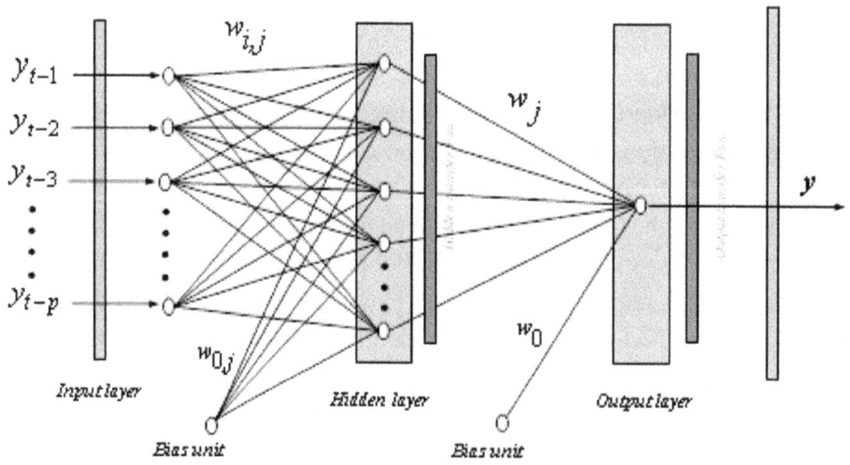

Thus, the neural network is the same as a nonlinear auto-regression model. The simple network given in (10) is extremely powerful in that it is able to approximate the arbitrary function as the number of the hidden nodes, q, is appropriately large (Khashei and Bijari, 2010). There is no exact method that can be used to choose parameter q since it is data-dependent. The nonlinear autocorrelation structure of the time series is determined by the lagged observations, p, which is estimated from artificial neural networks (ANN) model.

However, pruning algorithm, canonical decomposition method and the network information criteria can be used to find the optimal structure of the artificial neural networks (ANN) model. Furthermore, there are many other methods that can be employed in finding the optimal structure of the ANN model but these methods are somehow very difficult to implement. In general, these methods rely on testing various networks with different numbers of inputs and hidden nodes (p,q) then estimate generalization error and select the network with the most minimal generalization error. The parameters are evaluated such that the cost function (aggregate accuracy criterion) of neural network is limited, for example, mean squared error:

$$E = \frac{1}{N}\sum_{n=1}^{N}\left(e_i\right)^2 = \frac{1}{N}\sum_{n=1}^{N}\left(y_t - \left(w_0 + \sum_{j=1}^{q} w_j g\left(w_{0j} + \sum_{i=1}^{p} w_{ij} y_{t-i}\right)\right)\right)^2 \tag{14}$$

Given that N is the number of error terms, an efficient nonlinear optimization algorithm are employed but not the basic back-propagation training algorithm (Khashei and Bijari, 2010) where neural network parameters, w_{ij}, are altered by an amount Δw_{ij}, according to the following formula:

$$\Delta w_{ij} = -\tau \frac{\partial E}{\partial w_{ij}} \tag{15}$$

where (τ) is the learning rate and $\frac{\partial E}{\partial w_{ij}}$ is the partial derivative of the function E with respect to the weight, w_{ij}. This derivative is generally figured in two passes; forward pass, an input vector from the preparation set is connected to the input units of the network and is spread through the network layer by layer creating the last output, and backward pass, the output of the network is compared with the optimal output and the subsequent error is then propagated in reverse through the network modifying the network. Mcclelland et al., (1986) presented a momentum term δ to equation (15) to accelerate the learning procedure, while keeping away from the instability of the algorithm.

$$\Delta w_{ij}\left(t+1\right) = -\tau \frac{\partial E}{\partial w_{ij}} + \delta \Delta w_{ij}\left(t\right) \tag{16}$$

The momentum term may likewise be useful in prohibiting the learning process from being trapped into poor local minima, and the chapter uses the interval of [0,1] for training the neural network model.

Structural Vector Autoregressive (SVAR): To evaluate the possible drivers of portfolio inflows into emerging markets, this chapter applies the structural identification methodology of vector autoregressive models (SVAR) developed by the purported AB-model of Amisano and Giannini (1997). SVAR model allows researchers to impose limitations on structural relationships by using the theoretical assumptions on the model of which unrestricted vector autoregressive model has a shortcoming. This can be realized by presenting a theoretical and some auxiliary limitations to achieve econometric identification issue, assuming that $\Sigma = E[\varepsilon_t \varepsilon_t]$ presents the covariance matrix of the residuals. Thereby, the structural analysis uses the reduced form of the model given by:

$$A\varepsilon_t = B\mu_t \tag{17}$$

Given that the disturbance vector is of the form εt whereas a structural innovation vector in an unobserved form is denoted by μt equation (17) takes into account the reduced form disturbance and underlying structural shocks. Both A and B need some limitations with a dimension in the structural vector auto-regression (SVAR) analysis k*x*k to be added, noting that covariance matrix are present in the structural innovations given by $E[\mu_t \mu_t]=I$ where the identity matrix is given by I so that μ_t can enforce the following limitations on A and B.

$$A\Sigma A = BB \tag{18}$$

for the recognizable proof of the AB model, at least $k^2 + \frac{k(k-1)}{2} = k(3k-1)/2$ limitations are required. Hence, the need to consider variables used as part of vector autoregressive procedure to execute innovative accounting strategies, for instance, impulse responses require not to be stationary. Sims (1980) spearheaded paper on the VAR procedure contends against differencing regardless of whether the time series utilized takes after a unit root process while Sims et al., (1990) demonstrated that parameters that can be composed as coefficients on mean zero, non-integrated regressors, have normal asymptotic distributions and proposed that the basic practice with regards to transforming models to stationary form by difference operators appear likely that the series are of integrated form is unnecessary. Otherwise, some vital knowledge contained in the series would perhaps not be used by the researcher.

The log-likelihood function of the structural vector auto-regression (SVAR) model can be measured by way of a function of Π and Σ (Sims, 1986), and assuming that there are no cross limitations on **Π** and **Σ** or on the other hand, in more broad terms, that there are no limitations at all on **Π** meanwhile some form of limitations are enforced on **Σ**, the identification and analysis of the likelihood function can help to estimate the parameters of models:

$$L = c - \frac{T}{2} log|\Sigma| - \frac{T}{2} tr\left(\Sigma^{-1} \widehat{\Sigma}\right) \tag{19}$$

Given $\widehat{\Sigma} = T^{-1}\widehat{V}\widehat{V}'$ is the log-likelihood concentrated with respect to **Π**. The estimation of **Π** comparing to the concentration of the log likelihood unmistakably concurs with the OLS estimator when the log likelihood is molded on the principal *p* observations of the sample. Other reliable estimators would yield asymptotically identical results with respect to the resulting estimation of the **Σ** matrix. Following Amisano and Giannini (1997), impulse response functions for structural VARs can be calculated by:

$$\Theta_0 = K^{-1} = \left(B^{-1}A\right)^{-1} = A^{-1}B, \Theta_i = JM^iJ'\Theta_0, \quad i = 1,2,3\ldots \tag{20}$$

Given that:

$$M = \begin{bmatrix} A_1 & A_2 & \cdots & A_{k-1} & A_k \\ I_p & 0 & \ldots & 0 & 0 \\ 0 & I_p & \ldots & 0 & 0 \\ \vdots & \vdots & \vdots & \vdots & \vdots \\ 0 & 0 & \cdots & I_p & 0 \end{bmatrix} \tag{21}$$

M is a $(pk*pk)$ matrix, and J is a $(p*pk)$ matrix and

$$\Theta_i = \Theta_i.\Theta_0, \Theta_0 = I_p, \Theta_i = \sum_{j=1}^{i} \Theta_{i-j} A_j, \quad i = 1,2,3\ldots \tag{22}$$

where A_i=0 for $i>k$. The impulse responses are predicted by substituting the theoretical variables by the respective estimators. The estimated impulse response that is stored up to horizon H in vector $[\Theta_1, \Theta_2, \ldots, \Theta_H]$ is unbiased and asymptotically normally independently distributed with covariance matrix $\frac{1}{T}\Sigma_\Theta\left(H\right)$.

RESULTS AND DISCUSSIONS

Descriptive Statistics

Schumacker and Lomax (2004), and Kline (2005) suggested that prior to doing the actual analysis of any study; important issues must be addressed such as normality. In this section, besides the normality test, other descriptive statistics are discussed such as the mean and standard deviation of the series. These are summarized in Table 1.

The variables of interest have a multimodal distribution as shown in figure 4. Theoretical values of skewness and kurtosis based on the normality assumption must assume values of 0.00 and 3.00, respectively. However, the values of the two measures deviated from the theoretical values of both skewness and kurtosis; thereby the normality assumption is violated and the series is nonlinear. These results were quantified by employing the Jarqua Bera test and its probability value is 0.000, which rejects the null hypothesis of normality at all levels of significance 1%, 5% and 10% for all the included series. In terms of the normality test, with assumption being violated, Cziraky and Cumpek (2002) believe that by applying transformation such as the normal scores the series can be normally distributed. However, with the current study all the series still violates the normality assumption even after the appropriate transformation and this can be because all the variables are nonlinear in nature.

Table 1. Descriptive statistics

Variable	Portfolio flows	Exchange rate	Real GDP	Inflation-linked bonds
Mean	9.425058	6.704596	13.15178	11.73551
Median	9.641215	6.637533	13.16707	11.76145
Std. dev.	1.390921	0.227674	0.129336	0.762516
Skewness	-5.922360	0.858337	-0.694901	-0.005083
Kurtosis	39.31928	3.044956	2.470845	1.534455
Jarqua Bera	189173.1	382.2633	286.6726	278.4249
P-value	0.000000	0.000000	0.000000	0.000000

Figure 4. Graphical representation of the portfolio flows into South Africa and its key drivers

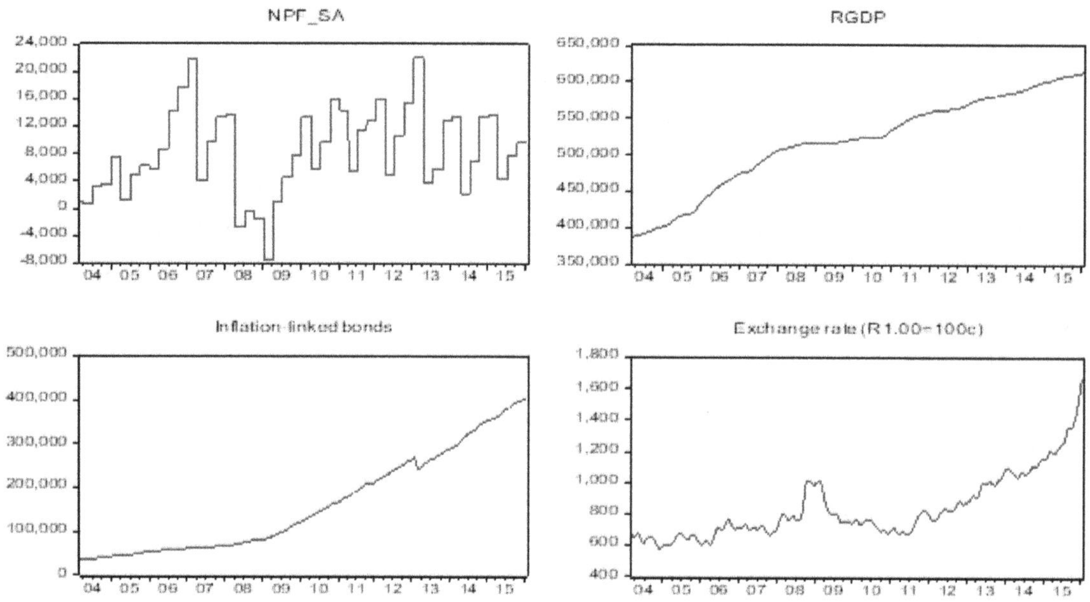

Knowing the characteristics of the series is of the paramount significant. For this purpose, the plot of the series is normally the one used to show any movements of the series from the start date to the end date of the series. Figure 4 shows the series plot of portfolio inflows into South Africa together with its key drivers covering the period of 01-March-2004 to 01-February-2016 consisting of 3111 observations.

A visual examination of the series plot shows that there are those periods in which South Africa received more portfolio inflows in their balance of payments. These periods are made up of those spikes in the data, which show volatility of the series. The series has several volatility periods, for assessing the stationarity of the series, this chapter employed the test of Bierens nonlinear unit root test (BNADF) and the results revealed that the series is nonlinear unit root.

Nonlinearity test results using the Brock-Dechert-Scheinkman (BDS) test and the CUSUM test of stability are presented in table 2 and figure 5, respectively.

Table 2. The Brock-Dechert-Scheinkman (BDS) test results

Variable(s)	Dimension	BDS Statistic	Std. Error	Z-statistic	P-value
Portfolio flows	2	0.205356	0.000794	258.5316	0.0000
	3	0.348782	0.001256	277.6763	0.0000
	4	0.448666	0.001488	301.6065	0.0000
	5	0.517933	0.001542	335.9499	0.0000
	6	0.565674	0.001478	382.6946	0.0000
Real GDP	2	0.208568	0.001094	190.7141	0.0000
	3	0.355180	0.001731	205.2321	0.0000
	4	0.458269	0.002051	223.4017	0.0000
	5	0.530777	0.002128	249.4462	0.0000
	6	0.581791	0.002042	284.9178	0.0000
Exchange rate	2	0.206652	0.001505	137.2934	0.0000
	3	0.351737	0.002392	147.0623	0.0000
	4	0.453641	0.002848	159.3021	0.0000
	5	0.525224	0.002968	176.9860	0.0000
	6	0.575494	0.002861	201.1239	0.0000
Inflation-linked bonds	2	0.208844	0.000952	219.4497	0.0000
	3	0.355196	0.001502	236.5584	0.0000
	4	0.457813	0.001774	258.0323	0.0000
	5	0.529800	0.001835	288.7783	0.0000
	6	0.580320	0.001755	330.6671	0.0000

Figure 5. Cumulative Sum (CUSUM) test results

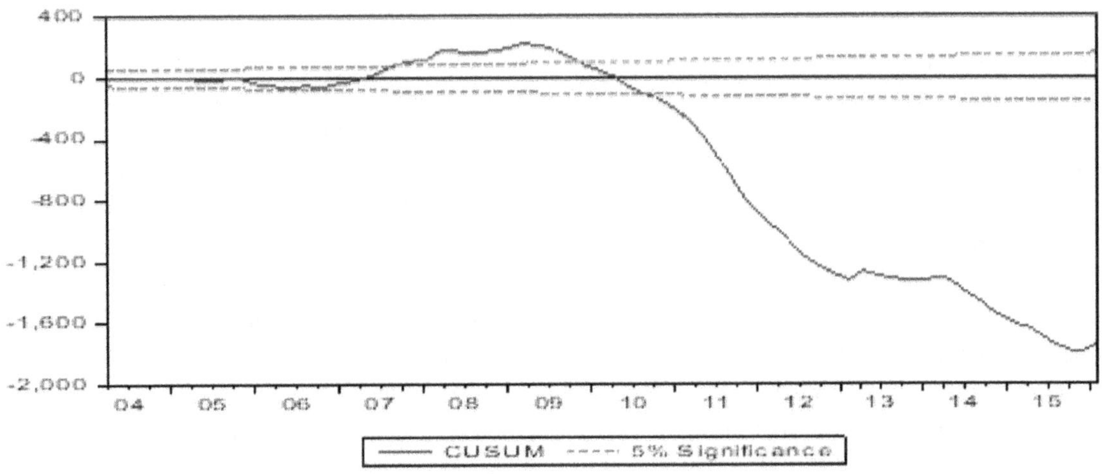

Based on the BDS test results in table 2, the null hypothesis of linearity is rejected in favour of alternative of nonlinearity at all significance level for all dimensions. Furthermore, the CUSUM test of stability is employed to support the results of BDS test, it shows that the series is unstable since the blue line of the graph goes outside the critical red lines of the graph, this confirms the results of the BDS test that the series is nonlinear.

Modelling and Estimation of Portfolio Inflows

Here, the researchers outline the estimation of the models in this chapter and results of Support Vector Regression (SVR), Artificial Neural Network (ANN) and Structural Vector Auto-Regression (SVAR) models are presented.

Support vector regression (SVR): For building support vector regression forecasting model, the *e*1071*R package* proposed by Meyer et al. (2015) is adopted. The time series data is firstly scaled into the interval of [0,1] to avoid large input variables that affect the smaller input variables which may biased the results. In searching for the optimal parameters in the model of Support Vector Regression (SVR), the study set an interval of the cost (*C*) and epsilon (ε) to be (1:100) and *seq*(0,1,0.1), respectively. Thus the optimal parameters chosen by the model are 100 and 0.4 for both the cost and epsilon parameters, respectively. These model parameters give the best forecasting results (minimum testing MSE) with the best value of 0.018477 and are set as the model parameters in forecasting portfolio inflows. The coefficients of the estimated model function of the support vector regression are given as:

$$w=\left[1,1:3\right]\begin{Bmatrix} -6.0605 & 10.5350 & -22.6916 \\ RGDP & exchange\ rate & inflation\ linked \end{Bmatrix}.$$

b=0.738

Figure 3, presents the two sets of parameters namely the cost parameter and the epsilon of the model. The model has the best performance in the darker blue area were the best performing parameters are (epsilon and cost parameters).

Artificial Neural Networks (ANNs): In building artificial neural networks (ANNs) model, the *neuralnet R package*of Fritsch et al. (2016) is used, the data set was scaled based on the interval of [0,1] so that the widely used activation function of logistic function could be used. Backpropagation algorithm is employed in training the neural network model. Since the response variable is numeric in nature the sum of squared error was used as the measuring error of the network. The threshold used is of the default of 0.01. The results of neural networks are presented in table 3.

In reaching absolute partial derivative, it took the training process of 13488 steps and that obtained smallest error function which is smaller than 0.01 (the default value of the threshold). The predicted weights have interval of -16.92 to 15.79, for instance, the intercept of the first hidden layer are 4.605 and -4.212 and the three weights leading to the first hidden neuron are estimated as -16.918, -5.498 and -7.410 for the covariates of RGDP, exchange rate and inflation linked. Figure 7 presents the architecture of the neural network model estimated for this study and it is made up of four input layer, 2 hidden layers and one output layer.

Figure 6. performance of the SVM regression model

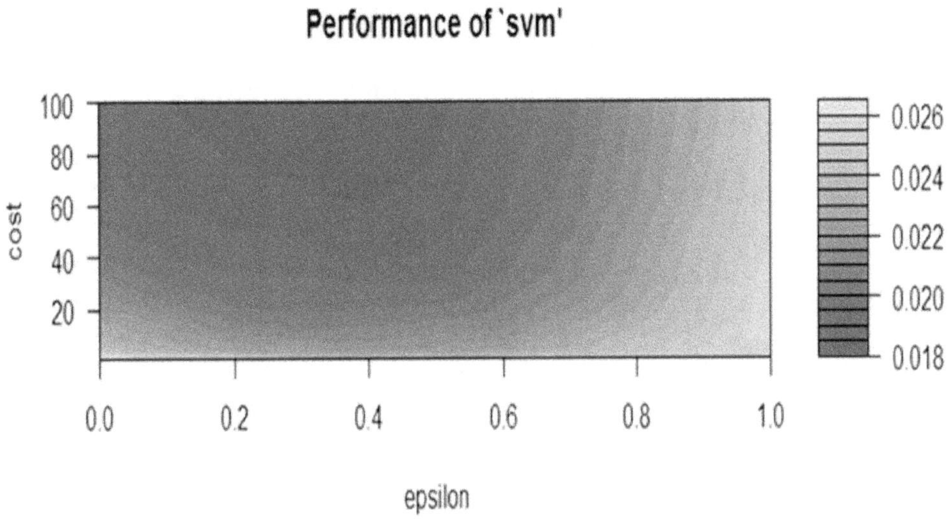

Table 3. The Artificial Neural Networks (ANNs) matrix

Error	25.7554
Reached threshold	0.0097
Steps	13488
Intercept to 1layhid1	4.6054
RGDP to 1layhid1	-16.9177
Exchange rate to 1layhid1	-5.4977
Inflation linked bonds to 1layhid1	15.7947
Intercept to 1layhid2	-4.2122
RGDP to 1layhid2	8.7387
Exchange rate to 1layhid2	2.5802
Inflation linked bonds to 1layhid2	-7.4104
Intercept to portfolio flows	1.4089
1layhid.1 to portfolio flows	-1.0273
1layhid.2 to portfolio flows	-1.8104

Structural vector autoregressive (SVAR): In estimating SVAR model, the unrestricted VAR model is estimated on the four stationary variables. The unrestricted VAR (8) model is initially estimated using AIC, SBC and HQC selected lag length (8) for the model. This chapter applies structural restrictions and assumed that portfolio flows are affected by all the shocks on the other variables. It is also assumed that exchange rate is responsive to shocks on the real gross domestic product. Inflation is also assumed to be responsive to shocks on both exchange rate and real gross domestic product. Thus the study used the AB model given as:

Figure 7. The Artificial Neural Networks architectural structure

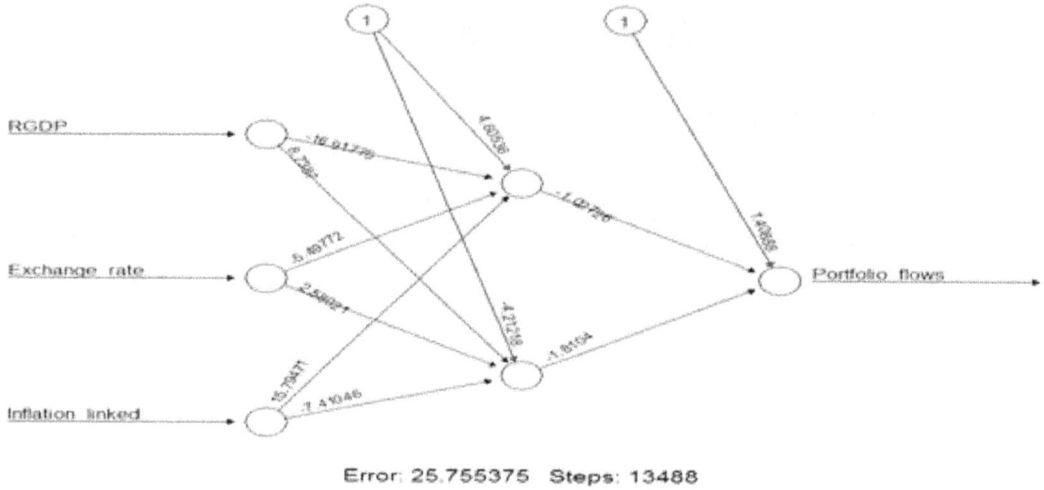

$$\begin{bmatrix} 1 & 0 & 0 & 0 \\ c(1) & 1 & 0 & 0 \\ c(2) & 0 & 1 & 0 \\ c(3) & 0 & 0 & 1 \end{bmatrix} \begin{bmatrix} \mu_t npf\_sa \\ \mu_t rgdp \\ \mu_t inflation \\ \mu_t exchangerate \end{bmatrix} = A$$

$$\begin{bmatrix} c(4) & 0 & 0 & 0 \\ 0 & c(5) & 0 & 0 \\ 0 & 0 & c(6) & 0 \\ 0 & 0 & 0 & c(7) \end{bmatrix} \begin{bmatrix} \varepsilon_t npf\_sa \\ \varepsilon_t rgdp \\ \varepsilon_t inflation \\ \varepsilon_t exchangerate \end{bmatrix} = B$$

Structural vector auto-regression (SVAR) estimated model is presented in figure 8, according to the results the model is over identified with 3 degrees of freedom. The LR test statistic estimated for the system identification restrictions under the null hypothesis is $\chi^2(3)$= 8.39E+09 with a probability value of 0.0000. The structural parameters are estimated by a method of scoring (analytic derivatives). Furthermore, the SVAR impulse response function (IRF) is represented in table 4 with periods of 10 days using 95% confidence interval.

Since there are no large margins or the confidence intervals, this gives a margin of certainty that the results estimated is reliable and unbiased. As expected, the portfolio inflows have a positive and negative response to its own shocks for the first period of the shock. Accordingly, it is seen that the 'push' factors that affect portfolio inflows are the real gross domestic product together with pull factor of exchange rate. The response of portfolio inflows due to shocks on inflation is along the zero line, thereby there is a positive shock to portfolio inflows that has been brought by inflation as a push factor. After taking into consideration the structural factorization of the impulse responses, table 4 presents the results of the variance decomposition of structural vector auto-regression (SVAR) model.

Figure 8. Structural Vector Auto-Regression (SVAR) Impulse Response Function

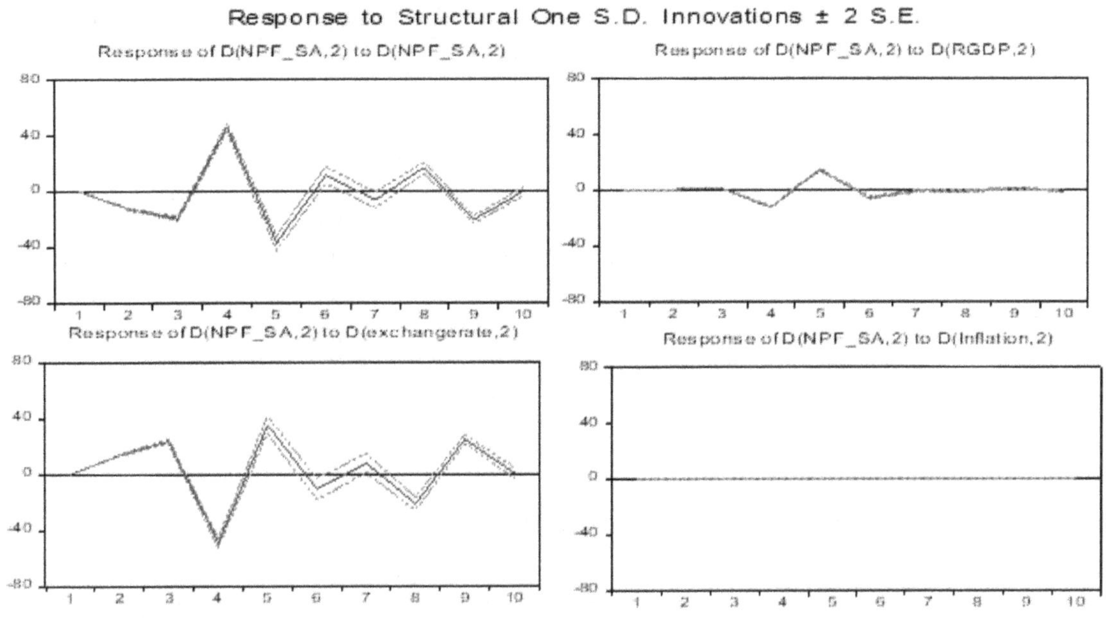

Variance decomposition analysis indicates that over the 10 periods of variation, forecast error variance of portfolio inflows of 99% can be attributed to its own shocks. The results also show that the variable which shows the best explanation (forecast error variance) is the real gross domestic product with a shock variable explaining nearly 55% of the variation to portfolio inflows. A shock to the inflation variable explains nearly 14% of variation to portfolio inflows while the least variable which explains the minimum variation of 9% is the exchange rate to portfolio inflows. Overall, the pull factors in the study explain nearly 69% of variation to portfolio inflows. All the estimation results indicate that the pull factors are indeed dominant over the push factors in acquiring the portfolio inflows as a form of investment into the South African economy one which is considered as an emerging market economy.

Performance criteria: According to the chapter objectives forecasting accuracy of the two machine learning (Support Vector Regression (SVR) and Artificial Neural Networks (ANNs)) models are compared. Forecasting of the future value of any economic variable(s) is of great importance to the academia, potential investors, decision making and policy formulation individuals. These models (SVR and ANNs) are here used to model and forecast portfolio inflows together with its determinants or drivers. Also, determine whether these models could be of use to financial forecasting of economic variable in terms of portfolio inflows. In terms of measuring the accuracy of the included models the study used the five criteria to assess the accuracy of the models. The five criteria are Mean Squared Error (MSE), Root Mean Squared Error (RMSE), Mean Absolute Error (MAE), Mean Absolute Squared Error (MASE) and Root Mean Squared Log Error (RMSLE) used as measuring error matrices. The results of these measurements are presented in table 5.

Table 4. Structural Vector Auto-Regression (SVAR) Variance Decomposition

Variance period	Portfolio flows	RGDP	Inflation-linked bonds	Exchange rate
1	100.0000	0.000000	0.000000	0.000000
2	99.93822	0.000747	0.057904	0.003133
3	99.85827	0.003058	0.126535	0.012141
4	99.61583	0.202605	0.130053	0.051513
5	99.28494	0.510752	0.132892	0.071419
6	99.23974	0.550461	0.136562	0.073234
7	99.23805	0.550547	0.137191	0.074209
8	99.22872	0.550732	0.138911	0.081636
9	99.21513	0.552498	0.140840	0.091535
10	99.21462	0.552880	0.141001	0.091503

Table 5. Forecasting Accuracy of SVR, ANNs and SVAR

Performance criterion	SVAR	SVR	ANN (3-2-1) model
MSE	N/A	0.0174	0.2637
RMSE	N/A	0.1561	0.5135
MAE	N/A	0.1028	0.2637
MASE	N/A	14.3975	12.9335
RMSLE	N/A	0.0851	0.3559

Table 6. Comparison of Model Accuracy: Diebold-Mariano Test

Variable	Hypothesis	Forecast horizon	Loss function power	t-statistic	P-value
Portfolio flows	ANN vs. SVR	1	2	8.5086376	< 0.0001

From the results, four of the five criteria select the Support Vector Regression (SVR) model as having the highest accuracy compared to the other model while only one criterion (MASE) chooses the artificial neural networks model. Thereby it can be concluded that the support vector regression model simulates and forecasts portfolio inflows better than the artificial neural networks (ANNs) model.

Comparing predictive accuracy of the models: According to Diebold (2015) the question that Diebold-Mariano (DM) test is answering is whether one model is truly superior to another model or is it mere luck. Lower error matrices do not necessarily mean that the forecasting ability of one model is superior to another one, as the difference between the error matrices may not be statistically significant. Hence, the need to assess the predictive accuracy of the models is apparent and the Diebold-Mariano test was employed to measure the predictive accuracy of the models. Summary of the results are presented in table 6.

The results in table 6 show that the null hypothesis of equal predictive accuracy of the models; Artificial Neural Network (ANN) and Support Vector Regression (SVR) is rejected since the respective probability value (< 0.0001) is less than the default significant level of 5% or 0.05. Furthermore, since the t-statistics (8.5086376) is positive it can be concluded that the estimated SVR model is more accurate than the estimated artificial neural networks model for portfolio flows.

FUTURE RESEARCH DIRECTIONS

machine learning techniques have been shown to estimate/model portfolio inflows better than conventional econometric models. Therefore, this work may be extended to cover the use of these machine learning techniques in modelling portfolio inflows of other banks in South Africa and other nations of the world.

CONCLUSION

This chapter showed that machine learning techniques can model portfolio inflows better than the econometric conventional model. In assessing the modelling and forecasting performance of the models, the error metrics; MSE, RMSE, MAE, MASE and RMSLE were utilized. The results showed that the support vector regression (SVR) model has the best modelling performance in modelling and forecasting portfolio inflows. In case of determining the factors that affect the allocation of portfolio inflows into South Africa based on the structural vector auto-regression (SVAR) model, it was shown that 69% of variation in the portfolio inflows is explained by pull factors (real gross domestic product and inflation linked bonds) and 9% was explained by push factor (exchange rate). However, 99% of variation in the portfolio inflows was attributed to its own shocks. Therefore, it is concluded that support vector regression (SVR) model is more accurate than its counterpart, artificial neural network (ANNs). Hence, this chapter recommends that the banking sector particularly the RMB bank should use these machine learning techniques in modelling portfolio inflows when embarking on their research.

LIMITATIONS

The study is restricted to using the historic data of portfolio inflows and its key drivers sourced from RMB covering the period of 01st March 2004 to 01st February 2016 on the basis of accessibility of data. There exist numerous models that can be used but for the current study, only three models; SVR, ANNs and SVAR are considered based on their empirical literature. Due to limited empirical studies on machine-learning models, the study used some sources which are older than ten years. Also, for the lack of appropriate syntax, the KSS-NADF is not employed in the current study.

ACKNOWLEDGMENT

The authors would like to express their sincere thanks to the editor and anonymous reviews for their time and valuable suggestions.

This research received no specific grant from any funding agency in the public, commercial, or not-for-profit sectors.

REFERENCES

Adebiyi, A.A., Adewumi, A.O. & Ayo, C.K., (2014). Comparison of ARIMA and artificial neural networks models for stock price prediction. *Journal of Applied Mathematics.* . doi:10.1155/2014/614342

Adebiyi, A. A., Ayo, C. K., Adebiyi, M. O., & Otokiti, S. (2012). Stock price prediction using neural network with hybridized market indicators. *Journal of Emerging Trends in Computing and Information Sciences, 3*, 1–9.

Ahlquist, J. S. (2006). Economic policy, institutions, and capital flows: Portfolio and direct investment flows in developing countries. *International Studies Quarterly, 50*(3), 681–704. doi:10.1111/j.1468-2478.2006.00420.x

Ajide, K. B., & Raheem, A. J. (2015). Determinants of foreign capital flows into Nigeria: A structural var analysis. *Business and Management Research Journal, 5*, 1–13.

Amisano, G., & Giannini, C. (1997). *From VAR models to Structural VAR models. In Topics in Structural VAR Econometrics*. Springer-Verlag Berlin. doi:10.1007/978-3-642-60623-6

Bing, Y., Hao, J. K., & Zhang, S. C. (2012). Stock Market Prediction Using Artificial Neural Networks. *Advanced Engineering Forum, 6-7*, 1055-1060.

Cao, L. (2003). Support vector machines experts for time series forecasting. *Neurocomputing, 51*, 321–339. doi:10.1016/S0925-2312(02)00577-5

Chen, K. Y., & Wang, C. H. (2007). Support vector regression with genetic algorithms in forecasting tourism demand. *Tourism Management, 28*(1), 215–226. doi:10.1016/j.tourman.2005.12.018

Çulha, A. (2006). A structural VAR analysis of the determinants of capital flows into Turkey. *Central Bank Review, 2*, 11–35.

Cziraky, D., & Cumpek, T. (2002). Multivariate analysis of the European economic and defence structure. *Croatian International Relations Review, 8*, 23–35.

Diebold, F. X. (2015). Comparing predictive accuracy, twenty years later: A personal perspective on the use and abuse of Diebold–Mariano tests. *Journal of Business & Economic Statistics, 33*(1), 1–9. do i:10.1080/07350015.2014.983236

Egly, P. V., Johnk, D. W., & Liston, D. P. (2010). Foreign Portfolio Investment Inflows to the United States: The Impact of Investor Risk Aversion and US Stock Market Performance. *North American Journal of Finance and Banking Research, 4*(4), 25–41.

Ekeocha, P.C., Ekeocha, C., Malaolu, V. & Oduh, M., (2012). Modelling the long run determinants of foreign portfolio investment in Nigeria. *Journal of Economics and Sustainable Development, 3*, 194-205.

Fritsch, S., Guenther, F., & Guenther, M. F. (2016). *Package 'neuralnet'*. The Comprehensive R Archive Network.

Hong, W. C. (2009). Chaotic particle swarm optimization algorithm in a support vector regression electric load forecasting model. *Energy Conversion and Management*, *50*(1), 105–117. doi:10.1016/j.enconman.2008.08.031

Hong, W. C., & Pai, P. F. (2007). Potential assessment of the support vector regression technique in rainfall forecasting. *Water Resources Management*, *21*(2), 495–513. doi:10.100711269-006-9026-2

Kaastra, I., & Boyd, M. (1996). Designing a neural network for forecasting financial and economic time series. *Neurocomputing*, *10*(3), 215–236. doi:10.1016/0925-2312(95)00039-9

Kazem, A., Sharifi, E., Hussain, F. K., Saberi, M., & Hussain, O. K. (2013). Support vector regression with chaos-based firefly algorithm for stock market price forecasting. *Applied Soft Computing*, *13*(2), 947–958. doi:10.1016/j.asoc.2012.09.024

Khashei, M., & Bijari, M. (2010). An artificial neural network (p, d, q) model for timeseries forecasting. *Expert Systems with Applications*, *37*(1), 479–489. doi:10.1016/j.eswa.2009.05.044

Kline, R. B. (2005). *Methodology in the social sciences. In Principles and practice of structural equation modeling*. Guilford Press.

Korap, L. (2010). *Identification of 'pull'& 'push' factors for the portfolio flows: SVAR evidence from the Turkish economy. MPRA Paper 24275*. University Library of Munich.

Lo Duca, M. (2012). *Modelling the time varying determinants of portfolio flows to emerging markets*. Working Paper Series of European Central Bank, Kaiserstrasse 29, D-60311 Frankfurt am Main, Germany.

Lu, C. J., Lee, T. S., & Chiu, C. C. (2009). Financial time series forecasting using independent component analysis and support vector regression. *Decision Support Systems*, *47*(2), 115–125. doi:10.1016/j.dss.2009.02.001

Mcclelland, J. L., Rumelhart, D. E., & Hinton, G. E. (1986). A general framework for parallel distributed processing. In *Parallel distributed processing: Explorations in the microstructure of cognition. Bradford Book*.

Meyer, D., Dimitriadou, E., Hornik, K., Weingessel, A. & Leisch, F. (2015). *e1071: misc functions of the department of statistics, probability theory group (formerly: E1071), TU Wien*. R package version 1.6-7.

Raghavan, M., Churchill, A., & Tian, J. (2014). *The Effects of Portfolio Capital Flows and Domestic Credit on the Australian Economy* (Unpublished manuscript). University of Tasmania, Launceston, Australia.

Ramani, P. & Murarka, P. (2013). *Stock market prediction using artificial neural network*. Doi:10.21090/ijaerd.c25

Schumacker, R. E., & Lomax, R. G. (2004). A beginner's guide to structural equation modeling. Psychology Press.

Schwarz, G. (1978). Estimating the dimension of a model. *Annals of Statistics*, *6*(2), 461–464. doi:10.1214/aos/1176344136

Sims, C. A. (1980). Macroeconomics and reality. *Econometrica*, *48*(1), 1–48. doi:10.2307/1912017

Sims, C. A. (1986). Are forecasting models usable for policy analysis? Quarterly Review, 2-16.

Sims, C. A., Stock, J. H., & Watson, M. W. (1990). Inference in linear time series models with some unit roots. *Econometrica*, *58*(1), 113–144. doi:10.2307/2938337

KEY TERMS AND DEFINITIONS

Bank: An institution dealing in financial issues, which is licenced to accepts deposits and make loans available to public.

Exchange Rate: A value of someone nation's currency in relation to the currency of another nation or economic zone.

Inflation Linked Bonds: These are securities designed to help protect investors from inflation.

Machine Learning: A science of applying artificial intelligence that provides systems the ability to automatically learn and improve without being explicitly programmed.

Mean Absolute Error: A measure of average magnitude of the errors in predictions without giving preference to their direction.

Portfolio Inflows: An influx of a group of financial assets as well as their equivalent funds.

Real Gross Domestic Product: A macroeconomic measure of country's total economic output taking into account the impact of inflation.

Chapter 15
Meta-Heuristic Approaches for the Optimization of Hydropower Energy: A Review

Abdus Samad Azad
Universiti Teknologi PETRONAS, Malaysia

Pandian Vasant
Universiti Teknologi PETRONAS, Malaysia

Junzo Watada
Universiti Teknologi PETRONAS, Malaysia

Rajalingam Al Sokkalingam
Universiti Teknologi PETRONAS, Malaysia

ABSTRACT

The concept of a multireservoir systems in hydropower introduces the function of multiple units simultaneously to reach the peak requirements. The reservoir optimized operation is a complex, extremely nonlinear, high dimensional, and multimodal task. The options that can be evaluated manually are generally limited in numbers, which made it difficult to identify the most appropriate option and should be taken into account while making decisions. Presently, for solving the optimization problems in multireservoir system, many modern heuristic stochastic search algorithms were established. That is possible because of the aspects of artificial and computational intelligence technologies. By connecting metaheuristic algorithms, the decision options can be identified to make the most suitable utilization of the scarce resources, best natural results for a given allotment can be attained, and the best trade-offs between contending goals can be established. In this chapter, the authors review the latest meta-heuristic optimization technique and their applications to maximize the economic factors.

DOI: 10.4018/978-1-7998-3645-2.ch015

INTRODUCTION

The power can be classified into two groups, the renewably and non-renewably generated ones as shown in Figure 1(Karaeren, 2014). A sustainable energy source like hydropower is depending on regular water cycle. It is most matured, biggest, dependable and profitable renewable power generation technology which is easily accessible. It creates the world's electricity about 16% of and the world's renewable electricity over 80%. At present, 90% of the electricity supply and more than 25 nations in the world rely upon hydropower.(Irena, 2012). Hydropower technology is flexible in meeting peak load demand with both higher and lower installed capacities. Although hydropower is considered the most economical power source but it is not true in case of multireservoir system by reason of the operation of many units at one time. Peak demand may vary from time to time and to meet this demand other alternative energy sources, for instance, the more valuable nuclear or thermal energy, have to be used. It increases the further cost of power generation. Hence It is needed to maximize the power generation by each operating unit while minimizing the use of other energy sources, as, in turn, the total operational cost will minimize (Srianan & Sangsawang, 2019). The plant operations can help in obtaining following benefits:

1. The settings of separate units.
2. The co-ordination of multiple unit operations.
3. Release patterns from multiple reservoirs.

The energy price, power production and power generated through the alternative source of energy are the most significant determinants of revenue in the hydropower system and need to be optimized (Personal, Archive, & Fleten, 2010). The method of finding the optimum solution of an objective equation under the defined constraint is known as optimization (Zelinka, 2019). It targets to maximize or minimize the objective function (Edenhofer et al., 2012).

Latterly, several novel meta-heuristic approaches were presented for high dimensional optimization problems. These techniques are promised for solving difficult real-world optimization problems, like engineering design problems

(Suresh & Lal, 2016, Naveen, Chandel, Vedik, & Topwal, 2016; Suresh & Lal, 2017). The NFL (No Free Lunch theorem) demonstrates that all optimization problems cannot be solved by a single meta-heuristic algorithm (Fister, Yang, Brest, & Fister, 2013). It has been considered as foundation of many studies in the field of optimization. Some widely-known meta-heuristic algorithms are Genetic Algorithm (GA) (Snaselova & Zboril, 2015), Particle Swarm Optimization (PSO) (Xin-She Yang, n.d.) and Elephant herding Optimization (EHO) (Wang, 2015). Monarch butterfly Optimization (MBO) (Wang, Deb, & Cui, 2015), Kidney-inspired Algorithm (KA) (Jaddi, Alvankarian, & Abdullah, 2016), Bat Algorithm (BA) (X. Yang, 2014), Firefly Algorithm (FA) (X. Yang, 2010), Krill Herd (KH) (Gandomi & Alavi, 2012) and Crow Search Algorithm (CSA) (Askarzadeh, 2016). Nonlinear dynamic techniques, exclusively chaos, can attract more consideration in many fields (Askarzadeh, 2016; Pan & Xu, 2016) high dimensional optimization problems. It is also observed that these techniques have some limitations in particular dimensionality problem, requirements of vast memory and lack of ability of handling nonlinear cost function. Hybrid methods combined with two or more methods were presented to get control of these problems (Mo, Lu, Wang, & Zhou, 2013).

The metaheuristic algorithms (MAs) have shown the capability of dealing with extremely nonlinear, high dimension and multimodal tasks (Pan & Xu, 2016, X. Yang & Press, 2010) . They can provide

Figure 1. Energy Resources (Karaeren, 2014)

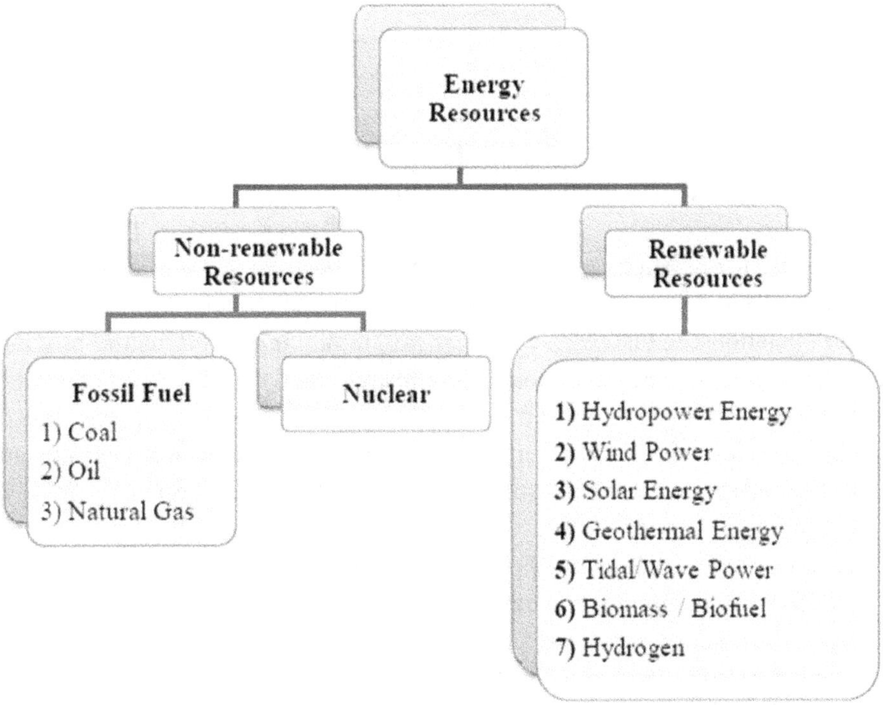

global or near-global solutions by using its few features. In spite of that NFL theorem (Fister et al., 2013, X. Yang & Press, 2010) state that no single meta-heuristic algorithm is not able to resolve all optimization problems. Otherwise speaking an algorithm performing well for a bunch of problems may perform poorer in set of different problems. Hence developing a modified algorithm is a real challenge. Therefore, few strategies have been established for obtaining worthwhile performance. Also, problem related customization of algorithm is very important task. Random matrices were developed to specifically model complex systems which contain universality - i.e. large and complex systems with highly interconnected components (Ganesan, Vasant, Thomas, Litvinchev, & Bank, 2020). This review includes the discussion on areas such as algorithms in optimization, multi-reservoirs and recent development in algorithms. It will help in identifying the gaps in this domain of research work and identify the future paths of research in optimization techniques.

MOTIVATION

The authors are motivated to do this work due to fast growing trend towards the development of computational techniques in handling multiple complex objective functions. Previously traditional methods undergoing large computational times high convergence rate and falling in local optima problems were not able to handle the hydropower optimization problem. These short comings in previous methods inspired the author to work on hybrid algorithms handling multi-objective functions. Combining two or more algorithms may offer increasing efficiency and reducing convergence time, which is motivation for this work.

BACKGROUND

Development in Algorithms and Optimization

First research associated to operations of single-reservoir had been carried out during 1980. For instance, in 1985 Yeh completed a look at to study the trendy mathematical models. The models had been developed for reservoir operations. In that research, linear-programming (LP), dynamic-programming (DP), nonlinear-programming (NLP) and simulations were illustrated and the necessities of using those models and the difficulties that scientists and engineers face on the optimization system of reservoirs. Yeh achieved this unique literature review of usual mathematical optimization models and their characteristics and stated the superiority and some future work suggestions for each optimization methodology(Yeh, 1985).

During 1990, Simonovic carried out a research to illustrate the knowledge gap in mathematical representation for reservoir optimizations in 1992. The closing of gap was depicted by two examples. The first instance was described respecting to a model for reservoir sizing using the method of simple simulation optimization. The other example was single multipurpose reservoir analysis which illustrated the benefits of knowledge-based technology(Simonovic, 1992). In 1993 Wurbs found out reservoir system evaluation models efficiently by deciding which techniques were almost all suitable in different kinds of decision making conditions (Wurbs, Ralph A., Member, 1993).

During 2000, in 2003 Barros et al. accomplished a research, in which 75 hydropower plants having capacity of 69,375 MW were involved in developing NLP model for the Brazilian hydropower system. The model was resolved by utilizing previous method which was linearized also resolved with linear programming (LP) and successive linear programming (SLP). Distinguishing between three methods revealed that SLP reached the desired convergence in shorter time while LP was a good way to resolve the problem. However, in real time operation, NLP was declared most accurate and suitable programming. Finally, it has being agreed that the nonlinear programming model fulfilled the demand and supply more energy and the NLP model was stated as the most useful technique for real-time operation(Mario T.L. Barros, Tsai, Yang, Lopes, & Yeh, 2003). In 2004, the research goal of Wang et al. was to boom the energy production of large scale hydropower system by developing a short-term hydro generation optimization model. There were two types of module one was cycling module and other was transition module. Discharging delay releases and end-of-study content of each reservoir had been covered by the cycling module. On the other hand, the transition module determined the optimal generation scheduling. Additionally, a DSP (direct search procedure) was stated also in the research. It was asserted that the DSP can get control the non-linearity and a large number of reservoirs. The engineers of Fujian Electric Power Company Ltd. (FEPCL) planned the hourly or half-hourly generation by using the DSP model (Chun Tian Cheng, Wang, Xu, & Chau, 2008). In 2005, Barros et al. carried out an examine to compare dissimilar objective functions for the optimization of complex hydropower systems. Objective function can directly influence the operation systems and should be cautiously defined. Therefore, six inconsistent functions were analyzed in this research they were minimizing the stored potential energy's loss, storage deviations from targets, spilled energy and energy complementation, while maximizing the production of energy and secondary energy's profit. In view of results, minimizing the complementation of energy was selected as the quality objective function for the studied hydropower system; It targeted on minimizing the use of alternative energy resources (M T L Barros, Ros, & Lopes, 2005).

The aim of the research of Cheng et al. (2008) was to illustrate that to reach optimal solutions Chaos Genetic Algorithm (CGA) was more effective in its application as compared to Genetic Algorithm (GA). In the research, CGA was used to overcome premature local optimum and the premature convergence problem of GA, in addition to how to aggravate the convergence speed. CGA desegregated the powerful global and local searching capability of GA and Chaos Optimization Algorithm (COA) respectively. In respect of their study, they proposed a result that CGA made better convergence speed and solution accuracy (Chun Tian Cheng et al., 2008). The goal of Yoo's study (2009) was to generate a linear objective function as a replacement of linear function to maximize hydropower energy production.

The operation of Yong dam Dam and HEPP was optimized by linear objective function model which has been explored also. Although the probable maximum energy generation was 214 GWh/per annum from the HEPP; it can be summarized that this model could generate 184GWh annual energy (Yoo, 2009). In Sulek's research (2012) a novel method was proposed for solving hydrothermal coordination problem named as hybrid optimization method which solved the hydro sub-problem. The combination of the traditional numerical methods and Genetic Algorithms were in hybrid optimization method. The research implemented on the Slovak power system and the outcome expressed the performance of hybrid optimization method. But in the opinion of same consequences, the given method required so much hardware equipment and also the execution time, so not considered as quick as required (Šulek, 2012).

Lu et al. (2013) accomplished a research for the Zhelin reservoir to find an optimization technique. The new method was needed to take into account the decrease in power production as a result of reduced rainfalls and water level. For this purpose, three different methods e.g. progressive optimization algorithm (POA), particle swarm optimization (PSO) and genetic algorithm (GA) were applied. Additionally, objective functions for minimization of water consumption rate were chosen. In the study of Zhang et al. (2013), the application of a novel method in short term joint optimal operation of Qing River cascade hydropower stations using improved Adaptive Genetic Algorithm (AGA) was revealed. The results acquired from the research provided that the improved Adaptive Genetic Algorithm (AGA) ought to produce more incredible solutions inside the same algebra. Besides, it could also be comprehended that the profit of maximum power generation was unequal to the amount of maximum power generation(K. Yang, Zheng, Yang, Zhou, & Liu, 2013). Yang et al. (2013) used different techniques for the premature convergence problem for the characteristics of cascade reservoirs optimization of the Genetic Algorithm. To generate solution space generation for producing the applicable initial population was the first strategy. After that, to find chaos optimization for optimizing the initial population was second strategy. Thirdly, suggesting new selective, trigonometric operators to sustain population variety was third strategy. Eventually, the last strategy was to admit the applicable possibility of crossing and mutation to promote the performance of convergence speed (K. Yang et al., 2013).

Development in Algorithms for Multi-reservoirs

First research associated with multi-reservoir operations have been performed at the end of the 20th century. With illustrations, Wardlaw et al. (1999) performed a research to reveal that accurate solutions were obtained by use of the Genetic Algorithm (GA) technique. It has followed from the research that the GA application might be implemented to a big finite horizon in multi-reservoir system. Furthermore, in this research, the advantages of the GA technique were revealed. The advantages of the GA technique provided nearly optimal solution and accurate results for a longer time frame (Millia & Delhi, 1999). Hınçal et al. carried out a research in 2010 to symbolize the performance of the GA for optimal solution of multi reservoirs. In this research a computer code named as Fortran Programming language was developed. Three reservoirs were used for the application of the code including the Morrow Point, Blue Mesa and the Crystal Reservoirs in the Colorado River. Comparison was made between operational data and achieved outcomes. The results proved GA as an efficient optimization tool (Hinçal, Altan-Sakarya, & Ger, 2011). Wu and Chen (2013) conducted a study to achieve the goal of fulfilling variable water demands d/s of a reservoir and optimize the dam to make sure its efficiency while the economic value of the dam was maximized. The hydrological simulation with the aid of the soil and water assessment tool (SWAT) were hired in attaining the objectives. The optimization tool used by the evolutionary algorithm had name as SCE. The data utilized was discovered for the facts that were reservoir outflows, inflows, water stages and hydropower generation. 22 years of data was at hand. By evaluating the simulation results of SWAT model for better crop water demand, Irrigation demand was determined. To improve the generation of hydropower of the reservoir the model was used.

By ensuring availability of additional water while assisting in optimized irrigation release in the reservoir, SWAT was used in reservoir operation. Precipitation, evapotranspiration, and inflow the components of hydrological were stochastically formulated using simulation models. Long time-series were produced subsequently.

After that, the parameterization of the operating regulations of the reservoir was done. To determine the management policy hydro system was simulated using the time series generated stochastically. The expected performance metric was done by utilizing describing significant objective functions. Three multi-objective surrogate-based optimizations (MOSBO) and two multi-objective evolutionary algorithm (MOEA)) were used for the optimization of two objective functions. At the end, by using performance indicators and methods the benchmarking of all the algorithms was done. Additionally, in the reservoir operation model the algorithms performed very well. Therefore, to solve a generated multi-objective reservoir model progressive optimality algorithm e dynamic programming successive approximation (POA-DPSA) was used. A long series of data were estimated below the gains of the reservoirs. To summarize, hydropower generation can be improved by joining operation of two reservoirs comparing to operating the reservoirs separately.

Another research was carried out to optimize the spillways operation on the three Gorges reservoir. The calculation of flood and design flood hydrograph was occurred by using progressive optimality algorithm for finding optimized spillways operation of a multi-reservoir system with a multidimensional DP model. For individual reservoirs there were many constraints and uncounted for the entire reservoir system that were possible to be solved by the two algorithms in the absence of any obstacle (Feng et al., 2018,Tospornsampan & Kita, 2005, Talbi, 2016). Additionally, Hakimi-Asiabaret al. (2010) presented a function that was resolved previously by using SOM-based multiobjective GA (SBMOGA) that overwhelmed current multi-objective optimization techniques (NSGA-II) (Hakimi-Asiabar et al., 2009). The performance of SBMOGA was limited by the SOM neuron number. Tangible developments were made in solution diversity, convergence rate, qualitative solution and running time by using SLGA. It illustrated that SLGA was able to solve large scale multi-reservoir optimization problems (Talbi, 2016, Kao & Zahara, 2008).

Moreover, Tsoukalas and Makropoulos (2015) conducted a study to show that large decision variables were there to be solved for multi-reservoir systems with the optimization of long-frame operating rules. That made the method complex. The objective functions were subjected to hydrological uncertainty and also nonlinear. To solve the stochasticity of the parameters, the solution was usually a coupling simulation model to the optimization model. Their study uses multi-objective surrogate-based optimization (MOSBO) to reduce the computational effort. Three MOSBO algorithms with two multi-objective evolutionary algorithms(EAs) was compared in that research. It was found that MOSBO performed better resulting in robust and faster uncertainty-aware operating rules. Tsoukalas and Makropoulos (2015) narrated the PSO model for hydrological uncertainty using stochastic simulation and developed reservoir operating rules which were aware of uncertainty. The confliction of multiple objectives was handled by multi-objective model. The results had an adjustable tool for helping in decision making to the reservoir water users (Tsoukalas & Makropoulos, 2015).

Rajabioun (2011) conducted a study that in the same approach, the cuckoo optimization algorithm (COA) which was inspired by lifestyle of cuckoo bird could solve nonlinear optimization. To illustrate, Atashpaz-Gargari and Lucas conducted a study in 2007, the imperialist competitive algorithm (ICA) was presented and showed population by population technique same as other Evolutionary algorithms. The algorithms were relatively new and were successfully verified on various basic problems. The results obtained were very competitive comparing to GA and NLP in the reservoir operation. The results generated by COA were better than ICA. Besides, ICA had the convergence ability to a global optimum. Following the same pattern, an improved result using a real-world reservoir system was introduced by Al-jawad et al. in 2017. Borg MOEA performed well to minimize the imbalances of total annual supply and demand using their EAs (Talbi, 2016, Kao & Zahara, 2008, Jitkongchuen & Initialization, 2015).

Many EA had shown efficient and produced competitive results for reservoir operation. Either single reservoir or multi-reservoirs, evolutionary algorithms were suitable for complex reservoir operation problems. In series or parallel, they were adapted well for different reservoir arrangements. Comparing to classical algorithms evolutionary algorithms had shown the superiority to able to provide abundant solutions in single simulation run. For this reason, the algorithms were recommended for difficult water resources problems. Yu Feng et al, 2018 used Global climate models (GCM) to find out the future climate changes impacts on cascade reservoir, the Xinjiang model (XAJ) was used to show the streamflow and then gravitational search algorithm (GSA) was taken to solve the LTHG problem (Feng et al., 2018).

Hydropower System

The coordinated operation of a multiple reservoir system for efficient management of available water to maximize the total energy production or minimize the total cost of the system, while satisfying various constraints, is a complex decision-making process (Tospornsampan & Kita, 2005). It is a mathematical optimization problem that can identify how to optimally manage one or more hydropower units over a specified time horizon. The evolutionary algorithms can produce satisfying solutions in op, regardless of the problem features (like continuity or no convexity). However, due to the premature convergence, evolutionary algorithms often fall into local optima, which have limited their widespread applications in practical engineering. Hence, it is necessary to find some effective modified strategies to enhance the performances of the evolutionary algorithms.

As metaheuristic algorithms (MAs) have exhibited promising performance when dealing with complex tasks that are extremely nonlinear, high dimension and multimodal (Mario T L Barros et al., 2003, Chun-tian Tian Cheng, Wang, Xu, & Chau, 2008). A combination of two or more algorithms is proposed in the literature as an effective strategy for hydropower operation problems. In this study, a hybrid approach is recommended to optimize the power generation while maintaining the cost and other energy sources. Grey Wolf Optimizer (GWO) is a new metaheuristic optimization. Since the GWO algorithm is easy to fall into local optimum especially when it is used in the high-dimensional data, an improved GWO-CS algorithm and GWS-CSA are proposed in this paper. These algorithms can actually tackle a quite general class of dynamic optimization problems, including the ones with nonlinear constraints.

Table 1. Hybrid metaheuristic techniques in solving optimization problem

Author	Year	Hybrid CI Techniques	Remarks
S. Arora et al.	2019	BBO-GWO	A hybrid algorithm based on GWO-BBA Algorithm was proposed for function optimization and feature selection problem. The limitations of GWO were minimized by hybridizing it with BBA.
A. Ella et al.	2018	CSA - RSS	Proposed algorithm provide initial optimal solution of optimization problem. In the study RSS phase was introduced as second stage to enhance the exploitation search and faster converge to the optimal solution.
Prashant J. et al.	2018	GWO – ABC	The proposed algorithm was applied to design an optimal fractional order PID (FOPID) controllers for variety of typical benchmark complex transfer functions and trajectory tracking problem of 2 degree-of-freedom (DOF) robotic manipulator. All simulation results, illustrations, and comparative analysis establish the GWO-ABC as viable alternative to design a controller with optimal parameters and enhance the performance of complex systems.
M. A. Tawhid et al.	2017	GWO - GA	Presented algorithm was optimistic, proficient, and have the ability to find global minimum or near global minimum of the molecular energy function. The proposed algorithm faster than its comparative algorithms.
N. Singh et al.	2017	GWO-PSO	The main idea was to improve the ability of exploitation in PSO with the ability of exploration in GWO to produce both variants' strength. The numerical and statistical solutions had shown that the hybrid variant outperforms significantly the PSO and GWO variants in terms of solution quality, solution stability, convergence speed, and ability to find the global optimum.
E. Nabil	2016	FPA-CSA	The proposed algorithm was able to find more accurate solutions than the standard FPA and the other techniques. The superiority of the proposed algorithm nominates it for being a part of intelligent and expert systems.
T. Jayabarathi et al.	2016	GWO-Crossover and Mutation Operators of GA	Given paper presented the application of the HGWO algorithm for solving the economic dispatch problem that is nonlinear, non-convex and discontinuous in nature, with numerous equality and inequality constraints. Four economic dispatch problems with prohibited operating zones, valve point effect and ramp rate limits were solved with and without transmission losses by HGWO.
D. Jitkongchuen et al.	2015	GWO – HDE	The proposed algorithm introduces a new improved mutation schemes. The performance of the well-known algorithm was proposed and evaluated on nine benchmark functions and it was compared to particle swarm optimization, the traditional differential evolution algorithm and the self-adaptive differential evolution algorithm besides it performed effectively to solving complex optimization problems.
Y. Kao et al.	2008	GA - PSO	An efficient approach was taken for solving global optimization of multimodal functions. 17 multimodal test functions were shown which described the superiority of the hybrid GA-PSO approach over the other search techniques in terms of solution quality and convergence rates.
J. Tsai et al.	2004	GA-Taguchi method	The proposed HTGA was effectively applied to solve 15 benchmark problems of global optimization with 30 or 100 dimensions and very large numbers of local minima. The computational experiments show that the proposed HTGA not only can find optimal or close-to-optimal solutions but also can obtain both better and more robust results. Therefore, it possesses the merits of global exploration, fast convergence, robustness, and statistical soundness.

Discussions

During last few decades hybrid algorithms had been center of attention for researchers to solve optimization problems. Furthermore, these algorithms showed better performance in solving complex problems compared to their counterparts (Talbi, 2016). In (Kao & Zahara, 2008), the authors had hybridized GA-PSO and had utilized the hybridization to generate individuals using crossover and mutation operators and also added global and local search operators of PSO. After that, the Taguchi method was also hybridized with GA for global numerical optimization (Kao & Zahara, 2008). The Taguchi method was added as an extra step in GA to obtain better genes and performance. In (Jitkongchuen & Initialization, 2015), for solving continuous global optimization problems, hybrid of grey wolf and hybrid differential evolution algorithm was developed. The conventional FPA solution was improved by developing a hybrid algorithm of clonal selection algorithm (CSA) and flower pollination algorithm (FPA) by Nabil (Nabil, 2016). Various benchmark test problems were solved by the proposed hybrid algorithm.

Tawhid et al. proposed a hybrid algorithm of GWO and GA to minimize the energy function (Tawhid & Ali, 2017). Jayabarathi et al. (Jayabarathi, Raghunathan, Adarsh, & Nagaratnam, 2016) developed the hybridization of GWO and crossover added with mutation operators of GA to solve economic dispatch problem. In order to improve the performance in another research, the exploration capability of GWO and exploitation ability of PSO was hybridized (Singh & Singh, 2017). The performance of complicated systems was enhanced using GWO and artificial bee colony hybrid and applied in parameters optimization (Singh & Singh, 2017). In (Arora, Singh, Sharma, Sharma, & Anand, 2019), hybridization of BBO and GWO were able to maintain exploration and exploitation and obtained enhanced performance comparing to individual performance of this two algorithms. Hassanien et al. hybridized CSA and rough searching scheme. Performance of CSA was improved and information of global optimal solution was handled smoothly. The studies proved that the hybrid algorithms are superior search algorithm. In the similar area, hybridization of ant colony and GA was also applied (Talbi, 2016) and cuckoo search (Kao & Zahara, 2008) were proposed. Furthermore, hybridization of DE and ABC (Jitkongchuen & Initialization, 2015), WOA and SA were employed for feature selection (Nabil, 2016). Presently, a technique was developed on grasshopper optimization algorithm (GHO) (Tawhid & Ali, 2017) and evolutionary population dynamics (EPD) (Jayabarathi et al., 2016) and applied for selecting the optimal solution set (Singh & Singh, 2017).

Objective Function

The selection of objective function plays vital role in optimized operation of reservoir system. Reservoir operation is a complex system and needs to be optimized using different objective funcions.

For optimizing complex hydropower operation systems different objective functions will be used. Linear or Nonlinear methods can be used for the optimization purpose.In this study non-linear method is selected to gain valid results.

A schematic diagram of hydropower system is shown in Figure 2. Where

H = Reservoir upstream water level;

HT = Water level at tailrace of reservoir and is function of $\left(R' + R''\right)$.

H – HT = Hb (*head*).

Figure 2. Schematic diagram of Hydropower System

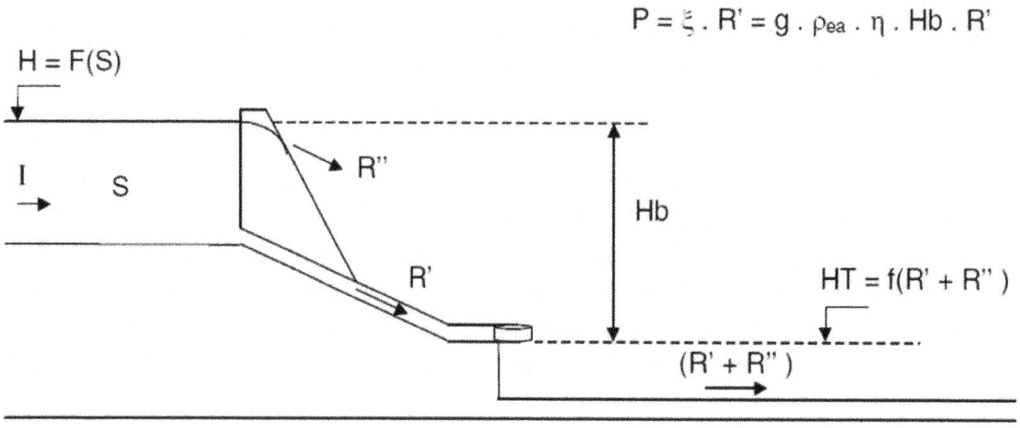

I = inflow
S = Storage in reservoir and it is a function of inflow
R' = Power release
R'' = Non-power release
P = Power = $R' * \xi$
ξ = Energy production function

Production of Energy is Given as Following

$$P = g \cdot \rho_{ea} \cdot \eta \cdot Hb \cdot R' \tag{1}$$

$$P = 9.81 \cdot 10^{-3} \cdot Hb \cdot R'$$

Where;

P = power (MW)
g = gravity (m/s^2)
ρ_{ea} = water specific weight (kg/cm^3)
R' = power release (m^3/s)
Hb = Head (m)
η = Efficiency of turbine.

The Energy Production Function is Given as Following

$$\xi = 9.81 \cdot 10^{-3} \cdot \eta \cdot Hb \tag{2}$$

So $P = \xi \cdot R'$

A schematic diagram of mathematical hydropower system is shown in Figure 3. The system is linked by spatial and time variations.

A general model of optimization can be represented as following:

{f(x)} = max or min
Constraints: g_1 (x) £ b_1
g_2 (x) ^{3} b_ (2)
g_3 (x) = b_3

where

x = Decision variables
f(x) = Objective function to be optimized
g_1 (x), g_2 (x) and g_3 (x) = Constraints
B= (b_1, b_2 and b_3) = Resource vector.

Figure 3. Hydro system Representation

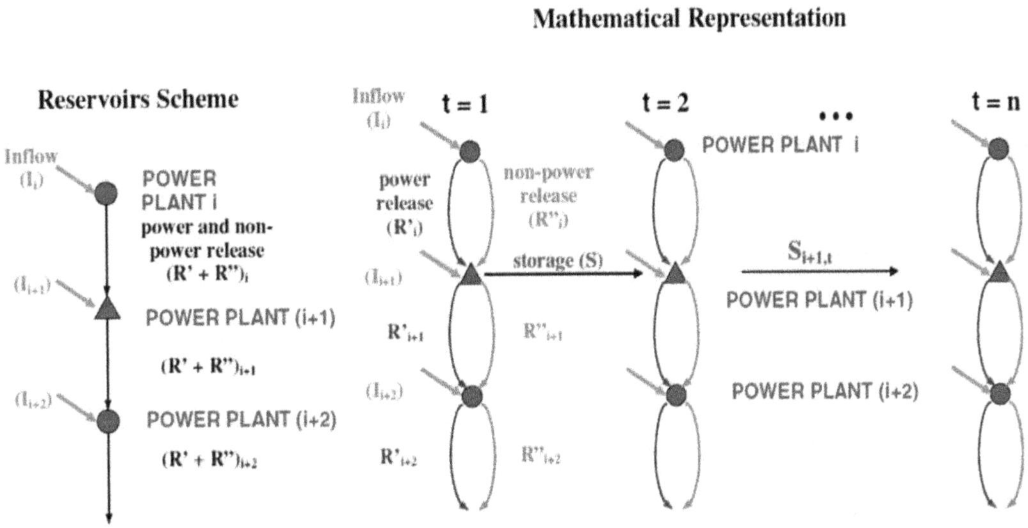

Objective Function 1 (FO1)

Maximization of total energy production:

$$\max f_1 = \sum_t \sum_i \left(\xi_{i,t} R^{'}_{i,t} \right) \tag{3}$$

Where $\xi_{i,t}$ = Function of energy production at the ith reservoir and at the end of time period t [$MW/(m^3/s)$].

Objective Function 2 (FO2)

Minimization of energy complementation:

$$\min f_2 = \sum_t \left[D_T - \sum_i \left(\xi_{i,t} R'_{i,t} \right) \right]^2 \tag{4}$$

where D_T = Total energy demand (avg/MW) including hydropower and alternative sources of energy at the end of time period t, in order to obtain the minimized total operational cost.

Objective Function 3 (FO3)

Maximization of profit obtained from secondary energy.

$$\max f_3 = \sum_t p_t \left(\sum_i \xi_{i,t} R'_{i,t} - D_C \right) \tag{5}$$

Where

p_t = Price of energy (avg*MW*) during time period t
D_C = System firm energy (avgMW) during time period t

Firm energy is known as the average long term produced energy having a risk of 5%. Secondary energy is the energy produced above firm energy.

Combined Objective Function:

Combined objective function is represented as follows:

$$\min Z = -w_1 f_1 + w_2 f_2 - w_3 f_3 \tag{6}$$

The objective functions are weighed by the coefficients ($w's$) depending on the importance of the objectives.

Constraints:

The following constraints are considered:

1) *Energy demand*:

$$\sum_i \xi_{i,t} R^{'}_{i,t} \geq d_t, \forall t \tag{7}$$

2) *T urbine capacity*:

$$\xi_{i,t} R^{'}_{i,t} \leq \overline{P_i}, \forall i, \forall t \tag{8}$$

3) *Flow continuity*:

$$S_{i,t} = S_{i,t-1} + \lambda I_{i,t} + \lambda \sum_{i \in IN} \left(R^{'}_{i,t} + R^{''}_{i,t} \right) - \lambda \left(R^{'}_{i,t} + R^{''}_{i,t} \right) \tag{9}$$

4) *Storage variation (Monthly)*:

$$S_{i,t} - S_{i,t+1} \leq \delta S_{i,t}, 0 \leq \delta \leq 1, \forall i, \forall t \tag{10}$$

1) *Minimum and maximum storages*:

$$S_i^{min} \leq S_{i,t} \leq S_i^{max}, \forall i, \forall t \tag{11}$$

2) *Minimum and maximum power releases*:

$$R_i^{'min} \leq R^{'}_{i,t} \leq R_i^{'max}, \forall i, \forall t \tag{12}$$

3) *Constraints on non-power release*:

$$0 \leq R^{''}_{i,t} \leq \infty, \forall i, \forall t \tag{13}$$

Where

$I_{i,t}$ = Inflow into the ith reservoir at the end of time period t (m^3/s)
$S_{i,t-1}$ = Initial storage into the ith reservoir (hm^3)
d_t = Demand of energy during time period t (*MW/month*)
$\overline{P_i}$ = Capacity of turbine for i^{th} power plant, the period of maintenance of turbine is also included (*MW*)
λ = Conversion factor

The decision variables in the nonlinear programming model are $R^{'}_{i,t}$, $S_{i,t}$ and $R^{''}_{i,t}$. The energy production function, $\xi_{i,t}$ can be expressed as follows:

$$\xi_{i,t} = \varepsilon_i \left(H_{i,t} - HT_{i,t} \right) \tag{14}$$

$$H_{i,t} = a_{0i} + a_{1i} S_{i,t} + a_{2i} S_{i,t}^2 + a_{3i} S_{i,t}^3 + a_{4i} S_{i,t}^4 \tag{15}$$

$$HT_{i,t} = b_{0i} + b_{1i} q_{i,t} + b_{2i} q_{i,t}^2 + b_{3i} q_{i,t}^3 + b_{4i} q_{i,t}^4 \tag{16}$$

Where

$H_{i,t}$ = i^{th} reservoir upstream water level (m)
$HT_{i,t}$ = i^{th} reservoir water level at tailrace (m)
$q_{i,t}$ = Total outflow (m^3/s)
ε_i = Producibility ($MW/(m^3/s)/m$)
$a_{0i}, a_{1i}, a_{2i}, a_{3i}, a_{4i}$ = polynomial coefficients of order four for the upstream water level equation
$b_{0i}, b_{1i}, b_{2i}, b_{3i}, b_{4i}$ = Polynomial coefficients of order four for water level equation at tailrace.

METHODOLOGY

The GWO Algorithm

Overview

Grey wolf optimizer (GWO) which based on meta-heuristics algorithm is a population based simulates. Mirjalili et al. was proposed this algorithm in 2014 which replicated to gray wolves in nature. The algorithm transcribed the leadership hierarchy and hunting mechanism of gray wolves. On account of gray wolves' top of the food chain, they are considered as apex predators. In average, grey wolves preferred to live 5-12 members in a groups (packs). A very strict social dominant hierarchy consists of four levels followed by all the members of the group. Alpha (α) is the first level who are the leaders of the pack. There are two groups one is male and the other one is female. The make decisions for hunting, when they have to walk, where they have to sleep and so forth. The group members have been dictated by the alpha's orders and by holding their tails down they respect the alpha. Beta (β) is in the second level. The beta works as supporting wolves that assist alpha in decision making. It may be either male or female. When the alpha passed away or be transformed into very old, the beta wolves judged as the best candidate. Moreover, it gives feedback to alpha by reinforcing alpha's commands throughout the pack. Delta (δ) is in third position that are not alpha or beta wolves as they called as subordinates. They submit to the alpha and beta but they dominate the omega (the lowest level in wolves' social hierarchy). There are different categories of delta as follows:

- **Scouts:** In case of any danger the scout wolves are responsible for observing the boundaries of the territory and warning the groups.
- **Sentinels:** For protecting the groups sentinel wolves are accountable for.
- **Elders:** The experienced wolves which used to be either alpha or beta are called elder.

- **Hunters:** The hunters' wolves look after for helping the alpha and beta wolves' in hunting and providing food for the groups.
- **Caretakers:** For taking care of the ill, weak and wounded wolves in the group's caretakers are responsible.

The final (lowest) level is called omega (ω). Omega wolves are evaluated as the scapegoat in the groups, they have to submit to all the other dominant wolves. They are the last allowed wolves to eat which seemed are not important in the groups. The entire groups fight in order to avoid the lose of omega.

Mathematical Model

The level determined the fitness function. The alpha_wolf, the beta_wolf, and the delta_wolf can be considered as the best fitness solution, according to the fitness value.

The fitness function is being determined by the level. The alpha_wolf, the beta_wolf, and the delta_wolf can be considered as the best fitness solution, according to the fitness value.

The grey wolves encircle the prey throughout the time of hunts. Following equations represented the encircling behavior in mathematically:

i. Encircling prey

$$\vec{D} = \left|\vec{C} \cdot \vec{X_p}(t) - \vec{X}(t)\right| \tag{17}$$

$$\vec{X}(t+1) = \vec{X_p}(t) - \vec{A} \cdot \vec{D} \tag{18}$$

Where current iteration is represented by t, the coefficient vectors as A and C, the position vector of the prey is X_p and the position vector of a grey wolf is presented by X. Furthermore, next iteration is being represented by $t+1$. The $\vec{A}$ and $\vec{C}$ are calculated as follows:

$$\vec{A} = 2\vec{a} \cdot \vec{r_1} - \vec{a} \tag{19}$$

$$\vec{C} = 2 \cdot \vec{r_2} \tag{20}$$

Here $\vec{r_1}$ and $\vec{r_2}$ are random number in [0,1] and the components of Vector $\vec{a}$ are linearly dropped from 2 to 0 over the course of iterations.

ii. Hunting prey

The alpha generally leads the way of hunting operation. The participation of beta and delta in hunting maybe on occasions. The hunting behavior of grey wolves can be described in the mathematical model. It can assume like that the alpha, beta and delta have better knowledge about the potential location of prey. Three leading solutions are saved and the others are obliging to update their positions as maintained by the position of the best search agents as shown in the following equations:

$$\vec{X}(t+1)=\frac{\vec{X_1}+\vec{X_2}+\vec{X_3}}{3} \tag{21}$$

$$\vec{X_1}=\left|\vec{X_\alpha}-\vec{A_1}\cdot\vec{D_\alpha}\right| \tag{22}$$

$$\vec{X_2}=\left|\vec{X_\beta}-\vec{A_2}\cdot\vec{D_\beta}\right| \tag{23}$$

$$\vec{X_3}=\left|\vec{X_\delta}-\vec{A_3}\cdot\vec{D_\delta}\right| \tag{24}$$

During the iteration process, the best three solutions can be presented as,, (key-group). the following equations are representing other parameters:

$$\vec{D_\alpha}=\left|\vec{C_1}\cdot\vec{X_\alpha}-\vec{X}\right| \tag{25}$$

$$\vec{D_\beta}=\left|\vec{C_2}\cdot\vec{X_\beta}-\vec{X}\right| \tag{26}$$

$$\vec{D_\delta}=\left|\vec{C_3}\cdot\vec{X_\delta}-\vec{X}\right| \tag{27}$$

iii. Attacking prey

Grey wolves generally attack at the time when the prey stops moving. The behavior of grey wolves approaching the prey is modeled by the below equation:

$$A=2-2\left(\frac{t}{Max}\right) \tag{28}$$

Here t represents the running number of times of the current algorithm, where an integer value between 0 and Max (max number of iteration).

Improving the Original GWO Algorithm

Integration With Cuckoo Search

By the unique nesting way of cuckoo birds and levy-flight-style the CS algorithm is inspired. Some cuckoos do not hatch their own by offspring. while the host goes out they lay down their eggs in the host's nest. Besides, they also remove host bird's eggs sometimes. For many animals Levy-flight-style is a typical characteristic of flight behaviors. To sum up, three idealized rules needed to be fit by CS algorithm as follows:

1) Cuckoo birds choose the nest randomly and they only place one egg at once.
2) Only the best nests will remain to be the next generations.
3) The number of bird's nests and the probability that the eggs are discovered are fixed. If an outsider's egg is discovered by the host bird, the host bird will abandon the nest and built a new one.

The nests are updated by complying with these three rules above. The iteration can be represented by following equations:

$$x_i\left(t+1\right)=x_i\left(t\right)+\alpha\oplus Levy\left(\lambda\right),\ i=1,2,\ldots,n \quad (29)$$

The entry-wise multiplication can be represented by $\oplus$. The new solutions for cuckoo i is presented by $x_i(t+1)$. The current solutions is $x_i(t)$. Since the step size can be controlled by $\alpha > 0$, in max situations, it's set to 1. The following equation can describe the levy-flight:

$$Levy \propto u = t^{-\lambda}, \left(1 \le \lambda \le 3\right) \quad (30)$$

Proposed CS-GWO Algorithm for Improvement

It is described from equation (21) that position of the GWO algorithm updates by trend search to individuals along with high fitness values, which is a key-group. If the search path is long or short with almost the same probabilities and the direction is highly random, the CS algorithm updates the positions of nest using random walk and levy-flights. In consequence, the jump from current area to other areas is easier. As stated by this feature of CS, improved GWO algorithm was integrated by the CS algorithm. The flow chart shown in Figure 4

Pseudo Code

The aim of this work is to evaluate the meta-heuristic approach established on the CS-GWO algorithm to generate an utmost solution for the optimization problem of hydropower. Furthermore, analyzing the performance of the CS-GWO algorithm in the optimization of Hydropower system.

```
initializing of the grey wolf position.
initialize a, A, C, P_a Compute the fitness of each search agent in the pack
Set the X_α,X_β,X_δ according to the fitness
t=1
While (t < Max)
        for each wolf
                Update the position by Equation (21)
        end for
        Initialize a, A and C
        Compute the fitness of each search agent in the pack
        Update the X_α,X_β,X_δ
        for X_α,X_β,X_δ
        Update the position by Equation (29)
        If random number > P_a
        Random change wolf's position
Compute the fitness and Update it according to the fitness
t=t+1
end While
Output X_α
```

Crow Search Algorithm

In 2016, Askarzadeh presented a nature-inspired algorithm that mimics the food hiding mechanism of crows. first of all, a crow can be considering as a clever bird that has sharpened brain. It possesses the ability that can be utilized to warn its species about some dangerous situations. In particular, the intelligence of crow is that they can convincingly hide and easily recall its food location. The working of CSA can be determined by four major principles as follows:

1. They live in a flock.
2. Remember the location of the hidden food.
3. Follow another member of their species and finally.
4. Shield their caches from being pilfered stochastically.

On account of its simplicity and efficiency, CSA has been used to solve different problems related to the selection of conductor size, feature selection image segmentation and electromagnetic optimization.

ADVANTAGES AND DISADVANTAGES

Algorithms have advantages and disadvantages. For instance, the genetic algorithm gets trapped by local optimization and finds local optima instead of global optimum. The particle swarm algorithm has an early convergence in solving some problems that cause immature answers instead of the main answer (Ehteram, Karami, & Farzin, 2018). Some of the evolutionary algorithms are slow in convergence and

Figure 4. Proposed flow chart of GWO-CS algorithm

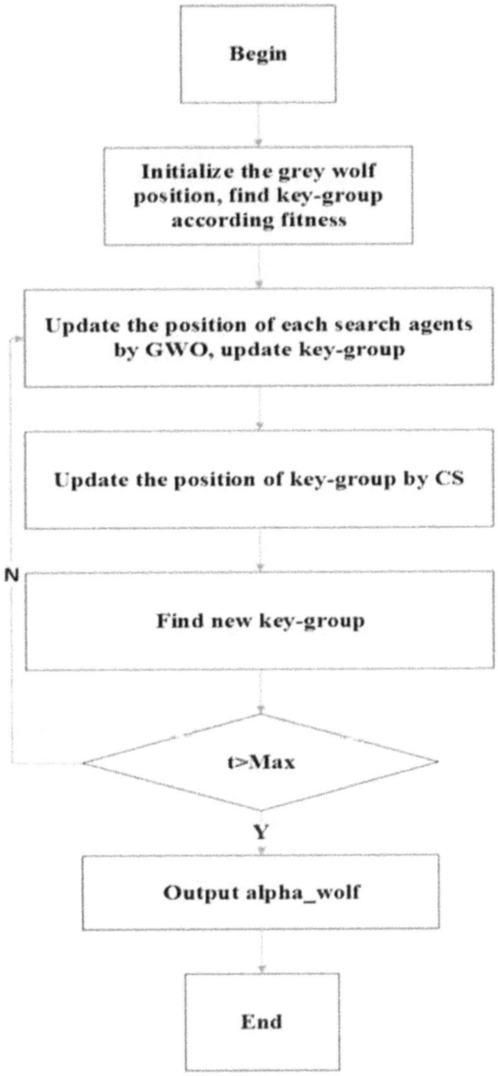

achieving the optimum answer. Also, it is difficult to determine the initial random parameters in other algorithms (Ehteram, Allawi, et al., 2017). Two important features in the evolutionary algorithms are exploration and exploitation capabilities. The exploration capability shows the ability of the algorithm to search freely without considering any achievements during the search process. In contrast, the exploitation is called the algorithm's attention to its achievements during the search process. Obviously, the more the exploration capability is in a search algorithm, the more random and unpredictable behavior can be seen for the algorithm (Ehteram, Karami, Mousavi, Farzin, & Kisi, 2017). On the contrary, enhancing the exploitation in an algorithm causes this algorithm to be more accountable and more cautious. The proposed algorithm has many adherent advantages that can be summarized as follows:

1) It is a free-derivative algorithm;
2) It has lower parameters to be adjusted;

Figure 5. Flow chart of the proposed GWO-CSA algorithm

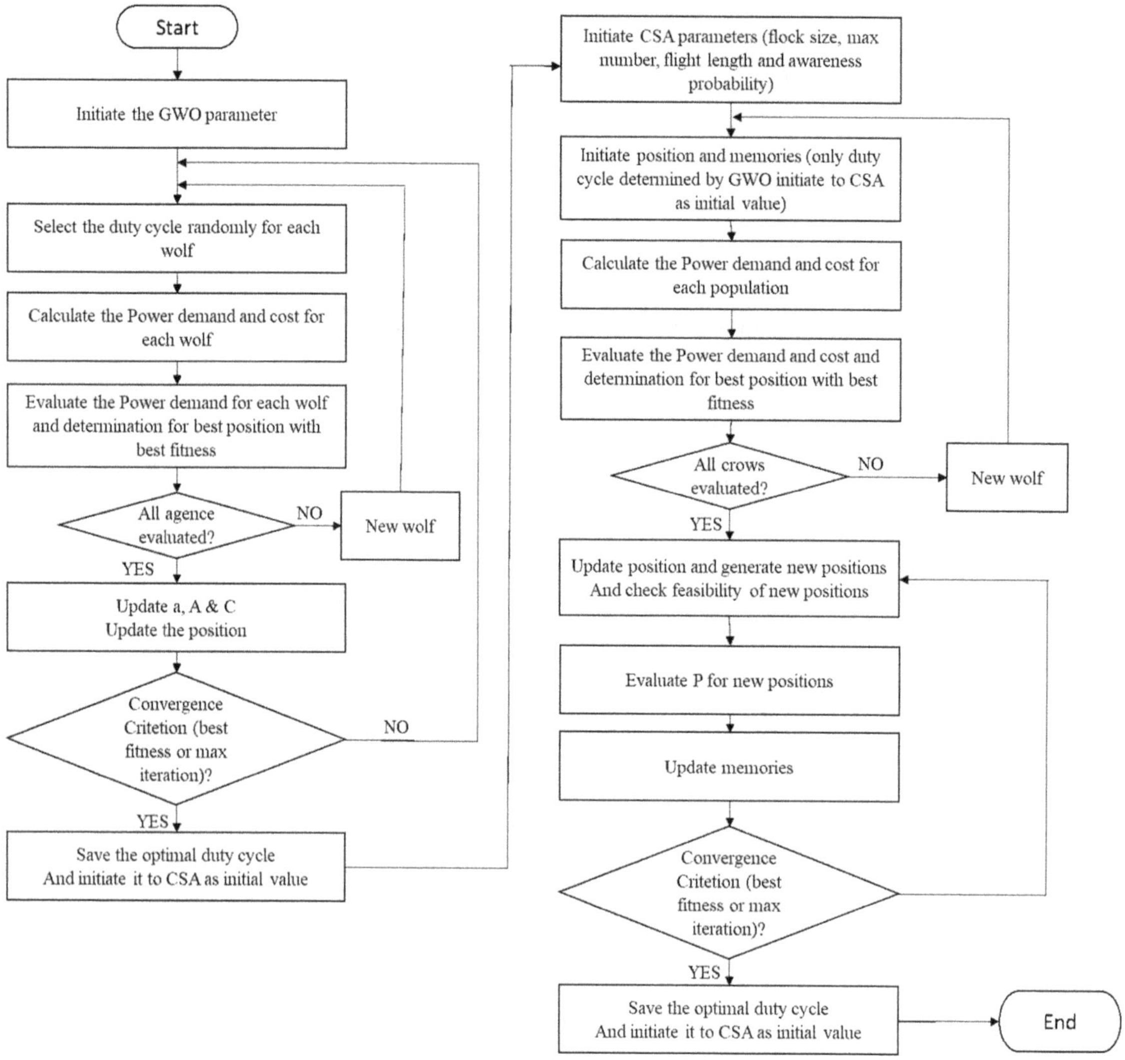

3) It has lower operators compared with other evolutionary algorithms;
4) It can preserve information about the search space with iterations and saves the best solution obtained;
5) It is simple and easy to implement .

LIMITATION OF THE STUDY

For this study, operational limitations should be taken into account e.g., mean global temperatures or mean annual discharges over long time periods, available water resources, water, evaporation rate, storage capacity, turbine efficiency, upstream and downstream water level etc. There is, therefore, a need to establish a balance between several different factors influencing a good path's design (Lehner, Czisch,

& Vassolo, 2005). Actually, there are so many constraints that can impact the efficiency of algorithm, energy production, cost minimization. But in this study, we don't consider everything. we just focused on a few constraints. Above these, there are so many constraints, so many variables which also impact the selection of all uncertainties, long-term average values are considered the most reliable outputs of climate and large-scale hydrological models but here we avoid it. Authors only searched in Scopus and Web of Science database. Authors also recognize that, in future, researchers should consider articles from other databases.

CONCLUSION

The paper constructs the review of a hybrid metaheuristic technique combined with GWO-CS and GWO-CSA. Several applications of hybrid GWO has provided more efficiency than any classical metaheuristic approach. In the future, the proposed CS-GWO algorithm will be applied to domain network security. The application of the algorithm indicates the initial results of experiments and its further applications will be verified by more experimental datasets particularly with multi-dimension datasets. So, the authors are hopeful that this methodology will provide more efficiency in multi-reservoir optimization problem.

FUTURE RESEARCH

The present review study can just incompletely make up for this requirement for future researchers to concentrate in more prominent profundity on the issues brought up in this paper. Our future perspective is to create an extreme solution based on the Hybrid algorithms employed and incorporate the novel operators and techniques in the Series Division Method to enhance its exploration and exploitation capabilities for the solution of optimization problem in hydropower generation.

ACKNOWLEDGMENT

The research leading to this review article has (a) been supported by both of the Graduate Assistantship (GA) (b) Fundamental and Applied Sciences Department (FASD), Universiti Teknologi PETRONAS.

REFERENCES

Arora, S., Singh, H., Sharma, M., Sharma, S., & Anand, P. (2019). A New Hybrid Algorithm Based on Grey Wolf Optimization and Crow Search Algorithm for Unconstrained Function Optimization and Feature Selection. *IEEE Access: Practical Innovations, Open Solutions*, *7*, 26343–26361. doi:10.1109/ACCESS.2019.2897325

Askarzadeh, A. (2016). A novel metaheuristic method for solving constrained engineering optimization problems: Crow search algorithm. *Computers & Structures*, *169*, 1–12. doi:10.1016/j.compstruc.2016.03.001

Barros, M T L, Ros, D. A., & Lopes, J. E. G. (2005). *Objective Functions For Hydropower System Operation*. Academic Press.

Barros, M. T. L., Tsai, F. T.-C., Yang, S., Lopes, J. E. G., & Yeh, W. W.-G. (2003). Optimization of large-scale hydropower system operations. *Journal of Water Resources Planning and Management, 129*(3), 178–188. doi:10.1061/(ASCE)0733-9496(2003)129:3(178)

Cheng, C. T., Wang, W.-C., Xu, D.-M., & Chau, K. W. (2008). Optimizing hydropower reservoir operation using hybrid genetic algorithm and chaos. *Water Resources Management, 22*(7), 895–909. doi:10.100711269-007-9200-1

Cheng, Tian, Wang, Xu, & Chau. (2008). Optimizing hydropower reservoir operation using hybrid genetic algorithm and chaos. *Water Resources Management, 22*(7), 895–909.

Edenhofer, O., Pichs-Madruga, R., Sokona, Y., Seyboth, K., Eickemeier, P., Matschoss, P., … Stechow, C. Von. (2012). Intergovernmental Panel On Climate Change - Summary for Policymakers. IPCC Special Report on Renewable Energy Sources and Climate Change Mitigation. Cambridge University Press.

Edition, S. (n.d.). *Advances in Metaheuristic Algorithms for Optimal Design of Structures*. Academic Press.

Ehteram, M., Allawi, M. F., Karami, H., Mousavi, S. F., Emami, M., El-Shafie, A., & Farzin, S. (2017). Optimization of Chain-Reservoirs' Operation with a New Approach in Artificial Intelligence. *Water Resources Management, 31*(7), 2085–2104. doi:10.100711269-017-1625-6

Ehteram, M., Karami, H., & Farzin, S. (2018). Reducing Irrigation Deficiencies Based Optimizing Model for Multi-Reservoir Systems Utilizing Spider Monkey Algorithm. *Water Resources Management, 32*(7), 2315–2334. doi:10.100711269-018-1931-7

Ehteram, M., Karami, H., Mousavi, S. F., Farzin, S., & Kisi, O. (2017). Optimization of energy management and conversion in the multi-reservoir systems based on evolutionary algorithms. *Journal of Cleaner Production, 168*, 1132–1142. doi:10.1016/j.jclepro.2017.09.099

Feng, Y., Zhou, J., Mo, L., Yuan, Z., Zhang, P., Wu, J., … Wang, Y. (2018). Long-term hydropower generation of cascade reservoirs under future climate changes in Jinsha River in Southwest China. Water, 10(2).

Fister, I., Yang, X. S., Brest, J., & Fister, D. (2013). A brief review of nature-inspired algorithms for optimization. *Elektrotehniski Vestnik/Electrotechnical Review, 80*(3), 116–122.

Gandomi, A. H., & Alavi, A. H. (2012). *Krill Herd Algorithm : A New Bio-Inspired Optimization Algorithm Krill Herd : A New Bio-Inspired Optimization Algorithm*. Academic Press.

Ganesan, T., Vasant, P., Thomas, J., Litvinchev, I., & Bank, R. (2020). *Random Matrix Generators for Optimizing a Fuzzy Biofuel Supply Chain System*. Academic Press.

Hinçal, O., Altan-Sakarya, A. B., & Ger, A. M. (2011). Optimization of Multireservoir Systems by Genetic Algorithm. *Water Resources Management, 25*(5), 1465–1487. doi:10.100711269-010-9755-0

IRENA. (2012). Renewable Energy Technologies: Hydropower. *Renewable Energy Technologies: Cost Analysis Series, 1*(3/5), 44.

Jaddi, N. S., Alvankarian, J., & Abdullah, S. (2016). PT. *Communications in Nonlinear Science and Numerical Simulation.*

Jayabarathi, T., Raghunathan, T., Adarsh, B. R., & Nagaratnam, P. (2016). Economic dispatch using hybrid grey wolf optimizer. *Energy*, *111*, 630–641. doi:10.1016/j.energy.2016.05.105

Jitkongchuen, D., & Initialization, A. (2015). *A Hybrid Differential Evolution with Grey Wolf Optimizer for Continuous Global Optimization*. Academic Press.

Kao, Y., & Zahara, E. (2008). *A hybrid genetic algorithm and particle swarm optimization for multi-modal functions*. Academic Press.

Karaeren, V. (2014). *Maximizing energy generation of a cascade hydropower system*. Middle East Technical University.

Lehner, B., Czisch, G., & Vassolo, S. (2005). The impact of global change on the hydropower potential of Europe: A model-based analysis. *Energy Policy*, *33*(7), 839–855. doi:10.1016/j.enpol.2003.10.018

Mo, L., Lu, P., Wang, C., & Zhou, J. (2013). Short-term hydro generation scheduling of Three Gorges-Gezhouba cascaded hydropower plants using hybrid MACS-ADE approach. *Energy Conversion and Management*, *76*, 260–273. doi:10.1016/j.enconman.2013.07.047

Nabil, E. (2016). A Modified Flower Pollination Algorithm for Global Optimization. *Expert Systems with Applications*, *57*, 192–203. doi:10.1016/j.eswa.2016.03.047

Naveen, P., Chandel, A. K., Vedik, B., & Topwal, K. (2016). *Economic dispatch with valve point effect using symbiotic oragnisms search algorithm*. Academic Press.

Pan, G., & Xu, Y. (2016). *Chaotic Glowworm Swarm Optimization Algorithm Based on Gauss Mutation*. Academic Press.

Personal, M., Archive, R., & Fleten, S. (2010). *Evaluation of static hedging strategies for hydropower producers in the Nordic market*. Academic Press.

Simonovic, S. P. (1992). Reservoir systems analysis: Closing gap between theory and practice. *Journal of Water Resources Planning and Management*, *118*(3), 262–280. doi:10.1061/(ASCE)0733-9496(1992)118:3(262)

Singh, N., & Singh, S. B. (2017). *Hybrid Algorithm of Particle Swarm Optimization and Grey Wolf Optimizer for Improving Convergence Performance*. Academic Press.

Snaselova, P., & Zboril, F. (2015). Genetic Algorithm using Theory of Chaos. *Procedia Computer Science*, *51*, 316–325. doi:10.1016/j.procs.2015.05.248

Srianan, T., & Sangsawang, O. (2019). *Path-Relinking for Fire Station Location* (Vol. 866). Advances in Intelligent Systems and Computing.

Šulek, P. (2012). A hydro power system operation using Genetic Algorithms and mixed-integer nonlinear programming. *Slovak Journal of Civil Engineering*, *20*(1), 1–9. doi:10.2478/v10189-012-0001-z

Suresh, S., & Lal, S. (2016). An efficient cuckoo search algorithm based multilevel thresholding for segmentation of satellite images using different objective functions. *Expert Systems with Applications, 58*, 184–209. doi:10.1016/j.eswa.2016.03.032

Talbi, E. (2016). *A Taxonomy of Hybrid Metaheuristics A Taxonomy of Hybrid Metaheuristics*. Academic Press.

Tawhid, M. A., & Ali, A. F. (2017). *A Hybrid Grey Wolf Optimizer And Genetic Algorithm For Minimizing Potential Energy Function*. Memetic Computing.

Tospornsampan, J., & Kita, I. (2005). Optimization of a multiple reservoir system operation using a combination of genetic algorithm and discrete differential dynamic programming : a case study in Mae Klong system. Academic Press.

Tsoukalas, I., & Makropoulos, C. (2015). Multiobjective optimisation on a budget: Exploring surrogate modelling for robust multi-reservoir rules generation under hydrological uncertainty. *Environmental Modelling & Software, 69*, 396–413. doi:10.1016/j.envsoft.2014.09.023

Wang, G. (2015). *Elephant Herding Optimization Elephant Herding Optimization*. Academic Press.

Wang, G., Deb, S., & Cui, Z. (2015). Monarch butterfly optimization. *Neural Computing and Applications*.

Yang, K., Zheng, J., Yang, M., Zhou, R., & Liu, G. (2013). Adaptive Genetic Algorithm for Daily Optimal Operation of Cascade Reservoirs and its Improvement Strategy. *Water Resources Management, 27*(12), 4209–4235. doi:10.100711269-013-0403-3

Yang, X. (2010). *Firefly algorithm, stochastic test functions and design optimisation*. Academic Press.

Yang, X., & Press, L. (2010). Nature-Inspired Metaheuristic Algorithms (2nd ed.). Academic Press.

Yeh, W. W. (1985). Reservoir Management and Operations Models: A State-of-the-Art Review. *Water Resources Research, 21*(12), 1797–1818. doi:10.1029/WR021i012p01797

Yoo, J. H. (2009). Maximization of hydropower generation through the application of a linear programming model. *Journal of Hydrology (Amsterdam), 376*(1–2), 182–187. doi:10.1016/j.jhydrol.2009.07.026

Zelinka, I. (2019). *Intelligent Computing and Optimization* (Vol. 2019). Academic Press.

APPENDIX: NOMENCLATURE

ABC: Artificial Bee Colony
ANN: Artificial Neural Network
AGA: Adaptive Genetic Algorithm
BA: Bat Algorithm
CGA: Chaos Genetic Algorithm
COA: Chaos Optimization Algorithm
CS: Cuckoo Search Algorithm
CSA: Crow Search Algorithm
DE: Differential Evolution
DP: Dynamic Programming
DSP: Direct Search Procedure
EA: Evolutionary Algorithm
EHO: Elephant herding Optimization
FA: Firefly Algorithm
FEPCL: Fujian Electric Power Company Ltd.
GA: Genetic Algorithm
GOA: Grasshopper Optimization Algorithm
GSA: Gravitational Search Algorithm
GWO: Grey Wolf Optimizer
KA: Kidney-inspired Algorithm
KH: Krill Herd
LP: Linear Programming
MA: Metaheuristic algorithms
MBO: Monarch butterfly Optimization
NFL: No Free Lunch theorem
NLP: Non Linear Programming
NSGA-II: Non-Dominated Sorting Genetic Algorithm II
PSO: Particle Swarm Optimization
POA: Progressive Optimization Algorithm
SE: Simulated Annealing
SLP: Successive Linear Programming
WOA: Whale Optimization Algorithm

Chapter 16
Wireless Technology of Electric Power Transmission Using Non-Metal Conductive Media

Dmitry Strebkov
https://orcid.org/0000-0002-2572-801X
Federal Scientific Agroengineering Centre VIM, Russia

Alexey Nekrasov
https://orcid.org/0000-0001-6141-984X
Federal Scientific Agroengineering Centre VIM, Russia

Anton Nekrasov
https://orcid.org/0000-0001-5816-8939
Federal Scientific Agroengineering Centre VIM, Russia

ABSTRACT

Over two million kilometers of power grids in Russia exist. They have to be replaced in the coming 15 years. However, nowadays we witness and participate in the development of advanced technologies. New wireless resonant electric-power systems for different power consumers are considered including stationary single-conductor transmission lines and single-trolley and noncontact high-frequency electric transport, using non-metal conductive media. The trends of the future development and application of wireless resonant systems for electric power transmission are described. In the future, electrified mobile robots with external wireless electric power supplies will allow the organization of agricultural production with full automation of technological processes. The chapter describes the method and means for electric power transmission without metal wires. In this case, several components of conventional transmission lines, such as metal wires, insulators, and cables, are excluded, which results in considerable reduction in the equipment cost.

DOI: 10.4018/978-1-7998-3645-2.ch016

INTRODUCTION

History of science demonstrates that very occasionally global inventions come to life that considerably change our perception of the outside world and the prospects of humanity development. As an example we can mention the discovery of electricity, the emergence of nuclear and solar energy, of aircraft and rocket engineering, computers and telecommunication technologies.

BACKGROUND

However, nowadays we witness and participate in the development of advanced technologies that change the world making it better, cleaner and safer. First and foremost we should mention technologies proposed by N. Tesla a hundred years ago. N. Tesla developed electric engineering solutions with the use of alternate current. (Tesla N. 1999, Tesla N. 1978, Tesla N. Lectures. 1956, Nikola Tesla. 2008, Sarbon S., M.Sc. 2013, Tesla N. 1999, Nikola Tesla 1999).

The known methods for electric power transmission are based on delivering active power by conduction currents through a closed circuit. (Pandian Vasant, Nikolai Voropai 2016, Pandian Vasant, Gerhard-Wilhelm Weber, Vo Ngoc. Dieu. 2016, Valeriy Kharchenko, Pandian Vasant. 2018, Valeriy Kharchenko, Pandian Vasant. 2020). In this case, the electromagnetic energy propagates along transmission lines in the form of traveling waves of electromagnetic field or charges. The line conductors, manufactured of aluminum or copper, serve as conductive channels, guiding the electromagnetic energy stream from the generator to the receiver and back to the generator. Maximum power delivered through a three-phase transmission line, is limited by the loss in the line resistance, maximum voltage (determined by the insulation electric strength) and electromagnetic stability of the line. The present-day approach to ensuring electromagnetic stability lies in strict regulation of line parameters by means of rapid-operating shunt reactors and series compensating capacitors, aimed at suppressing the streams of reactive power and the line resonant properties.

A method for power transmission through a single metal conductor has been considered in. (Strebkov D.S., Nekrasov A.I. 2005, Strebkov D.S., Nekrasov A.I., Trubnikov V.Z., Nekrasov A.A. 2018, Strebkov D.S. 2014, D.S. Strebkov, A.I. Nekrasov, A.A. Nekrasov. (2019). In this case, the power is transmitted using stationary waves of reactive current and resonant properties of the line. The purpose of the present Chapter is to study the possibility of using non-metal conductive mediums for power transmission. (Strebkov D.S. 2002, RF Patent No. 2172546. 2001, RF Patent No. 2143775. 1999, Strebkov D.S., Nekrasov A.I., Avramenko S.V. 2002, Strebkov D.S., Avramenko S.V., Nekrasov A.I. 2002).

The present chapter describes the method and means for electric power transmission without metal wires. In this case, several components of conventional transmission lines, such as metal wires, insulators and cables, are excluded, which results in considerable reduction of the equipment cost.

New resonant electric-power system for different power consumers are considered including stationary single-conductor transmission lines and single-trolley and noncontact high-frequency electric transport (RF Patent No. 2172546. 2001, RF Patent No. 2143775. 1999).

The trends of the future development and application of resonant systems for electric power transmission are described. In future electrified mobile robots with external wireless electric power supply will allow to organize agricultural production on the principle “Industrial factories on the fields” with full automation of technological processes.

Motivation

Continued work of N. Tesla in the field of electric power transmission through non-metallic conductive materials. Reducing energy losses during electricity transmission; saving non-ferrous metals for the construction of power lines and the manufacture of power cables.

Main Part

The power source and the load are connected to a channel (or several channels), containing solid, liquid or gaseous conductive substance. Resonant electromagnetic oscillation at a frequency of 0.3 to 300 kHz is generated in the conductive channel. The amplitude of the resonant oscillation may exceed the source voltage by factor of 2 to 50. The transmitted electric field energy (mainly reactive) is converted to active energy consumed by the load. If necessary, the electric energy may be converted to the mechanical energy or heat.

An example of the power transmission system using a non-metal conductive channel is shown in Figure 1. The power source *1* is connected to the high-frequency Tesla transformer *2*, which is coupled to the conductive channel *3* via the high-voltage cable *4*. The channel is represented by the water-filled pipe *5*, manufactured of an insulating material (for example, of polyethylene). The pipe external diameter is *D*, its internal diameter is *d*, the length is *l*. The water (in its natural state, or tap water) demonstrates ionic conductivity due to various salts dissolved in it. The water has a specific resistance of 0.5 to 10 Ω cm, which enables the streams of electromagnetic energy to flow through it.

The pipe butts *6* are connected to the high-voltage cable *4*. The conductive channel *3* is connected to the load *7* via the step-down Tesla transformer *8* or the diode-capacitor units *9*. The diode-capacitor unit *9* represents a voltage-doubling circuit. It contains two diodes connected to the capacitor; the common point of the diodes is coupled to the power source. The positive half-wave of the alternating current flows to one capacitor electrode, the negative half-wave flows to the other electrode. The capacitor is charged until its voltage reaches double amplitude of the input alternating voltage. (RF Patent No. 2172546. 2001).

Using the resonant method, transmission lines of arbitrary length may be used. In addition, power grid systems, connecting multiple power sources and consumers through non-metal lines may be realized.

An example of multi-channel power transmission system using optical waveguides with a conductive coating is illustrated in Figure 2; the system includes several sources and receivers. The first conductive channel *3* is connected to the second conductive channel *10*. The second conductive channel *10* is connected to the third conductive channel *11* which receives the electric power from the energy source *12* via the high-voltage Tesla transformer *13*. In order to transmit the electric power to the consumer, the second channel *10* is connected to the load *14* through the step-down Tesla transformer *15*.

The conductive channel *15* coupled to the conductive channel *11* is used to transmit the electric power from the source *16*, located apart from the channels *3*, *10* and *11*. The source *16* is connected to the channel *15* via the high-voltage Tesla transformer *17*. The power consumer *18* coupled to the channel *15* receives the electric power through the diode-capacitor unit *19*.

The conductive channels *3*, *10*, *11* and *15* are manufactured of optical waveguides *20* whose surface is coated with the conductive film *21* and the insulation layer *22*. In the single-conductor electric power system, the optical waveguides *20* are used to transmit the information and control signals between the receivers and the generators, while the electric power is transmitted via the conductive film *21*. The input-output devices *23*, the optical signal amplifiers *24*, as well as the power transmitters and receivers

Figure 1. System for electric power transmission to stationary consumers through a non-metal conductive channel

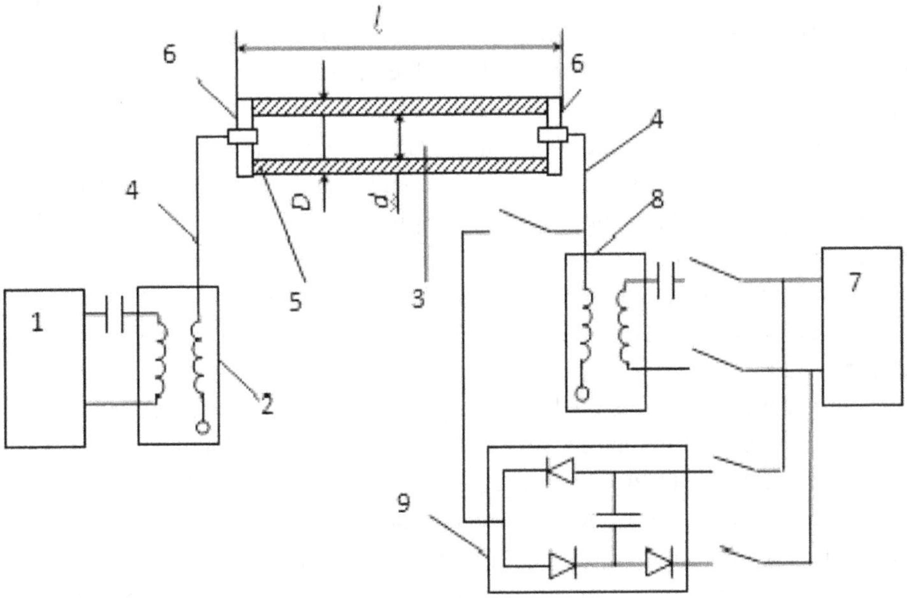

Figure 2. Block diagram of the power transmission system using several sources and receivers connected by optical waveguides

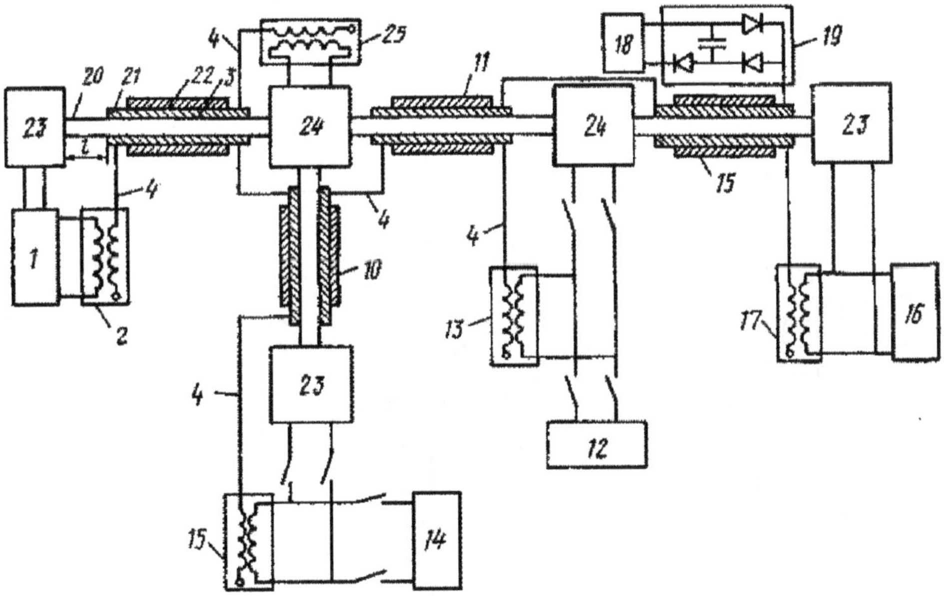

2, 13, 15, 17 and *19*, are distributed along the conductive channels. The optical waveguide ends are free from the conductive coating at a certain length, which is used to enter, to read or to amplify the optical information signals. In these places the conductive channels are connected to an insulated single-wire cable. The amplifiers *24* are located at regular intervals, which is necessary to keep the required level of the optical signals. The amplifiers are powered through conductive channels *3, 10, 11, 15* via the step-down Tesla transformers *25*, rectifiers and inverters, installed close to the amplifiers.

The electric power may be efficiently transmitted through a pipeline mains used to transport liquid conductive substance, for example, through the water-supply system. The block diagram of the considered system is shown in Figure. 3. The insulated water-supply pipes *26, 27, 28* are also used as conductive channels for the power transmission system. In order to meet the requirements of electrical safety, the input and output ends of the conductive channels are kept at zero potential. The water potential in the pipeline may be equal to the output potential of the Tesla transformer *29* (or may exceed this value) (RF Patent No. 2172546. 2001).

The power source *1* is connected to the low-voltage winding *30* of the Tesla transformer *29*. Its high-voltage multi-layer winding *31* is manufactured of insulated hollow pipes wound as a coil round the transformer ferrite core *32*. The external terminal *33* of the hollow high-voltage winding is joined to the electric pump *35* and the water source *36* by means of the tubular junction *34*. The electric pump *35* is powered by the source *1*. The internal terminal *37* of the winding *31* is connected through the pipe *38* to the conductive channel *26* of the water supply system. This terminal is located near the winding axis, while the terminal *33* is located near the low-voltage winding *30*. Due to such structure, the terminal *33* has a low potential (near zero relative to the ground), so the water potential at the transformer entrance is also near zero.

Figure 3. System transmitting electric power through a pipeline mains, used to transport liquid conductive substance

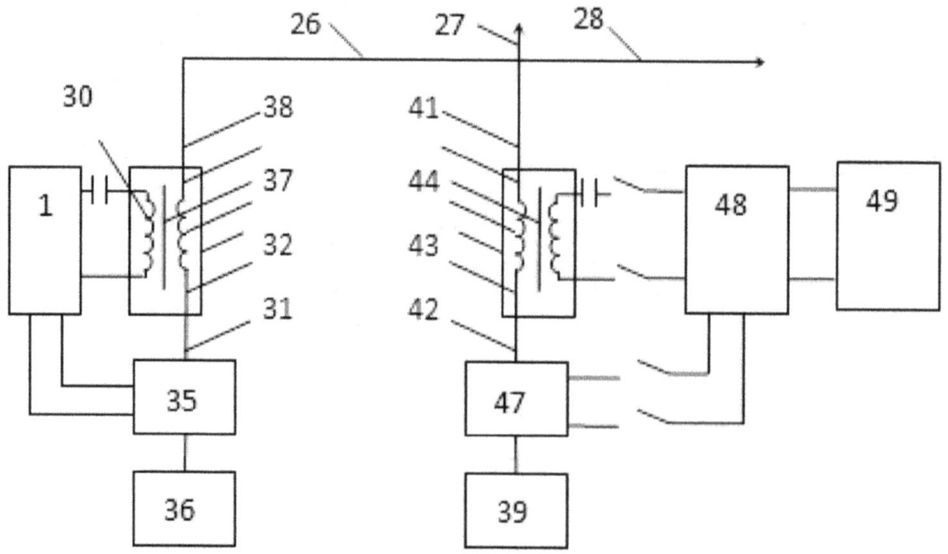

The water pump *35* is connected to the winding *31* at the terminal *33* having zero potential, so the water entering the transformer is safe for handling and transporting. Maximum potential value is reached in the conductive channels *26, 27, 28*.

The step-down high-voltage Tesla transformer *40*, installed at the pipeline end, is located close to the water consumer *39*. The water from the insulated conductive channel *26* is transported through the pipe *41* into the coil *42* (consisting of insulated pipes wound round the ferrite core *43*), which forms the high-voltage winding of the transformer *40*.

The pipeline *41* of the conductive channel *26* is joined to the first tubular turn *44* of the coil *42*, located on its internal surface, close to the axis. The last (external) tubular coil *45* is joined to the electric pump *47* by means of the tubular junction *46*. The pump is used to deliver the water with zero potential to the consumer *39*. The electric power transmitted through the conductive channels *26, 27, 28,* is delivered to the power consumer *49* via the Tesla transformer *40* and the inverter *48*.

The proposed method allows to use the water supply and irrigation systems to transmit the electric power to the water consumers. In case of using municipal water supply systems, the mains pipelines insulated from the ground may be also used to supply the electric power to houses, industrial buildings or to regional distributive stations (RF Patent No. 2172546. 2001).

The system shown in Figure 4 contains the conductive channel *50*, which is an open water line insulated from the ground. The water and the electric energy are supplied into the channel via the high-voltage winding *31* of the high-frequency Tesla transformer *29*. The water is taken from the channel *50* by the electric sprinkler *51* which is also powered from the conductive channel. In this case, the step-down Tesla transformer *40* is installed on the sprinkler *51* which moves along the conductive channel *50* and sprinkles water to the adjacent ground territory. The tubular water receiver *53* is installed on the insulators *52*; it receives also the electric energy, which is delivered to the high-voltage winding *42* of the step-down Tesla transformer *40*. The winding is manufactured as a multi-layer coil with tubular turns, similar to the case, illustrated in Fig. 3. Then the water having zero potential is sprinkled by means of the electric pump *47*. The electric energy from the Tesla transformer *40* is delivered to the inverter *48*; it is used to power the electric pump *47* and the electric drive system *54* of the sprinkler *51*.

In the system shown in Figure 5, the conductive channel *55* is formed by separating a part of the water area *56* (which is an irrigation canal or a river) by means of the open insulating gutter *52*. The electric vehicle *57* (for example, a sprinkler) receives the electric power from the conductive fluid using the insulated flexible trolley *58*. In order to prevent water freezing in winter, it is additionally salted. The electric energy is supplied into the conductive channel *55* from stationary power sources *1* and high-frequency high-voltage Tesla transformers *2*, installed with certain intervals along the channel. The water having zero potential is delivered from the irrigation canal (or river) *56* to the sprinkler *57* through the pipeline *59*.

The vehicle *57* contains the step-down Tesla transformer *60*. The internal terminal *61* of its high-voltage winding *62* is connected to the fluid of the conductive channel *55* by means of the trolley *58*. The external winding terminal *63* is connected to the natural capacitor *64*. The low-voltage transformer winding *65* is connected through the rectifier *66* and the inverter *67* to the electric drive and control system *68* and to the electric pump *69*. The pump is used to take the water from the irrigation canal *56* through the pipeline *59* and to sprinkle it. (RF Patent No. 2172546. 2001).

Figure 4. System used to power a sprinkler, moving along the water channel

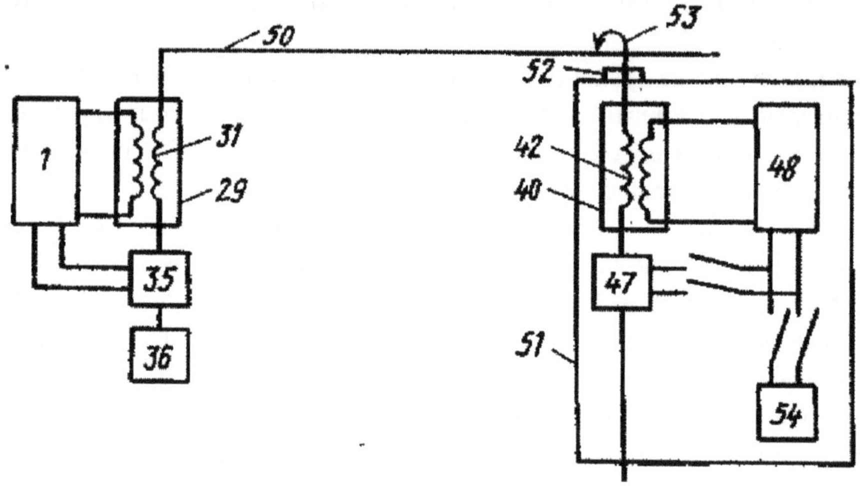

Figure 5. System used to power a sprinkler through a conductive channel, installed on the surface of an irrigation canal or a river

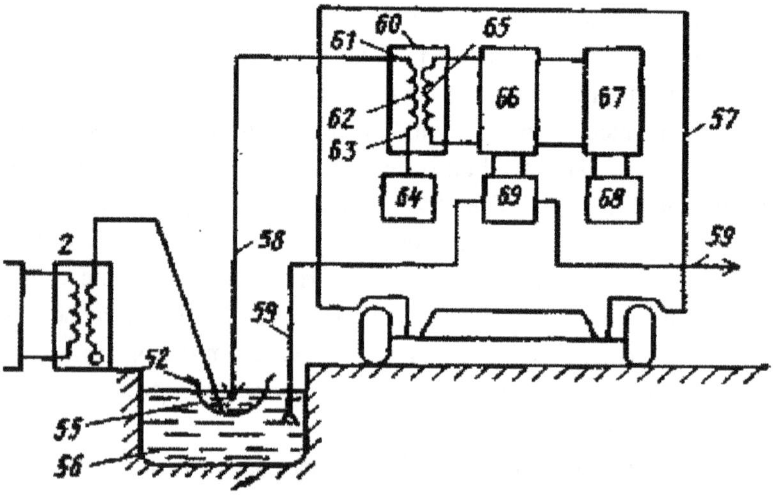

In the power transmission system shown in Figure 6, conductive channels are formed along each lane of the roadway 66, using conductive strips *70*. The strip, whose width L_0 ranges from 1 to 2 m, is laid over the insulating layer *71*, separating the conductive channel from the ground. Each vehicle *57* moving along the roadway has a flexible conductive trolley *59* under the bottom, touching the conductive coating layer *70* between the wheels. The protective barriers *73* along the roadway *66* prevent from penetration of people and animals into the dangerous zone. Special-purpose areas *74*, having the earth potential, are provided for parking the vehicles.

In the system shown in Figure 7, the conductive channel *75* is realized as the continuous open insulated gutter *76*, having the conductive coating *77* at the bottom. The channel is installed above the water surface on the floats *78*, joined together by means of the rope *79*, along the coast, canal or river *80*. The ship *81* moves along the conductive channel *75*; it receives the electric energy through the trolley *82*, touching the water in the gutter *76*. The power transmission equipment (including the high-frequency Tesla transformer *83* installed on the ship) is similar to that shown in Fig. 5. (Pandian Vasant, Gerhard-Wilhelm Weber, Vo Ngoc. Dieu. 2016).

Figure 6. System used to power electric vehicles moving along a roadway with conductive strips

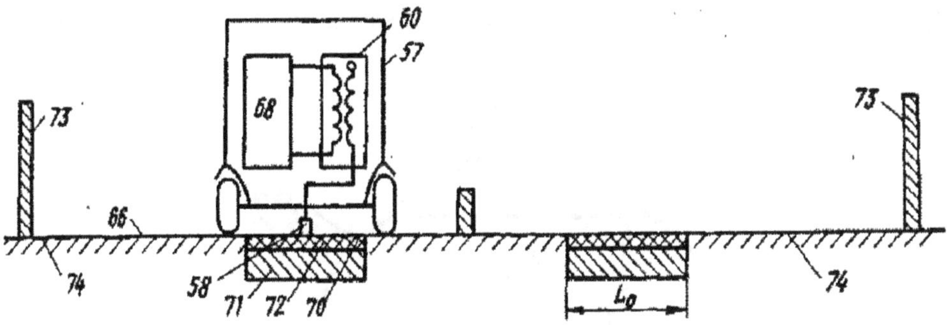

Figure 7. System used to power water vehicles through a water-filled conductive channel

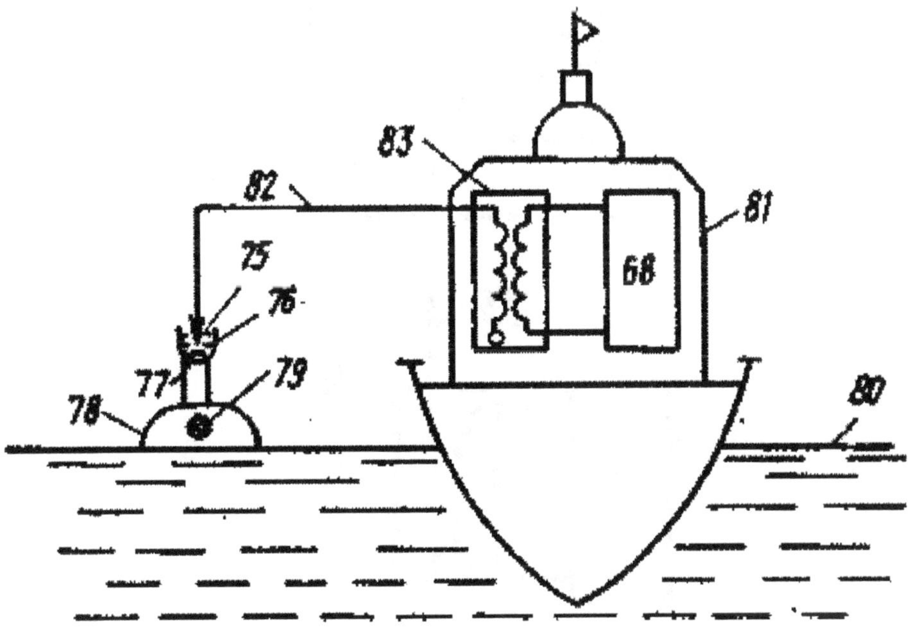

The system shown in Figure 8 is used to power the flying vehicle *89*, for example a balloon, a dirigible, a helicopter, a missile or a vertical takeoff airplane. In this case, the conductive channel *84* is realized as a vertical non-metal thread *85* with a conductive coating. As an example, a carbon-plastic, glass-carbon or quartz filament covered with an indium-tin oxide (ITO) film may be used. As an alternative, the conductive coating may be manufactured of a carbon, copper or aluminum film, 0.1 to 40 μm in thickness.

The lower terminal *86* of the conductive channel *84* is connected to the power source *1* via the high-voltage high-frequency Tesla transformer *2*. The upper filament terminal *87* is connected to the step-down high-frequency Tesla transformer *88*, installed on the vehicle *89*. The transformer *88* is connected to the vehicle drive and control system *90* through the inverter *48*.

Figure 8. System used to power a flying vehicle through a non-metal conductive channel

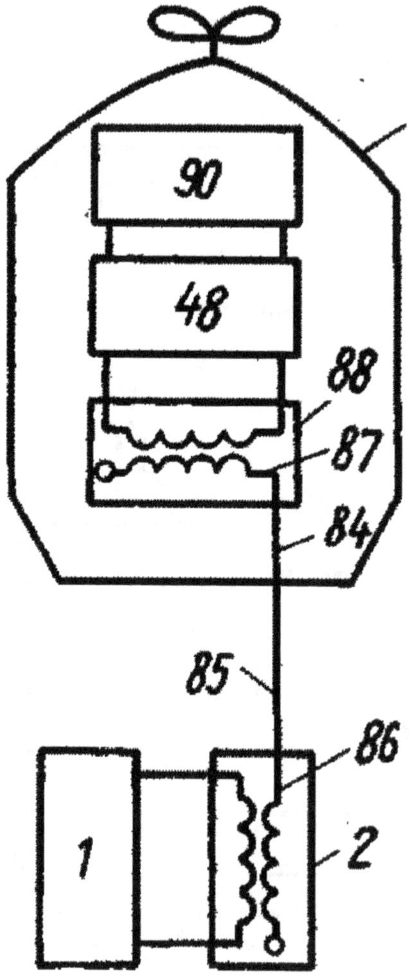

In Figure 9, the conductive channel *91* is represented by the insulated pipeline *92* used to transport water, coal, gas, oil or other materials. Solid or liquid substances may be mixed with water. If the transported substance is a dielectric (for example, oil products or dry natural gas), the pipeline *92* may be manufactured of a conductive material and used as the energy transmitting channel. The pipeline manufactured of a dielectric material may be covered with a conductive coating *93*. The electric energy of the source *1* is transmitted to the consumer *7* through the high-frequency Tesla transformers *2*, *8* and the conductive channel *91*. (RF Patent No. 2172546. 2001)

Figure 9. Power transmission system with a conductive channel represented by the pipeline 92 used to transport coal, gas, oil or other solid and liquid substances

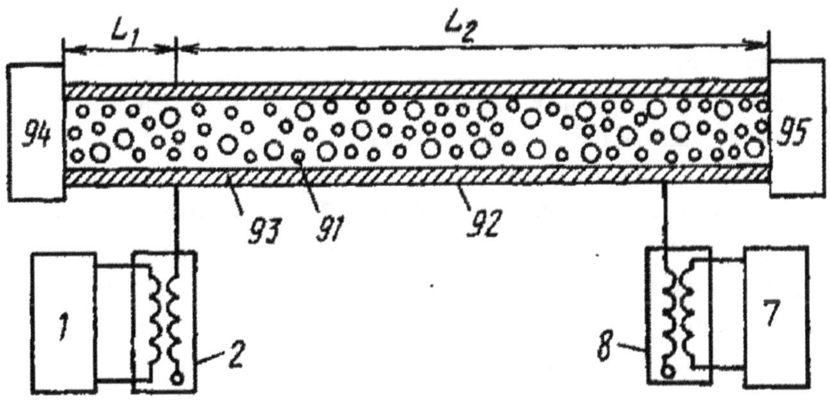

In order to load the transported substance, the loading unit *94* is installed at a distance L_1 from the high-frequency Tesla transformer *2*. This distance is

$$L_1 = \lambda/2 - A, \tag{1}$$

where λ - is the wavelength in the single-conductor transmission line, corresponding to the operating frequency; *A* is the length of the transformer high-voltage winding (Strebkov D.S. 2014).

The unloading unit *95* is installed at a distance L_2 from the Tesla transformer 2:

$$L_2 = \lambda/2\,(n+1) - A, \tag{2}$$

where $n = 0, 1, 2, \ldots$.

The considered structure ensures zero potential of the conductive channel 91 at the places, where the transported substance is loaded or unloaded. The pipeline manufactured of an insulating material may be filled with damp air or with vapor of a conductive substance (for example, mercury or alkaline metal). In this case, a high concentration of ions and electrons is generated in the conductive channel using a high-frequency discharge. Enhanced channel conductance may be also reached using a generator of charged aerosol or a microwave generator.

As an example, the output voltage of the step-up Tesla transformer *2* is taken 10 kV, the operating frequency is equal to 5 kHz. The conductive channel length depends on the source power and the operating frequency, it ranges usually from 2 m to 10 000 km. The electric power transmitted through the conductive channel depends on the source power, the frequency, the voltage transfer ratio and the recharging energy for the capacitance of the receiver and the line. The power P delivered to the receiver, having the recharged capacitance C, is found using the following equation:

$$P = \alpha\, 2\pi f\, C V_0^2\, K_u^2, \tag{3}$$

where α - is the efficiency of converting the transmitted (mainly reactive) power to the active load power; its value ranges from 0 to 1;

f - is the operating frequency;
V_0 - is the voltage across the high-voltage winding of a step-up high-frequency Tesla transformer *2*;
K_u - is the voltage transfer ratio of the conductive channel.

Assuming for the considered example $C = 1$ μF, $K_u = 30$ and $\alpha = 0.5$, we obtain $P = 141.3$ MW. The potential of the conductive channel end is equal to zero, if the channel length is corresponding (2).

The reference point for measuring the channel length coincides with the place, where the Tesla transformer *2* is installed. In case of $f = 5$ kHz, the half-wave channel length is equal to 30 km. The channel points with zero potential may be used to load and unload the transported substance. Maximum voltage values correspond to the channel points, whose distance L from the Tesla transformer equals to

$$L = \lambda/4\,(2\,n + 1) - A, \tag{4}$$

where $n = 0, 1, 2, \ldots$

In the frequency range from 1 to 10 kHz, the radiation loss of the electromagnetic energy is insignifican.

When the resonant single-conductor power transmission systems are used, the consumption of non-ferrous metals (Al or Cu) by the line conductors may be reduced by factor of 10; sometimes non-ferrous metals may be completely eliminated from the transmission lines. The energy transmission loss may be also considerably reduced.

The experimental investigations has been carried out using the resonant single-conductor electric system (RSCES), whose diagram is shown in Figures 10 - 11.

The RSCES consists of the 1-kW voltage-regulated high-frequency generator *1*, the transmitter and receiver Tesla transformers *2*, *3* with the conductive channel *4* between them, the rectifier *5* and the load *7*. The load is represented by incandescent lamps or by an 1-kW 220-V electromotor. The high-voltage winding of the transmitter transformer contains 4 000 to 6 000 turns; it is located on a ferrite core, 50 mm in diameter. Its internal terminal is connected to the conductive channel, the external terminal remains free. The input low-voltage winding contains 40 to 60 turns. The transformer feed circuit and the circuit containing the high-voltage windings and the conductive channel, operates in the resonant regime. The transformer output voltage ranges from 1 to 10 kV at a frequency of 1 to 25 kHz. In addition, the transformer (as a helical antenna) radiates an electromagnetic signal with a wavelength, corresponding

Figure 10. Diagram of a single-conductor electric system with a non-metal conductive channel: RSCES with symmetric location of Tesla transformers

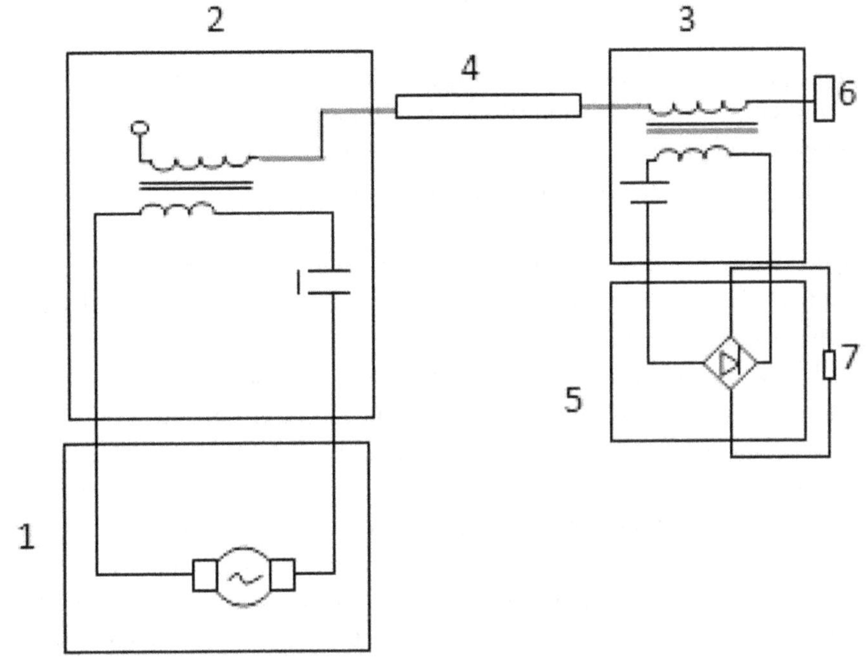

Figure 11. Diagram of a single-conductor electric system with a non-metal conductive channel: RSCES with a diode-capacitor unit at the end of the conductive channel

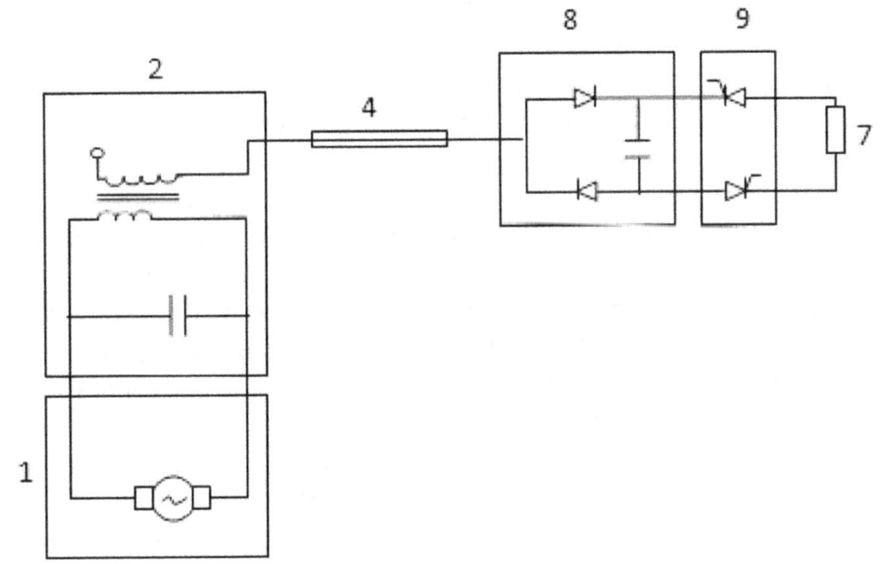

to the length of the high-voltage winding turn. The high-voltage winding of the receiver transformer *3* is terminated by the natural capacitor *6*. (Strebkov D.S. 2002).

The following versions of the conductive channel have been used:

- a polyethylene pipe, 10 mm in diameter and 1.5 m in length, filled with tap water or sea water;
- a plastic gutter with a wet soil layer (150 × 10 × 400 mm in dimension);
- a 300-Ω indium-tin oxide film, 0.3 μm in thickness, on a glass substrate;
- a 100-Ω graphite filament, 0.1 mm diameter and 500 mm in length.

Steel and copper wires, 0.1 mm diameter and 5 m in length, have been also used for the sake of comparison (see Figure 12). The voltage, current and power at the line terminal and across the load are measured by means of standard measuring devices.

If a metal conductive channel is replaced by a channel of non-metal conductive material (or series connection of channels), the transmitted power is not reduced; the channel material is not heated.

If an air gap is made in the conductive channel between a conductor and the water surface, a reactive plasma discharge appears in the gap (see Figure 13). However, the water temperature does not rise after transmitting a 100-W power within 1 hr, which confirms low energy loss in the conductive channel. The water temperature variation does not influence the transmitted power value. The water pH level has varied from 6 to 4. If the sea-salt content in the water is enhanced to 7 g/l, the transmitted power remains the same, as in the case of using tap water. However, if the tap water is replaced by deionized water, 16 MΩ in resistance, the power transmission is suppressed (Strebkov D.S. 2002, RF Patent No. 2172546. 2001, RF Patent No. 2143775. 1999)

It may be concluded, that the source active power is consumed by the loss in the feed circuit conductors and in the core of the Tesla transformer (which does not exceed 2%), as well as by generating reactive current in the conductive channel. The loss in the resistance of the conductive channel, caused by reactive current in the resonant regime, is insignificant. The loss caused by the corona discharge and leakage current may be reduced by insulating the conductive channel. The active current component and the line magnetic field are low, while the line electric field has maximum value. The maximum transmitted power is limited by the line charging power (as well as in the conventional transmission line). The phase angle between the voltage vectors at the line terminals is equal to zero. The quality factor of the RSCES, operating at a frequency of 5 kHz, is 100 times higher, than that of the conventional 50-Hz transmission line, which results in considerable increase of the line voltage and the power, transmitted through the conductive channel in the resonant regime. In contrast to it, the conventional transmission line is characterized by insignificant voltage variation along the line, while the phase angle between the voltage vectors at the line terminals is proportional to its length (related to the wavelength).

The carried out investigations have resulted in proposing new methods for power transmission to stationary and mobile consumers. The following types of conductive channel are used: plastic water pipes or irrigation canals, insulated from the ground; pipelines transporting gas, oil, hot or cold water; fiber-optic cables, covered with a conductive film; carbon-composite cable; insulated sections of soil or water surface (including highway sections). The requirements for electric safety and limitation for the use of potable and hot water from pipelines under high voltage have been developed. The key requirement lies in grounding the pipelines in the points, whose distance from the generator equals to an integer number of half-wave lengths (where the line voltage is equal to zero). If a lateral pipeline branch is used, it shall be grounded in the point, whose distance from the main pipeline equals to an odd number of quarter-wave

Figure 12. Sections of the conductive channel, manufactured of various materials: 1 – bath with tap water; 2 – gutter with wet ground; 3 – polyethylene tube with water; 4 – graphite filament; 5 – step-down transformer; 6 – incandescent lamp (220 V, 60 W)

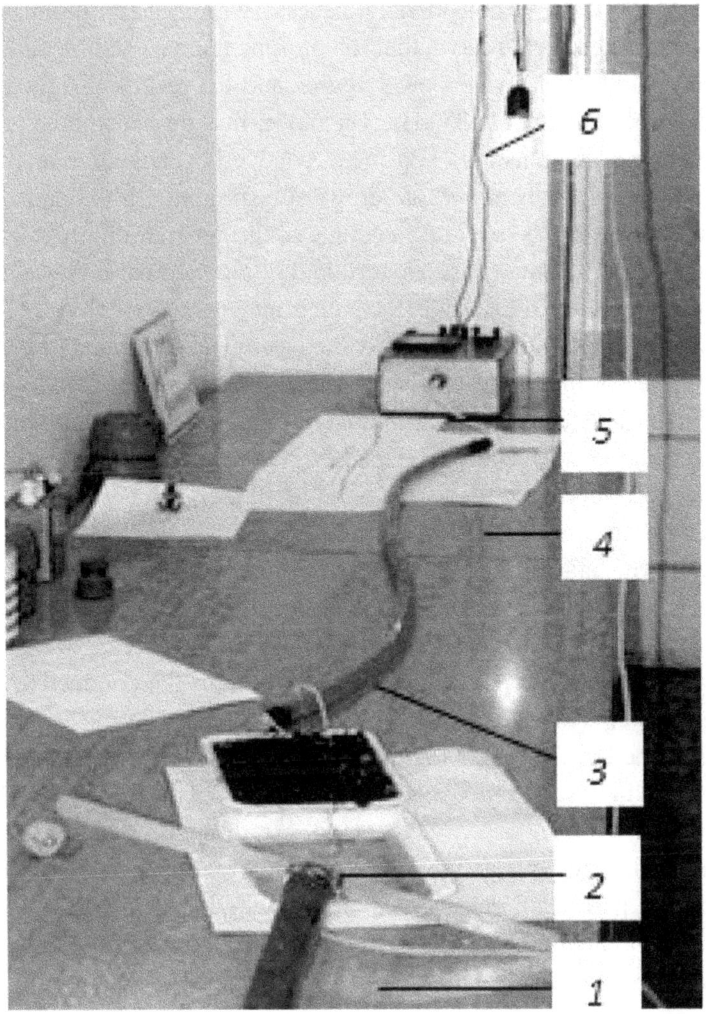

Figure 13. Discharge of a "cold" reactive plasma in a single-conductor line gap between the conductor 1 and the water surface 2 (the voltage is 4.5 kV; the power is 100 W)

lengths. If the frequency equals to 5 kHz, the quarter-wave length is equal to 15 km. (Strebkov D.S., Nekrasov A.I., Avramenko S.V. 2002, Strebkov D.S., Avramenko S.V., Nekrasov A.I. 2002)

N. Tesla grounded one terminal of the transformer high-voltage windings at the transmitter and receiver ends of the resonant system; he considered it necessary for transmitting power along the ground. The results of our investigations have shown, that the presence of a closed metal conductor and of current flow through the ground between the power source and the receiver is not necessary for electric power transmission at a frequency of 1 to 25 kHz. The power may be transmitted to the receiver through a single-conductor guiding system, using a non-metal conductive channel, just as the electromagnetic energy is transmitted along a laser beam or a microwave beam, but with higher efficiency due to low attenuation and radiation loss. In this case, one terminal of the generator high-voltage winding has zero potential and remains free, while the symmetric terminal of the receiver high-voltage winding shall be connected to the natural capacitor *6* (see Fig.10), which may be represented by an aerostat shell or a tractor frame. In our experiments, a metal case was used as the natural capacitor.

In an alternative version of the power transmission system, the receiver terminal of the conductive channel is coupled to the diode-capacitor unit *8*, containing the conventional voltage-doubling circuit (see Figure 11). Its capacitor is connected to the load *7* via the electronic switch *9*. In this case, the total length of the conductive channel *4* and the output winding of the source transformer *2* shall be equal to an odd number of quarter-wave lengths.

A non-metal conductive channel, for example, a fiber-optic or a carbon-plastic cable, may be used to transmit electric power not only along the ground, but also perpendicularly to it: for example, to an aerostat with a retransmitted or to a balloon.

In laser – based wireless resonant power transmission systems the conductive channel between the power source and the receiver is formed using the method of photoionization (or impact ionization). The conductive channel is formed by the power source or by the receiver of the electromagnetic energy. The energy may be transmitted in the continuous-wave form or in a pulsed regime (RF Patent No. 2143775. 1999).

The wireless transmission systems may be used to power stationary electric equipment or mobile object of various types. In particular, electric vehicles moving along a roadway or along an arbitrary pathway may be powered. In addition, the electric power may be transmitted to various types of flying vehicles. In order to increase the total distance to the power consumer, two or more radiation generators may be installed in sequence.

A wireless transmission system used to power stationary consumers is shown in Figure 14. The energy source *1* is connected to the radiation generator (laser) *2* and to the high-voltage Tesla transformer *3*.

The Tesla transformer *3* is coupled to the device, forming the conductive channel *5* (channel former *4*). The channel former is a pipe manufactured of a conductive material and installed coaxially with the laser *2*. The diameter *D* of the channel former *4* is equal to the diameter of the beam *8* radiated by the laser (or a little more than it); it may range from 0.1 mm to several tens millimeters.

The insulating shield *7* is installed between the channel former *4* and the laser *2*. It insulates the laser from the high voltage applied to the channel former. The shield represents a hermetic evacuated dielectric cylinder whose butt-end windows are transparent for the laser beam *8*. The cylinder diameter exceeds the beam diameter by factor of 5 to 50; its length shall be 150 mm for each 10 kV of the channel former voltage. In order to enhance the power transmission efficiency, the shield surface opposite to the laser is covered with a transparent conductive coating *6* (for example, an ITO film), connected to the channel former.

Figure 14. Wireless transmission system used to power stationary consumers

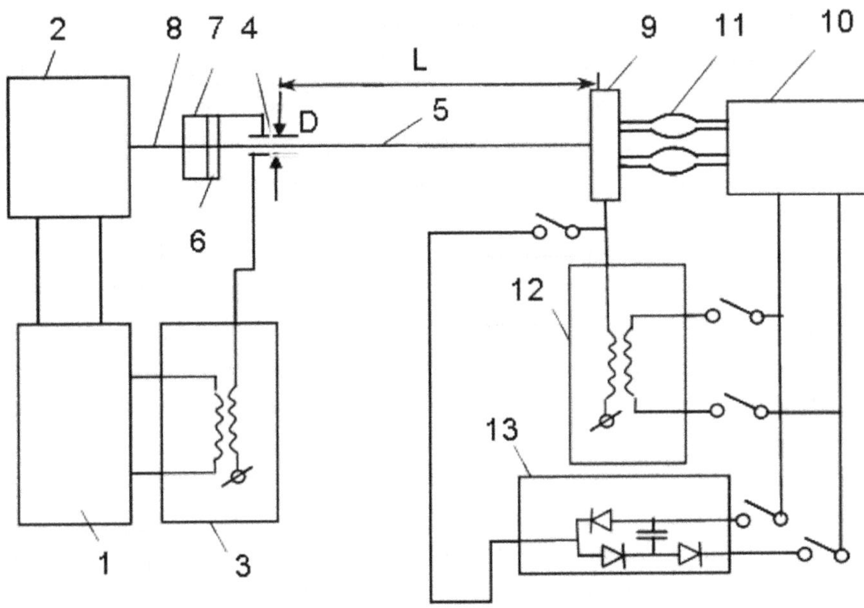

The shield may be also manufactured of a bulk dielectric material, transparent for the radiation. Its butt-end windows are polished and covered with an antireflecting coating.

The receiver *9*, located at the end of the conductive channel *5*, is manufactured of a conductive material (for example of steel). It is insulated from the housing of the power consumer *10* by means of high-voltage insulators *11*. The receiver *9* is connected to the power consumer via the step-down Tesla transformer *12* or the diode-capacitor units *13*. The diode-capacitor unit *13* is similar to that used in voltage doubling circuits. It consists of two diodes and a capacitor; the common point of the diodes is connected to the power source.

The laser radiation is used to generate free carriers within the beam *8* by means of photoionization. The beam enters the pipe of the channel former *4*, which is connected to the high-voltage winding of the Tesla transformer *3*. Due to additional impact ionization by the high electric field, the high-conductance channel *5* is formed inside the pipe, it is connected to the transformer output. Electric streamers are generated in the pipe, they propagate along the channel *5* at a high speed (about 1 km/s). The conductive channel *5* connects the power source *1* to the receiver *9* of the power consumer *10*. Due to the insulating shield *7*, the radiation generator *2* is decoupled of the high voltage applied to the channel former *4*.

The inductance of the Tesla transformer *3*, combined with the capacitance of the line *5* and the load, form a resonant circuit, which results in increasing the line voltage. The power transmission through the single-conductor line *5* in the resonant regime is characterized by a high efficiency (up to 96 or 99%).

The radiation generator *2* is used to form the conductive channel only; its power is less than the transmitted electric power by factor of 50 to 100. Therefore its low efficiency (10 to 15%) does not considerably degrade the total efficiency of power transmission (RF Patent No. 2143775. 1999)

The laser may operate in a pulsed regime, when its power is considerably higher. In this case, the pulse repetition frequency shall be high enough, so that the channel conductance does not disappear during the intervals because of carrier diffusion and recombination.

The length L of the conductive channel *5* is limited by the power of the radiation generator (laser) *2*. If the distance between the source and the power consumer exceeds maximum permissible length of the conductive channel, two or more radiation generators may be used.

An example of a system transmitting electric power over a long distance and containing several power sources and receivers is illustrated in Figure 15. The receiver *9* of the first conductive channel *5* is connected to the channel former *14* of the second conductive channel *15*. The second conductive channel is formed using the second radiation generator *16*, which is insulated from the second channel former *14* by means of the second insulating shield *17*. The generator *16* obtains the electric power from the step-down Tesla transformer *12*, connected to the receiver *9* of the first conductive channel *5*. The receiver *18* of the second conductive channel *15* is connected to the former *19* of the third conductive channel *20*. The receiver of the second conductive channel *18* is connected to the power consumer *21* via the high-frequency step-down Tesla transformer *22*. If the consumer requires dc power or industrial frequency (50 Hz) voltage, a rectifier or an inverter is added to the equipment, following the step-down Tesla transformer.

The radiation generator *24* of the conductive channel *20* is powered from the step-down Tesla transformer *22*. The shield *25* insulates the generator *24* from the high voltage applied to the former *19* of the conductive channel *20*. The insulating shields *7*, *17* and *25* are manufactured of a bulk dielectric material transparent for the radiation (RF Patent No. 2143775. 1999)

Using the resonant method and wireless transmission lines, a system of arbitrary length may be realized, connecting the required number of power sources and consumers (operating at the same frequency in the connection points).

The power source *29* is connected to the conductive channel *28* via the high-voltage Tesla transformer *30* and the channel former *31*. The conductive channel *28* is formed using the radiation generator *32* and the insulating shield *33*. The generator *32* is powered by the source *29*. The receiver *34* of the conductive channel *20* is fastened on the housing of the power consumer *36* by means of the insulators *35*. The consumer *36* is powered from the receiver *34* via the diode-capacitor unit *37*.

The radiation generator *2* is represented by a CO_2 laser, 1 kW in power, operating at a wavelength of 10.6 μm (a high-power microwave generator may be also used). The insulating shield *7* is transparent for the laser radiation. The channel former *4* is powered from the Tesla transformer *3* having an output voltage of 35 kV at a frequency of 30 kHz. The receiver *9* at the end of the conductive channel *5* is a plate, 0.5 m in diameter, manufactured of a refractory metal, for example, of titanium (RF Patent No. 2143775. 1999).

The electric power transmitted through the conductive channel depends on the source voltage, the capacitance of the line and the receiver, and the recharging frequency. Assuming the recharged capacitance equal to 10 000 pF and the line voltage of 35 kV at a frequency of 30 kHz, we obtain maximum transmitted power about 5 MW. If the voltage is increased to 1000 kV, the maximum transmitted power will reach 5 000 MW.

The length of a conductive channel depends on the power of the radiation generator and on the beam divergence angle. If the laser power is 1 to 10 kW, and the beam divergence angle is 1^2 to 2^2, the of conductive channel length will range from 100 m to 10 km. Using several conductive channels connected in series, the power transmission distance may be enhanced to 100 km and more.

Figure 15. System transmitting electric power over a long distance and containing several power sources and receivers

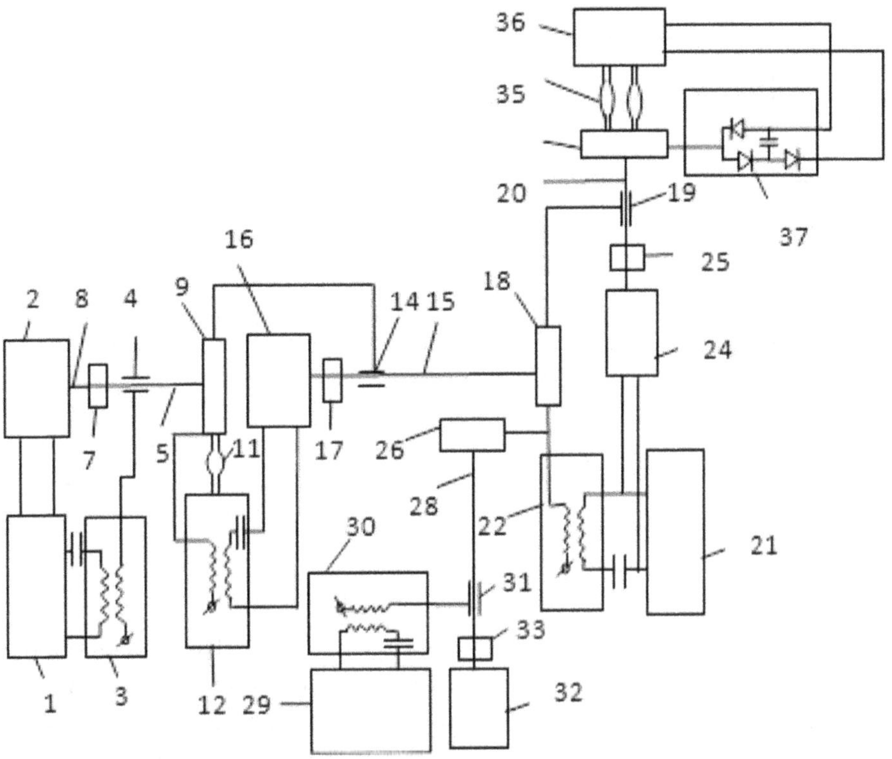

In the system shown in Figure 16, the receiver *9* of the conductive channel *5* is fastened on the vehicle *38* (for example, an electric tractor) by means of high-voltage insulators *11*. The electric drive and control system *10* of the tractor is connected to the receiver *9* via the diode-capacitor unit *13* (RF Patent No. 2143775. 1999).

The transmitter unit, containing the power source *1*, the radiation generator *2*, the insulating shield *7* and the channel former *4*, is located at a certain distance from the vehicle *38*. The unit is equipped with a tracking device *39*, which ensures pointing the conductive channel *5* to the receiver *9* of the vehicle *38*, in the course of its arbitrary displacement. The tracking system contains an laser locator or a radar used to determine the vehicle coordinates, and an actuating mechanism turning the platform with the transmitter unit. If the electric power shall be transmitted in the pulsed form, the synchronizing device *40* is added to the system equipment. It ensures that the power and laser pulses are generated at the same periods. The stationary power source may have a number of transmitter units used to power several mobile vehicles simultaneously.

A system used to power electric vehicles moving along a highway is shown in Figure 17. The stationary power source *1* contains the high-frequency Tesla transformer *3*, which is connected through the cable *41* to the V-shaped metal strip *42*, installed on the insulators *11* along the highway *43*. The strip *42* represents a receiver for the conductive channel, used to power the vehicles *44* which are equipped with both an electric motor, and a combustion engine.

Figure 16. Block diagram of a system used to power an electric vehicle moving along an arbitrary pathway

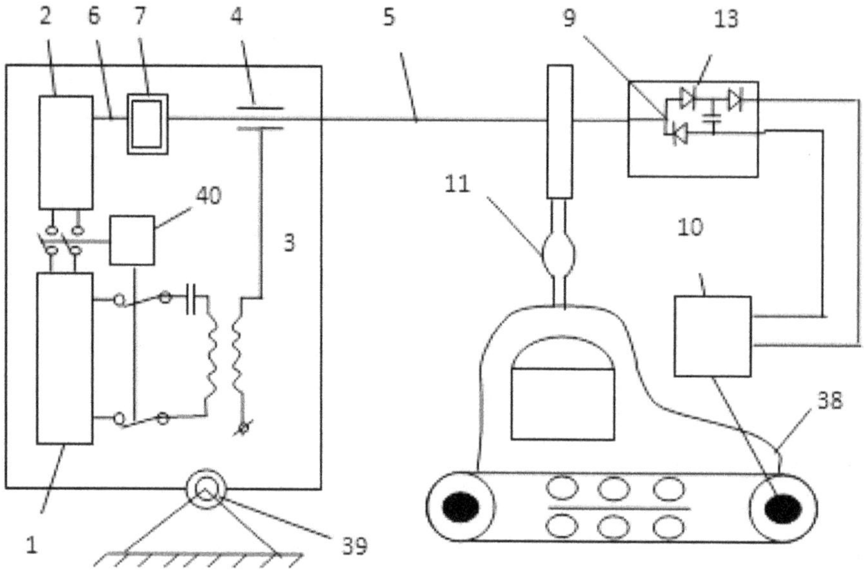

The block diagram of the vehicle equipment is shown in Figure 18. The auxiliary low-power source *46* is used to power the radiation generator *2*, which is a laser based on neodymium glass. The laser, 0.5 kW in power, operates at a wavelength of 0.53 μm in the frequency doubling regime. An alternative radiation generator may be used to form a high-conductance channel along the beam axis (for example, a microwave, an X-ray or an aerosol generator). The equipment includes the insulating shield *7* and the channel former *4*, connected to the auxiliary power source *46* via the high-frequency Tesla transformer *47*. The channel former is also connected to the vehicle drive and control system *10* via the diode-capacitor unit *13*.

The orientation device *45* ensures pointing the laser beam to the *V*-shaped tape receiver *42* (Figure 18), installed along the highway on the insulators *11*, at a height of *5* to *6* m. The tape width is *40* to *60* cm, its length is equal to the highway length. As the height of the *V*-shaped tape receiver *42* is constant along the road *43*, the automobile moving in the same lane keeps an invariable orientation of the beam pointing to the receiver. In case of changing the lane, the angular orientation of the beam shall be corrected (RF Patent No. 2143775. 1999).

Let us consider a highway having *8* traffic lanes, 4 m in width, in each direction. If the tape receiver *42* is installed at a height of 6 m above the highway medial line, minimum length of the conductive channel is about 8 m, while its maximum required length is equal to 32 m. Starting from this result, the requirements to the radiation generator power and to the beam angular orientation may be established.

If the line operating voltage is 110 kV at a frequency of 30 kHz, and the capacitance of a line and the load is 2 000 pF the transmitted power will reach 60 MW. Assuming the automobile drive power equal to 60 kW, a single power source is capable to supply electric power to 1 000 automobiles simultaneously. In order to increase the number of powered automobiles, several stationary power sources may be installed along the highway in regular intervals and connected to the receiver.

Figure 17. System used to power objects moving along a definite pathway (for example, electric vehicles on a highway)

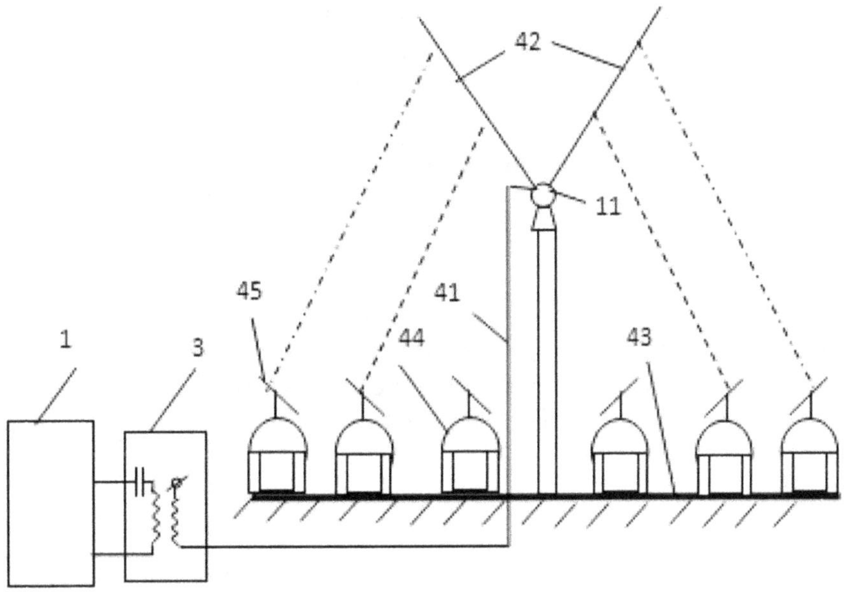

Figure 18. Vehicle equipment used to form the conductive channel and to receive the electric energy

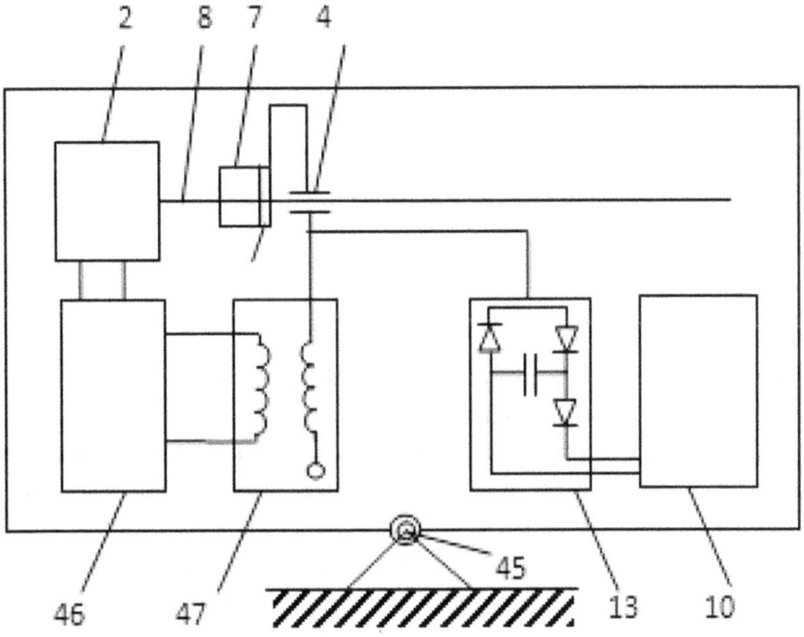

Benefits of

Wireless power transmission is carried out by the wave method in resonant mode at an increased frequency without the use of conductors of metal power lines. As a result of the research, the features of the distribution of the intensity of the magnetic component of the reactive magnetic fields in the space between the transmitting and receiving transformers were revealed, as well as the intensity of the electromagnetic field in the zone of the receiving transformer was revealed.

Originality

The proposed method makes it possible to demonstrate the possibility of wireless power transmission in resonance mode, to study the physical processes accompanying wireless power transmission, to investigate the property of standing waves of a reactive electromagnetic field, interference of electric and magnetic fields, and also to provide the most efficient mode of electric power transmission.

Main Contribution

A circuit diagram has been developed, original elements and components have been manufactured, an experimental laboratory bench has been mounted to study the wave properties of the resonant system for wireless transmission of electric power by high-frequency currents.

FUTURE RESEARCH DIRECTIONS

The ability to transfer electricity to consumers by the wave method in resonant mode at an increased frequency without the use of conductive metal power lines. Continued work of N. Tesla in the field of electric power transmission by high frequency currents on non-metallic conductive materials.

CONCLUSION

1. There are considerable advantages of the resonant single-conductor system for power transmission, as compared to wireless systems transmitting the electromagnetic waves through the air. First of all, it is high efficiency of power transmission (more than 85%) combined with rather simple electric equipment. The high efficiency is stipulated by presence of the resonant conductive channel between the generator and the receiver. The second important difference is that the resonant single-conductor system does not need transmitting and receiving antennas, it is only necessary to ensure resonance conditions for the high-voltage and low-voltage windings of the Tesla transformer. The third important conclusion is that power transmission can be carried out without using overhead lines. In this case, the most frequent reason for interrupting the power supply as a result of glaze, storm, fire or earthquake. The future energetics will use reliable, and safe underground or underwater single-conductor cables, as well as sea-water or ground as a resonant single-conductor line.

2. Resonant techniques for electric power transmission along underground high-frequency single-conductor waveguides is reducing the ground occupation and improving ecological conditions. Taking into consideration a reserve ground 15 to 30 m on both sides of the overhead transmission line, its replacement with the underground cable single-conductor line will reduce the ground occupation by 3 to 6 million hectares, in case of total line length of 10^6 km. This figure includes the following regions: St.-Petersburg (776 km of transmission lines): 40 000 hectares; Moscow (1000 km of transmission lines): 60 000 hectares. The cost of land works in the cities will make 1500 to 3 000 US dollars per 100 $м^2$.
3. Today, the energy security of Russia is a key component of the national security. The main directions for heightening the energy security are the following. Distributed energy production combined with a united power grid. Increasing the volume of decentralized energy production to 50% of the total produced energy. The following measures are necessary to ensure the energy security of Russia: development of modern energy technologies and state programs for using them, as well as attractive investment and tax policy aimed at developing new industrial power branches and manufacturing new materials, increasing the reliability of electric power supply under the conditions of natural and technogenic factors or terrorist attacks.
4. New wireless resonant electric-power system for different power consumers are considered including stationary single-conductor transmission lines and single-trolley and noncontact high-frequency electric transport, using non-metal conductive media. In this case, several components of conventional transmission lines, such as metal wires, insulators and cables, are excluded, which results in considerable reduction of the equipment cost.

REFERENCES

Kharchenko, V., & Vasant, P. (2018). *Handbook of Research on Renewable Energy and Electric Resources for Sustainable Rural Development*. doi:10.4018/978-1-5225-3867-7

Kharchenko, V., & Vasant, P. (2020). *Handbook of Research on Energy-Saving Technologies for Environmentally-Friendly Agricultural Development*. Academic Press.

RF Patent No. 2143775. (1999). Method and Apparatus for Electric Power Transmission. *Inventor's Bulletin, 36.*

RF Patent No. 2172546. (2001). Method and Apparatus for Electric Power Transmission. *Inventor's Bulletin, 23.*

Sarbon, S. (2013). *The unresolved patents of Nikola Tesla. The inventions that Tesla did not protect.* Nikola Tesla Museum.

Strebkov, D. S. (2002). Possibility of Transmitting Electric Power without Metal Wires. *Proceedings of the Russian Academy of Agricultural Sciences, 1*, 47–50.

Strebkov, D. S. (2014). Perspectives of using technologies of Nikola Tesla in up-to-datas power engineering. *Light & Engineering, 2*, 4–14.

Strebkov, D. S., Avramenko, S. V., & Nekrasov, A. I. (2002, July–August). Investigation of Electric Energy Transmission Process in non-Metallic Conducting Channels. *New Energy Technologies, 4*(7), 24–27.

Strebkov, D. S., & Nekrasov, A. I. (2005). Development of Resonant Methods for Electric Power Transmission in Russia. *Proceedings of the 8-th Symposium "Electric Engineering – 2010"*.

Strebkov, D. S., Nekrasov, A. I., & Avramenko, S. V. (2002). Investigation of Electric Power Transmission through Non-Metal Conducting Channels. *Proceedings of the 3-rd International Scientific-and-Practical Conference "Ecology and Agricultural Techniques", 1*, 45–52.

Strebkov, D. S., Nekrasov, A. I., & Nekrasov, A. A. (2019). *Renewable Energy and Power Supply Challenges for Rural Regions*. Resonant Power System for Electric Transport.

Strebkov, D. S., Nekrasov, A. I., Trubnikov, V. Z., & Nekrasov, A. A. (2018). Single-wire resonant electric power systems for renewable based electric grid. In Handbook of Research on Renewable Energy and Electric Resources for Sustainable Rural Development. doi:10.4018/978-1-5225-3867-7.ch019

Tesla, N. (1956). *Lectures. Patents. Articles*. N. Tesla Museum.

Tesla, N. (1978). *Colorado Springs Notes 1899 – 1900*. Nolit.

Tesla, N. (1999a). *My Inventions. Electrical Experimenter, February, March, June and October. 1919. In Nikola Tesla. Selected works*. Academic Press.

Tesla, N. (1999b). The Problem of Increasing Human Energy. The Century Illustrated Monthly Magazine. June, 1900. In *Nikola Tesla. Selected works*. Academic Press.

Tesla, N. (1999c). World System of wireless transmission of energy telegraph and telephone Age. Oct.16, 1927. In Nikola Tesla, Articles (vol. 1). Academic Press.

Tesla, N. (2008). *From Colorado to Long island: Research Notes – Colorado Springs 1899-1900 – New York. 1900-1901*. Nikola Tesla Museum.

Vasant, Weber, & Dieu. (2016). *Handbook of Research on Modern Optimization Algorithms and Applications in Engineering and Economics*. Academic Press.

Vasant & Voropai. (2016). *Sustaining Power Resources through Energy Optimization and Engineering*. Academic Press.

Chapter 17
Development of Box Behnken Design to Predict the Optimum Operating Condition of Rectangular Sheet Membrane to Increase Permeate Flux

Anirban Banik
https://orcid.org/0000-0002-2987-2713
National Institute of Technology, Agartala, India

Sushant Kumar Biswal
https://orcid.org/0000-0002-1413-7206
National Institute of Technology, Agartala, India

Tarun Kanti Bandyopadhyay
National Institute of Technology, Agartala, India

ABSTRACT

The chapter focuses on the implementation of Box Behnken Design (BBD) to increase permeate flux of rectangular sheet membrane. Box Behnken Design (BBD) was used to optimize the membrane operation by predicting the optimum conditions. The factors such as operating pressure, feed velocity, and pore size were selected as the input of the model. The study illustrates the optimum conditions of operating pressure, feed velocity, and pore size, which was found to be 14.5Pa, 0.179 m/s, and 0.59µm respectively. Analysis of variance was used to identify the significant terms in the model equation. The effect of input parameters on the model output evaluated using Pareto analysis. It shows that operating pressure is the most significant parameter in the developed model. The BBD predicted results follow the actual results with high accuracy.

DOI: 10.4018/978-1-7998-3645-2.ch017

INTRODUCTION

Membrane filtration method based on semi-permeable barrier implemented to treat feed stream without using any chemicals. The process divides the flow through the membrane into two parts, namely, permeate and concentrate the flux. The stream of feed which is allowed to flow through the membrane is said to permeate flux and collected in permeate tank. The stream of feed which is not allowed to flow through the membrane is said to be concentrate flux. Zhao et al. developed a ceramic membrane to improve the quality of municipal wastewater. The study shows that membrane can remove chemical oxygen demand and phosphorus by 90% and 99.9% respectively. The research reveals that it is required to develop effective cleaning strategies for the effective operation of the proposed membrane (Zhao, Li, Li, & Li, 2019). Zhang et al. proposed an integrated method of ozonation, membrane and biologically active carbon to treat and reuse wastewater. The study illustrates that the integrated method poses high performance in treating and improving the wastewater quality (Zhang et al., 2020). Song et al. developed a sponge moving bed membrane with pre-positioned anoxic biofilter to treat saline wastewater. The method illustrates a high total organic carbon (TOC) removal efficiency of 92.8% to 96.2% and excellent total nitrogen removal of 93.2%. Regarding the membrane filtration process, the filtration time is extended from 10days to 44 days by mitigating the cake layer resistance (Song et al., 2018). The chapter focuses on developing Box Behnken design (BBD) to predict the optimum condition of the rectangular sheet membrane.

MOTIVATION AND BACKGROUND

The Global scarcity of water resource and rapid degradation of water quality has advocated for water reclamation and reuse(Lyu, Chen, Zhang, Fan, & Jiao, 2016). So, the membrane is widely used for water reclamation and overcoming water scarcity.

The membrane was a well-known technique since the eighteen century and implemented for water reclamation. The membrane widely used in different industries like the food industry, paper industry, and rubber industry (Mohammadi, Man, Hassan, & Yee, 2013)(Zhou, Zhao, Bai, Zhang, & Tang, 2012). Membrane separation technique are classified as microfiltration(Lim & Bai, 2003), ultra filtration(Abdessemed, Nezzal, & Ben Aim, 2000) and reverse osmosis (Sahinkaya et al., 2019)(Tong et al., 2020) and implemented to improve quality of water feed stream. Membrane attracted many researchers and environmental scientists for its ability to produce high quality of permeate with high reuse potential, but it is limited in its application for its rapid fouling tendency. Fouling is due to partial and complete pore blocking of the membrane during the separation process (Tong et al., 2020)(Liu, Wu, Hou, & Wang, 2020). To predict optimum operating factors and enhance permeate flux, Box Behnken method used in the present study.

Thamaraiselvan et al. studied non-woven carbon nanotube for filtration of water, and the developed membrane investigated in terms of surface morphology, water permeability, selectivity and chemical resistance. The proposed membrane illustrates high permeability and high rejection of globular protein. The result also shows simplicity in fabrication, modification and support free configuration to enhance chemical stability(Thamaraiselvan, Lerman, Weinfeld-Cohen, & Dosoretz, 2018). Jeong et al. investigated the feasibility of anaerobic ceramic membrane bioreactor (AnCMBR) and compared it with conventional membrane bioreactor (CMBR). The AnCMBR shows lower fouling tendency compared to CMBR. Surface charge and hydrophobic behaviour were studied, it found that the sludge of AnCMBR was more hydrophobic compared to CMBR (Jeong, Kim, Jin, Hong, & Park, 2018). Banik et al. studied the hydrodynamic property of disc-shaped membrane using computational fluid dynamics (CFD) based solver. Hydrodynamic properties and permeates quality of stationary and rotating disc membrane compared. The study demonstrates the superior performance of rotating over the stationary membrane(Banik, Bandyopadhyay, & Biswal, 2017). Li et al. implemented composite flocculant polymerization hydrolysis product of aluminium, iron, sulfate, silicate, and boron (PAFSSB) as pre-treatment technique for reverse osmosis (RO) membrane to improve quality of pulp and paper wastewater. Use of PAFSSB before membrane separation shows higher membrane efficiency and higher chemical oxygen demand (COD) removal rate (Li & Zhang, 2011). Changmai et al. studied the electro-coagulation combined with microfiltration membrane to improve the quality of oily wastewater. The study reported a considerable reduction in oil and grease from 35 mg/L to 10.2 mg/L within 20 min of membrane operation.(Changmai, Pasawan, & Purkait, 2018). Kusworo et al. developed nano-TiO2 loaded polysulfone membrane and studied the effectiveness of the membrane on the removal of the pollutant from natural rubber wastewater treatment. The study illustrates improved fouling resistance property and hydrophilic behaviour of the membrane(Kusworo, Ariyanti, & Utomo, 2020).

NOVELTY AND OBJECTIVE OF THE STUDY

To the best of our knowledge, application of Box Behnken design to predict permeate flux of rectangular sheet membrane not reported earlier. Most of the studies in the concerned field are experimental based study with limited application of soft computing. The present study illustrates the application of Box Behnken design an effective method to predict membrane performance and lowers the cost and time of experimentation.

The chapter focuses on predicting optimum conditions of rectangular sheet membrane using Box Behnken Design to increase permeate flux. The BBD model is constructed and implemented using operating pressure, inlet velocity and pore size as the input of the model and permeate flux as the model response. Analysis of variance (ANOVA) is used to predict the significant terms of the developed model equation. The effect of input parameters on the model outcome evaluated using Pareto analysis. The study also compares predicted results with experimental results to evaluate the adequacy of the developed model.

RESEARCH CONTRIBUTION

The study provides a detailed overview of the membrane separation process and its importance to overcome the water scarcity problem. Detail review of literature illustrates a clear picture among readers regarding the use of Box Behnken method to predict optimum operating factors of the membrane. In the present study, BBD model developed using a limited experimental run, thus minimizing the cost and time of experimentation. The study also includes Pareto analysis, and it used to demonstrate the terms of high significance in the developed model. The predicted results are compared with experimental results to validate the acceptability of the model.

EXPERIMENTAL DESCRIPTION

Schematic diagram of the experimental setup demonstrated using Figure 1. The feed (Rubber Industrial effluent) was initially stored in the storage tank to monitor the pH and temperature of the feed. It is essential because the cellulose acetate membrane used in the study operates in the pH range 4 to 6 and the operating temperature below 35^{0}C. The feed pumped inside the membrane module with the help of a centrifugal pump. The feed separated into two parts during the membrane separation technique, namely reject/concentrate and permeate flux. The reject is re-circulated into the feed tank whereas permeate collected inside the permeate tank. The feed samples collected from common effluent treatment plant situated in TIDC (Tripura Industrial Development Corporation LTD.) complex in Bodhjung Nagar, India. The characterizations of the sample conducted in the institute laboratory. Parameters like total dissolved solids, total suspended solids, total nitrogen, oil and grease, biochemical oxygen demand, and sulfide, tested and their respective values listed in Table 1.

Figure 1. Schematic illustration of experimental setup

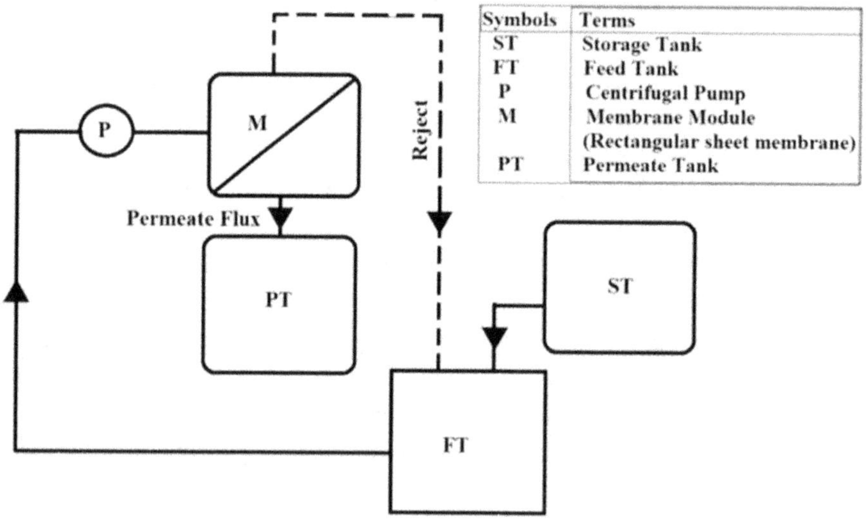

Table 1. Characterization of feed sample

Sl. No.	Parameters	Units	Values
1.	Total dissolved solids	mg/L	3120
2.	Total suspended solids	mg/L	352
3.	Sulfide	mg/L	20.7
4.	Biochemical oxygen demand	mg/L	725
5.	Oil and Grease	mg/L	11
6.	pH		5.5

BOX BEHNKEN DESIGN (BBD)

Experimental design of the optimization of rectangular sheet membrane was achieved by using Box Behnken design (BBD). It is a combination of both statistical and mathematical technique to predict the optimum conditions and study the effect of input parameters on the model response. BBD is widely used in industrial research to develop new methods. Thus leading to the determination of the boundary for the parameters in which the model response is present. The BBD model was developed by considering operating pressure, feed velocity, and pore size as independent parameters and permeate flux as dependent variables. The BBD is the ideal method which allows testing of the model using lack of fit when desired numbers of experimental design points provided. BBD defined as a rotatable second-order design based on incomplete factorial design. The experimental run required for BBD model estimated by using N=2G× (G-1) +C_0 (where G is the number of inputs and C_0 represents the number of central points) (Swati, Ghosh, & Thakur, 2019)(Ghelich, Jahannama, Abdizadeh, Torknik, & Vaezi, 2019). The results predicted from experiments implemented to determine the coefficients of the regression equation and predict optimum conditions to achieve a maximum model response. The effects of parameters on permeate flux and interactions among the parameters also investigated. Equation 1 illustrates the second-order regression equation consists of three linear, quadratic and interaction terms implemented to fit the design points of the experiments and to predict model response.

$$Y = \alpha_0 + \alpha_1 X_1 + \alpha_2 X_2 + \alpha_3 X_3 + \alpha_{11} X_1^2 + \alpha_{22} X_2^2 + \alpha_{33} X_3^2 + \alpha_{12} X_1 X_2 + \alpha_{13} X_1 X_2 + \alpha_{23} X_2 X_3 \quad (1)$$

Where, α represents the regression coefficients of the developed equation whereas X_1, X_2, X_3 represents the input parameters of the proposed model. The F- test and p-value also used to find the significant parameters in the developed equation. Two dimensional and three-dimensional plots used to investigate the individual and their corresponding effect of parameters on the model outcome.

RESULTS AND DISCUSSIONS

Development of Model Equation

From experimental investigation, it observed that independent factors such as operating pressure, feed velocity and pore size of the membrane affect the model response (permeate flux). Box Behnken model developed using parameters mentioned above, such as operating pressure, pore size, and inlet velocity was selected as the input of the model and permeate flux as model output. BBD is used to establish the relationship between the independent and dependent parameters of the model. Table 2 shows the complete design matrix with their corresponding responses. The centre design points repeated three times to estimate the quadratic terms of the constructed BBD model. Table 3 shows the determined values of all the coefficients of the developed regression equation. The results of the table 3 evaluated using the experimental design point to construct model equation 2 for the response (permeate flux).

$$\text{Permeate flux} = 75.86 - [8.297\times(X_1)] - [2.664\times(X_2)] - [2.326\times(X_3)] - [4.98\times(X_1)^2] - [3.96\times(X_2)^2] - [0.82\times(X_3)^2] + [2.34\times(X_1)\times(X_2)] + [0.64\times(X_1)\times(X_3)] + [0.61\times(X_2)\times(X_3)] \quad (2)$$

In equation 2, X_1, X_2, and X_3 are operating pressure, pore size, and feed velocity, respectively. From equation 2, it found that response of the model depends on the linear, square and 2-way interaction of the independent factors such as operating pressure, pore size, and inlet velocity. Two-dimensional plots and three-dimensional plots are very useful for collecting information regarding experimental design and provide insight into the developed model. The interaction between the input and output of the model illustrated using the plots. Figure 2(a-c) shows the contour plot of the permeate flux of the rectangular sheet membrane. Figure 2(a) demonstrate the plot of permeate flux at constant feed velocity. The plot shows that increases with an increase in operating pressure initially, but as the operation progresses, permeate flux reduces due to fouling of the membrane. Figure 2(b) shows the plot of permeate flux at constant operating pressure. It shows that permeates of the membrane increases initially but reduces with gradually as operation progress due to high resistance offered by the membrane pore size. Figure 2(c) shows the permeate flux plot at constant pore size where the trends of the plot illustrate that permeate of the membrane reduces with the progress of the filtration. The surface plot is shown by Figure 3(a-c), the highest point in the plot demonstrates the responses of the design points which estimated from input factors such as operating pressure, pore size, and inlet feed velocity.

Table 2. Design matrix for Box Behnken Model

Sl. No.	Coded Parameters			Decoded Parameters			Model Response (%)
	Operating pressure (Pa)	Pore Size (μm)	Feed velocity (m/s)	Operating Pressure (Pa)	Pore Size (μm)	Feed Velocity (m/s)	
1	-1	0	1	14.5	1.6	5.126	77.61
2	0	0	0	472.75	1.6	2.6525	75.86
3	0	-1	-1	472.75	0.2	0.179	78.12
4	-1	0	-1	14.5	1.6	0.179	79.86
5	0	0	0	472.75	1.6	2.6525	75.86
6	1	0	-1	931	1.6	0.179	61.23
7	-1	-1	0	14.5	0.2	2.6525	80.24
8	1	-1	0	931	0.2	2.6525	59.73
9	1	1	0	931	3	2.6525	58.27
10	-1	1	0	14.5	3	2.6525	69.44
11	0	0	0	472.75	1.6	2.6525	75.86
12	0	1	-1	472.75	3	0.179	72.38
13	1	0	1	931	1.6	5.126	61.54
14	0	1	1	472.75	3	5.126	65.26
15	0	-1	1	472.75	0.2	5.126	68.57

Figure 2. Contour plots of permeate flux at (a) constant feed velocity; (b) operating pressure; and (c) constant operating pressure

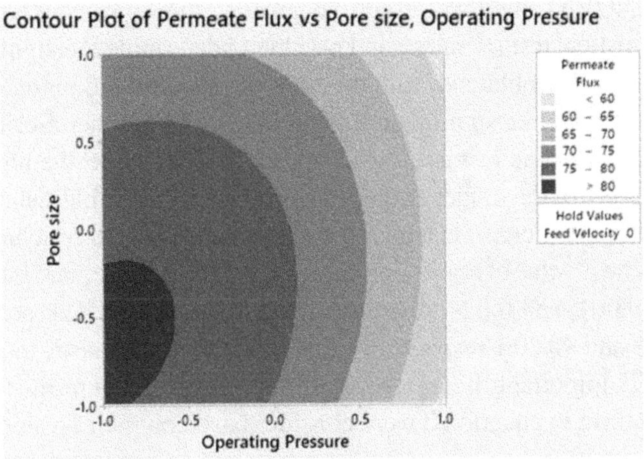

(a)

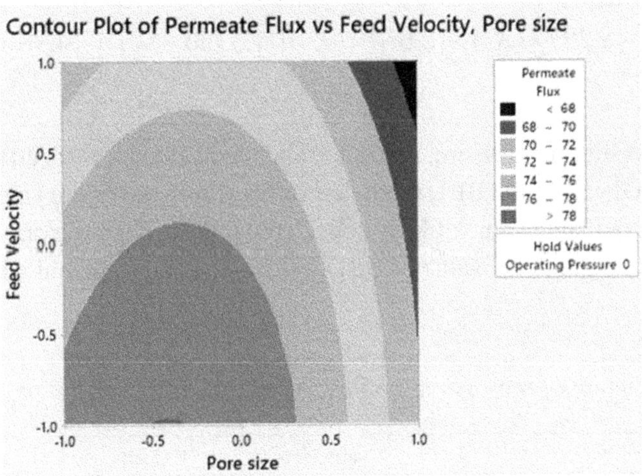

(b)

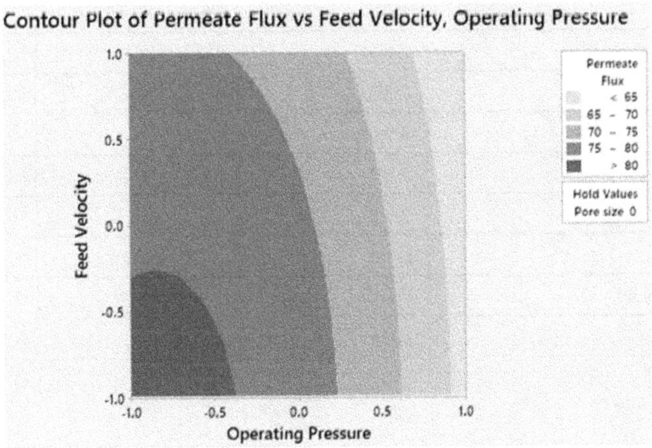

(c)

Analysis of Variance (ANOVA) Study

Analysis of variance (ANOVA) study was used to determine the significant terms of the model equations. It uses Fisher's statistical test (F-value) and p-value to determine the significance of each term of the developed equation. F-value obtained from the ratio of mean square regression value to error. The terms in equation 2 considered to be significant if the f-value is higher than f-critical and corresponding p-value less than 0.05. The F-value test used to calculate the value of R^2 the regression equations. The acceptability and workability of the model evaluated using lack of fit. Table 4 shows the regression coefficient, linear and square interactions of equation 2 are significant, which concluded from f-value and its corresponding p-value. The p-value of regression coefficient, linear and square interaction was calculated and found to be 0.004, 0.001, and 0.024, respectively (less than 0.05). The predicted and adjusted R^2 were found to be 96.50% and 90.20% respectively. From ANOVA analysis it was found that all terms in equation 2 are not equally important. It helps to separate the significant terms from insignificant ones. From table 3 significant term in equation 2 were consider based on their f-value and p-value. Equation 3 shows the new model equation.

$$\textit{Permeate flux} = 75.86 - [8.297 \times (X_1)] - [2.664 \times (X_2)] - [2.326 \times (X_3)] - [4.98 \times (X_1)^2] - [3.96 \times (X_2)^2] \tag{3}$$

Here, X_1, X_2, and X_3 are operating pressure, pore size, and feed velocity, respectively. Figure 4 shows the plot between experimental values and BBD predicted values. From the figure value of R^2 is estimated and found to be 0.932. The calculated value of R^2 is considered to be acceptable one as the value is closer to unity. The value shows that predicted value of BBD follows the experimental value with high accuracy.

Table 3. Estimated coefficient of regression for Box Behnken Design (BBD) model

Factors	Coefficients	SE Coefficient	T-Value	P-Value
Constant	75.86	1.40	54.05	0.000
Operating pressure	-8.297	0.860	-9.65	0.000
Pore size	-2.664	0.860	-3.10	0.027
Feed velocity	-2.326	0.860	-2.71	0.042
$(\text{Operating pressure})^2$	-4.98	1.27	-3.94	0.011
$(\text{Pore Size})^2$	-3.96	1.27	-3.13	0.026
$(\text{Feed velocity})^2$	-0.82	1.27	-0.65	0.546
Operating pressure × Pore size	2.34	1.22	1.92	0.113
Operating pressure × Feed velocity	0.64	1.22	0.53	0.621
Pore Size × Feed velocity	0.61	1.22	0.50	0.638

Figure 3. Surface plot of permeate flux at (a) constant pore size; (b) constant operating pressure; and (c) constant feed velocity

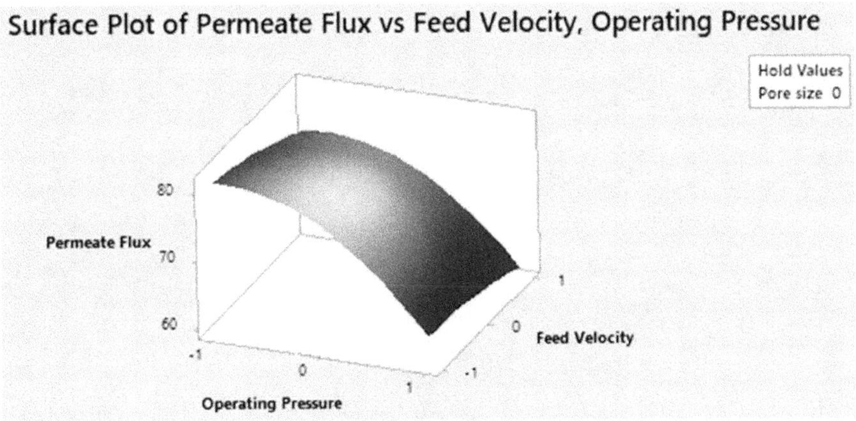

(a)

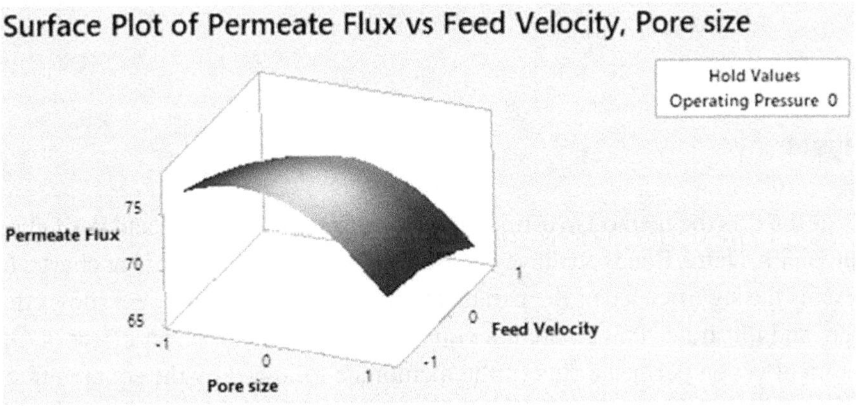

(b)

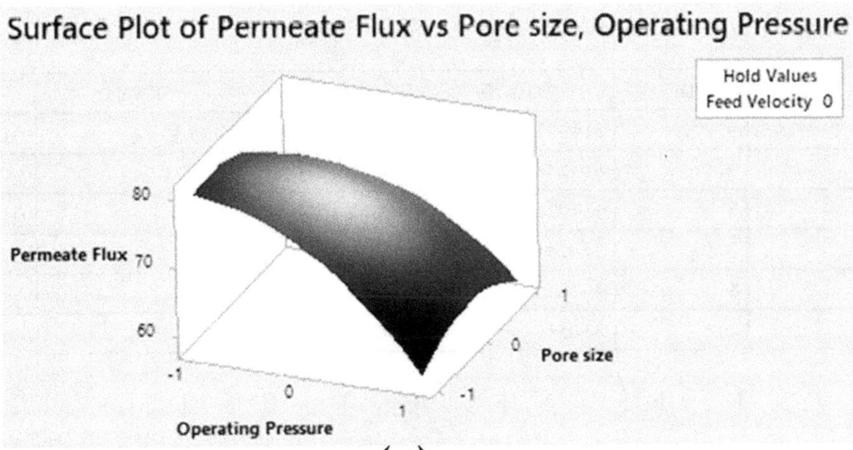

(c)

Figure 4. Experimental versus predicted results

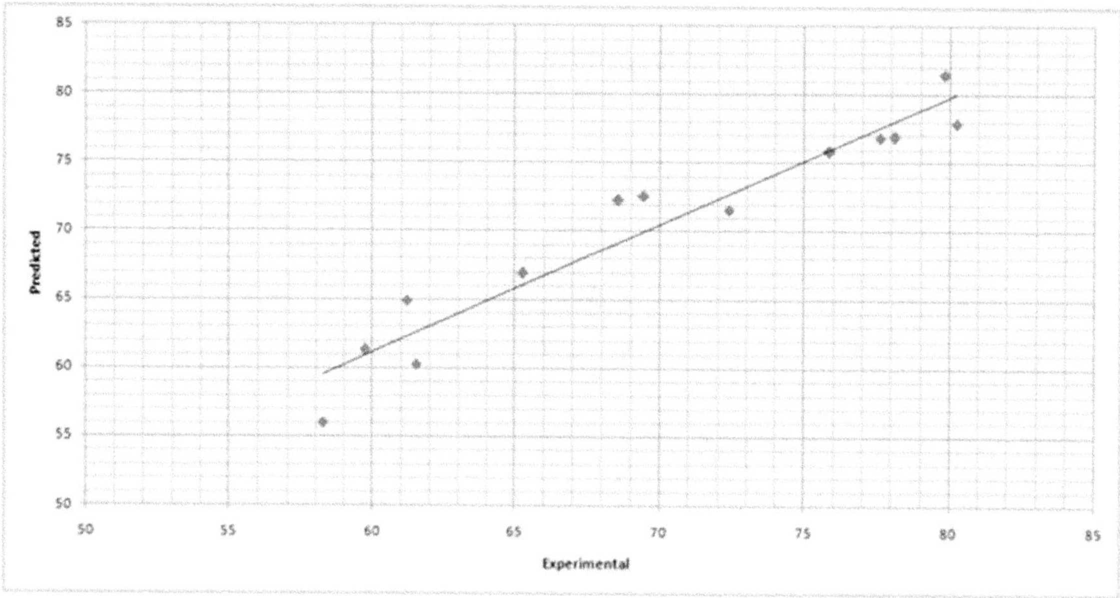

Pareto Analysis

Pareto Analysis defined as the method to estimate the significant parameters and their cumulative effect on model output(Banik, Dutta, Bandyopadhyay, & Biswal, 2018). It consists of bar chart where the length of the bar represents the significance of the parameter on model output. Figure 5 shows the significance of the parameters and illustrates that all are not equally important. The linear effect of Operating pressure has maximum effect on permeate flux of the membrane followed by the square effect of operating pressure and pore size, respectively.

Table 4. Analysis of Variance (ANOVA) Study

Source	DF	ADJ. SS	ADJ. MS	F-Value	P-Value
Model	9	814.992	90.555	15.32	0.004
Linear	3	650.844	216.948	36.71	0.001
Square	3	139.224	46.408	7.85	0.024
2-way Interaction	3	24.924	8.308	1.41	0.344
Error	5	29.550	5.910		
Total	14	844.541			

Figure 5. Pareto analysis

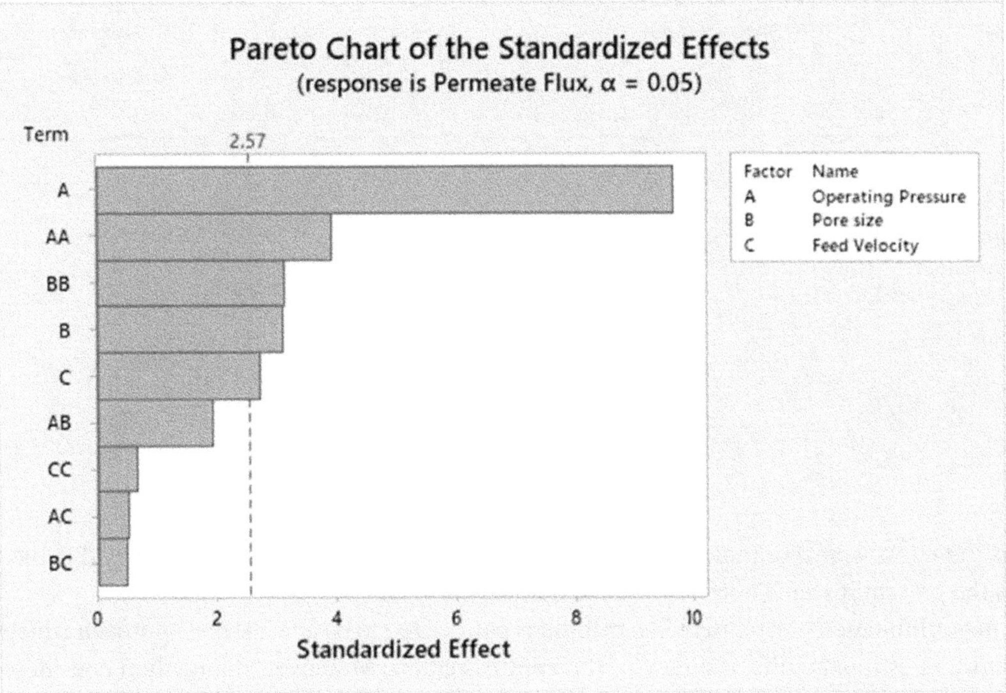

Process Optimization, and Optimum condition

Optimization is a mathematical technique of searching the optimum conditions within the defined boundary(Dutta, 2013)(Banik, Biswal, Majumder, & Bandyopadhyay, 2018)(Ganesan, Vasant, & Elamvazuthi, 2016)(Thomas, Karagoz, Ahamed, & Vasant, 2019)(Vasant, Zelinka, & Weber, 2019)(Vasant, Zelinka, & Weber, 2020)(Vasant, Marmolejo, Litvinchev, & Aguilar, 2019)(Zelinka, Tomaszek, Vasant, Dao, & Hoang, 2018). In the study, parameters such as operating pressure, feed velocity and pore size are used as model input and permeate flux as model output. Model output predicted using maximization objective function. Figure 6 shows the optimization plot of the developed model. The used membrane illustrates permeate flux of 83.31% corresponding to the optimum parameters (Pore size=0.59μm; operating pressure= 14.5Pa; feed velocity=0.179 m/s). When the experiments conducted by implementing optimum conditions, the membrane exhibits permeate flux of 80.56%. The experimental value holds a good agreement with the predicted value, thus justifying the acceptability of the model.

Advantages, Disadvantages, and Limitation

Box Behnken design (BBD) is widely applicable for optimization with a limited number of runs, thus limiting the experimental cost and time. It implemented for a minimum number of the experimental run, estimation of transformation, constant check on variance, the sequential building of higher-order design from a simple design, satisfactory graphical analysis, and ability to estimate internal errors. One of the significant disadvantages of BBD is that it only recommended for 3 numbers of factors. When the

Figure 6. Optimization plot

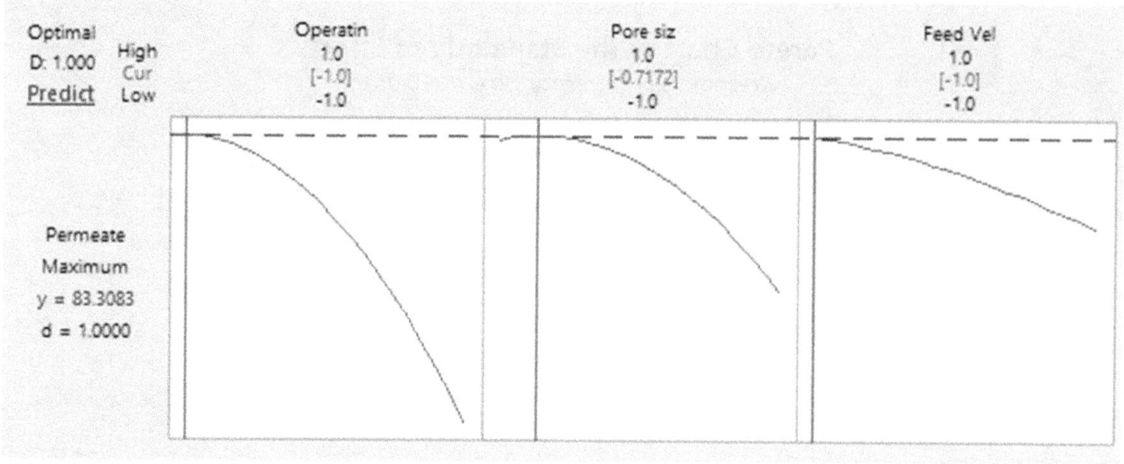

numbers of factors were 4 or more, the advantages of BBD disappear. BBD implemented in the present study as the present research consists of three numbers of factors.

The major limitation of Box Behnken method is considered to be a black-box approach which results in difficulty in estimating the accuracy of the approximation. Moreover, the method considered to be local analysis limiting its workability within the predefined boundary of the factors.

CONCLUSION

The study was carried out to increase permeate flux of rectangular sheet membrane using Box Behnken Design (BBD). The study shows that permeate flux of the membrane get affected by the factors such as pore size, feed velocity, and operating pressure. BBD was constructed to optimize the membrane operation and to predict the optimum conditions. BBD was used to evaluate the effect of independent factors on model outcome. Analysis of variance (ANOVA) analysis illustrate that all the terms in the model equation was not equally significant. According to the ANOVA, the square interaction of feed velocity and 2-way interaction of pore size, feed velocity and operating pressure were found to be insignificant. The Pareto study shows that operating pressure was the most significant one followed square interaction of operating pressure and pore size. The actual values were compared with the predicted values to evaluate the workability of the constructed model. It shows that predicted values follow the actual values with high accuracy. The optimum conditions for operating pressure, feed velocity and pore size were estimated and found to be 14.5 Pa, 0.179 m/s, and 0.59µm respectively. The permeate flux corresponding to the optimum conditions was found to be 83.31%. The optimum conditions can be used to increase the permeate flux and antifouling property of the membrane.

FUTURE RESEARCH DIRECTION

The research focuses on increasing permeate flux of rectangular sheet membrane using Box Behnken method. The study shows that different parameters have a different degree of significance on the outcome of the model. Future research will focus on implementing the outcome of the studies in the real field. Moreover, future research will be conducted to lower the cost membrane operation to creating a product for price-sensitive markets of Tripura. Thus, a cost optimization study will be carried out in future to lower the operational cost of the membrane.

REFERENCES

Abdessemed, D., Nezzal, G., & Ben Aim, R. (2000). Coagulation-adsorption-ultrafiltration for wastewater treatment and reuse. *Desalination*, *131*(1–3), 307–314. doi:10.1016/S0011-9164(00)90029-8

Banik, A., Dutta, S., Bandyopadhyay, T. K., & Biswal, S. K. (2018). Prediction of maximum permeate flux (%) of disc membrane using Response Surface Methodology (RSM). *Canadian Journal of Civil Engineering*, *46*(6), 299–307.

Banik, A., Bandyopadhyay, T. K., & Biswal, S. K. (2017). Computational fluid dynamics simulation of disc membrane used for improving the quality of effluent produced by the rubber industry. *International Journal of Fluid Mechanics Research*, *44*(6), 499–512. doi:10.1615/InterJFluidMechRes.2017018630

Banik, A., Biswal, S. K., Majumder, M., & Bandyopadhyay, T. K. (2018). Development of an adaptive non-parametric model for estimating maximum efficiency of disc membrane. *Int. J. Convergence Computing*, *3*(1), 3–19. doi:10.1504/IJCONVC.2018.091111

Changmai, M., Pasawan, M., & Purkait, M. K. (2018). Treatment of oily wastewater from drilling site using electrocoagulation followed by microfiltration. *Separation and Purification Technology*, *210*(February), 463–472. doi:10.1016/j.seppur.2018.08.007

Dutta, S. (2013). Optimization of Reactive Black 5 removal by adsorption process using Box-Behnken design. *Desalination and Water Treatment*, *51*(40–42), 7631–7638. doi:10.1080/19443994.2013.779597

Ganesan, T., Vasant, P., & Elamvazuthi, I. (2016). *Advances in Metaheuristics: Applications in Engineering Systems (1st Editio.)*. CRC Press. doi:10.1201/9781315297651

Ghelich, R., Jahannama, M. R., Abdizadeh, H., Torknik, F. S., & Vaezi, M. R. (2019). Central composite design (CCD)-Response surface methodology (RSM) of effective electrospinning parameters on PVP-B-Hf hybrid nanofibrous composites for synthesis of HfB 2 -based composite nanofibers. *Composites Part B: Engineering, 166*(Ccd), 527–541. doi:10.1016/j.compositesb.2019.01.094

Jeong, Y., Kim, Y., Jin, Y., Hong, S., & Park, C. (2018). Separation and Puri fi cation Technology Comparison of fi ltration and treatment performance between polymeric and ceramic membranes in anaerobic membrane bioreactor treatment of domestic wastewater. *Separation and Purification Technology, 199*(January), 182–188. doi:10.1016/j.seppur.2018.01.057

Kusworo, T. D., Ariyanti, N., & Utomo, D. P. (2020). Effect of nano-TiO2 loading in polysulfone membranes on the removal of pollutant following natural-rubber wastewater treatment. *Journal of Water Process Engineering, 35*(November), 101190. doi:10.1016/j.jwpe.2020.101190

Li, S., & Zhang, X. (2011). The study of PAFSSB on RO pre-treatment in pulp and paper wastewater. *Procedia Environmental Sciences, 8*, 4–10. doi:10.1016/j.proenv.2011.10.003

Lim, A. L., & Bai, R. (2003). Membrane fouling and cleaning in microfiltration of activated sludge wastewater. *Journal of Membrane Science, 216*(1–2), 279–290. doi:10.1016/S0376-7388(03)00083-8

Liu, X., Wu, J., Hou, L. an, & Wang, J. (2020). Fouling and cleaning protocols for forward osmosis membrane used for radioactive wastewater treatment. *Nuclear Engineering and Technology, 52*(3), 581–588. doi:10.1016/j.net.2019.08.007

Lyu, S., Chen, W., Zhang, W., Fan, Y., & Jiao, W. (2016). Wastewater reclamation and reuse in China: Opportunities and challenges. *Journal of Environmental Sciences (China), 39*, 86–96. doi:10.1016/j.jes.2015.11.012

Mohammadi, M., Man, H., Hassan, M., & Yee, P. (2013). Treatment of wastewater from rubber industry in Malaysia. *African Journal of Biotechnology, 9*(38), 6233–6243. http://www.ajol.info/index.php/ajb/article/view/92237

Sahinkaya, E., Tuncman, S., Koc, I., Guner, A. R., Ciftci, S., Aygun, A., & Sengul, S. (2019). Performance of a pilot-scale reverse osmosis process for water recovery from biologically-treated textile wastewater. *Journal of Environmental Management, 249*(August), 109382. doi:10.1016/j.jenvman.2019.109382

Song, W., Li, Z., Ding, Y., Liu, F., You, H., Qi, P., Wang, F., Li, Y., & Jin, C. (2018). Performance of a novel hybrid membrane bioreactor for treating saline wastewater from mariculture: Assessment of pollutants removal and membrane filtration performance. *Chemical Engineering Journal, 331*, 695–703. doi:10.1016/j.cej.2017.09.032

Swati, G. P., & Thakur, I. S. (2019). Biodegradation of pyrene by Pseudomonas sp. ISTPY2 isolated from landfill soil: Process optimisation using Box-Behnken design model. *Bioresource Technology Reports, 8*(September), 100329. doi:10.1016/j.biteb.2019.100329

Thamaraiselvan, C., Lerman, S., Weinfeld-Cohen, K., & Dosoretz, C. G. (2018). Characterization of a support-free carbon nanotube-microporous membrane for water and wastewater filtration. *Separation and Purification Technology, 202*, 1–8. doi:10.1016/j.seppur.2018.03.038

Thomas, J. J., Karagoz, P., Ahamed, B. B., & Vasant, P. (2019). *Deep Learning Techniques and Optimization Strategies in Big Data Analytics*. IGI Global.

Tong, X., Cui, Y., Wang, Y. H., Bai, Y., Yu, T., Zhao, X. H., Ikuno, N., et al. (2020). Fouling properties of reverse osmosis membranes along the feed channel in an industrial-scale system for wastewater reclamation. *Science of the Total Environment, 713*, 136673. doi:10.1016/j.scitotenv.2020.136673

Vasant, P., Marmolejo, J. A., Litvinchev, I., & Aguilar, R. R. (2019). *Nature-inspired meta-heuristics approaches for charging plug-in hybrid electric vehicle*. Wireless Networks., doi:10.100711276-019-01993-w

Vasant, P., Zelinka, I., & Weber, G.-W. (2019). *Intelligent Computing & Optimization*. Springer International Publishing. doi:10.1007/978-3-030-00979-3

Vasant, P., Zelinka, I., & Weber, G.-W. (2020). *Intelligent Computing and Optimization*. Springer International Publishing. doi:10.1007/978-3-030-33585-4

Zelinka, I., Tomaszek, L., Vasant, P., Dao, T. T., & Hoang, D. V. (2018). A novel approach on evolutionary dynamics analysis – A progress report. *Journal of Computational Science, 25*, 437–445. doi:10.1016/j.jocs.2017.08.010

Zhang, K., Zhang, Z., Wang, H., Wang, X., Zhang, X., & Xie, Y. F. (2020). Synergistic effects of combining ozonation, ceramic membrane filtration and biologically active carbon filtration for wastewater reclamation. *Journal of Hazardous Materials, 382*(August), 121091. doi:10.1016/j.jhazmat.2019.121091

Zhao, Y., Li, P., Li, R., & Li, X. (2019). Characterization and mitigation of the fouling of flat-sheet ceramic membranes for direct filtration of the coagulated domestic wastewater. *Journal of Hazardous Materials*, 121557. PMID:31735469

Zhou, Y., Zhao, H., Bai, H., Zhang, L., & Tang, H. (2012). Papermaking Effluent Treatment: A New Cellulose Nanocrystalline/Polysulfone Composite Membrane. *Procedia Environmental Sciences, 16*, 145–151. doi:10.1016/j.proenv.2012.10.020

Chapter 18
Computational Intelligence in the Reversible Data-Hiding Processes Using Radon Transform

B. Bazeer Ahamed
https://orcid.org/0000-0003-1559-8386
Al Musanna College of Technology, Oman

ABSTRACT

Data hiding has emerged as a major research area due to the phenomenal growth in internet and multimedia technologies. Securing data transmitted over the internet becomes a challenging issue caused in digitization and networking over the past decade. Data hiding schemes have been adopted to protect digital media content which involves confidential data such as text, video, audio, images, and compression coding. A good reversible data hiding scheme is characterized by the possession of attributes like reversible, imperceptible, high payload capacity, and robustness. By reversible, it's meant that the extraction of the payload as well as the restoration of the host image perfectly from the stego image. Secondly, the imperceptible stego image resemblance against the cover/host image. Finally, robustness counts for the ability to sustain the secret payload against both intentional and unintentional attacks; it has been observed that all the proposed algorithms are more robust and reversible against various attacks in lower bit error rate and higher normalization coefficient.

INTRODUCTION

An efficient reversible data hiding method must be robust against a wide range of image processing operations such as image enhancement, cropping, rotation, scaling, compression and signal processing operation such as addition of noise (Abdulfetah et al, 2010). However, conventional data hiding algorithms are more sensitive to geometric distortions. Hence, radon transform is introduced to perform rotation, scaling and translation operations on the cover image. These operations change the positioning of the

DOI: 10.4018/978-1-7998-3645-2.ch018

secret bits (Bedi et al, 2013). Without taking inverse Radon Transform, it is very difficult to detect the embedded data and subsequently this increases the security of embedded payload.

In this chapter, the data embedding is performed in a hybrid domain. The cover image is first transformed from spatial domain to radon domain and then this radon image is applied with integer lifting wavelet transform (Chakraborty et al, 2013). Various orthogonal and bi-orthogonal wavelets are used for the implementation of reversible integer lifting wavelet transform. Then the middle bit planes of high frequency lifting coefficients are compressed using arithmetic coding to provide space for embedding secret payload (Cheddad et al, 2010).Radon transform (RT) was introduced in 1917 by the Austrian mathematician Johann Radon. The Radon Transformation is a fundamental tool, which is used in various applications such as radar imaging, geophysical imaging, nondestructive testing and medical imaging (Coltuc, D 2011). The Radon transform is applied to transform the image from the spatial domain to the Radon domain and the inverse radon transform does the reverse. Radon transform is a linear transform and an effective method to analyze signal between the spatial domain and its projection space. It represents the image as a collection of projections along various directions (Gerami et al 2012). It computes the projection of the image intensity along a radial line oriented at a specific angle as shown in Figure 1.

Figure 1. Projection of image at a specific angle of rotation

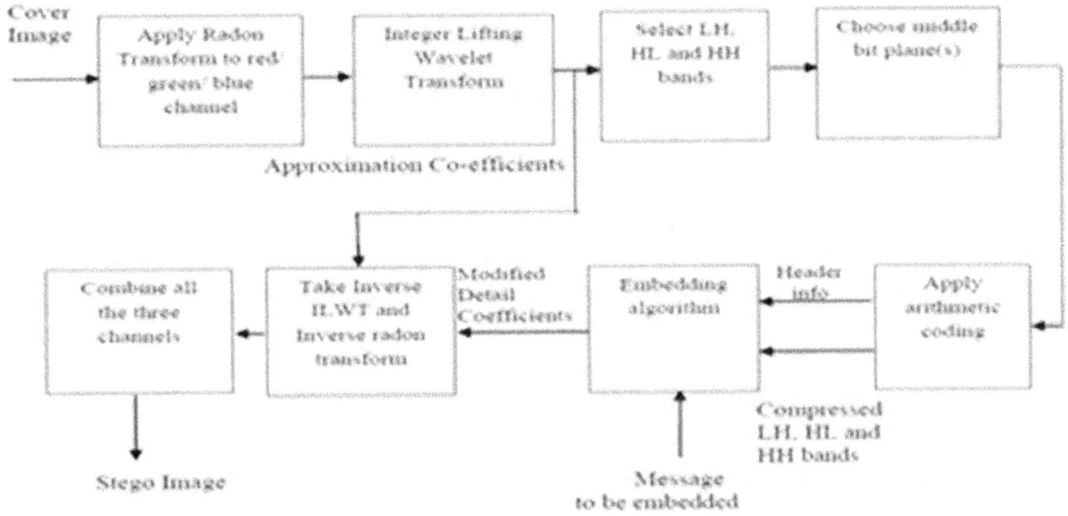

For each angle θ and at each distance φ, the intensity of a ray perpendicular to the ρ axis is summed up at R (ρ,θ). Radon transform converts rectangular coordinates (x, y) into polar coordinates R (ρ,θ). The simplest form of discrete Radon transform is to select finite number of the angular variable of projection, then take the summation on the discrete image along the projection line.

LITERATURE REVIEW

Basically reversible data hiding techniques are classified into three categories. They are spatial domain, frequency domain and compressed domain techniques. In spatial domain, pixel values are modified directly to embed the confidential data. Most of the spatial domain techniques are developed based on two principles, i.e., Difference expansion (DE) (Ahamed, B. B., & Ramkumar, T.,2018) and histogram modification (Ahamed, B. B., & Ramkumar, T, 2015). The algorithms based on difference expansion provide high embedding capacity (Hu et al 2008) while the algorithms based on histogram shifting contribute good visual quality to the extracted image (Huang, HC et al, 2010). Frequency domain techniques involve the calculation of frequency coefficients of the image using some reversibility guaranteed transforms like discrete wavelet transform, discrete fourier transform, discrete cosine transform, integer lifting transform, slantlet transform, curvelet transform, discrete radon transform, etc., Then the frequency coefficients are modified to embed secret data. In compression domain, the cover image is compressed to minimize the memory required and the bandwidth space of the embedded image.

MAIN FOCUS OF THE CHAPTER

Robustness Based Schemes

Some systems were designed to address the robustness property, where the payload capacity and imperceptibility are treated with a less concern. (Ahamed, B. B., & Ramkumar, T. 2016) suggested one such scheme that is based on patchwork theory, permutation scheme and the distribution features of pixel groups. The stego images generated were robust to JPEG2000/JPEG compression and had no salt-and-pepper noise. However the payload size was only 1,024 bits and picture quality did not exceed 38 dB. Another lossless, robust data hiding scheme based on shifting the mathematical difference values in a block was proposed by (Ishtiaq et al,2010). They embedded the secret payload into blocks. The method was resistant to inadvertent JPEG compression as the image was separated into two zones, namely bit-0-zone and the bit-1-zone. The resulting images exhibited an increased payload capacity, but at the cost of bit error rate and picture quality of the output image.

Radon Transform Based Schemes

Data embedding is also performed in radon domain, which shows a considerable improvement in bit error rate. (Lee, et al, 2007) have introduced a radon-based approach to incorporate translation invariance properties to the payload. (Li et al, 2013) have proposed an RST invariant watermarking technique by utilizing the Fourier transform and transforming them to log polar coordinates, which are quite flexible towards rotation, scaling and translation attacks.

Data Hiding Using Optimization Techniques

Evolutionary algorithms such as PSO (Lin et al, 2010), Genetic Algorithm (Liu et al 2014) and Bacterial foraging (Niu et al 2011) are widely used for data hiding. These evolutionary techniques are used to identify coefficients of the image in transform domain to embed data. But there is a problem in finding the optimal values of multiple scaling factors (MSFs). This problem can be overcome by combining evolutionary techniques with the transform domain techniques. (Saha, B & Sharma S, 2012), have combined the Genetic algorithm with SVD to find values of MSFs. However, (Thabit, R & Khoo, 2013) found that the algorithm proposed by Lee (2007) caused a false positive detection, even if the embedded watermark is different or nonexistent. (Venkatanarasimhulu, C & Satyaprasad 2011) have utilized multi-objective Ant colony optimization in LWT – SVD domain to find the values of MSFs.

(Yang et al, 2011) proposed a discrete wavelet transform based image watermarking technique with particle swarm optimization to search optimal parameters. (Li & Wang 2013) proposed a steganography technique, using particle swarm optimization algorithm to find the substitution matrix and used entropy coding for embedding secret data. Hung et al (2010) proposed a data hiding algorithm, which used ant colony algorithm to find the substitution matrix for the cover image, and the secret data is embedded in that matrix.

The Main motivation of this research is to implement novel reversible data hiding algorithms in spatial and frequency domains to augment the security level by enhancing the imperceptibility, robustness, embedding capacity and reversibility. This research work is an investigation of certain algorithms for reversible data hiding in spatial and frequency domains. The performance of any reversible data hiding system depends on four important criteria, namely, imperceptibility, data embedding capacity and robustness. As they are interlinked with one another and improvement in one or more factor is achieved at the cost of distortion of the other. Hence, the proposed works have established a balanced compromise between the three extremities for an optimal data hiding in the natural images.

Disadvantage is as the Radon transform performs rotation, scaling and translation operations on the cover image, it changes the positioning of the secret bits. Hence, it is very difficult to detect the embedded data without taking the inverse Radon transform and subsequently increases the security of the embedded payload. Efforts should either investigate the feasibility of new algorithms for enhancing the embedding capacity and achieving better image quality in the future scope of this research.

SOLUTIONS AND RECOMMENDATIONS

The majority of the methods discussed in the literature addressed only a few of the desired characteristics, namely lossless/reversible, imperceptible, high payload capacity and robustness, and not all. The proposed method of reversible data hiding is based on bit plane embedding in radon and integer lifting wavelet domain. It aims to meet all the desired characteristics to an optimal level. The block diagram of the proposed data embedding algorithm is shown in Figure 2.

Data Embedding Algorithm

Input: Cover image with M-rows and N-columns, Secret payload bits.

Output: Stego Image

Step 1: Read the cover image of size M x N.
Step 2: Separate color channels (Red, Green and Blue) of the cover image.
Step 3: Apply radon transform on any one of the channels.

Figure 2. Block diagram of embedding process

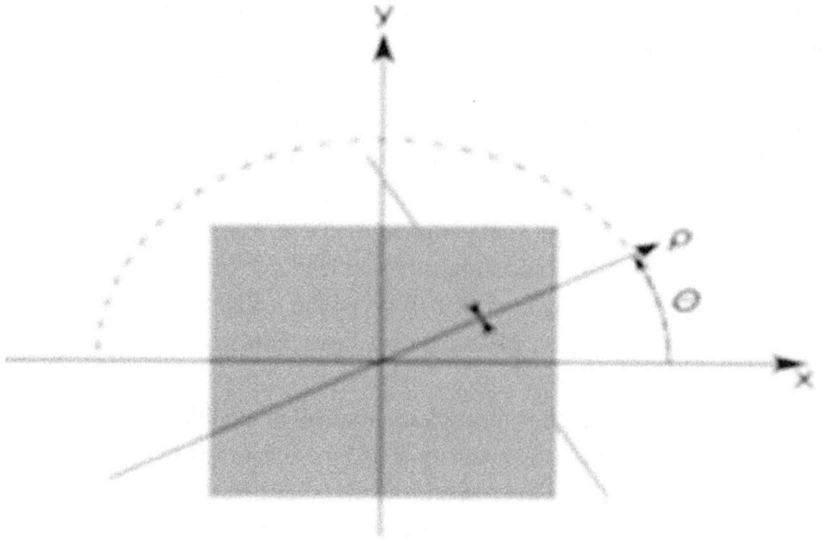

Step 4: Radon image undergoes single level integer lifting wavelet transform which results in 4 sub bands (LL, LH, HL, HH) of size M/2 x N/2, each.

Step 5: Construct binary images from the chosen bit planes of LH, HL and HH bands.

Step 6: Compress the original bits in the chosen bit plane of these bands using arithmetic coding and obtain the header information required for the arithmetic encoder and decoder.

Step 7: Read the secret payload and convert it into a bit string.

Step 8: Now, concatenate the header length, header information and compressed bit streams of CLH, CHL and CHH and the secret. Payload to get a single bit stream.

Step 9: Embed bit stream into chosen bit plane of LH band. If not enough, embed in HL band and then in HH band and it results in embedded LH, HL and HH components.

Step 10: Now, apply inverse integer lifting transform on LL coefficients and the modified LH, HL and HH coefficients.

Step 11: Compute inverse radon transform of the image obtained from

Step 12: Combine all the three color channels to get stego image.

Data Extraction Algorithm

Step 1: Read the stego image of size M x N.
Step 2: Separate color channels (Red, Green and Blue) of stego image.
Step 3: Apply radon transform to the channel in which the data is embedded.
Step 4: Radon image undergoes single level integer lifting wavelet transform which results in 4 sub bands (LL, LH, HL, HH) of size M/2 x N/2, each.
Step 5: Construct binary images from chosen bit planes of LH, HL and HH bands.
Step 6: Derive the header information and header length needed for arithmetic decoding.

Step 7: Extract the compressed bits from the chosen bit plane of these bands using arithmetic decoder and decompress the sub bands to get the reconstructed sub bands.
Step 8: Apply inverse integer lifting transform on the reconstructed sub bands LH, HL and HH along with LL sub band.
Step 9: Finally, compute inverse radon transform of the image obtained from Step 8.
Step 10: Combine all the three channels to get the original cover image.

EXPERIMENTAL RESULTS AND DISCUSSIONS

In order to investigate the performance of the proposed data hiding algorithm, several experiments are carried out in a computer system equipped with Intel core 2 duo processor with 2 GB memory and a clock speed of 2 GHz. Mat lab 8 (R2013a) platform is used for the digital simulation of the algorithm. Five standard 512 x 512 color images such as (a) Airplane, (b) Baboon, (c) Boats, (e) Lena and (e) Pepper, obtained from USC-SIPI (Image database), have been used as cover images. The performance of the algorithm is investigated in terms of imperceptibility, robustness and reversibility.

Imperceptibility

The metrics used to test the imperceptibility property of the proposed algorithm are PSNR, MSE and SSIM. Table.1 shows the PSNR and SSIM of cover images after embedding a payload of 10,000 bits using the wavelet cdf2.4 in the fourth bit plane of Red, Green and Blue channels. As the PSNR obtained from the stego images is greater than 42 dB, the embedded payload is highly imperceptible to the human eye, i.e., the perceptual quality of the resultant stego images is good. The red channel offers better PSNR and SSIM compared to green and blue channels. The red channel gives an improvement in PSNR of about 2.0 to 7.0 dB over green and blue channels. Among all cover images, the Airplane image yields better PSNR for the same payload.

The experimental results of the proposed scheme are compared with the various schemes discussed in Tsai & Sun (2007), Fu & Shen (2008) and Niu et al (2011) and summarized in Table 2. The PSNR of the cover images is measured after embedding a payload of 10,000 bits in Red channel using the wavelet cdf2.4. The PSNR offered by the proposed scheme is about 15% greater than Niu et al (2011) scheme. The proposed scheme gives better PSNR for Lena image compared to Baboon and Barbara images

Table 3 summarizes the quality of the Lena image under varying payload, after embedding in the bit plane 4, 5 and 6 of red channel using wavelet cdf2.4. The perceptual quality of cover image will get reduced if the data is embedded in higher bit planes and also PSNR drops down as more number of bits embedded in that plane. The bit plane 4 can accommodate only 90,000 bits as the bit plane 4 provides less space for data embedding compared to other bit planes. The maximum embedding capacity of bit planes 5 and 6 is 95,000 bits and 96,000 bits respectively. Beyond 95,000 bits, there is no space to accommodate the secret payload in both the bit planes. The embedding capacity completely depends upon cover image and is also based on the bit distribution of the chosen bit plane. The PSNR varies from 41.12 dB at the embedding rate of 0.004 bits per pixel to 37.04 dB at 0.362 bits per pixel for bit plane 6.

Table 1. Quality metrics after embedding a secret payload of 10,000 bits in bit plane 4 of Red, Green and Blue channels of cover images using wavelet cdf2.4

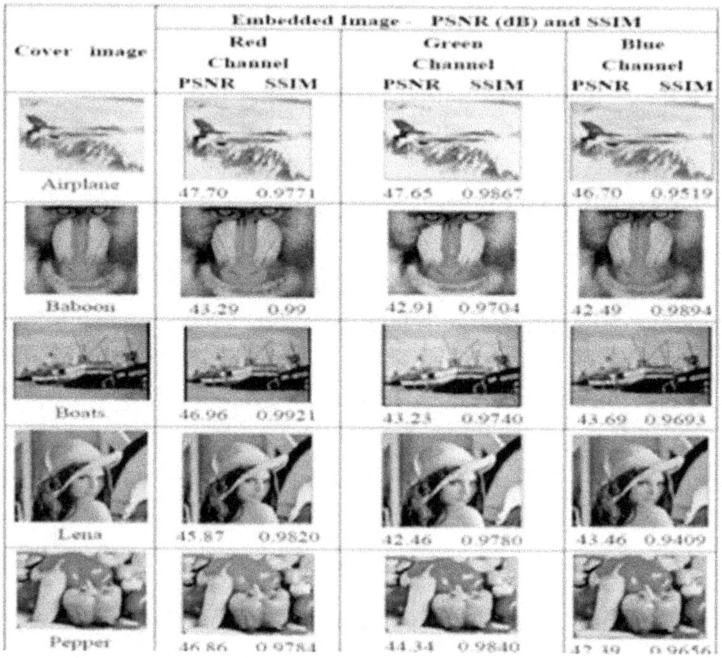

Cover image	Embedded Image - PSNR (dB) and SSIM: Red Channel PSNR	Red Channel SSIM	Green Channel PSNR	Green Channel SSIM	Blue Channel PSNR	Blue Channel SSIM
Airplane	47.70	0.9771	47.65	0.9867	46.70	0.9519
Baboon	43.29	0.99	42.91	0.9704	42.49	0.9894
Boats	46.96	0.9921	43.23	0.9740	43.69	0.9693
Lena	45.87	0.9820	42.46	0.9780	43.46	0.9409
Pepper	46.86	0.9784	44.34	0.9840	47.39	0.9656

Table 2. Comparison of proposed scheme with the schemes in the literature

Host Images	PSNR(dB)			
	Tsai and Sun Scheme(2007)	*Fu and Shen Scheme(2008)*	*Niu et.al Scheme(2011)*	*Proposed Sceme*
Lena	*39.43*	*38.10*	*40.57*	*45.87*
Baboon	*41.76*	*38.90*	*41.67*	*43.29*
Barbara	*42.49*	*39.72*	*40.71*	*44.03*

Table 3 is also illustrated in Figure 3. It shows the comparison of embedding rate and image quality in different bit planes. It is inferred from Figure 2 that the embedding capacity of bit plane 6 is higher than that of bit plane 4 and bit plane 5. Higher the bit plane, lower the quality of the stego image. However, the PSNR obtained after embedding data in bit plane 6 is about 4 dB less than that of bit plane 4 for the same embedded payload.

The robustness of the proposed data hiding technique against various image processing and signal processing attacks is measured using bit error rate. The stego images are added with Gaussian noise having a variance of 0.2 and 0.4, Poisson Noise, Impulse Noise with a variance of 0.05 and 0.1 and Speckle Noise with a variance of 0.05 and 0.1. Generally, addition of noise is responsible for the degradation of the image. The image processing operations such as Rotation (5 and 10 degrees), Scaling (200% and 400%), Blurring and Cropping (10% and 25%) are performed on embedded images.

Table 3. Embedding capacity vs. PSNR (dB) after embedding secret bits in bit plane 3, 4 and 5 of red channel of Lena image using wavelet cdf2.4

Secret payload (bits)	Embedding Rate (bpp	Bit Plane 6		
		Bit Plane 4	Bit Plane 5	Bit Plane 6
1000	*0.004*	45.27	43.08	41.42
3000	*0.011*	45.26	43.07	41.40
6000	**0.023**	45.14	43.04	41.20
10,000	**0.038**	45.10	43.02	41.06
20,000	**0.076**	44.68	42.54	40.83
40,000	**0.153**	42.21	40.41	38.33
50,000	**0.191**	42.20	39.80	38.33

Figure 3. Comparison of embedding rate and image quality in different bit planes

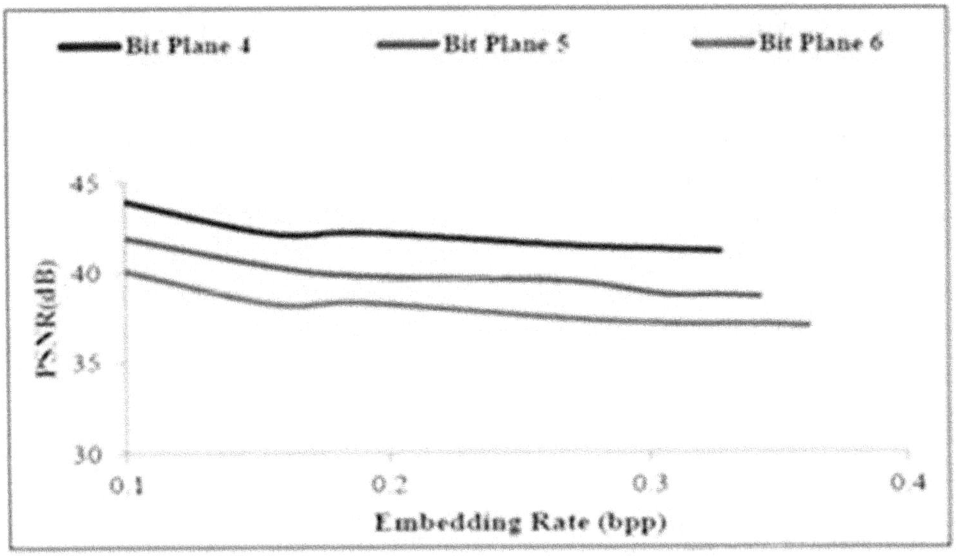

Reversibility

The reversibility of the proposed data hiding technique against various attacks is measured using Normalized Correlation Coefficient. Various attacks like Gaussian noise, Poisson Noise, Salt and pepper Noise, Speckle Noise, Rotation, Scaling, Blurring and Cropping are applied on embedded image. After subjecting to the attacks, the original cover image is extracted and Normalized correlation coefficient of extracted cover image against original cover image is computed.

CONCLUSION

Reversible data hiding has emerged as a major research area due to the phenomenal growth in internet and multimedia technologies. It involves concealing confidential data within another seemingly innocuous cover media such as text, video, audio, images, and compression coding. The embedded data can then be used as authentication codes for protecting intellectual copyrights (Water marking) or as confidential data for sharing information (Steganography). However, in those applications, permanent destruction of the cover media is normally unavoidable, even after the embedded media has been extracted. The reversible data hiding is implemented in a hybrid domain, i.e., Radon and ILWT domain to enhance the security of the system. The cover image is first transformed from spatial domain to radon domain. Then the radon image is applied with integer lifting wavelet transform. Haar, Daubechies, Symlets, Coiflets, Bi-orthogonal and CDF wavelets are used for the implementation of reversible integer lifting wavelet transform. Then the middle bit planes of high frequency lifting wavelet coefficients are compressed using arithmetic coding. As the arithmetic coding is a lossless compression method, it guarantees the recovery of original payload. To achieve better embedding capacity, higher bit planes have to be used. However, it affects the visual quality of the embedded image. Lower bit planes offer lesser embedding capacity. Hence, bit plane coding embeds secret payload into the middle bit planes in order to achieve the reasonable embedding capacity and higher imperceptibility. The results have been compared with the existing works in the literature and the proposed method gives 10 to 15% improvement in PSNR. The embedding capacity of bit plane 6 is about 95,000 bits whereas for bit plane 4, it is only 86,000 bits. i.e., Higher the bit plane, better the embedding capacity. Image quality rapidly changes when different wavelets are used. When the image intensity is projected at different angles using radon transform, the perceptual quality of the embedded image also gets changed. The original cover is exactly restored at the destination as the NCC values are greater than 0.97 and hence the algorithm is said to be reversible. As the proposed algorithms have low computational complexity, they can also be used for embedding data in a video file.

REFERENCES

Abdulfetah, A. A., Sun, X. M., Yang, H., & Mohammad, N. (2010). Robust Adaptive Image Watermarking using Visual Models in DWT and DCT Domain. *International Journal of Information Technology*, *9*(3), 470–488.

Ahamed, B. B., & Ramkumar, T. (2015). Deduce User Search Progression with Feedback Session. *Advances in Systems Science and Applications*, *15*(4), 366–383.

Ahamed, B. B., & Ramkumar, T. (2016). An intelligent web search framework for performing efficient retrieval of data. *Computers & Electrical Engineering*, *56*, 289–299. doi:10.1016/j.compeleceng.2016.09.033

Ahamed, B. B., & Ramkumar, T. (2018). Proficient Information Method for Inconsistency Detection in Multiple Data Sources. *International Journal of Artificial Intelligence*, *16*(1), 182–194.

Bedi, P., Bansal, R., & Sehgal, P. (2013). Using PSO in a spatial domain based image hiding scheme with disortion tolerance. *International Journal of Computers and Electrical Engineering*, *39*(2), 640–654. doi:10.1016/j.compeleceng.2012.12.021

Chakraborty, S., Samanta, S., Biswas, D., & Dey, N. (2013), Particle Swarm Optimization based parameter optimization technique in medical information hiding, *IEEE International Conference on Computational Intelligence and Computing Research*, 1-6. 10.1109/ICCIC.2013.6724173

Cheddad, A., Condell, J., Curran, K., & McKevitt, P. (2010). Digital image steganography survey and analysis of current methods. *Journal of Signal Processing*, *90*(3), 727–752. doi:10.1016/j.sigpro.2009.08.010

Coltuc, D. (2011). Improved embedding for prediction-based reversible watermarking. *IEEE Transactions on Information Forensics and Security*, *6*(3), 873–882. doi:10.1109/TIFS.2011.2145372

Gerami, P., Subariah, I., & Morteza, B. (2012). Least Significant Bit Image Steganography using Particle Swarm Optimization and Optical Pixel Adjustment. *International Jisuanji Yingyong*, *55*(2), 20–25.

Huang, Chen, & Abraham. (2010). Optimized Watermarking Using Swarm-Based Bacterial Foraging. *Journal of Information Hiding and Multimedia Signal Processing*, *1*(1), 51–58.

Ishtiaq, M., Bushra, S., Jaffar, A. M., & Aziz, K. (2010). Adaptive Watermark Strength Selection Using Particle Swarm Optimization. *ICIC Express Letters*, *4*(5), 1–6.

Lee, S., Yoo, C. D., & Kalker, T. (2007). Reversible image watermarking based on integer-to-integer wavelet transform. *IEEE Transactions on Information Forensics and Security*, *2*(3), 321–330. doi:10.1109/TIFS.2007.905146

Li, J., Li, X., & Yang, B. (2013). Reversible data hiding scheme for color image based on prediction-error expansion and cross-channel correlation. *Signal Processing*, *93*(9), 2748–2758. doi:10.1016/j.sigpro.2013.01.020

Lin, S. D., Shie, S. C., & Guo, J. Y. (2010). Improving the robustness of DCT based image watermarking against JPEG compression. *International Journal of Computer Standards and Interfaces*, *32*(1), 54–60. doi:10.1016/j.csi.2009.06.004

Liu, L., Chen, T., Zhu, S., Hong, W., & Si, X. (2014). A Reversible Data Hiding Method Using Improved Neighbor Mean Interpolation and Random-Block Division. *Information Technology Journal*, *13*(15), 2374–2384. doi:10.3923/itj.2014.2374.2384

Niu, P. P., Wang, X. Y., Yang, Y. P., & Ming, Y. (2011). A novel color image watermarking scheme in non-sampled contourlet-domain. *Journal of Expert Systems with Applications*, *38*(3), 2081–2098. doi:10.1016/j.eswa.2010.07.147

Saha, B., & Sharma, S. (2012). Steganographic Techniques of Data Hiding using Digital Images. *Defence Science Journal*, *62*(1), 11–18. doi:10.14429/dsj.62.1436

Thabit, R., & Khoo, B. E. (2014, February). Robust reversible watermarking scheme using Slantlet transform matrix. *Journal of Systems and Software*, *88*, 74–86. doi:10.1016/j.jss.2013.09.033

Venkatanarasimhulu, C., & Satyaprasad, K. (2011). A new SVD based hybrid color image watermarking for copyright protection using contourlet transform. *International Journal of Computers and Applications*, *20*(8), 18–27. doi:10.5120/2455-3314

Yang, C. H., Huang, C. T., & Wang, S. J. (2011). Reversible steganographybased on side match and hit pattern for VQ compressed images. *Information Sciences*, *181*(11), 2218–2230. doi:10.1016/j.ins.2011.01.015

Zeng, X. T., Ping, L. D., & Pan, X. Z. (2010). A lossless robust data hiding scheme. *Journal of Pattern Recognition*, *43*(4), 1656–1667. doi:10.1016/j.patcog.2009.09.016

Chapter 19
Genetic Algorithm and Particle Swarm Optimization Techniques in Supply Chain Design Problems:
A Survey

Md. Ashikur Rahman
Universiti Teknologi PETRONAS, Malaysia

Pandian M. Vasant
Universiti Teknologi PETRONAS, Malaysia

Junzo Watada
Universiti Teknologi PETRONAS, Malaysia

Rajalingam Sokkalingam
Universiti Teknologi PETRONAS, Malaysia

ABSTRACT

Metaheuristics has become a top research area. Numerous optimization problems have been solved by metaheuristics as they showed comprehensive improvements to solve these intractable optimization problems. Complex problems like supply chain design problems need strategic decisions, and metaheuristics can intensify the decisions while designing supply chain network. In this chapter, the authors have introduced how nature memetic algorithms (e.g., genetic algorithm and particle swarm algorithms) are implemented to solve supply chain network design problem. A discussion about the recent research in this field shows an important direction to the future research.

DOI: 10.4018/978-1-7998-3645-2.ch019

INTRODUCTION

Supply chain management (SCM) has become one of the major Industrial Engineering application and research areas over the last two decades. "A supply chain, which is the subject of interest in SCM, is a complex network of business entities (e.g., suppliers, manufacturers, distributors, service providers, warehouses, customers) involved in the upstream and downstream flows of material, information and finances" (Altiparmak et al., 2016). Efficient and effective management of these flows is a key to the success of supply chain. Therefore, the management of supply chain flows need to be supported by a coherent supply chain strategy and an appropriate design and planning of the supply chain for execution of the flows.

Traditionally, strategic decisions related to the design and planning of the supply chain revolve around design of the network. A supply chain network consists of a series of linked levels, such as suppliers, production facilities (plants), warehouses, distribution centers and customers, that are involved in processes associated with the flow of products/services.

Supply chain network involves decisions about the configuration of the network, locating facilities, deciding their capacity optimizing the transportation of material among them. The network design problem is often solved using the transportation of material among them.

OBJECTIVE OF STUDY

The network design problem is often solved using large optimization models.

Supply Chain Design (SCND) problems are combinatorial, complex in nature and thus not well suited to be handled by conventional optimization techniques. Metaheuristic optimization such as genetic algorithms, often outperform conventional optimization methods when applied to these problems. This chapter illustrates the practice of Genetic Algorithms (GAs) and Particle Swarm Optimization (PSO) as one of the important tools to optimize SCND problems. GA is an evolutionary search algorithm that uses mechanics of natural selection and genetics (e.g. crossover, mutation) to find near-optimal solutions to problems. Moreover, PSO has a wide area of application in optimizing NP hard problems as like as SCND problems.

The remainder of the chapter is organized into four sections. The next section describes the basic characteristics of the SCND problem. The third section examines how GA and Particle Swarm Optimization (PSO) techniques are applied to solve SCND problems. The fourth section classifies the studies in the recent literature and outlines research opportunities. Finally, the last section provides a summary of the key points of this chapter and an outlook.

BACKGROUND STUDY ABOUT GA AND PSO

Altiparmak et al.(2006) discussed a distribution network chain combined with three stage. Initially they introduced a mathematical model, consequently their solution-based approach was defined by Genetic Algorithm. Sue et al. (2008) solved a complex product distribution problem through GA, Tabu search and improved Path relinking method. Thanh et al. (2008) discussed a multi- product and multi planning based distribution network. They first formulated a cost reduction model through MILPM. Calvete and

Gale (2010) proposed a multi-objective model for supply chain planning. Their model was formulated with multi producer, multi distribution centres along with multi retailers. Wang et al. (2011) elaborated a single echelon supply chain network through a Mixed Integer Linear Programming (MILP) model. The considered factor of the model was stochastic demand. Particularly that profit based model was solved by Genetic algorithm along with efficient greedy heuristic. Amorim et al. (2012) introduced a hybrid GA model to solve a distribution network problem for the perishable products. Kadadevaramath et al. (2012) designed a 2-echelon distribution network. Their solution approach was a comparative study between GA and PSO. Varthanan et al. (2012) presented a stochastic demand-based distribution network. A MILP model was applied to reduce the cost of that supply chain network. Zamarripa et al. (2012) designed a Multi period supply chain network. They formulated a cost reduction model by MILP and found the solution by GA. Raa et al. (2013) introduced a MILP model for an established Product Distribution Network (PDN). The model was optimized by a metaheuristic algorithm which has a wide use of expounding large instance problems. Nasiri et al. (2014) discussed a multistage distribution planning considering demand uncertainty. Alike majority of the distribution network model they proposed a cost reduction model. Later, a heuristic method based on Genetic Algorithm.

Several researchers have developed cost-based modelling with metaheuristic approach to solve the problems. Ma et al. (2016) presented a distribution network considering uncertain transportation cost. Their research focus on to lowering the total cost of that model. A hybrid GA was adopted to solve that proposed model. Govindan and Fattahi (2017) concentrated on the uncertain demand in supply chain network and developed a MILP for the cost reduction. Khalifehzadeh et al. discussed a two-echelon distribution network designed a cost-based modelling. Their multi-objective model was concentrated on minimizing cost along with maximizing reliability of transportation system. Their proposed methodology was segregated with portions. For small instance example they used LINGO repeatedly for large instance they applied a novel heuristic called Ranking Genetic Algorithm. [24]

Wang and Liu (2010) addressed that Chaotic Particle Swarm Optimization (CPSO) can visualize the issues of assembly line balancing problem, using the chaos method to boost solution and convergence efficiency.

Kanthavel et al. (2012) established nested particle swarm optimization as the fusion of multiple other mechanisms, Master Particle Swarm Optimization (MPSO) and Slave Particle Swarm Optimization (SPSO), to resolve the Vehicle Routing Planning (VRP) with simultaneous distribution and pick up. Shanker et al. (2012) have introduced the bi-objective distribution scheduling with hybrid multi-objective PSO algorithm. Another approach named Discrete Particle Swarm Algorithm (MODPSA) was followed by Venkatesan and Kumanan (2012) for designing of a supply chain architecture. They used two individual global guide selection technique for that optimization.

Varamath et al (2012) suggested a three-echelon supply chain network model utilizing the particle swarm optimization to tackle the market uncertainty and constraints encountered by particular echelon in the supply chain design operators.

In *Table 1* recent development and metaheuristic approaches are enlisted to optimize supply chain problems.

Table 1. Recent development of GA and PSO for solving Supply Chain Design Problems

Hiassat et al. (2017)	Solving a Vehicle Routing Planning (VRP) model for perishable products with divisible delivery and pickup	GA
Kannan et al. (2018)	Comparative analysis of different metaheuristics on vehicle routing problem optimization	PSO
Alshamsi and Diabat (2017)	Considering of a practical scenario of a network design for Reverse Logistics (RL)	GA
Ting et al. (2017)	A multi -vehicle selective pickup and delivery problem study and optimization	GA
Fathollahi Fard et al. (2017	Architecture of a bi-objective stochastic closed-loop supply chain network problem with a downside risk factor	GA and PSO
Kaya and Urek (2016)	Constructing a design of the network for a closed-loop supply chain with the collection of used products from the customers	GA
Zhalechian et al. (2016)	Outline of a closed-loop location-routing-inventory supply chain network in consideration of mixed uncertainty	GA
Ghezavati and Beigi (2016)	Study of a VRP model of the location-routing problem considering time windows	GA
Behmanesh and Pannek (2016)	Constructing a model which includes flexible and multistage both way (forward/reverse) logistic network	GA
Yang et al. (2016)	Description of a CLSC model which can optimize facility allocation problem	GA
Shi et al. (2016)	Illustration of a model which can modify a Closed Loop Supply Chain design problem	GA
Jiang et al. (2016)	A study how general acyclic supply chain networks deals with an inventory location problem	GA
Guo and Ya (2015)	Analysing a production planning model of a manufacturing/ recycling system	GA and PSO
Farrokhi-Asl et al. (2017)	Regarding the model of waste management including consumer waste collection	PSO
Subulan et al. (2015)	Optimizing the Reverse Logistics (RL) problems through metaheuristic techniques	PSO
Asl-Najafi et al. (2015)	Considering the case of a dynamic closed-loop location-inventory problem	GA and PSO
Soleimani and Kannan (2014)	A description regarding the planning and design of Close Loop Supply Chain (CLSC)	GA and PSO

SUPPLY CHAIN NETWORK DESIGN PROBLEM

SCND problem has also been studied under the term's logistics network design or distribution network design and has been referred to as facility location models in the literature. The facility location problem in the context of SCND is primarily concerned with minimizing the total costs of opening and operating a facility and shipping materials/goods while satisfying demands of the customers. Chopra and Meindl (2010) stated that the term facility is used in its broadest sense and includes entities operating at different levels of the supply chain such as plants, warehouses, distribution centres, retail stores, customers and the like. Ideally when modelling a network design, information about the location of supply sources and customers, location of potential facility locations, demand forecast by customer/market, facility, labour and material costs by facility, inventory costs by facility, transportation costs between each pair of facilities, sales price of the product, taxes, tariffs, desired customer response time, and other required service factors is available.

The contemporary form of the facility models is the uncapacitated facility location problem (ULFP) that illustrates how to lower the network cost for a single commodity provided with unlimited facility and flow capacities and linear costs. The described model can be formulated as

$$\text{Min} \sum_{j\epsilon D}\sum_{i\epsilon} c_{ij}x_{ij} \,.$$

$$\text{subject to} \sum_{i\in F} x_{ij} = 1, \quad j \in D. \tag{1}$$

$$x_{ij} \leq y_j, \ \ i \in F, j \in D$$

$$x_{ij} \geq 0, \ \ i \in F, j \in D$$

$$y_i \in \{0,1\}, i \in F$$

The formulation can be denoted by following description. F is the set of potential facility locations and building/operating a facility at location i ∈ F has a fixed cost of fi. Any open facility can provide an unlimited amount of the certain commodity. D is the set of customers (demand points) and customer j ∈ D has a demand for the commodity. The decision variable x_{ij} is the fraction of the demand of jth customer supplied from i^{th} .facility. Due to the single assignment property of UFLP, we can define x_{ij} as x_{ij} $\{0,1\}$, i ∈ F, j ∈ D; since a customer is always entirely supplied from its nearest facility. Hence, x_{ij}=1 means that j^{th} customer is entirely supplied from i^{th} facility. y_i is a binary location variable: y_i indicates a facility is located at i ∈ F, 0 otherwise. The total cost of service or distribution c_{ij} is incurred for supplying j^{th} customer from i^{th} facility; c_{ij} is a variable cost that includes the cost of moving the product between different levels in the network. The core goal is to open a subset of facilities and assign customers to these facilities at minimal total cost, integrating the total cost of service plus the fixed cost of opening the facilities.

Models of the SCND problem defined through various attributes such as: objective function (single/ multiple), capacity limitation (capacitated/ un capacitated), number of stages/echelons (single/ multiple) number of commodities/products (single/ multiple), number of sourcing facilities (single/ multiple), planning horizon (single-period/multi-period), capacity flexibility (fixed/expandable), and data type (deterministic/ stochastic). The basic model is extended to accommodate the varying attributes, by introducing additional parameters, variables and constraints. A broad variety of literature on the creation and testing of algorithms searching for a solution to variants of the facility location problem, for a review of contributions and classifications the reader may refer to studies of Revelle et al. (2008), Klose and Drexl (2005), Melo et al. (2009), Aikens (1985), among others.

APPLYING GENETIC ALGORITHMS IN SUPPLY CHAIN

Stochastic optimization methods just like Evolutionary Algorithms (EA) can be implemented for finding new paths for searching as they opt natural evolution technique. Even though all contemporary evolutionary techniques, e.g. Genetic algorithms (GA), Memetic algorithms (MA), Cultural algorithms (CA), Differential search strategies (DS), Genetic programming (GP) etc., have owned their respective strengths and weaknesses, for the SCND problem the most common algorithms studied in the literature are variants of GAs (Whitley, 2001). Targeted random search techniques such as Genetic Algorithms (GA) can be utilized to find a near-optimal solution in multi-dimensional search spaces for many problems (Guner et al., 2010).

The fundamental driver underlying GA is natural selection which is inspired by a natural evolution process. The process exactly works for biological organism. The process exactly works for biological organism. The key term which is transferred from one generation to another is the changing chromosome combinations to survive in the living world respect to Darwinian principle "survival of the fittest". GA has a special characteristic is that if it implemented through a proper coding it can iteratively enhance the probability of finding global optima rather than being confined into local optima. Genetic agents exploit the search for globally optimal individuals(chromosomes) in a population to boost their fitness values. Unlike many other heuristics or metaheuristics, GA are much more independent of the problem and more flexible in respect to formulation of objectives and constraints.

The basic steps of GAs may be listed as follows. Notable fact is that advanced hybrid approaches can impact the flow in respective way -

1. Designing of a representation flowsheet for a possible solution
2. Preparing scheme for the initialization of particles
3. Defining the fitness function
4. Several genetic operators such as (reproduction, mutation, crossover, elitism) illustrations
5. Preferences of the parameters such as population size, genetic operators probabilities etc
6. Terminate rule denotation

To initialize GA a collection of starting population needs to be established. Chromosomes are treated as related solutions in the search space at this period. Fitness function valuate these solutions as well as chosen with pre-defined techniques. Crossover and mutation these two operators act to guide the total population into global optimum. Eventually diversification of population and information exchange can be performed through crossover stage. Consequently, a solution would be made with new population. And this preferred method is continued until it meets a stopping criterion.

PARTICLE SWARM OPTIMIZATION

To find best solution, the PSO algorithm induces the rotation of a series of particles in search space under predetermined laws. Eberhart and Kennedy (1995) first suggested PSO, which was influenced by bird flocking, fish schooling, and even human social behaviour. Reynolds (1987) schemed a bird inspired model to mimic the behaviour of bird flocking. Three basic rules such as collision avoidance, velocity matching and flock centring are implemented to every individual on that model. Individuals represent-

Figure 1. Schematic flowsheet of Genetic Algorithm (Altiparmak et al., 2006)

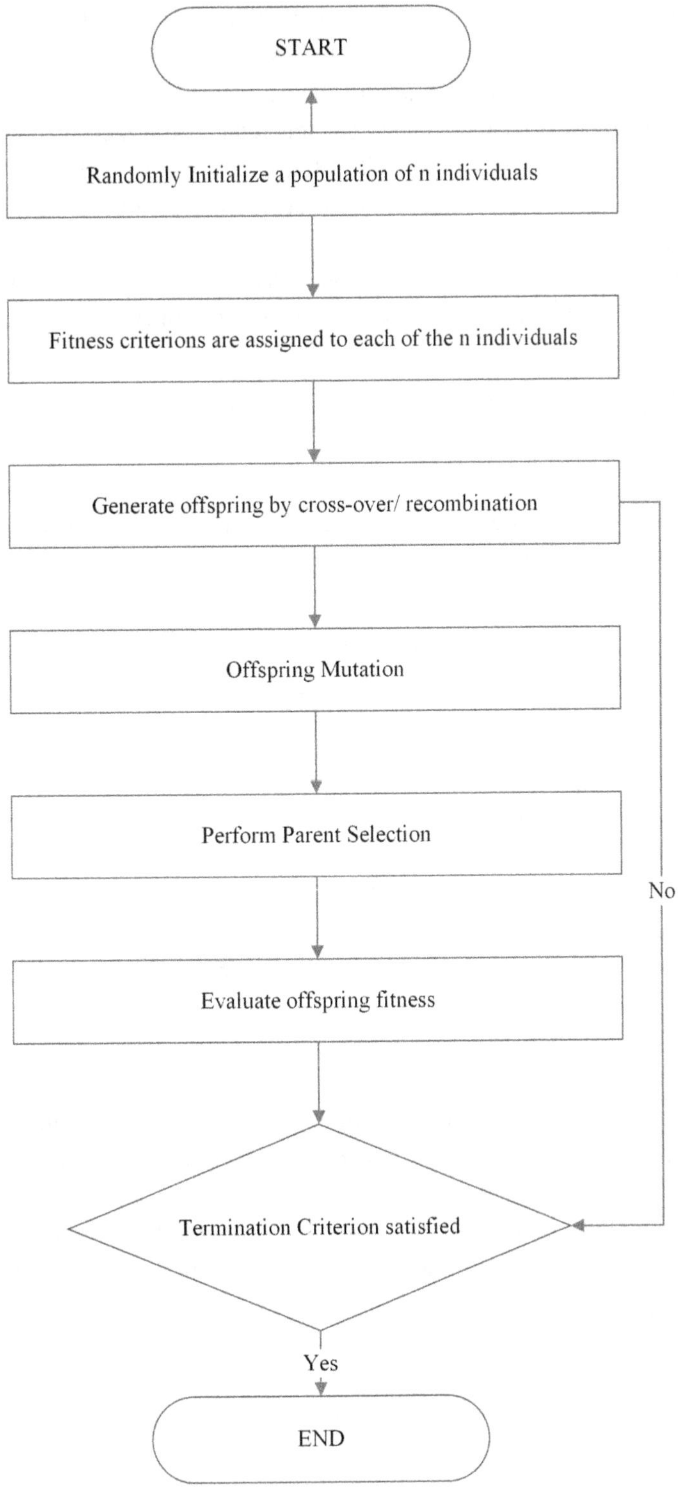

ing solutions are viewed as particles in PSO derived from the bird-oid model and each particle is defined by its corresponding fitness value, position vector and velocity vector. Beside the three inner attributes, each particle also recollects its historical best position (local best position) x_i^* and global best position g^* and refers to those two positions. If a particle reaches a better location than any previously reached location, it changes the position as particle i .new best current. Consequently, there is a new maximum for all n paryicles at any time t during iterations. The target is to discover the global best among all the current best until the objective no longer upgrades or after a certain number of iterations. Particle motion can be depicted in the figure (2) below.

Assume x_i and v_i are denoting position vector and velocity for particle i respectively. The new velocity vector for particle i respectively. Following equation (3) can exemplify the new velocity vector:

$$v_i^{t+1} = v_i^t + \alpha\varepsilon_1\left[g^* - x_i^t\right] + \beta\varepsilon_2\left[x_i^{*(t)} - x_i^t\right]. \tag{2}$$

Figure 2. Graphical diagram of a PSO particle approaching toward the global best g^* and the current best x_i^* for each particle i (Yang, 2014)

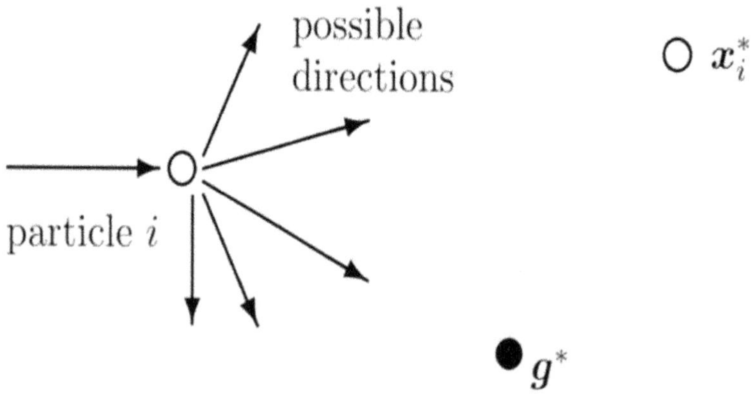

In equation (2), ε_1 and ε_2 are representing two random vectors and each entry has value range of (0 – 1). α and β can be defined as the learning parameters or acceleration constants, which can be usually be assumed as α≈β≈2.

Several multimodal problems required a uniformed distribution of the particles as they can cover up a large search space of those problems. Initially, the velocity of a particle can be arbitrated as zero, that is $v_i^{t=0} = 0$. Updated new position can be illustrated by

$$x_i^{t+1} = x_i^t + v_i^{t+1}. \tag{3}$$

Figure 3. Flowchart for the PSO Algorithm (Yang, 2014)

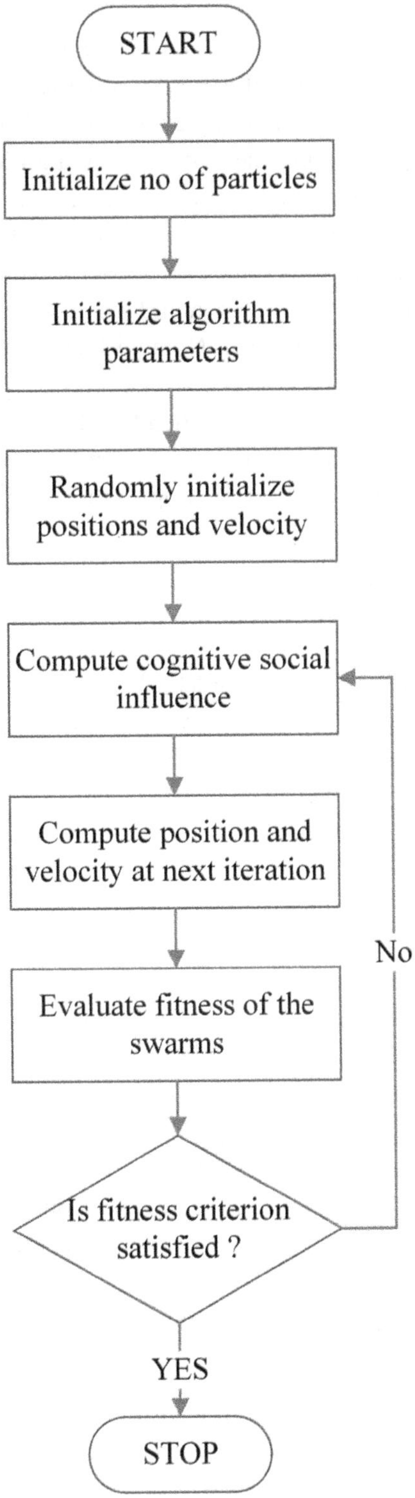

PARAMETER TUNING FOR GA & PSO

Parameter tuning has a significant impact on any type of optimization. For example, GA consists of 3 operators crossover, mutation, selection. Normally crossover helps to exploit and increase the chance of convergence. From several empirical results researchers describe that higher probability for crossover could be range of 0.6-0.95 (Yang, 2014). On the contrary, mutation probability is relatively very low probably 0.001-0.05 (Yang,2014). Another important issue has found on the literature as the population size would be efficient from n=40 to 200 for large portion of true life challenging problems (Yang, 2014).

Like most unimodal functions the typical values for APSO (accelerated PSO) could be $\alpha \approx 0.1 \sim 0.4$ and $\beta \approx 0.1 \sim 0.7$. It is noteworthy that the parameters α and β can be general as per as the search domain accepts as well as scale of independent variables. Notable fact is that simplified APSO can provide global convergence (Yang, 2014).

Numerous studies show that PSO has outperformed GA and other conventional methods for solving real life problems. As GA has better broadcasting ability to reach quicker convergence. Among all the particles the information can easily transferred as its impact on neighbourhood is very good.

LIMITATIONS

Though GA and PSO both algorithms have been used for better optimality, these algorithms have some drawbacks also. Many scientists depicted that not every time PSO and GA will ensure optimum solution (Kolias et al., 2011). So, several modifications are needed to maintain greater optimality as different objective function demands. Moreover, PSO has shown premature convergence which is a decisive case for optimization. So, several researches are ongoing to reach at better optimality through these algorithms.

FUTURE RESEARCH PROSPECT

Considering the drawbacks of GA and PSO, some decision can be taken for optimizing real life problems. As PSO is a generic category of swarm intelligence, authors suggest hybridizing this with any evolutionary algorithm. Genetically modified operators are the root element of hybrids of GA. Whenever adopted operators are intercorporate with PSO, they trend to produce more fit applicants (Angeline and P.J, 1998). Swarm algorithm has proven to reach better exploration and the evolutionary algorithms have better exploitation, so their hybridization can be a better approach to solve supply chain network problems. Moreover, hybrid metaheuristics have been recognised tool for optimizing stochastic problems (Gen, Hao & Zhang, 2017). So, the research prospect can be wider if these steps are in consideration for solving complex problems like supply chain optimization.

CONCLUSION

Supply chain network optimization is earning a vast concentration form all over the world. In this discussion authors have tried to conclude the application the most popular algorithms e.g. GA and PSO. The authors believe if the parameters can be tuned as the objective function demands, the convex problems like supply chain network problems could be solved with better optimality. Even more, convenient hybridization can provide better convergence the literature describes. The authors believe that this approach can be efficient in supply chain optimization.

ACKNOWLEDGMENT

The research leading to this review article has (a) been assist by the Graduate Assistantship (GA) (b) been supported by Fundamental and Applied Sciences Department (FASD), Universiti Teknologi PETRONAS.

REFERENCES

Abraham, A. J., Kumar, K. R., Sridharan, R., & Singh, D. (2015). A genetic algorithm approach for integrated production and distribution problem. *Procedia: Social and Behavioral Sciences*, *189*, 184–192. doi:10.1016/j.sbspro.2015.03.213

Afrouzy, Z. A., Nasseri, S. H., Mahdavi, I., & Paydar, M. M. (2016). A fuzzy stochastic multi-objective optimization model to configure a supply chain considering new product development. *Applied Mathematical Modelling*, *40*(17-18), 7545–7570. doi:10.1016/j.apm.2016.03.015

Aikens, C. H. (1985). Facility location models for distribution planning. *European Journal of Operational Research*, *22*(3), 263–279. doi:10.1016/0377-2217(85)90246-2

Alshamsi, A., & Diabat, A. (2017). A Genetic Algorithm for Reverse Logistics network design: A case study from the GCC. *Journal of Cleaner Production*, *151*, 652–669. doi:10.1016/j.jclepro.2017.02.096

Altiparmak, F., Gen, M., Lin, L., & Paksoy, T. (2006). A genetic algorithm approach for multi-objective optimization of supply chain networks. *Computers & Industrial Engineering*, *51*(1), 196–215. doi:10.1016/j.cie.2006.07.011

Amorim, P., Günther, H. O., & Almada-Lobo, B. (2012). Multi-objective integrated production and distribution planning of perishable products. *International Journal of Production Economics*, *138*(1), 89–101. doi:10.1016/j.ijpe.2012.03.005

Angeline, P. J. (1998, May). Using selection to improve particle swarm optimization. In *1998 IEEE International Conference on Evolutionary Computation Proceedings. IEEE World Congress on Computational Intelligence (Cat. No. 98TH8360)* (pp. 84-89). IEEE. 10.1109/ICEC.1998.699327

Balinski, M. L. (1965). Integer programming: Methods, uses, computations. *Management Science*, *12*(3), 253–313. doi:10.1287/mnsc.12.3.253

Beamon, B. M. (1998). Supply chain design and analysis: Models and methods. *International Journal of Production Economics*, *55*(3), 281–294. doi:10.1016/S0925-5273(98)00079-6

Behmanesh, E., & Pannek, J. (2016). Modeling and random path-based direct encoding for a closed loop supply chain model with flexible delivery paths. *IFAC-PapersOnLine*, *49*(2), 78–83. doi:10.1016/j.ifacol.2016.03.014

Calvete, H. I., & Galé, C. (2010). A multiobjective bilevel program for production-distribution planning in a supply chain. In *Multiple criteria decision making for sustainable energy and transportation systems* (pp. 155–165). Springer. doi:10.1007/978-3-642-04045-0_13

Chan, F. T., Jha, A., & Tiwari, M. K. (2016). Bi-objective optimization of three echelon supply chain involving truck selection and loading using NSGA-II with heuristics algorithm. *Applied Soft Computing*, *38*, 978–987. doi:10.1016/j.asoc.2015.10.067

Chopra, S., & Meindl, P. (2013). *Supply Chain Management: Strategy, Planning, and Operation*. Academic Press.

Diabat, A., & Deskoores, R. (2016). A hybrid genetic algorithm based heuristic for an integrated supply chain problem. *Journal of Manufacturing Systems*, *38*, 172–180. doi:10.1016/j.jmsy.2015.04.011

Fard, A. F., Gholian-Jouybari, F., Paydar, M. M., & Hajiaghaei-Keshteli, M. (2017). A bi-objective stochastic closed-loop supply chain network design problem considering downside risk. *Industrial Engineering & Management Systems*, *16*(3), 342–362. doi:10.7232/iems.2017.16.3.342

Farrokhi-Asl, H., Tavakkoli-Moghaddam, R., Asgarian, B., & Sangari, E. (2017). Metaheuristics for a bi-objective location-routing-problem in waste collection management. *Journal of Industrial and Production Engineering*, *34*(4), 239–252. doi:10.1080/21681015.2016.1253619

Fathian, M., Jouzdani, J., Heydari, M., & Makui, A. (2018). Location and transportation planning in supply chains under uncertainty and congestion by using an improved electromagnetism-like algorithm. *Journal of Intelligent Manufacturing*, *29*(7), 1447–1464. doi:10.100710845-015-1191-9

Gen, M., Hao, X., & Zhang, W. (2017). Advances in hybrid metaheuristics for stochastic manufacturing scheduling: Part I models and methods. In *Proceedings of the Tenth International Conference on Management Science and Engineering Management* (pp. 1063-1077). Springer. 10.1007/978-981-10-1837-4_88

Ghezavati, V. R., & Beigi, M. (2016). Solving a bi-objective mathematical model for location-routing problem with time windows in multi-echelon reverse logistics using metaheuristic procedure. *Journal of Industrial Engineering International*, *12*(4), 469–483. doi:10.100740092-016-0154-x

Goren, H. G., Tunali, S., & Jans, R. (2010). A review of applications of genetic algorithms in lot sizing. *Journal of Intelligent Manufacturing*, *21*(4), 575–590. doi:10.100710845-008-0205-2

Govindan, K., & Fattahi, M. (2017). Investigating risk and robustness measures for supply chain network design under demand uncertainty: A case study of glass supply chain. *International Journal of Production Economics*, *183*, 680–699. doi:10.1016/j.ijpe.2015.09.033

Guo, J., & Ya, G. (2015). Optimal strategies for manufacturing/remanufacturing system with the consideration of recycled products. *Computers & Industrial Engineering*, *89*, 226–234. doi:10.1016/j.cie.2014.11.020

Hiassat, A., Diabat, A., & Rahwan, I. (2017). A genetic algorithm approach for location-inventory-routing problem with perishable products. *Journal of Manufacturing Systems*, *42*, 93–103. doi:10.1016/j.jmsy.2016.10.004

Jabbarzadeh, A., Fahimnia, B., & Sheu, J. B. (2017). An enhanced robustness approach for managing supply and demand uncertainties. *International Journal of Production Economics*, *183*, 620–631. doi:10.1016/j.ijpe.2015.06.009

Jiang, D., Li, H., Yang, T., & Li, D. (2016). Genetic algorithm for inventory positioning problem with general acyclic supply chain networks. *European Journal of Industrial Engineering*, *10*(3), 367–384. doi:10.1504/EJIE.2016.076385

John, T., DeWitt, W., Keebler, J. S., Min, S., Nix, N. W., Smith, C. D., & Zacharia, Z. G. (2001). Defining Supply Chain Management. *Journal of Business Logistics*.

Kadadevaramath, R. S., Chen, J. C., Shankar, B. L., & Rameshkumar, K. (2012). Application of particle swarm intelligence algorithms in supply chain network architecture optimization. *Expert Systems with Applications*, *39*(11), 10160–10176. doi:10.1016/j.eswa.2012.02.116

Kadadevaramath, R. S., Chen, J. C., Shankar, B. L., & Rameshkumar, K. (2012). Application of particle swarm intelligence algorithms in supply chain network architecture optimization. *Expert Systems with Applications*, *39*(11), 10160–10176. doi:10.1016/j.eswa.2012.02.116

Kaya, O., & Urek, B. (2016). A mixed integer nonlinear programming model and heuristic solutions for location, inventory and pricing decisions in a closed loop supply chain. *Computers & Operations Research*, *65*, 93–103. doi:10.1016/j.cor.2015.07.005

Klose, A., & Drexl, A. (2005). Facility location models for distribution system design. *European Journal of Operational Research*, *162*(1), 4–29. doi:10.1016/j.ejor.2003.10.031

Lambert, D. M., Cooper, M. C., & Pagh, J. D. (1998). Supply chain management: Implementation issues and research opportunities. *International Journal of Logistics Management*, *9*(2), 1–20. doi:10.1108/09574099810805807

Liu, S., & Papageorgiou, L. G. (2013). Multi-objective optimisation of production, distribution and capacity planning of global supply chains in the process industry. *Omega*, *41*(2), 369–382. doi:10.1016/j.omega.2012.03.007

Ma, Y., Yan, F., Kang, K., & Wei, X. (2016). A novel integrated production-distribution planning model with conflict and coordination in a supply chain network. *Knowledge-Based Systems*, *105*, 119–133. doi:10.1016/j.knosys.2016.05.007

Melo, M. T., Nickel, S., & Saldanha-Da-Gama, F. (2009). Facility location and supply chain management–A review. *European Journal of Operational Research*, *196*(2), 401–412. doi:10.1016/j.ejor.2008.05.007

Nasiri, G. R., Zolfaghari, R., & Davoudpour, H. (2014). An integrated supply chain production–distribution planning with stochastic demands. *Computers & Industrial Engineering*, *77*, 35–45. doi:10.1016/j.cie.2014.08.005

Raa, B., Dullaert, W., & Aghezzaf, E. H. (2013). A matheuristic for aggregate production–distribution planning with mould sharing. *International Journal of Production Economics*, *145*(1), 29–37. doi:10.1016/j.ijpe.2013.01.006

Revelle, C. S., Eiselt, H. A., & Daskin, M. S. (2008). A bibliography for some fundamental problem categories in discrete location science. *European Journal of Operational Research, 184*(3), 817–848. doi:10.1016/j.ejor.2006.12.044

Shi, J., Liu, Z., Tang, L., & Xiong, J. (2017). Multi-objective optimization for a closed-loop network design problem using an improved genetic algorithm. *Applied Mathematical Modelling, 45*, 14–30. doi:10.1016/j.apm.2016.11.004

Soleimani, H., & Kannan, G. (2015). A hybrid particle swarm optimization and genetic algorithm for closed-loop supply chain network design in large-scale networks. *Applied Mathematical Modelling, 39*(14), 3990–4012. doi:10.1016/j.apm.2014.12.016

Su, S., Zhan, D., & Xu, X. (2008). An extended state task network formulation for integrated production-distribution planning in supply chain. *International Journal of Advanced Manufacturing Technology, 37*(11-12), 1232–1249. doi:10.100700170-007-1063-3

Subulan, K., Taşan, A. S., & Baykasoğlu, A. (2015). A fuzzy goal programming model to strategic planning problem of a lead/acid battery closed-loop supply chain. *Journal of Manufacturing Systems, 37*, 243–264. doi:10.1016/j.jmsy.2014.09.001

Thanh, P. N., Bostel, N., & Péton, O. (2008). A dynamic model for facility location in the design of complex supply chains. *International Journal of Production Economics, 113*(2), 678–693. doi:10.1016/j.ijpe.2007.10.017

Ting, C. K., Liao, X. L., Huang, Y. H., & Liaw, R. T. (2017). Multi-vehicle selective pickup and delivery using metaheuristic algorithms. *Information Sciences, 406*, 146–169. doi:10.1016/j.ins.2017.04.001

Varthanan, P. A., Murugan, N., & Kumar, G. M. (2012). A simulation based heuristic discrete particle swarm algorithm for generating integrated production–distribution plan. *Applied Soft Computing, 12*(9), 3034–3050. doi:10.1016/j.asoc.2012.05.001

Wang, K. J., Makond, B., & Liu, S. Y. (2011). Location and allocation decisions in a two-echelon supply chain with stochastic demand–A genetic-algorithm based solution. *Expert Systems with Applications, 38*(5), 6125–6131. doi:10.1016/j.eswa.2010.11.008

Whitley, D. (2001). An overview of evolutionary algorithms: Practical issues and common pitfalls. *Information and Software Technology, 43*(14), 817–831. doi:10.1016/S0950-5849(01)00188-4

Yang, S. S., Ong, S. K., & Nee, A. Y. C. (2016). A decision support tool for product design for remanufacturing. *Procedia CIRP, 40*, 144–149. doi:10.1016/j.procir.2016.01.085

Yang, X. S. (2014). *Nature-inspired optimization algorithms*. Elsevier.

Zamarripa, M., Silvente, J., & Espuña, A. (2012). Supply chain planning under uncertainty using genetic algorithms. *Computer-Aided Chemical Engineering, 30*, 457–461. doi:10.1016/B978-0-444-59519-5.50092-7

Zhalechian, M., Tavakkoli-Moghaddam, R., Zahiri, B., & Mohammadi, M. (2016). Sustainable design of a closed-loop location-routing-inventory supply chain network under mixed uncertainty. *Transportation Research Part E: Logistics and Transportation Review, 89*, 182-214.

Compilation of References

Abadi, J., & Brunnermeier, M. (2018). *Blockchain economics* (Working paper). Princeton University.

Abakouy, R., En-Naimi, E. M., & El Haddadi, A. (2017). Classification and Prediction Based Data Mining algorithms to Predict Email Marketing Campaigns. *Proceedings of the 2nd International Conference on Computing and Wireless Communication Systems - ICCWCS'17*, 1–5. 10.1145/3167486.3167520

Abdelwahab, S., Hamdaoui, B., Guizani, M., & Rayes, A. (2014). Enabling smart cloud services through remote sensing: An internet of everything enabler. *IEEE Internet of Things Journal*, *1*(3), 276–288. doi:10.1109/JIOT.2014.2325071

Abdessemed, D., Nezzal, G., & Ben Aim, R. (2000). Coagulation-adsorption-ultrafiltration for wastewater treatment and reuse. *Desalination*, *131*(1–3), 307–314. doi:10.1016/S0011-9164(00)90029-8

Abdulfetah, A. A., Sun, X. M., Yang, H., & Mohammad, N. (2010). Robust Adaptive Image Watermarking using Visual Models in DWT and DCT Domain. *International Journal of Information Technology*, *9*(3), 470–488.

Abraham, A. J., Kumar, K. R., Sridharan, R., & Singh, D. (2015). A genetic algorithm approach for integrated production and distribution problem. *Procedia: Social and Behavioral Sciences*, *189*, 184–192. doi:10.1016/j.sbspro.2015.03.213

Abrukov, V. S., Kochakov, V. D., Abrukov, S. V., Anufrieva, D. A., Vasilev, A. I., & Smirnov, A. V. (2017). Intelligent Forecasting System for Solar Power Plants. *ISJAEE*, *16*, 30–42. doi:10.15518/isjaee.2017.16-18.030-042

Adebiyi, A.A., Adewumi, A.O. & Ayo, C.K., (2014). Comparison of ARIMA and artificial neural networks models for stock price prediction. *Journal of Applied Mathematics*. . doi:10.1155/2014/614342

Adebiyi, A. A., Ayo, C. K., Adebiyi, M. O., & Otokiti, S. (2012). Stock price prediction using neural network with hybridized market indicators. *Journal of Emerging Trends in Computing and Information Sciences*, *3*, 1–9.

Adelantado, F., Vilajosana, X., Tuset-Peiro, P., Martinez, B., Melià-Seguí, J., & Watteyn, T. (2017). Understanding the limits of LoRaWAN. *IEEE Communications Magazine*, *55*(9), 34–40. doi:10.1109/MCOM.2017.1600613

Ade-Omowaye, B. I. O., Rastogi, N. K., Angersbach, A., & Knorr, D. (2002). Osmotic dehydration of bell peppers: Influence of high intensity electric field pulses and elevated temperature treatment. *Journal of Food Engineering*, *54*(1), 35–43. doi:10.1016/S0260-8774(01)00183-2

Adhami, J., Giudici, G., & Martinazzi, S. (2018). Why do businesses go crypto? An empirical analysis of initial coin offerings. *Journal of Economics and Business*, 20–32.

Adya, M., & Collopy, F. (1998). How effective are neural networks at forecasting and rediction? A review and evaluation. *Journal of Forecasting*, *17*(5-6), 481–495. doi:10.1002/(SICI)1099-131X(1998090)17:5/6<481::AID-FOR709>3.0.CO;2-Q

Afrouzy, Z. A., Nasseri, S. H., Mahdavi, I., & Paydar, M. M. (2016). A fuzzy stochastic multi-objective optimization model to configure a supply chain considering new product development. *Applied Mathematical Modelling*, *40*(17-18), 7545–7570. doi:10.1016/j.apm.2016.03.015

Agrawal, S. (2019). *What the heck is Word Embedding*? Retrieved 25 September 2019, from https://towardsdatascience.com/what-the-heck-is-word-embedding-b30f67f01c81

Ahamed, B. B., & Ramkumar, T. (2015). Deduce User Search Progression with Feedback Session. *Advances in Systems Science and Applications*, *15*(4), 366–383.

Ahamed, B. B., & Ramkumar, T. (2016). An intelligent web search framework for performing efficient retrieval of data. *Computers & Electrical Engineering*, *56*, 289–299. doi:10.1016/j.compeleceng.2016.09.033

Ahamed, B. B., & Ramkumar, T. (2018). Proficient Information Method for Inconsistency Detection in Multiple Data Sources. *International Journal of Artificial Intelligence*, *16*(1), 182–194.

Ahire, J. (2019). *The Artificial Neural Networks Handbook: Part 1*. Retrieved 19 November 2019, from https://medium.com/coinmonks/the-artificial-neural-networks-handbook-part-1-f9ceb0e376b4

Ahlquist, J. S. (2006). Economic policy, institutions, and capital flows: Portfolio and direct investment flows in developing countries. *International Studies Quarterly*, *50*(3), 681–704. doi:10.1111/j.1468-2478.2006.00420.x

Aikens, C. H. (1985). Facility location models for distribution planning. *European Journal of Operational Research*, *22*(3), 263–279. doi:10.1016/0377-2217(85)90246-2

Ajide, K. B., & Raheem, A. J. (2015). Determinants of foreign capital flows into Nigeria: A structural var analysis. *Business and Management Research Journal*, *5*, 1–13.

Akpakwu, G. A., Silva, B. J., Hancke, G. P., & Abu-Mahfouz, A. M. (2018). A survey on 5G networks for the Internet of Things: Communication technologies and challenges. *IEEE Access : Practical Innovations, Open Solutions*, *6*, 3619–3647. doi:10.1109/ACCESS.2017.2779844

Akyildiz, I. F., Wang, X., & Wang, W. (2005). Wireless mesh networks: A survey. *Computer Networks*, *47*(4), 445–487. doi:10.1016/j.comnet.2004.12.001

Alanne, K., & Saari, A. (2006). Distributed energy generation and sustainable development. *Renewable & Sustainable Energy Reviews*, *10*(6), 539–558. doi:10.1016/j.rser.2004.11.004

Al-Fuqaha, A., Guizani, M., Mohammadi, M., Aledhari, M., & Ayyash, M. (2015). Internet of Things: A survey on enabling technologies, protocols, and applications. *IEEE Communications Surveys and Tutorials*, *17*(4), 2347–2376. doi:10.1109/COMST.2015.2444095

Alikaniotis, D., Yannakoudakis, H., & Rei, M. (2016). *Automatic Text Scoring Using Neural Networks*. Retrieved from https://arxiv.org/pdf/1606.04289.pdf

Alkhasov, A. B. (2010). *Vozobnovlyaemaya energetika* [Renewable energy]. FIZMATLIT.

Allen, R. (1997). Self-calibrating method for estimating solar radiation from air temperature. *Journal of Hydrologic Engineering*, *2*(2), 56–67. doi:10.1061/(ASCE)1084-0699(1997)2:2(56)

Almorox, J., & Hontoria, C. (2004). Global solar radiation estimation using sunshine duration in Spain. *Energy Conversion and Management*, *45*(9-10), 1529–1535. doi:10.1016/j.enconman.2003.08.022

Alshamsi, A., & Diabat, A. (2017). A Genetic Algorithm for Reverse Logistics network design: A case study from the GCC. *Journal of Cleaner Production*, *151*, 652–669. doi:10.1016/j.jclepro.2017.02.096

Altiparmak, F., Gen, M., Lin, L., & Paksoy, T. (2006). A genetic algorithm approach for multi-objective optimization of supply chain networks. *Computers & Industrial Engineering, 51*(1), 196–215. doi:10.1016/j.cie.2006.07.011

Amentie, M. D., Sheng, M., Song, J., & Liu, J. (2016, October). Minimum delay guaranteed cooperative device-to-device caching in 5G wireless networks. In *8th International Conference on Wireless Communications & Signal Processing (WCSP)* (pp. 1-5). IEEE.

American Society and Heating, Refrigerating and Airconditioning Engineers (ASHRAE). (1972). Handbook of Fundamentals. ASHRAE.

Amisano, G., & Giannini, C. (1997). *From VAR models to Structural VAR models. In Topics in Structural VAR Econometrics*. Springer-Verlag Berlin. doi:10.1007/978-3-642-60623-6

Amorim, P., Günther, H. O., & Almada-Lobo, B. (2012). Multi-objective integrated production and distribution planning of perishable products. *International Journal of Production Economics, 138*(1), 89–101. doi:10.1016/j.ijpe.2012.03.005

Andia, P., Sant, R. V., & Whiteley, N. (2018). A Comprehensive Real-Time Data Analysis Tool for Fluid Gains and Losses. *IADC/SPE Drilling Conference and Exhibition, 2018*(March), 6–8. doi:10.2118/189576-MS

Andolfatto, D. (2018). Blockchain: What it is, what it does, and why you probably don't need one. *Federal Reserve Bank of St. Louis Review, 100*(2).

Androulaki, E., Karame, G. O., Roeschlin, M., Scherer, T., & Capkun, S. (2013). Evaluating user privacy in Bitcoin. In A. R. Sadeghi (Ed.), *Proceedings of the 17th International Conference on Financial Cryptography and Data Security, LNCS 7859* (pp. 34-51). Springer.

Andrychowicz, M., Dziembowski, S., Malinowski, D., & Mazurek, L. (2015a). On the malleability of bitcoin transactions. In M. Brenner, N. Christin, B. Johnson, & K. Rohloff (Eds.), Lecture notes in computer science: Vol. 8976. *Financial cryptography and data security (FC 2015)* (pp. 1–18). Springer. doi:10.1007/978-3-662-48051-9_1

Angeline, P. J. (1998, May). Using selection to improve particle swarm optimization. In *1998 IEEE International Conference on Evolutionary Computation Proceedings. IEEE World Congress on Computational Intelligence (Cat. No. 98TH8360)* (pp. 84-89). IEEE. 10.1109/ICEC.1998.699327

Angstrom, A. (1924). Solar and terrestrial radiation. *Quarterly Journal of the Royal Meteorological Society, 50*(210), 121–126. doi:10.1002/qj.49705021008

Annandale, J. G., Jovanic, N. Z., Benade, N., & Allen, R. G. (2002). Software for missing data error analysis of Penman–Monteith reference evapotranspiration. *Irrigation Science, 21*(2), 57–67. doi:10.1007002710100047

An, S. C., Leung, M. T., & Daouk, H. (2003). Application of neural networks to an emerging Financial market: Forecasting and trading the Taiwan Stock Index. *Computers & Operations Research, 30*(6), 901–923. doi:10.1016/S0305-0548(02)00037-0

Antonopoulos, A. M. (2015). *Mastering Bitcoin*. O'Reilly Media.

Arafat, A., & Raihanuzzaman, M. (2016). *Automated Essay Grading with Recommendation*. Retrieved from https://pdfs.semanticscholar.org/7bb7/22d9f01706bf3da54faae9b763fa8fd64f76.pdf

Arora, S., Singh, H., Sharma, M., Sharma, S., & Anand, P. (2019). A New Hybrid Algorithm Based on Grey Wolf Optimization and Crow Search Algorithm for Unconstrained Function Optimization and Feature Selection. *IEEE Access: Practical Innovations, Open Solutions, 7*, 26343–26361. doi:10.1109/ACCESS.2019.2897325

Asadi, S., Hadavandi, E., Mehmanpazir, F., & Nakhostin, M. M. (2012). Hybridization of evolutionary Levenberg–Marquardt neural networks and data pre-processing for stock market prediction. *Knowledge-Based Systems, 35*, 245–258. doi:10.1016/j.knosys.2012.05.003

Askarzadeh, A. (2016). A novel metaheuristic method for solving constrained engineering optimization problems: Crow search algorithm. *Computers & Structures, 169*, 1–12. doi:10.1016/j.compstruc.2016.03.001

Atsalakis, G. S., & Valavanis, K. P. (2009). Surveying stock market forecasting techniques –Part II: Soft computing methods. *Expert Systems with Applications, 36*(3), 5932–5941. doi:10.1016/j.eswa.2008.07.006

Atzori, L., Iera, A., & Morabito, G. (2010). The Internet of Things: A survey. *Computer Networks, 54*(15), 2787–2805. doi:10.1016/j.comnet.2010.05.010

Ayele, E. D., Das, K., Meratnia, N., & Havinga, P. J. (2018, February). Leveraging BLE and LoRa in IoT network for wildlife monitoring system (WMS). In *2018 IEEE 4th World Forum on Internet of Things (WF-IoT)* (pp. 342-348). IEEE.

Ayele, E. D., Meratnia, N., & Havinga, P. J. (2018, February). Towards a new opportunistic iot network architecture for wildlife monitoring system. In *2018 9th IFIP International Conference on New Technologies, Mobility and Security (NTMS)* (pp. 1-5). IEEE. 10.1109/NTMS.2018.8328721

Ayodele, T. R., & Ogunjuyigbe, A. S. (2015). Prediction of monthly average global solar radiation based on statistical distribution of clearness index. *Energy, 90*, 1733–1742. doi:10.1016/j.energy.2015.06.137

Babaev, B. D. (2015). Resursy vozobnovlyaemyh istochnikov energii Respubliki Dagestan: Uchebno-spravochnoe posobie [Resources of renewable energy sources of the Republic of Dagestan: manual]. Mahachkala: Izdatel'stvo "Raduga".

Babaev, B. D. (2016b). *Razrabotka i issledovanie energosistem na osnove vozobnovlyaemyh istochnikov s fazoperekhodnym akkumulirovaniem tepla* [Development and study of energy systems based on renewable sources with phase transition heat storage]. Dis. dok. tekhn. nauk. Mahachkala.

Babaev, B. D., & Sheverdiev, R. P. (2017b). Optimizaciya sistemy kompleksnogo energosnabzheniya avtonomnogo potrebitelya na osnove mestnyh vozobnovlyaemyh energoistochnikov [Optimization of the integrated energy supply system of an autonomous consumer based on local renewable energy sources]. In *Materialy V Mezhdunarodnoj konferencii "Vozobnovlyaemaya energetika: problemy i perspektivy". Tom 2. Mahachkala* [Materials of the V International Conference "Renewable Energy: Problems and Prospects". Volume 2. Makhachkala]. Academic Press.

Babaev, B. D., Kharchenko, V. V., & Panchenko, V. (2019). Development and Research of Phase-Transition and Thermochemical Materials for Heat Accumulation. Handbook of Research on Smart Computing for Renewable Energy and Agro-Engineering, 1 – 26. Doi:10.4018/978-1-7998-1216-6.ch001

Babaev, B.D. & Sheverdiev, R.P. (2017a). Certificate of state registration of a computer program No. 2017663797. In *Optimizaciya sostava energokompleksa* [Optimization of the energy complex]. Registered in the Register of Computer Programs.

Babaev, B.D. & Sheverdiev, R.P. (2018). Optimizaciya ugla naklona solnechnyh modulej v zavisimosti ot mesta raspolozheniya i avtomatizirovannyj raschet solnechnoj radiacii [Optimization of the angle of inclination of solar modules depending on the location and automated calculation of solar radiation]. *Energetik, 12*, 29 - 31.

Babaev, B.D. (2003). *Kriterii vybora al'ternativnyh kombinirovannyh sistem energosnabzheniya* [Criteria for the selection of alternative combined energy supply systems]. *Izv. vuzov. Sev. -Kavk. region. Tekhn. nauki*, 49 - 50.

Babaev, B.D. (2016a). Metodika sravnitel'nogo analiza variantov kombinirovannogo energosnabzheniya po neskol'kim kriteriyam [Technique for a comparative analysis of the options for combined energy supply according to several criteria]. *Energetik, 5*, 19 - 22.

Babaev, B.D., Kharchenko, V., Panchenko, V. & Vasant, P. (2019). Materials and Methods of Thermal Energy Storage in Power Supply Systems. *Renewable Energy and Power Supply Challenges for Rural Regions*, 115 – 135. DOI: . doi:10.4018/978-1-5225-9179-5.ch005

Badescu, V. (1999). Correlations to estimate monthly mean daily solar global irradiation: Application to Romania. *Energy, 24*(10), 883–893. doi:10.1016/S0360-5442(99)00027-4

Badescu, V. (2002). 3D isotropic approximation for solar diffuse irradiance on tilted surfaces. *Renewable Energy, 26*(2), 221–233. doi:10.1016/S0960-1481(01)00123-9

Baev V.I., Petrukhin V.A. (2014) Kachestvennoye opisaniye protsessa prirastaniya privivok rasteniy pri elektricheskoy stimulyatsii [A qualitative description of the process of growth of plant grafts during electrical stimulation]. *Izvestiya Nizhnevolzhskogo agrouniversitetskogo kompleksa: Nauka i vyssheye professional'noye obrazovaniye, 4*(36), 241–245.

Bahel, V., Bakhsh, H., & Srinivasan, R. (1987). A correlation for estimation of global solar radiation. *Energy, 12*(2), 131–135. doi:10.1016/0360-5442(87)90117-4

Bahrammirzaee, A. (2010). A comparative survey of artificial intelligence applications in finance: Artificial neural networks, expert system and hybrid intelligent systems. *Neural Computing & Applications, 19*(8), 1165–1195. doi:10.100700521-010-0362-z

Bai, L., Yang, D., Wang, X., Tong, L., Zhu, X., Bai, C., & Powell, C. A. (2020). Chinese experts' consensus on the Internet of Things-aided diagnosis and treatment of coronavirus disease 2019. *Clinical eHealth.*

Bakirci, K. (2009). Correlations for estimation of daily global solar radiation with hours of bright sunshine in Turkey. *Energy, 34*(4), 485–501. doi:10.1016/j.energy.2009.02.005

Balaji, S., Swaminathan, V., & Kannan, K. (2010). An effective algorithm for minimum weighted vertex cover problem. *International Journal of Mathematical, Computational, Physical, Electrical and Computer Engineering, 4*(7), 770–774.

Balinski, M. L. (1965). Integer programming: Methods, uses, computations. *Management Science, 12*(3), 253–313. doi:10.1287/mnsc.12.3.253

Bamert, T., Decker, C., Wattenhofer, R., & Welten, S. (2014). Bluewallet: The secure bitcoin wallet. In *Proceedings of the 10th International Workshop Security Trust Management*, (pp. 65-80). Academic Press.

Bamert, T., Decker, C., Elsen, L., Wattenhofer, R., & Welton, S. (2015). Have a snack, pay with bitcoins. *Proceedings of the 13th IEEE International Conference on Peer-to-Peer Computing*. doi: 10.1109/P2P.2013.6688717

Bandyopadhyay, D., & Sen, J. (2011). Internet of Things: Applications and challenges in technology and standardization. *Wireless Personal Communications, 58*(1), 49–69. doi:10.100711277-011-0288-5

Bangert, P. (2019). Predicting and Detecting Equipment Malfunctions using Machine Learning. *SPE Middle East Oil and Gas Show and Conference, MEOS, Proceedings, 2019-March*. Retrieved from https://www.onepetro.org/conference-paper/SPE-195149-MS

Banik, A., Dutta, S., Bandyopadhyay, T. K., & Biswal, S. K. (2018). Prediction of maximum permeate flux (%) of disc membrane using Response Surface Methodology (RSM). *Canadian Journal of Civil Engineering, 46*(6), 299–307.

Banik, A., Bandyopadhyay, T. K., & Biswal, S. K. (2017). Computational fluid dynamics simulation of disc membrane used for improving the quality of effluent produced by the rubber industry. *International Journal of Fluid Mechanics Research, 44*(6), 499–512. doi:10.1615/InterJFluidMechRes.2017018630

Banik, A., Biswal, S. K., Majumder, M., & Bandyopadhyay, T. K. (2018). Development of an adaptive non-parametric model for estimating maximum efficiency of disc membrane. *Int. J. Convergence Computing, 3*(1), 3–19. doi:10.1504/IJCONVC.2018.091111

Barros, M T L, Ros, D. A., & Lopes, J. E. G. (2005). *Objective Functions For Hydropower System Operation*. Academic Press.

Barros, M. T. L., Tsai, F. T.-C., Yang, S., Lopes, J. E. G., & Yeh, W. W.-G. (2003). Optimization of large-scale hydropower system operations. *Journal of Water Resources Planning and Management, 129*(3), 178–188. doi:10.1061/(ASCE)0733-9496(2003)129:3(178)

Barski, C., & Wilmer, C. (2015). *Bitcoin for the befuddled*. No Starch Press.

Bartels, A. H., Daley, E., Parker, A., Evelson, B., & Muteba, C. (2009). *Smart computing drives the new era of IT growth*. Forrester Inc.

Bartolozzi, M., Leinweber, D. B., & Thomas, A. W. (2006). Scale-free avalanche dynamics in the stock market. *Physica A, 370*(1), 132–139. doi:10.1016/j.physa.2006.04.024

Beamon, B. M. (1998). Supply chain design and analysis: Models and methods. *International Journal of Production Economics, 55*(3), 281–294. doi:10.1016/S0925-5273(98)00079-6

Becker, E. (n.d.). *Barcelona: Taming tourism*. Available at: https://www.travelweekly.com/Europe-Travel/Barcelona-Taming-tourism

Bedi, P., Bansal, R., & Sehgal, P. (2013). Using PSO in a spatial domain based image hiding scheme with disortion tolerance. *International Journal of Computers and Electrical Engineering, 39*(2), 640–654. doi:10.1016/j.compeleceng.2012.12.021

Behmanesh, E., & Pannek, J. (2016). Modeling and random path-based direct encoding for a closed loop supply chain model with flexible delivery paths. *IFAC-PapersOnLine, 49*(2), 78–83. doi:10.1016/j.ifacol.2016.03.014

Beigel, O. (2019). *Bitcoin historical price & events*. Retrieved from https://99bitcoins.com/bitcoin/historical-price/

Bera, S., Misra, S., & Rodrigues, J. (2015). Cloud computing applications for smart grid: A survey. *IEEE Transactions on Parallel and Distributed Systems, 5*(5), 1477–1494. doi:10.1109/TPDS.2014.2321378

Berentsen, A., & Schär, F. (2018). *A short introduction to the world of cryptocurrencies*. Academic Press.

Bertoldi, P. (2019). Overview of the European Union policies to promote more sustainable behaviours in energy end-users. In M. Lopes, C. H. Antunes, & K. B. Janda (Eds.), *Energy and Behaviour* (pp. 451–477). Academic Press.

Besharat, F., Dehghan, A. A., & Faghih, A. R. (2013). Empirical models for estimating global solar radiation: A review and case study. *Renewable & Sustainable Energy Reviews, 21*, 798–821. doi:10.1016/j.rser.2012.12.043

Bezrukikh, P. P., Degtyarev, V. V., & Elistratov, V. V. (2007). *Spravochnik po resursam vozobnovlyaemyh istochnikov energii Rossii i mestnym vidam topliva/pokazateli po territoriyam* [Handbook of Russian renewable energy resources and local fuels/indicators by territory]. IAC Energiya.

Bhardwaj, S., & Bhola, P. (2016). Solar Energy Estimation Techniques: A Review. *Conference Paper*.

Bhaskar, B., Mahesh, D., Kumaraswamy, P., & Nagender, Y. (2018). Caching Strategies Based on Data Density Estimation in Mobile Ad-hoc Networks. *International Journal of Engineering & Technology*, *7*(1.8), 240-243.

Bhatoolaul, D., Ghosh, A., Mangalvedhe, N., & Ratasuk, R. (2017). *LTE-M evolution towards 5G massive MTC. In 2017 IEEE Globecom Workshops (GC Workshops)*. IEEE.

Bing, Y., Hao, J. K., & Zhang, S. C. (2012). Stock Market Prediction Using Artificial Neural Networks. *Advanced Engineering Forum*, *6-7*, 1055-1060.

Bishop, C. M. (2006). Pattern Recognition and Machine Learning. *Information Science and Statistics*. Retrieved from https://www.springer.com/gp/book/9780387310732

BitLegal. (2016). *BitCoin legal status around the world.* Retrieved May 9, 2016 from http://bitlegal.io/; http://money.cnn.com/interactive/technology/where-is-bitcoin-legal/

Black, B. T. V. (2019). *White Box Testing: Key Differences*. Retrieved 3 December 2019, from https://www.guru99.com/back-box-vs-white-box-testing.html

Black, J. N. (1956). The distribution of solar radiation over the earth's surface. *Arch. Met. Geoph. Biokl,* (7), 165-189.

Black, J. D., Forsyth, F. R., Fensom, D. S., & Ross, R. B. (1971). Electrical stimulation and its effects on growth and ion accumulation in tomato plants. *Canadian Journal of Botany*, *49*(10), 1809–1815. doi:10.1139/b71-255

Bloomberg NEF. (2019). *New Energy Outlook 2019*. Retrieved from https://about.bnef.com/new-energy-outlook/

Boes, K., Buhalis, D., & Inversini, A. (2015). Conceptualising Smart Tourism Destination Dimensions. In *Information and Communication Technologies in Tourism 2015* (pp. 391–403). Springer International Publishing. doi:10.1007/978-3-319-14343-9_29

Böhm, T. (2018). *The Three Main Branches of Word Embeddings*. Retrieved 25 September 2019, from https://towardsdatascience.com/the-three-main-branches-of-word-embeddings-7b90fa36dfb9

Boland, J., Scott, L., & Luther, M. (2001). Modelling the diffuse fraction of global solar radiation on a horizontal surface. *Environmetrics*, *12*(12), 103–116. doi:10.1002/1099-095X(200103)12:2<103::AID-ENV447>3.0.CO;2-2

Bornstein, A. (2018). *Beyond Word Embeddings Part 2- Word Vectors & NLP Modeling from BoW to BERT.* Retrieved 11 November 2019, from https://towardsdatascience.com/beyond-word-embeddings-part-2-word-vectors-nlp-modeling-from-bow-to-bert-4ebd4711d0ec

Bos, J. W., Halderman, J. A., Heninger, N., Moore, J., Naehrig, M., & Wustrow, E. (2014). Elliptic curve cryptography in practice. *Proceedings of the 18th International Conference on Financial Cryptography and Data Security (FC'14)*, 157-175.

Boyd, D., & Crawford, K. (2012). Critical questions for big data: Provocations for a cultural, technological, and scholarly phenomenon. *Information Communication and Society*, *15*(5), 662–679. Advance online publication. doi:10.1080/1369118X.2012.678878

Bramwell, B., Henry, I., Jackson, G., Prat, A. G., Richards, G., & van der Straaten, J. (1996). *Sustainable tourism management: principles and practice* (B. Bramwell, Ed.). Tilburg University Press.

Breazeale, E. L., & McGeorge, W. T. (1953). Cation uptake by plants as affected by an applied potential. *Soil Science*, *75*(6), 443–448. doi:10.1097/00010694-195306000-00004

Bristow, C. L., & Campbell, G. S. (1984). On the relationship between incoming solar radiation and daily maximum and minimum temperature. *Agricultural and Forest Meteorology*, *31*(2), 159–166. doi:10.1016/0168-1923(84)90017-0

Brito, J., & Castillo, A. (2013). *Bitcoin: A primer for policymakers*. Mercatus Center and George Mason University.

Brock, W., Lakonishok, J., & Lebaron, B. (1992). Simple technical trading rules and the stochastic properties of stock returns. *The Journal of Finance*, *47*(5), 1731–1764. doi:10.1111/j.1540-6261.1992.tb04681.x

Brown, D. R. L. (2010). *SEC 2: Recommended elliptic curve domain parameters, standards for efficient cryptography group (SECG)*. Retrieved March 22, 2017 from http://www.secg.org/sec2-v2.pdf

Brunekreeft, G., Buchmann, M., Dänekas, C., Guo, X., Mayer, C., Merkel, M., Rehtanz, C., Göring, A., Herrmann, A., Kodali, R., Stadler, M., Uslar, M., Vogel, N., Luhmann, T., Menz, T., Müller, S.-U., & Recknagel, P. (2015). *China's way from conventional power grids towards smart grids. Regulatory Pathways for Smart Grid Development in China*. Springer Vieweg. doi:10.1007/978-3-658-08463-9

Buhalis, D. (2002). *eTourism: Information technology for strategic tourism management*. Financial Times/Prentice Hall.

Buhalis, D., & Amaranggana, A. (2013). Smart Tourism Destinations. In *Information and Communication Technologies in Tourism 2014* (pp. 553–564). Springer International Publishing. doi:10.1007/978-3-319-03973-2_40

Buhalis, D., & Amaranggana, A. (2015). Smart Tourism Destinations Enhancing Tourism Experience Through Personalisation of Services. In *Information and Communication Technologies in Tourism 2015* (pp. 377–389). Springer International Publishing. doi:10.1007/978-3-319-14343-9_28

Burmistrov, A. A., Vissarionov, V. I., & Deryugina, G. V. (2009). Metody rascheta resursov vozobnovlyaemyh istochnikov energii [Methods for calculating renewable energy resources]. Moskva: Izdatel'skij dom MEI.

Burton, T., Sharpe, D., Jenkins, N., & Bossanyi, E. (2001). *Wind Energy Handbook*. John Wiley & Sons, Ltd. doi:10.1002/0470846062

Buterin, V. (2014). *A next-generation smart contract and decentralized application platform* (White paper). Toronto: Ethereum.

Butler, R. W. (1999). Sustainable tourism: A state-of-the-art review. *Tourism Geographies*, *1*(1), 7–25. doi:10.1080/14616689908721291

Calvete, H. I., & Galé, C. (2010). A multiobjective bilevel program for production-distribution planning in a supply chain. In *Multiple criteria decision making for sustainable energy and transportation systems* (pp. 155–165). Springer. doi:10.1007/978-3-642-04045-0_13

Canchumuni, S. A., Emerick, A. A., & Pacheco, M. A. (2017). Integration of ensemble data assimilation and deep learning for history matching facies models. *OTC Brasil*, *2017*, 385–397. doi:10.4043/28015-MS

Cao, L. (2003). Support vector machines experts for time series forecasting. *Neurocomputing*, *51*, 321–339. doi:10.1016/S0925-2312(02)00577-5

Capizzi, G., Napoli, C., & Bonanno, F. (2012). Innovative second-generation wavelets construction with recurrent neural networks for solar radiation forecasting. *IEEE Transactions on Neural Networks and Learning Systems*, *23*(11), 1805–1815. doi:10.1109/TNNLS.2012.2216546 PMID:24808074

Carstens, A. (2019). Money in the digital age: What role for central banks? In Central Bank of Ireland. Whitaker Lecture.

Celik, A. N., & Muneer, T. (2013). Neural network based method for conversion of solar radiation data. *Energy Conversion and Management*, *67*, 117–124. doi:10.1016/j.enconman.2012.11.010

Chait, G. (2015). *Distributed crypto identity as a mechanism for legal empowerment of the poor and stimulating local economic development* (Working paper). Sidney: Whythawk.

Chakraborty, S., Samanta, S., Biswas, D., & Dey, N. (2013), Particle Swarm Optimization based parameter optimization technique in medical information hiding, *IEEE International Conference on Computational Intelligence and Computing Research*, 1-6. 10.1109/ICCIC.2013.6724173

Chand, N., Joshi, R. C., & Misra, M. (2007). Cooperative caching in mobile ad hoc networks based on data utility. *Mobile Information Systems*, *3*(1), 19–37. doi:10.1155/2007/193641

Chan, F. T., Jha, A., & Tiwari, M. K. (2016). Bi-objective optimization of three echelon supply chain involving truck selection and loading using NSGA-II with heuristics algorithm. *Applied Soft Computing*, *38*, 978–987. doi:10.1016/j.asoc.2015.10.067

Changmai, M., Pasawan, M., & Purkait, M. K. (2018). Treatment of oily wastewater from drilling site using electrocoagulation followed by microfiltration. *Separation and Purification Technology*, *210*(February), 463–472. doi:10.1016/j.seppur.2018.08.007

Chao, H., Li-li, H., & Ting-ting, H. (2012). Financial time series forecasting based on wavelet kernel support vector machine. *International Conference on Natural Computation, 8,* 79-83.

Charan, P., Usmani, T., Paulus, R., & Saeed, S. H. (2019). Performance of Distributed Energy Aware Routing (DEAR) Protocol with Cooperative Caching for Wireless Sensor Networks. *Wireless Sensor Network*, *11*(3), 35–45. doi:10.4236/wsn.2019.113003

Chauhan, N., Awasthi, L. K., Chand, N., Joshi, R. C., & Misra, M. (2010). Energy efficient cooperative caching in mobile ad hoc networks. *International Journal of Applied Engineering Research*, *1*(3), 360–371.

Cheddad, A., Condell, J., Curran, K., & McKevitt, P. (2010). Digital image steganography survey and analysis of current methods. *Journal of Signal Processing*, *90*(3), 727–752. doi:10.1016/j.sigpro.2009.08.010

Chen, Y., Zhao, D.J., Wang, Z.Y., Wang, Z.Y., Tang, G.L., & Huang, L. (2016). Plant Electrical Signal Classification Based on Waveform Similarity. *Algorithms, 9*(4), 70. doi:10.3390/a9040070

Chen, B., Shen, H., & Cao, X. (2016). Cooperative caching placement in cognitive device-to-device networks. *Journal of Communication*, *11*(1), 42–49.

Cheng, Tian, Wang, Xu, & Chau. (2008). Optimizing hydropower reservoir operation using hybrid genetic algorithm and chaos. *Water Resources Management, 22*(7), 895–909.

Cheng, C. T., Wang, W.-C., Xu, D.-M., & Chau, K. W. (2008). Optimizing hydropower reservoir operation using hybrid genetic algorithm and chaos. *Water Resources Management*, *22*(7), 895–909. doi:10.100711269-007-9200-1

Chen, K. Y., & Wang, C. H. (2007). Support vector regression with genetic algorithms in forecasting tourism demand. *Tourism Management*, *28*(1), 215–226. doi:10.1016/j.tourman.2005.12.018

Chen, R., Ersi, K., Yang, J., Lu, S., & Zhao, W. (2004). Validation of five global radiation models with measured daily data in China. *Energy Conversion and Management*, *45*(11-12), 1759–1769. doi:10.1016/j.enconman.2003.09.019

Chen, Y. (2018). Blockchain tokens and the potential democratization of entrepreneurship and innovation. *Business Horizons*, *61*(4), 567–575. doi:10.1016/j.bushor.2018.03.006

Chi-Jie, L., & Jui-Yu, W. (2011). An efficient CMAC neural network for stock index forecasting. *Expert Systems with Applications*, *38*(12), 15194–15201. doi:10.1016/j.eswa.2011.05.082

Chi-Jie, L., Tian-Shyug, L., & Chih-Chou, C. (2009). Financial time series forecasting using independent component analysis and support vector regression. *Decision Support Systems*, *47*(2), 115–125. doi:10.1016/j.dss.2009.02.001

Chollet, F. (2018). *Deep learning with Python* (1st ed.). Manning Publications Co.

Chopra, S., & Meindl, P. (2013). *Supply Chain Management: Strategy, Planning, and Operation*. Academic Press.

Chow, C. Y., Leong, H. V., & Chan, A. T. (2007). GroCoca: Group-based peer-to-peer cooperative caching in mobile environment. *IEEE Journal on Selected Areas in Communications*, *25*(1), 179–191. doi:10.1109/JSAC.2007.070118

Chowdhury, G. (2003). Natural language processing. *Annual Review Of Information Science And Technology*, *37*, 51-89.

Cocchia, A. (2014). *Smart and digital city: A systematic literature review. In Smart city*. Springer.

Coindesk. (2017). *Bitcoin - Daily number of transactions*. Retrieved January 10, 2017 from http://www.coindesk.com/data/bitcoin-daily-transactions/

CoinMarketCap. (2019). *Top 100 cryptocurrencies by market capitalization*. Retrieved August 8, 2019 from https://coinmarketcap.com/

Collins. (n.d.). *Collins Overtourism*. Available at: https://www.collinsdictionary.com/submission/19794/Overtourism

Coltuc, D. (2011). Improved embedding for prediction-based reversible watermarking. *IEEE Transactions on Information Forensics and Security*, *6*(3), 873–882. doi:10.1109/TIFS.2011.2145372

Corbet, S., Lucey, B., Urquhart, A., & Yarovaya, L. (2018). *Cryptocurrencies as a financial asset: A systematic analysis*. International Institute of Financial Analytics Journal.

Cowles, A. III. (1933). Can stock market forecasters forecast? *Econometrica*, *1*(3), 309–324. doi:10.2307/1907042

Crnkovic-Friis, L., & Erlandson, M. (2015). Geology driven EUR prediction using deep learning. *Proceedings - SPE Annual Technical Conference and Exhibition,* 1062–1071. 10.2118/174799-MS

Crouch, G. I., & Ritchie, J. R. B. (2000). The competitive destination: A sustainability perspective. *Tourism Management*. Advance online publication. doi:10.1080/1461668042000280264

Çulha, A. (2006). A structural VAR analysis of the determinants of capital flows into Turkey. *Central Bank Review*, *2*, 11–35.

Cziraky, D., & Cumpek, T. (2002). Multivariate analysis of the European economic and defence structure. *Croatian International Relations Review*, *8*, 23–35.

Dai, W., Jui-Yu, W., & Chi-Jie, L. (2012). Combining nonlinear independent component analysis and neural network for the prediction of Asian stock market indexes. *Expert Systems with Applications*, *39*(4), 4444–4452. doi:10.1016/j.eswa.2011.09.145

Dannehl, D. (2018). Effects of electricity on plant responses. *Scientia Horticulturae*, *234*, 382–392. doi:10.1016/j.scienta.2018.02.007

Dannehl, D., Huyskens-Keil, S., Eichholz, I., Ulrichs, C., & Schmidt, U. (2009). Effects of intermittent-direct-electric-current (IDC) on polyphenols and antioxidant activity in radish (Raphanus sativus L.) during growth. *Journal of Applied Botany and Food Quality*, *83*, 54–59.

Dannehl, D., Huyskens-Keil, S., Eichholz, I., Ulrichs, C., & Schmidt, U. (2011). Effects of direct-electric-current on secondary plant compounds and antioxidant activity in harvested tomato fruits (Solanum lycopersicon L.). *Food Chemistry*, *126*(1), 157–165. doi:10.1016/j.foodchem.2010.10.092

Dannehl, D., Huyskens-Keil, S., Wendorf, D., Ulrichs, C., & Schmidt, U. (2012). Influence of intermittent-direct-electric-current (IDC) on phytochemical compounds in garden cress during growth. *Food Chemistry*, *131*(1), 239–246. doi:10.1016/j.foodchem.2011.08.069

Davidson, R., & Maitland, R. (1997). *Tourism Destinations*. Hodder & Stoughton.

Davies, J. (2020). 5G Conspiracy Theories; What They Are, Why They Are Wrong and What Can Be Done. *Telecoms.com*. https://telecoms.com/503845/5g-conspiracy-theories-what-they-are-why-they-are-wrong-and-what-can-be-done/

De Bondt, W. F., & Thaler, R. (1985). Does the stock market overreact? *The Journal of Finance*, *40*(3), 793–805. doi:10.1111/j.1540-6261.1985.tb05004.x

De Faria, E. L., Albuquerque, M. P., Gonzalez, J. L., Cavalcante, J. T. P., & Albuquerque, M. P. (2009). Predicting the Brazilian stock market through neural networks and adaptive exponential smoothing methods. *Expert Systems with Applications*, *36*(10), 12506–12509. doi:10.1016/j.eswa.2009.04.032

De Filippi, P. (2014). Bitcoin: A regulatory nightmare to a libertarian dream. *Internet Policy Review*, *3*(2), 43. doi:10.14763/2014.2.286

De Oliveira, F. A., Nobre, C. N., & Zárate, L. E. (2013). Applying artificial neural networks to prediction of stock price and improvement of the directional prediction index –Case study of PETR4, Petrobras, Brazil. *Expert Systems with Applications*, *40*(18), 7596–7606. doi:10.1016/j.eswa.2013.06.071

Decker, C., & Wattenhofer, R. (2014). Bitcoin transaction malleability and MtGox. *Proceedings of the 19th European Symposium for Computer Security Research (ESORICS'14)*. Retrieved July 10, 2016 from https://link.springer.com/chapter/10.1007/978-3-319-11212-1_18

Deloitte. (2019). *Beyond the energy transition. When the demand-side is demanding change*. Retrieved from https://www2.deloitte.com/au/en/pages/energy-and-resources/articles/beyond-energy-transition.html

Denko, M. K., Nkwe, T., & Obaidat, M. S. (2010). Efficient cooperative caching with improved performance in wireless mesh networks. In *IEEE International Conference on Communications* (ICC) (pp. 1-5). IEEE. 10.1109/ICC.2010.5501921

Deryugina, G. V., & Tyagunov, M. G., & Lin, Z. I. (2017). Data verification for usage in regional Geographic Information System «Renewable energy sources». *Energetik*, *5*, 36–41.

Diabat, A., & Deskoores, R. (2016). A hybrid genetic algorithm based heuristic for an integrated supply chain problem. *Journal of Manufacturing Systems*, *38*, 172–180. doi:10.1016/j.jmsy.2015.04.011

Diebold, F. X. (2015). Comparing predictive accuracy, twenty years later: A personal perspective on the use and abuse of Diebold–Mariano tests. *Journal of Business & Economic Statistics*, *33*(1), 1–9. doi:10.1080/07350015.2014.983236

Dikli, S. (2006). Automated Essay Scoring. *Turkish Online Journal of Distance Education*, *7*(1). Retrieved from https://www.researchgate.net/publication/26415982_Automated_Essay_Scoring

Director, L. B., Zaichenko, V. M., & Maykov, I. L. (2012). Intellektual'nye sistemy upravleniya avtonomnymi energeticheskimi kompleksami v sostave lokal'nyh raspredelitel'nyh setej maloj energetiki [Intelligent control systems for autonomous energy complexes as part of local distribution networks of small energy]. *Izvestiya Rossiiskoi Akademii Nauk*, *1*, 38 – 48.

Director, L.B. & Maykov, I.L. (2010). Reshenie zadach optimizacii slozhnyh energeticheskih sistem [The solution of problems of optimization of complex energy systems]. *Upravlenie bol'shimi sistemami*, *28*, 274 – 292.

Director, L.B., Maykov, I.L. & Ivanin, O.A. (2012). Zadacha optimizacii avtonomnyh energeticheskih kompleksov v sostave lokal'nyh raspredelitel'nyh setej [The optimization problem of autonomous energy complexes as part of local distribution networks]. *Energotekhnologii i resursosberezhenie, 4*, 33 - 41.

Di, X. (2014). *Stock trend prediction with technical indicators using SVM. Independent Work Report.* Stanford Univ.

Dogniaux, R., & Lemoine, M. (1983). Classification of radiation sites in terms of different indices of atmospheric transparency. *Solar energy research and development in the European Community, series F, 2.*

Donatell, M., & Campbell, G. S. (1998). A simple model to estimate global solar radiation. *Proceedings of the fifth European society of agronomy congress*, 133-134.

Dong, F., Zhang, Y., & Yang, J. (2017). Attention-based Recurrent Convolutional Neural Network for Automatic Essay Scoring. *Proceedings of the 21st Conference on Computational Natural Language Learning*, 153-162. Retrieved from http://aclweb.org/anthology/K17-1017

Dos Santos, G. A. M., Barnes, Z., Lo, E., Ritoper, B., Nishizaki, L., Tejeda, X., . . . Kastner, R. (2014, October). Small unmanned aerial vehicle system for wildlife radio collar tracking. In *2014 IEEE 11th International Conference on Mobile Ad Hoc and Sensor Systems* (pp. 761-766). IEEE. 10.1109/MASS.2014.48

Draeger, A., Engell, S., & Ranke, H. (1995). Model Predictive Control Using Neural Networks. *IEEE Control Systems, 15*(5), 61–66. doi:10.1109/37.466261

Duffie, J., Beckman, W., & Worek, W. (2003). *Solar Engineering of Thermal Processes.* Wiley.

Dutta, S. (2013). Optimization of Reactive Black 5 removal by adsorption process using Box-Behnken design. *Desalination and Water Treatment, 51*(40–42), 7631–7638. doi:10.1080/19443994.2013.779597

Edenhofer, O., Pichs-Madruga, R., Sokona, Y., Seyboth, K., Eickemeier, P., Matschoss, P., … Stechow, C. Von. (2012). Intergovernmental Panel On Climate Change - Summary for Policymakers. IPCC Special Report on Renewable Energy Sources and Climate Change Mitigation. Cambridge University Press.

Edition, S. (n.d.). *Advances in Metaheuristic Algorithms for Optimal Design of Structures.* Academic Press.

Egly, P. V., Johnk, D. W., & Liston, D. P. (2010). Foreign Portfolio Investment Inflows to the United States: The Impact of Investor Risk Aversion and US Stock Market Performance. *North American Journal of Finance and Banking Research, 4*(4), 25–41.

Ehteram, M., Allawi, M. F., Karami, H., Mousavi, S. F., Emami, M., El-Shafie, A., & Farzin, S. (2017). Optimization of Chain-Reservoirs' Operation with a New Approach in Artificial Intelligence. *Water Resources Management, 31*(7), 2085–2104. doi:10.100711269-017-1625-6

Ehteram, M., Karami, H., & Farzin, S. (2018). Reducing Irrigation Deficiencies Based Optimizing Model for Multi-Reservoir Systems Utilizing Spider Monkey Algorithm. *Water Resources Management, 32*(7), 2315–2334. doi:10.100711269-018-1931-7

Ehteram, M., Karami, H., Mousavi, S. F., Farzin, S., & Kisi, O. (2017). Optimization of energy management and conversion in the multi-reservoir systems based on evolutionary algorithms. *Journal of Cleaner Production, 168*, 1132–1142. doi:10.1016/j.jclepro.2017.09.099

EIA (U.S. Energy Information Administration). (2019). *Annual Energy Outlook 2019 with projections to 2050.* Retrieved from https://www.eia.gov/outlooks/aeo/

Ejimuda, C. (2019). Using deep learning and computer vision techniques to improve facility corrosion risk management systems 2.0. *Society of Petroleum Engineers - SPE Nigeria Annual International Conference and Exhibition 2019, NAIC 2019*. 10.2118/198863-MS

Ekeocha, P.C., Ekeocha, C., Malaolu, V. & Oduh, M., (2012). Modelling the long run determinants of foreign portfolio investment in Nigeria. *Journal of Economics and Sustainable Development, 3*, 194-205.

Ekici, B. (2014). A least squares support vector machine model for prediction of the next day solar insolation for effective use of PV systems. *Measurement, 50*, 255–262. doi:10.1016/j.measurement.2014.01.010

Eleven EU countries have already fulfilled their res development plans for 2020. (n.d.). http://www.renen.ru/eleven-eu-countries-have-already-fulfilled-their-res-development-plans-for-2020/

Elias, A. R., Golubovic, N., Krintz, C., & Wolski, R. (2017, April). Where's the bear?-automating wildlife image processing using iot and edge cloud systems. In *2017 IEEE/ACM Second International Conference on Internet-of-Things Design and Implementation (IoTDI)* (pp. 247-258). IEEE.

Elistratov, E. E., & Aronova, E. S. (2012). Solnechnye energoustanovki. Ocenka postupleniya solnechnogo izlucheniya [Solar power plants. Assessment of solar radiation]. Izdatel'stvo Politekhnicheskogo universiteta.

Elistratov, V. V. (2017). Tekhnologicheskie i tekhniko-ekonomicheskie osobennosti energokompleksov na osnove VIE dlya slozhnyh prirodno-klimaticheskih uslovij [Technological and technical and economic features of energy complexes based on renewable energy for complex climatic conditions]. In *Materialy V Mezhdunarodnoj konferencii "Vozobnovlyaemaya energetika: problemy i perspektivy". Tom 1. Mahachkala* [Materials of the V International Conference "Renewable Energy: Problems and Prospects". Volume 1. Makhachkala]. Academic Press.

Ellsmoor, J. (2019a, April 27). *Blockchain is the next big thing for renewable energy*. Retrieved from https://www.forbes.com/sites/jamesellsmoor/2019/04/27/blockchain-is-the-next-big-thing-for-renewable-energy/

Ellsmoor, J. (2019b, April 27). *Meet 5 Companies Spearheading Blockchain For Renewable Energy*. Retrieved from https://www.forbes.com/sites/jamesellsmoor/2019/04/27/meet-5-companies-spearheading-blockchain-for-renewable-energy/#5dac8ac7f2ae

El-Metwally, M. (2005). Sunshine and global solar radiation estimation at different sites in Egypt. *Journal of Atmospheric and Solar-Terrestrial Physics, 67*(14), 1331–1342. doi:10.1016/j.jastp.2005.04.004

EMD International A/S. (n.d.). *WindProMeteo Data Export* (Version 5) [Software]. Available from https://www.emd.dk/windpro/downloads/

emd.dk. (2013, June 6). *Blended Coastal Winds*. Retrieved from http://www2.emd.dk/admin/helpWiki/index.php/Blended_Coastal_Winds

emd.dk. (2013, October 18). *MERRA Data*. Retrieved from http://www2.emd.dk/admin/helpWiki/index.php/MERRA_Data

Energeticheskaya strategiya Rossii na period do 2035 goda [Energy strategy of Russia for the period until 2035]. (2014). Academic Press.

Ernie, H., & Wall, G. (1992). *Marketing Tourism Destinations: A Strategic Planning Approach*. Wiley.

ETS. (2013). *Practice Book for the Paper-delivered GRE General Test*. Retrieved 3 March 2019, from https://www.ets.org/s/gre/pdf/practice_book_GRE_pb_revised_general_test.pdf

EWF (Energy Web Foundation). (2019). *EW-DOS: The energy web decentralized operating system*. Retrieved from https://www.energyweb.org/reports/ewdos-vision-purpose/

Eyal, I., & Sirer, E. G. (2014). *How to disincentivize large Bitcoin mining pools.* Retrieved January 6, 2017 from http://hackingdistributed.com/2014/06/18/how-to-disincentivize-large-bitcoin-mining-pools/

Fabian, B., Ermakova, T., & Sander, U. (2016). Anonymity in Bitcoin – The users' perspective. *37th International Conference on Information Systems.*

Fairbrother, C., Guccione, L., Henchen, M., & Teixeira, A. (2017). *Pathways for Innovation: The Role of Pilots and Demonstrations in Reinventing the Utility Business Model.* Retrieved from Rocky Mountain Institute website: www.rmi.org/insights/reports/pathwaysforinnovation/

Fairfield, J. (2014). Smart contracts, Bitcoin bots, and consumer protection. *Washington and Lee Law Review Online, 71*, 35–50.

Fang, F., Nguyen, T. H., Sinha, A., Gholami, S., Plumptre, A., Joppa, L., ... Critchlow, R. (2017). Predicting poaching for wildlife protection. *IBM Journal of Research and Development, 61*(6), 3–1. doi:10.1147/JRD.2017.2713584

Farahmand, M. (2019). *Pre-trained Word Embeddings or Embedding Layer? A Dilemma.* Retrieved 25 September 2019, from https://towardsdatascience.com/pre-trained-word-embeddings-or-embedding-layer-a-dilemma-8406959fd76c

Fard, A. F., Gholian-Jouybari, F., Paydar, M. M., & Hajiaghaei-Keshteli, M. (2017). A bi-objective stochastic closed-loop supply chain network design problem considering downside risk. *Industrial Engineering & Management Systems, 16*(3), 342–362. doi:10.7232/iems.2017.16.3.342

Farrokhi-Asl, H., Tavakkoli-Moghaddam, R., Asgarian, B., & Sangari, E. (2017). Metaheuristics for a bi-objective location-routing-problem in waste collection management. *Journal of Industrial and Production Engineering, 34*(4), 239–252. doi:10.1080/21681015.2016.1253619

Farsangi, A. S. N., Hadayeghparast, S., Mehdinejad, M., & Shayanfar, H. (2018). A novel stochastic energy management of a microgrid with various types of distributed energy resources in presence of demand response programs. *Energy, 160*, 257–274. doi:10.1016/j.energy.2018.06.136

Fathian, M., Jouzdani, J., Heydari, M., & Makui, A. (2018). Location and transportation planning in supply chains under uncertainty and congestion by using an improved electromagnetism-like algorithm. *Journal of Intelligent Manufacturing, 29*(7), 1447–1464. doi:10.100710845-015-1191-9

Felsen, J. (1975). Learning pattern recognition techniques applied to stock market forecasting. *IEEE Transactions on Systems, Man, and Cybernetics, SMC-5*(6), 583–594. doi:10.1109/TSMC.1975.4309399

Femenia-sierra, F., Neuhofer, B., & Ivars-Baidal, J.A. (2018). Towards a conceptualisation of smart tourists and their role within the smart destination scenario. *The Service Industries Journal, 39*(2).

Feng, Y., Zhou, J., Mo, L., Yuan, Z., Zhang, P., Wu, J., … Wang, Y. (2018). Long-term hydropower generation of cascade reservoirs under future climate changes in Jinsha River in Southwest China. Water, 10(2).

Fernandez-Rodriguez, F., Gonzalez-Martel, C., & Sosvilla-Rivero, S. (2000). On the profitability of technical trading rules based on artificial neural networks: Evidence from the Madrid stock market. *Economics Letters, 69*(1), 89–94. doi:10.1016/S0165-1765(00)00270-6

Fister, I., Yang, X. S., Brest, J., & Fister, D. (2013). A brief review of nature-inspired algorithms for optimization. *Elektrotehniski Vestnik/Electrotechnical Review, 80*(3), 116–122.

Fortov, V. E., & Popel, O. S. (2011). Energetika v sovremennom mire: Nauchnoe izdanie [Energy in the modern world: Scientific publication]. Dolgoprudnyj: Izdatel'skij Dom "Intellekt".

Foster, R., Ghassemi, M., & Cota, A. (2010). *Solar Energy*. CRC Press.

Foundation, D. D. S. (2019). *How does DDS compare to other IoT technologies?* Accessed April 23, 2019 from https://www.dds-foundation.org/features-benefits/

Franco, P. (2015). *Understanding Bitcoin: Cryptography, engineering and economics*. John Wiley & Sons.

Fritsch, S., Guenther, F., & Guenther, M. F. (2016). *Package 'neuralnet'*. The Comprehensive R Archive Network.

Gallagher, P., & Kerry, C. (2013). *Federal Information Processing Standards (FIPS) Publication 186-4: Digital Signature Standard (DSS)*. Retrieved March 20, 2017 from http://nvlpubs.nist.gov/nistpubs/FIPS/NIST.FIPS.186-4.pdf

Gandomi, A. H., & Alavi, A. H. (2012). *Krill Herd Algorithm : A New Bio-Inspired Optimization Algorithm Krill Herd : A New Bio-Inspired Optimization Algorithm*. Academic Press.

Ganesan, T., Vasant, P., Thomas, J., Litvinchev, I., & Bank, R. (2020). *Random Matrix Generators for Optimizing a Fuzzy Biofuel Supply Chain System*. Academic Press.

Ganesan, T., Vasant, P., & Elamvazuthi, I. (2016). *Advances in Metaheuristics: Applications in Engineering Systems (1st Editio.)*. CRC Press. doi:10.1201/9781315297651

Gang, L., & Song, H. (2008). Tourism Demand Modelling and Forecasting. *Tourism Management*, *29*(2), 203–220. doi:10.1016/j.tourman.2007.07.016

Gardašević, G., Veletić, M., Maletic, N., Vasiljevic, D., Radusinovic, I., Tomovic, S., & Radonjic, M. (2017). The IoT architectural framework, design issues and application domains. *Wireless Personal Communications*, *92*(1), 127–148. doi:10.100711277-016-3842-3

Gariepy, J. (1980). *Estimation of global solar radiation. International report*. Service of Meteorology.

Gea, T., Paradells Aspas, J., Lamarca, M., & Roldan, D. (2013). Smart cities as an application of Internet of Things: Experiences and lessons learnt in Barcelona. In *2013 Seventh International Conference on Innovative Mobile and Internet Services in Ubiquitous Computing (IMIS)*. IEEE. 10.1109/IMIS.2013.158

Gen, M., Hao, X., & Zhang, W. (2017). Advances in hybrid metaheuristics for stochastic manufacturing scheduling: Part I models and methods. In *Proceedings of the Tenth International Conference on Management Science and Engineering Management* (pp. 1063-1077). Springer. 10.1007/978-981-10-1837-4_88

Gerami, P., Subariah, I., & Morteza, B. (2012). Least Significant Bit Image Steganography using Particle Swarm Optimization and Optical Pixel Adjustment. *International Jisuanji Yingyong*, *55*(2), 20–25.

Gervais, A., Karame, G. O., Capkun, V., & Capkun, S. (2014). Is Bitcoin a decentralized currency? *IEEE Security and Privacy*, *12*(3), 54–60. doi:10.1109/MSP.2014.49

Ghelich, R., Jahannama, M. R., Abdizadeh, H., Torknik, F. S., & Vaezi, M. R. (2019). Central composite design (CCD)-Response surface methodology (RSM) of effective electrospinning parameters on PVP-B-Hf hybrid nanofibrous composites for synthesis of HfB 2 -based composite nanofibers. *Composites Part B: Engineering*, *166*(Ccd), 527–541. doi:10.1016/j.compositesb.2019.01.094

Ghezavati, V. R., & Beigi, M. (2016). Solving a bi-objective mathematical model for location-routing problem with time windows in multi-echelon reverse logistics using metaheuristic procedure. *Journal of Industrial Engineering International*, *12*(4), 469–483. doi:10.100740092-016-0154-x

Global Wind Energy Council. (2019, April). *Global wind report 2018*. Retrieved from http://jbr.org/articles.html

Glover, J., & McCulloch, S. (1958). The empirical relation between solar radiation and hours of sunshine. *Quarterly Journal of the Royal Meteorological Society*, *84*(360), 172–175. doi:10.1002/qj.49708436011

Gogo, E. O., Huyskens-Keil, S., Krimlowski, A., Ulrichs, C., Schmidt, U., Opiyo, A., & Dannehl, D. (2016). Impact of direct-electric-current on growth and bioactive compounds of African nightshade (Solanum scabrum Mill.) plants. *Journal of Applied Botany and Food Quality*, *89*, 60–67.

Goldfeder, S., Bonneau, J., Felten, E. W., Kroll, J. A., & Narayanan, A. (2014). Securing bitcoin wallets via threshold signatures. *International Technical Review*. Retrieved June 20, 2016 from http://www.cs.princeton.edu/~stevenag/bitcoin_ threshold_signatures.pdf

Goldfeder, S., Gennaro, R., & Kalodner, H. (2015). *Securing bitcoin wallets via a new DSA/ECDSA threshold signature scheme* (Technical report). Retrieved February 5, 2017 from http://stevengoldfeder.com/papers/threshold_sigs.pdf

Gómez, V., & Casanovas, A. (2003). Fuzzy modeling of solar irradiance on inclined surfaces. *Solar Energy*, *75*(4), 307–315. doi:10.1016/j.solener.2003.08.006

Gordeev, A. M., & Sheshnev, V. B. (1991). *Elektrichestvo v zhizni rasteniy* [Electricity in plant life]. Nauka.

Goren, H. G., Tunali, S., & Jans, R. (2010). A review of applications of genetic algorithms in lot sizing. *Journal of Intelligent Manufacturing*, *21*(4), 575–590. doi:10.100710845-008-0205-2

Gorshenin, V.A., Adamovich, B.A., & Adamovich, A. B. (1994) *Sposob povysheniya urozhaynosti i snizheniya srokov vegetatsii rasteniy* [Method for increasing yield and reducing growing period of plants]. Patent RF, 2058717.

Govindan, K., & Fattahi, M. (2017). Investigating risk and robustness measures for supply chain network design under demand uncertainty: A case study of glass supply chain. *International Journal of Production Economics*, *183*, 680–699. doi:10.1016/j.ijpe.2015.09.033

Gretzel, U. (2011a). Intelligent systems in tourism. A Social Science Perspective. *Annals of Tourism Research*, *38*(3), 757–779. Advance online publication. doi:10.1016/j.annals.2011.04.014

Gretzel, U., Sigala, M., Xiang, Z., & Koo, C. (2015). Smart tourism: Foundations and developments. *Electronic Markets*, *25*(3), 179–188. doi:10.100712525-015-0196-8

Gridasov, M. V., Kiseleva, S. V., Nefedova, L. V., Popel, O. C., & Frid, S. E. (2010). *"Renewable energy sources in Russia" information system developing*. Retrieved October 15, 2019, from www.gis-vie.ru

Grijalva, S., & Tariq, M. U. (2011). Prosumer-based smart grid architecture enables a flat, sustainable electricity industry. In *Proceedings Innovative Smart Grid Technologies (ISGT)* (pp. 1–6). IEEE. doi:10.1109/ISGT.2011.5759167

Grinberg, R. (2012). Bitcoin: An innovative alternative digital currency. *Hastings Science & Technology Law Journal*, *4*, 159–174.

Gruber, S. (2013). Trust, identity and disclosure: Are bitcoin exchanges the next virtual havens for money laundering and tax evasion. *Quinnipiac Law Review*, *32*, 135–208.

Grushack, J. (2014). *Currency 3.0: Examining digital crypto currency markets* (Doctoral dissertation). Department of Computer Science and Economics, Union College, Schenectady, NY.

Guderjan, M., Topfl, S., Angersbach, A., & Knorr, D. (2005). Impact of pulsed electric field treatment on the recovery and quality of plant oils. *Journal of Food Engineering*, *67*(3), 281–287. doi:10.1016/j.jfoodeng.2004.04.029

Gueymard, C. (1993). Critical analysis and performance assessment of clear sky solar irradiance models using theoretical and measured data. *Solar Energy*, (51), 385–397. doi:10.1016/0038-092X(93)90059-W

Guo, J., & Ya, G. (2015). Optimal strategies for manufacturing/remanufacturing system with the consideration of recycled products. *Computers & Industrial Engineering*, *89*, 226–234. doi:10.1016/j.cie.2014.11.020

Guo, S., Qiang, M., Luan, X., Xu, P., He, G., Yin, X., ... Fang, D. (2015). The application of the Internet of Things to animal ecology. *Integrative Zoology*, *10*(6), 572–578. doi:10.1111/1749-4877.12162 PMID:26338071

Gureeva, M. A. (2018). The effects of electrical stimulation on plants through the root system. *Exact Science.*, *21*, 18–21.

Hajdarbegovic, N. (2014). *Bitcoin miners ditch Ghash.io pool over fears of 51% attack, Coindesk*. Retrieved December 12, 2016 from http://www.coindesk.com/bitcoin-miners-ditch-ghash-io-pool-51-attack/

Hancke, G., Silva, B., & Hancke, G. Jr. (2012). The role of advanced sensing in smart cities. *Sensors (Basel)*, *13*(1), 393–425. doi:10.3390130100393 PMID:23271603

Handbook on the Climate of the USSR (Vol. 32). (1989). Leningrad.

Han, W., Wang, G., & Ying, P. (2017, December). Energy-Efficiency Aware Cooperative Caching Strategy for Content-Centric Networks. In *International Conference on Security, Privacy and Anonymity in Computation, Communication and Storage* (pp. 389-398). Springer. 10.1007/978-3-319-72395-2_36

Hargreaves, G. H., & Samani, Z. A. (1982). Estimating potential evapotranspiration. *Journal of Irrigation and Drainage Engineering*, *108*(IR3), 223–230.

Harrison, C., Eckman, B., Hamilton, R., Hartswick, P., Kalagnanam, J., Paraszczak, J., & Williams, P. (2010). Foundations for Smarter Cities. *IBM Journal of Research and Development*, *54*(4), 1–16. Advance online publication. doi:10.1147/JRD.2010.2048257

Heidmann, L. (2019). *AI vs. Machine Learning vs. Deep Learning*. Retrieved 11 November 2019, from https://blog.dataiku.com/ai-vs.-machine-learning-vs.-deep-learning

Hendrickson, J., Hogan, T., & Luther, W. (2016). The political economy of bitcoin. *Economic Inquiry*, *54*(2), 925–939. doi:10.1111/ecin.12291

Hiassat, A., Diabat, A., & Rahwan, I. (2017). A genetic algorithm approach for location-inventory-routing problem with perishable products. *Journal of Manufacturing Systems*, *42*, 93–103. doi:10.1016/j.jmsy.2016.10.004

Hinçal, O., Altan-Sakarya, A. B., & Ger, A. M. (2011). Optimization of Multireservoir Systems by Genetic Algorithm. *Water Resources Management*, *25*(5), 1465–1487. doi:10.100711269-010-9755-0

Hochreiter, S., & Schmidhuber, J. (1997). Long Short-Term Memory. *Neural Computation*, *9*(8), 1735–1780. doi:10.1162/neco.1997.9.8.1735 PMID:9377276

Hoeink, T., & Zambrano, C. (2017). Shale discrimination with machine learning methods. *51st US Rock Mechanics / Geomechanics Symposium 2017*, *4*, 2840–2845. Retrieved from https://www.onepetro.org/conference-paper/ARMA-2017-0769

Hong, K. H. (2017). Bitcoin as an alternative investment vehicle. *Information Technology Management*, *18*(4), 265–275. doi:10.100710799-016-0264-6

Hong, S., & Kim, H. (2019). Analysis of Bitcoin exchange using relationship of transactions and addresses. In *Proceedings of 21st International Conference on Advanced Communication Technology (ICACT)* (pp. 67-70). 10.23919/ICACT.2019.8701992

Hong, W. C. (2009). Chaotic particle swarm optimization algorithm in a support vector regression electric load forecasting model. *Energy Conversion and Management*, *50*(1), 105–117. doi:10.1016/j.enconman.2008.08.031

Hong, W. C., & Pai, P. F. (2007). Potential assessment of the support vector regression technique in rainfall forecasting. *Water Resources Management, 21*(2), 495–513. doi:10.100711269-006-9026-2

Hsu, C. M. (2011). A hybrid procedure for stock price prediction by integrating self-organizing map and genetic programming. *Expert Systems with Applications, 38*(11), 14026–14036. doi:10.1016/j.eswa.2011.04.210

Huang, Chen, & Abraham. (2010). Optimized Watermarking Using Swarm-Based Bacterial Foraging. *Journal of Information Hiding and Multimedia Signal Processing, 1*(1), 51–58.

Huang, C., Li-li, H., & Ting-ting, H. (2012). Financial time series forecasting based on wavelet kernel support vector machine. *8th International Conference on Natural Computation.* 10.1109/ICNC.2012.6234569

Huang, W., Nakamori, Y., & Shou-Yang, W. (2005). Forecasting stock market movement direction with support vector machine. *Computers & Operations Research, 32*(10), 2513–2522. doi:10.1016/j.cor.2004.03.016

Hubert. (2017). *AI In Education — Automatic Essay Scoring.* Accessed 5 December 2019, from https://medium.com/hubert-ai/ai-in-education-automatic-essay-scoring-6eb38bb2e70

Huhtinen, T. (2014). *Bitcoin as a monetary system: Examining attention and attendance* (Thesis). Department of Finance, School of Business, Aalto University, Aalto, PA.

Hui, T. K. L., Sherratt, R. S., & Sánchez, D. D. (2017). Major requirements for building smart homes in smart cities based on Internet of Things technologies. *Future Generation Computer Systems, 76*, 358–369. doi:10.1016/j.future.2016.10.026

Hunter, W. C., Chung, N., Gretzel, U., & Koo, C. (2015). Constructivist Research in Smart Tourism. *Asia Pacific Journal of Information Systems, 25*(1), 105–120. doi:10.14329/apjis.2015.25.1.105

Hutchinson, J. M., Goodin, D. G., Vanderlip, R. L., & Knapp, M. C. (1999). Estimating solar irradiance for crop modeling using daily air temperature data. *Agronomy Journal, 91*(5), 845–851. doi:10.2134/agronj1999.915845x

IEA. (2017). *Prospects for Distributed Energy Systems in China.* Retrieved from https://webstore.iea.org/prospects-for-distributed-energy-systems-in-china

IEA. (2018). *Population without access to electricity falls below 1 billion.* Retrieved from https://www.iea.org/commentaries/population-without-access-to-electricity-falls-below-1-billion

IEA. (2019). *CO2 Emissions from Fuel Combustion.* Retrieved from https://www.iea.org/data-and-statistics

Ignatev, E. V. (2019). *Energy efficiency increasing assessment of the wind energy complex in the united power system* (Unpublished candidate of engineering sciences dissertation). Moscow Power Engineering Institute (National Research University), Moscow, Russia.

IPCC. (2019). *Global warming of 1.5 °C.* Retrieved from https://www.ipcc.ch/sr15/

Iqbal, M. (1979). Correlation of average diffuse and beam radiation with hours of bright sunshine. *Solar Energy, 23*(23(2)), 169–173. doi:10.1016/0038-092X(79)90118-X

Iqbal, M. (1983). *An Introduction to Solar Radiation.* Academic Press.

IRENA. (2012). Renewable Energy Technologies: Hydropower. *Renewable Energy Technologies: Cost Analysis Series, 1*(3/5), 44.

IRENA. (2019a). *Energy transition.* Retrieved from https://www.irena.org/energytransition

IRENA. (2019b). *Transforming the energy system – and holding the line on the rise of global temperatures.* Author.

IRENA. (2019c). *Tracking SDG7: the energy progress report.* Retrieved from https://www.irena.org/publications/2019/May/Tracking-SDG7-The-Energy-Progress-Report-2019

Ishikawa, H., & Evans, M. L. (1990). Electrotropism of maize roots. *Plant Physiology, 94*(3), 913–918. doi:10.1104/pp.94.3.913 PMID:11537481

Ishtiaq, M., Bushra, S., Jaffar, A. M., & Aziz, K. (2010). Adaptive Watermark Strength Selection Using Particle Swarm Optimization. *ICIC Express Letters, 4*(5), 1–6.

ITU. (2017). *Minimum requirements related to technical performance for IMT-2020 radio interface (s).* ITU.

Jabbarzadeh, A., Fahimnia, B., & Sheu, J. B. (2017). An enhanced robustness approach for managing supply and demand uncertainties. *International Journal of Production Economics, 183*, 620–631. doi:10.1016/j.ijpe.2015.06.009

Jaddi, N. S., Alvankarian, J., & Abdullah, S. (2016). PT. *Communications in Nonlinear Science and Numerical Simulation.*

James Ayre. (2012). *The Red List Of Threatened Species, Annual Report Released.* planetsave.com

Jasic, T., & Wood, D. (2004). Theprofitability of daily stock market indices trades based on neural network predictions: Case study for the S&P 500, the DAX, the TOPIX and the FTSE in the period 1965-1999. *Applied Financial Economics, 14*(4), 285–297. doi:10.1080/0960310042000201228

Jawalkar, M. S., Desale, N., Suratwala, F., Lamkhade, A., Mahalle, P. N., & Railkar, P. N. (2017). Intelligent wildlife tracking using ubiquitous technological suite. *International Journal of Synthetic Emotions, 8*(1), 44–59. doi:10.4018/IJSE.2017010104

Jayabarathi, T., Raghunathan, T., Adarsh, B. R., & Nagaratnam, P. (2016). Economic dispatch using hybrid grey wolf optimizer. *Energy, 111*, 630–641. doi:10.1016/j.energy.2016.05.105

Jeans, E. (2015). Funny money or the fall of fiat: Bitcoin and forward-facing virtual currency regulation. *Journal on Telecommunications & High Technology Law, 13*, 99–171.

Jegadeesh, N., & Titman, S. (1993). Returns to buying winners and selling losers: Implications for stock market efficiency. *The Journal of Finance, 48*(1), 65–91. doi:10.1111/j.1540-6261.1993.tb04702.x

Jeong, Y., Kim, Y., Jin, Y., Hong, S., & Park, C. (2018). Separation and Puri fi cation Technology Comparison of fi ltration and treatment performance between polymeric and ceramic membranes in anaerobic membrane bioreactor treatment of domestic wastewater. *Separation and Purification Technology, 199*(January), 182–188. doi:10.1016/j.seppur.2018.01.057

Jiang, D., Li, H., Yang, T., & Li, D. (2016). Genetic algorithm for inventory positioning problem with general acyclic supply chain networks. *European Journal of Industrial Engineering, 10*(3), 367–384. doi:10.1504/EJIE.2016.076385

Jiao, W. (2018). *What are the advantages and disadvantages of Word2vec and GloVe?* Retrieved 18 November 2019, from https://www.quora.com/What-are-the-advantages-and-disadvantages-of-Word2vec-and-GloVe

Jie, W., Jian-Zhou, W., Zhe-George, Z., & Shu-Po, G. (2012). Stock index forecasting based on a hybrid model. *Omega, 40*(6), 758–766. doi:10.1016/j.omega.2011.07.008

Jin, M., Feng, W., Marnay, C., & Spanos, C. (2018). Microgrid to enable optimal distributed energy retail and end-user demand response. *Applied Energy, 210*, 1321–1335. doi:10.1016/j.apenergy.2017.05.103

Jin, Z., Yezheng, W., & Gang, Y. (2005). General formula for estimation of monthly average daily global solar radiation in China. *Energy Conversion and Management, 46*(2), 257–268. doi:10.1016/j.enconman.2004.02.020

Jitkongchuen, D., & Initialization, A. (2015). *A Hybrid Differential Evolution with Grey Wolf Optimizer for Continuous Global Optimization*. Academic Press.

John, T., DeWitt, W., Keebler, J. S., Min, S., Nix, N. W., Smith, C. D., & Zacharia, Z. G. (2001). Defining Supply Chain Management. *Journal of Business Logistics*.

Joshua Thomas, J., Pinar Karagoz, B., & Bazeer Ahamed, P. V. (Eds.). (2020). Deep Learning Techniques and Optimization Strategies in Big Data Analytics. doi:10.4018/978-1-7998-1192-3

Jukan, A., Masip-Bruin, X., & Amla, N. (2017). Smart computing and sensing technologies for animal welfare: A systematic review. *ACM Computing Surveys*, *50*(1), 1–27. doi:10.1145/3041960

Kaastra, I., & Boyd, M. (1996). Designing a neural network for forecasting financial and economic time series. *Neurocomputing*, *10*(3), 215–236. doi:10.1016/0925-2312(95)00039-9

Kabalci, E., & Kabalci, Y. (2019). *From smart grid to internet of energy*. Academic Press.

Kadadevaramath, R. S., Chen, J. C., Shankar, B. L., & Rameshkumar, K. (2012). Application of particle swarm intelligence algorithms in supply chain network architecture optimization. *Expert Systems with Applications*, *39*(11), 10160–10176. doi:10.1016/j.eswa.2012.02.116

Kaimoyo, E., Farag, M. A., Sumner, L. W., Wasmann, C., Cuello, J. L., & VanEtten, H. (2008). Sub-lethal levels of electric current elicit the biosynthesis of plant secondary metabolites. *Biotechnology Progress*, *24*(2), 377–384. doi:10.1021/bp0703329 PMID:18331050

Kalogirou, S. A. (2001). Artificial neural networks in renewable energy systems applications: A review. 2001;5(4). *Renewable & Sustainable Energy Reviews*, *5*(4). Advance online publication. doi:10.1016/S1364-0321(01)00006-5

Kaminsky, D. (2011). Black ops of TCP/IP. In *Proceedings of Black Hat Technical Security Conference*. Retrieved November 15, 2016 from http://de.slideshare.net/dakami/black-ops-of-tcpip-2011-black-hat-usa-2011

Kamphuis, R. (2017). Integrating demand flexibility with distributed generation – renewable energy sources (DG-RES) at the residential household and commercial customer level in electricity grids. *CIRED - Open Access Proceedings Journal, 1*, 1827-1830.

Kao, Y., & Zahara, E. (2008). *A hybrid genetic algorithm and particle swarm optimization for multimodal functions*. Academic Press.

Karaeren, V. (2014). *Maximizing energy generation of a cascade hydropower system*. Middle East Technical University.

Karame, G. O., Androulaki, E., & Capkun, S. (2012). Double-spending attacks on fast payments in bitcoin. *Proceedings of the 2012 ACM Conference on Computer and Communications Security (CCS'12)*. doi: 10.1145/2382196.2382292

Karame, G. O., Androulaki, E., Roeschlin, M., Gervais, A., & Capkun, S. (2015). Misbehavior in bitcoin: A study of double-spending and accountability. *ACM Transactions on Information and System Security*, *18*(1), 1–32. Advance online publication. doi:10.1145/2732196

Karani, D. (2018). *Introduction to Word Embedding and Word2Vec*. Retrieved 25 September 2019, from https://towardsdatascience.com/introduction-to-word-embedding-and-word2vec-652d0c2060fa

Kara, Y., Boyacioglu, M. A., & Baykan, Ö. K. (2011). Predicting direction of stock price index movement using artificial neural networks and support vector machines: The sample of the Istanbul Stock Exchange. *Expert Systems with Applications*, *38*(5), 5311–5319. doi:10.1016/j.eswa.2010.10.027

Karlstrom, H. (2014). Do libertarians dream of electric coins? The material embeddedness of Bitcoin. *Distinktion: Scandinavian Journal of Social Theory*, *15*(1), 23–36. doi:10.1080/1600910X.2013.870083

Karpathy, A. (2015). *The Unreasonable Effectiveness of Recurrent Neural Networks*. Retrieved 17 March 2019, from http://karpathy.github.io/2015/05/21/rnn-effectiveness/

Kaya, O., & Urek, B. (2016). A mixed integer nonlinear programming model and heuristic solutions for location, inventory and pricing decisions in a closed loop supply chain. *Computers & Operations Research*, *65*, 93–103. doi:10.1016/j.cor.2015.07.005

Kays, R., Tilak, S., Crofoot, M., Fountain, T., Obando, D., Ortega, A., ... Hirsch, B. (2011). Tracking animal location and activity with an automated radio telemetry system in a tropical rainforest. *The Computer Journal*, *54*(12), 1931–1948. doi:10.1093/comjnl/bxr072

Kazem, A., Sharifi, E., Hussain, F. K., Saberi, M., & Hussain, O. K. (2013). Support vector regression with chaos-based firefly algorithm for stock market price forecasting. *Applied Soft Computing*, *13*(2), 947–958. doi:10.1016/j.asoc.2012.09.024

Khaidem, L., Saha, S., & Dey, S. R. (2016). *Predicting the direction of stock market prices using random forest.* arXiv preprint arXiv:1605.00003

Khan, A., Rehmani, M. H., & Rachedi, A. (2017). Cognitive-radio-based Internet of Things: Applications, architectures, spectrum related functionalities, and future research directions. *IEEE Wireless Communications*, *24*(3), 17–25. doi:10.1109/MWC.2017.1600404

Kharchenko, V., Panchenko, V., Tikhonov, P. V., & Vasant, P. (2018). Cogenerative PV Thermal Modules of Different Design for Autonomous Heat and Electricity Supply. Handbook of Research on Renewable Energy and Electric Resources for Sustainable Rural Development, 86 – 119. Doi:10.4018/978-1-5225-3867-7.ch004

Kharchenko, V., Nikitin, B., Tikhonov, P., Panchenko, V., & Vasant, P. (2019). Evaluation of the Silicon Solar Cell Modules. Intelligent Computing & Optimization. *Advances in Intelligent Systems and Computing*, *866*, 328–336. doi:10.1007/978-3-030-00979-3_34

Kharchenko, V., & Vasant, P. (2018). *Handbook of Research on Renewable Energy and Electric Resources for Sustainable Rural Development*. IGI Global. doi:10.4018/978-1-5225-3867-7

Kharchenko, V., & Vasant, P. (2019a). *Handbook of Research on Energy-Saving Technologies for Environmentally-Friendly Agricultural Development*. IGI Global.

Kharchenko, V., & Vasant, P. (2019b). *Handbook of Research on Smart Computing for Renewable Energy and Agro-Engineering*. IGI Global.

Khashei, M., & Bijari, M. (2010). An artificial neural network (p, d, q) model for timeseries forecasting. *Expert Systems with Applications*, *37*(1), 479–489. doi:10.1016/j.eswa.2009.05.044

Khokhlov, A., Melnikov, Y., Veselov, F., Kholkin, D., & Datsko, K. (2018). *Distributed energy resources in Russia: development potential*. SKOLKOVO Energy Centre.

Kholmansky A.S., Kozhevnikov Yu.M. (2016) Mekhanizm generatsii elektrokhimicheskogo potentsiala dereva [The mechanism of generation of the electrochemical potential of a tree]. *Izvestiya vysshikh uchebnykh zavedeniy. Lesnoy zhurnal*, *5*(353), 73–80.

Kimball, H. H. (1919). Variations in the total and luminious solar radiation with geographical position in the United States. *Monthly Weather Review*, *47*(47(11)), 769–793. doi:10.1175/1520-0493(1919)47<769:VITTAL>2.0.CO;2

Kim, K. J., & Lee, W. B. (2004). Stock market prediction using artificial neural networks with optimal feature transformation. *Neural Computing & Applications, 13*(3), 255–260. doi:10.100700521-004-0428-x

Kim, S. (2018). A novel Bitcoin mining scheme based on the multi-leader multi-follower Stackelberg Game Model. *IEEE Access: Practical Innovations, Open Solutions, 1*(6), 48902–48912. doi:10.1109/ACCESS.2018.2867631

Kiseleva, S. V., Popel, O. S., & Kolomiets, Y. G. (2015). Evaluation of solar resources in the Central Asia. *Geliotechnics*, 69-74.

Klein, S. A. (1977). Calculation of monthly average insolation on tilted surfaces. *Solar Energy, 19*(4), 325–329. doi:10.1016/0038-092X(77)90001-9

Kline, R. B. (2005). *Methodology in the social sciences. In Principles and practice of structural equation modeling.* Guilford Press.

Klose, A., & Drexl, A. (2005). Facility location models for distribution system design. *European Journal of Operational Research, 162*(1), 4–29. doi:10.1016/j.ejor.2003.10.031

Koblitz, N. (1987). Elliptic curve cryptosystems. *Mathematics of Computation, 48*(177), 203–209. doi:10.1090/S0025-5718-1987-0866109-5

Koblitz, N., & Menezes, A. (2016). Cryptocash, cryptocurrencies, and cryptocontracts. *Designs, Codes and Cryptography, 78*(1), 87–102. doi:10.100710623-015-0148-5

Komninos, N., Philippou, E., & Pitsillides, A. (2014). Survey in smart grid and smart home security: Issues, challenges and countermeasures. *IEEE Communications Surveys and Tutorials, 16*(4), 1933–1954. doi:10.1109/COMST.2014.2320093

Korap, L. (2010). *Identification of 'pull' & 'push' factors for the portfolio flows: SVAR evidence from the Turkish economy. MPRA Paper 24275.* University Library of Munich.

Korjani, M., Popa, A., Grijalva, E., Cassidy, S., & Ershaghi, I. (2016). A new approach to reservoir characterization using deep learning neural networks. *Society of Petroleum Engineers - SPE Western Regional Meeting.* doi:10.2118/180359-ms

Koshy, P., Koshy, D., & McDaniel, P. (2014). An analysis of anonymity in Bitcoin using P2P network traffic. In N. Christin & R. Safavi-Naini (Eds.), *Proceedings of the 18th International Conference on Financial Cryptography and Data Security, LNCS 8437* (pp. 469-485). Springer.

Kubilay, T., & Bingöl, O. (2014). Web-based smart home automation: PLC-controlled implementation. *Acta Polytechnica Hungarica, 11*(3), 51–63.

Kumar, M., & Thenmozhi, M. (2006). *Forecasting stock index movement: A comparison of support vector machines and random forest.* Indian Institute of Capital Markets 9th Capital Markets Conference Paper.

Kumaresan, R., & Bentov, I. (2014). How to use bitcoin to incentivize correct computations. *Proceedings of the 2014 ACM SIGSAC Conference on Computer and Communications Security*, 30-41. 10.1145/2660267.2660380

Kusworo, T. D., Ariyanti, N., & Utomo, D. P. (2020). Effect of nano-TiO2 loading in polysulfone membranes on the removal of pollutant following natural-rubber wastewater treatment. *Journal of Water Process Engineering, 35*(November), 101190. doi:10.1016/j.jwpe.2020.101190

Kyoung-Jae, J. (2003). Financial time series forecasting using support vector machines. *Neurocomputing, 55*(1-2), 307–319. doi:10.1016/S0925-2312(03)00372-2

Kyoung-Jae, J., & Ingoo, H. (2000). Genetic algorithms approach to feature discretization in artificial neural networks for the prediction of stock price index. *Expert Systems with Applications, 19*(2), 125–132. doi:10.1016/S0957-4174(00)00027-0

Laffont, J. J., & Maskin, E. S. (1990). The efficient market hypothesis and insider trading on the stock market. *Journal of Political Economy, 98*(1), 70–93. doi:10.1086/261669

Lambert, D. M., Cooper, M. C., & Pagh, J. D. (1998). Supply chain management: Implementation issues and research opportunities. *International Journal of Logistics Management, 9*(2), 1–20. doi:10.1108/09574099810805807

Lam, J. C., & Li, D. (1996). Correlation between global solar radiation and its direct and diffuse components. *Building and Environment, 31*(6), 527–535. doi:10.1016/0360-1323(96)00026-1

Lamsfus, C. (2015). Smart Tourism Destinations: An Extended Conception of Smart Cities Focusing on Human Mobility. In *Information and Communication Technologies in Tourism 2015* (pp. 363–375). Springer International Publishing. doi:10.1007/978-3-319-14343-9_27

Lauridsen, M., Berardinelli, G., Tavares, F. M., Frederiksen, F., & Mogensen, P. (2016, May). Sleep modes for enhanced battery life of 5G mobile terminals. In *2016 IEEE 83rd Vehicular Technology Conference (VTC Spring)* (pp. 1-6). IEEE.

Lazzús, J. A., Pérez Ponce, A. A., & Marín, J. (2011). Estimation of global solar radiation over the city of La Serena (Chile) using a neural network. *Applied Solar Energy, 47*(1), 66–73. doi:10.3103/S0003701X11010099

Lebedev, S. I., Baranskiy, P. I., Litvinenko, L. G., & Shiyan, L. T. (1977). Barley growth in superweak magnetic field. *Electronic Treatment of Materials, 3*, 71–73.

LeCun, Y., Bengio, Y., & Hinton, G. (2015). Deep Learning. *Nature, 521*(7553), 436–444. doi:10.1038/nature14539 PMID:26017442

Lee, S., Yoo, C. D., & Kalker, T. (2007). Reversible image watermarking based on integer-to-integer wavelet transform. *IEEE Transactions on Information Forensics and Security, 2*(3), 321–330. doi:10.1109/TIFS.2007.905146

Lehner, B., Czisch, G., & Vassolo, S. (2005). The impact of global change on the hydropower potential of Europe: A model-based analysis. *Energy Policy, 33*(7), 839–855. doi:10.1016/j.enpol.2003.10.018

Letendre, S., Makhyoun, M., & Taylor, M. (2014). *Predicting Solar Power Production: Irradiance Forecasting Models*. Applications and Future Prospects.

Leung, M. T., Daouk, H., & An-Sing, C. (2000). Forecasting stock indices: A comparison of classification and level estimation models. *International Journal of Forecasting, 16*(2), 173–190. doi:10.1016/S0169-2070(99)00048-5

Li, L., Jiuchun, R., & Zhu, Q. (2017). On the application of LoRa LPWAN technology in Sailing Monitoring System. In *2017 13th Annual Conference on Wireless On-demand Network Systems and Services (WONS)*.IEEE.

Liang, T. P., Chandler, J. S., & Han, I. (1990). Integrating statistical and inductive learning methods for knowledge acquisition. *Expert Systems with Applications, 1*(4), 391–401. doi:10.1016/0957-4174(90)90048-Y

Li, J., Li, X., & Yang, B. (2013). Reversible data hiding scheme for color image based on prediction-error expansion and cross-channel correlation. *Signal Processing, 93*(9), 2748–2758. doi:10.1016/j.sigpro.2013.01.020

Lim, A. L., & Bai, R. (2003). Membrane fouling and cleaning in microfiltration of activated sludge wastewater. *Journal of Membrane Science, 216*(1–2), 279–290. doi:10.1016/S0376-7388(03)00083-8

Ling-Jing, K., Chih-Chou, C., Chi-Jie, L., & Jung-Li, Y. (2013). Integration of nonlinear independent component analysis and support vector regression for stock price forecasting. *Neurocomputing, 99*, 534–542. doi:10.1016/j.neucom.2012.06.037

Lin, P., Song, Q., & Jamalipour, A. (2019). Multidimensional Cooperative Caching in CoMP-Integrated Ultra-Dense Cellular Networks. *IEEE Transactions on Wireless Communications, 19*(3), 1977–1989. doi:10.1109/TWC.2019.2960329

Lin, S. D., Shie, S. C., & Guo, J. Y. (2010). Improving the robustness of DCT based image watermarking against JPEG compression. *International Journal of Computer Standards and Interfaces*, *32*(1), 54–60. doi:10.1016/j.csi.2009.06.004

Li, S., & Zhang, X. (2011). The study of PAFSSB on RO pre-treatment in pulp and paper wastewater. *Procedia Environmental Sciences*, *8*, 4–10. doi:10.1016/j.proenv.2011.10.003

Lischke, M., & Fabian, B. (2016). Analyzing the Bitcoin network: The first four years. *Future Internet*, *8*(1), 7. doi:10.3390/fi8010007

Liu, B., & Jordan, R. (1961). Daily insolation on surfaces tilted towards equator. *ASHRAE,* (10), 53-59.

Liu, X., Wu, J., Hou, L. an, & Wang, J. (2020). Fouling and cleaning protocols for forward osmosis membrane used for radioactive wastewater treatment. *Nuclear Engineering and Technology, 52*(3), 581–588. doi:10.1016/j.net.2019.08.007

Liu, B. Y., & Jordan, R. C. (1960). The interrelationship and characteristic distribution of direct, diffuse and total solar radiation. *Solar Energy*, *4*(4(3)), 1–19. doi:10.1016/0038-092X(60)90062-1

Liu, L., Chen, T., Zhu, S., Hong, W., & Si, X. (2014). A Reversible Data Hiding Method Using Improved Neighbor Mean Interpolation and Random-Block Division. *Information Technology Journal*, *13*(15), 2374–2384. doi:10.3923/itj.2014.2374.2384

Liu, R., Vellaithurai, C., Biswas, S. S., Gamage, T., & Srivastava, A. (2015). Analyzing the cyber-physical impact of cyber events on the power grid. *IEEE Transactions on Smart Grid*, *6*(5), 2444–2453. doi:10.1109/TSG.2015.2432013

Liu, S., & Papageorgiou, L. G. (2013). Multi-objective optimisation of production, distribution and capacity planning of global supply chains in the process industry. *Omega*, *41*(2), 369–382. doi:10.1016/j.omega.2012.03.007

Liu, X., Yang, T., & Yan, B. (2015, November). Internet of Things for wildlife monitoring. In *2015 IEEE/CIC International Conference on Communications in China-Workshops (CIC/ICCC)* (pp. 62-66). IEEE.

Li, Y., Yang, W., He, P., Chen, C., & Wang, X. (2019). Design and management of a distributed hybrid energy system through smart contract and blockchain. *Applied Energy*, *248*, 390–405. doi:10.1016/j.apenergy.2019.04.132

Lo Duca, M. (2012). *Modelling the time varying determinants of portfolio flows to emerging markets.* Working Paper Series of European Central Bank, Kaiserstrasse 29, D-60311 Frankfurt am Main, Germany.

LO3 Energy. (2020). *Pando*. Retrieved from https://lo3energy.com/pando/

Loia, V., Terzija, V., Vaccaro, A., & Wall, P. (2015). An affine-arithmetic-based consensus protocol for smart-grid computing in the presence of data uncertainties. *IEEE Transactions on Industrial Electronics*, *62*(5), 2973–2982. doi:10.1109/TIE.2014.2363046

López, G., Battles, F. J., & Tovas-Pescador, J. (2005). Selection of input parameters to model direct solar irradiance by using artificial neural networks. *Energy*, *30*(9), 30. doi:10.1016/j.energy.2004.04.035

Lowe, G., Duffie, J., & Smith, C. (1965). *World Distribution of Solar Radiation.* Academic Press.

Lucero, S. (2016). *IoT platforms: Enabling the Internet of Things* [white paper].

Lugovic, S., Mrsic, L., & Korona, L. Z. (2019). Public WiFi Security Network Protocol Practices in Tourist Destination. In *International Symposium on Pervasive Systems, Algorithms and Networks* (pp. 321-332). Springer. 10.1007/978-3-030-30143-9_27

Lustig, C., & Nardi, B. (2015). Algorithmic authority: The case of Bitcoin. *48th Hawaii International Conference on System Sciences IEEE*, 743-752.

Luther, W. (2013). *Cryptocurrencies, network effects, and switching costs* (Working paper). Arlington, VA: Mercatus Center and George Mason University.

Luther, W. (2018). Is bitcoin intrinsically worthless? *J. Priv. Enterprise*, (33), 31-45.

Lyu, S., Chen, W., Zhang, W., Fan, Y., & Jiao, W. (2016). Wastewater reclamation and reuse in China: Opportunities and challenges. *Journal of Environmental Sciences (China), 39*, 86–96. doi:10.1016/j.jes.2015.11.012

Mabaso, M. (2018). *Community distributed energy systems: an end-user energy services approach* (Unpublished master's thesis). Stellenbosch University, South Africa.

Ma, C., & Iqbal, M. (1982). Statistical Comparison Of Solar Radiation. *Solar Energy, 33*(2).

Machler, M. A., & Iqbal, M. A. (1967). A modification of the ASHRAE clear sky irradiation model. *ASHRAE Transactions*, (91), 106–115.

Maghrabi, A. H. (2009). Parameterization of a simple model to estimate monthly global solar radiation based on meteorological variables, and evaluation of existing solar radiation models for Tabouk, Saudi Arabia. *Energy Conversion and Management, 50*(11), 2754–2760. doi:10.1016/j.enconman.2009.06.024

Mahajan Shubhrata, D., Deshmukh Kaveri, V., Thite Pranit, R., Samel Bhavana, Y., & Chate, P. J. (2016). Stock market prediction and analysis using Naïve Bayes. *Int. J. Recent Innov. Trends Comput. Commun., 4*(11), 121–124.

Makarov, A. A., Grigoryev, L. M., & Mitrova, T. A. (2016). Prognoz razvitiya energetiki mira i Rossii [Forecast of the development of the energy industry in the world and Russia]. Academic Press.

Makridakis, S. G., Wheelwright, S. C., & Hyndman, R. J. (1998). *Forecasting: Methods and Applications*. Academic Press.

Maleki, S. A., Hizam, H., & Gomes, C. (2017). Estimation of Hourly, Daily and Monthly Global Solar Radiation on Inclined Surfaces: Models Re-Visited. *Energies*, 10.

Malkiel, B. G. (1989). Efficient market hypothesis. In *Finance* (pp. 127–134). Palgrave Macmillan. doi:10.1007/978-1-349-20213-3_13

Mallick, P. K. (Ed.). (2015). *Research Advances in the Integration of Big Data and Smart Computing*. IGI Global.

Manjunath, M., Singh, P., Mandal, A., & Parihar, G. S. (2014). Consumer behaviour towards electricity–A field study. *Energy Procedia, 54*, 541–548. doi:10.1016/j.egypro.2014.07.295

Margarita, S. (1991). Neural networks, genetic algorithms and stock trading. In *Artificial neural networks* (pp. 1763–1766). North Holland. doi:10.1016/B978-0-444-89178-5.50183-4

Marzetta, T. L. (2010). Noncooperative cellular wireless with unlimited numbers of base station antennas. *IEEE Transactions on Wireless Communications, 9*(11), 3590–3600. doi:10.1109/TWC.2010.092810.091092

Matthew Knight. (2012). *Extinction threat 'a call to world leaders' at Rio Earth Summit*. edition.cnn.com

Mavromoustakis, A., Constandinos, X., Mastorakis, G., & Batalla, J. M. (Eds.). (2016). Internet of Things (IoT) in 5G Mobile Technologies: Vol. 8. *Modeling and Optimization in Science and Technologies*. Springer.

Ma, Y., Yan, F., Kang, K., & Wei, X. (2016). A novel integrated production-distribution planning model with conflict and coordination in a supply chain network. *Knowledge-Based Systems, 105*, 119–133. doi:10.1016/j.knosys.2016.05.007

Mayer-Shönberg, V., & Cukier, K. (2013). *Big Data: A Revolution That Will Transform How We Live, Work, and Think*. Hodder & Stoughton.

Maykov, I.L. & Director, L.B. (2011). Reshenie zadach optimizacii i upravleniya gibridnymi energeticheskimi kompleksami v strukture raspredelennoj generacii [Solving the problems of optimization and control of hybrid energy complexes in the structure of distributed generation]. *Upravlenie bol'shimi sistemami, 35*, 250 – 264.

Maykov, I.L., Director, L.B. & Zaichenko, V.M. (2010). Reshenie zadach optimizacii energeticheskih sistem s neskol'kimi avtonomnymi energoustanovkami [Solving the problems of optimizing energy systems with several autonomous power plants]. *Upravlenie bol'shimi sistemami, 31*, 110 – 129.

McAfee, A., & Brynjolfsson, E. (2012). *Big Data. The management revolution.* Harvard Buiness Review. doi:10.100712599-013-0249-5

Mcclelland, J. L., Rumelhart, D. E., & Hinton, G. E. (1986). A general framework for parallel distributed processing. In *Parallel distributed processing: Explorations in the microstructure of cognition. Bradford Book.*

McKinsey & Company. (2017). *Coping with success: managing overcrowding in tourism destinations.* Available at: https://www.wttc.org/-/media/files/reports/policy-research/coping-with-success---managing-overcrowding-in-tourism-destinations-2017.pdf?la=en

Meiklejohn, S., Pomarole, M., Jordan, G., Levchenko, K., McCoy, D., Voelker, G. M., & Savage, S. (2013). A fistful of Bitcoins: Characterizing payments among men with no names. *Proceedings of the 2013 Conference on Internet Measurement (IMC '13).* 10.1145/2504730.2504747

Meirs, I., Garman, C., Green, M., & Rubin, A. D. (2013). Zerocoin: Anonymous distributed e-cash from bitcoin. *Proceedings of the 2013 IEEE Symposium on Security and Privacy (S&P'13).* 10.1109/SP.2013.34

Mekki, K., Bajic, E., Chaxel, F., & Meyer, F. (2018). *A comparative study of LPWAN technologies for large-scale IoT deployment.* ICT Express.

Mellit, A. (2008). Artificial intelligence technique for modelling and forecasting of solar radiation data: A review. *International Journal of Artificial Intelligence and Soft Computing, 1*(1), 52–76. doi:10.1504/IJAISC.2008.021264

Mellit, A., Kalogirou, S. A., Hontoria, L., & Shaari, S. (2009). Artificial intelligence techniques for sizing photovoltaic systems: A review. *Sustainable Energy Reviews, 13*(2), 406–419. doi:10.1016/j.rser.2008.01.006

Mellit, A., & Pavan, A. (2010). A 24-h forecast of solar irradiance using artificial neural network: Application for performance prediction of a grid-connected PV plant at Trieste, Italy. *Solar Energy, 84*(5), 84. doi:10.1016/j.solener.2010.02.006

Melo, M. T., Nickel, S., & Saldanha-Da-Gama, F. (2009). Facility location and supply chain management–A review. *European Journal of Operational Research, 196*(2), 401–412. doi:10.1016/j.ejor.2008.05.007

Mendling, J., van der Aalst, W., vom Brocke, J., Cabanillas, C., Weidlich, M., Weske, M., ... Zhu, L. (2018). Blockchains for business process management - Challenges and opportunities. *ACM Transactions on Management Information Systems, 9*(1), 1–16. doi:10.1145/3183367

Mengelkamp, E., Gärttner, J., Rock, K., Kessler, S., Orsini, L., & Weinhardt, C. (2018). Designing microgrid energy markets: A case study: The Brooklyn Microgrid. *Applied Energy, 210*, 870–880. doi:10.1016/j.apenergy.2017.06.054

Mercator Research Group. (2015). *Understanding Bitcoin: Blockchain applications beyond payments.* Retrieved December 24, 2016 from https://www.mercatoradvisorygroup.com/Reports/Understanding-Bitcoin--Blockchain-Applications-Beyond-Payments/

Metcalf, G. E., & Malkiel, B. G. (1994). The Wall Street Journal contests: The experts, the darts, and the efficient market hypothesis. *Applied Financial Economics, 4*(5), 371-374.

Meteonorm. (2019). Retrieved October 7, 2019, from http://meteonorm.ru

Meteotest, A. G. (2020). *Meteonorm* (Version 7.3.4) [Software]. Available from https://meteonorm.com/en/download

Meyer, D., Dimitriadou, E., Hornik, K., Weingessel, A. & Leisch, F. (2015). *e1071: misc functions of the department of statistics, probability theory group (formerly: E1071), TU Wien.* R package version 1.6-7.

Michie, D., Spiegelhalter, D. J., & Taylor, C. C. (1994). Machine learning. *Neural and Statistical Classification, 13.*

Mikolov, T., Chen, K., Corrado, G., & Dean, J. (2013). *Efficient Estimation of Word Representations in Vector Space.* Retrieved 25 September 2019, from https://arxiv.org/pdf/1301.3781.pdf

Milano, C., Cheer, J. M., & Novelli, M. (n.d.). *Overtourism is becoming a major issue for cities across the globe.* Available at: https://www.weforum.org/agenda/2018/07/overtourism-a-growing-global-problem

Militano, L., Araniti, G., Condoluci, M., Farris, I., & Iera, A. (2015). Device-to-device communications for 5G Internet of Things. *EAI Endorsed Transactions on Internet of Things, 15*(1), 1–15.

Miller, V. S. (1985). Use of elliptic curves in cryptography. *Proceedings of the Fifth Conference of Advanced Cytology.* doi: 10.1007/3-540-39799-X_31

Mingyue, Q., Cheng, L., & Yu, S. (2016). Application of the artificial neural network in predicting the direction of stock market index. *10th International Conference on Complex, Intelligent, and Software Intensive Systems.*

Miraz, M. H., Ali, M., Excell, P. S., & Picking, R. (2018). Internet of nano-things, things and everything: Future growth trends. *Future Internet, 10*(8), 68. doi:10.3390/fi10080068

Mironyuk, S. S., Smirnov, A., Sokolov, A., & Proshkin, Y. (2020) Optimization of spectral composition and energy economy effectiveness of phyto-irradiators with use of digital technologies. In Handbook of Research on Energy-Saving Technologies for Environmentally-Friendly Agricultural Development. IGI-Global.

Mohammadi, K., Shamshirband, S., Anisi, M. H., Alam, K. A., & Petković, D. (2015). Support vector regression based prediction of global solar radiation on a horizontal surface. *Energy Conversion and Management, 91*, 433–441. doi:10.1016/j.enconman.2014.12.015

Mohammadi, M., Man, H., Hassan, M., & Yee, P. (2013). Treatment of wastewater from rubber industry in Malaysia. *African Journal of Biotechnology, 9*(38), 6233–6243. http://www.ajol.info/index.php/ajb/article/view/92237

Mo, L., Lu, P., Wang, C., & Zhou, J. (2013). Short-term hydro generation scheduling of Three Gorges-Gezhouba cascaded hydropower plants using hybrid MACS-ADE approach. *Energy Conversion and Management, 76*, 260–273. doi:10.1016/j.enconman.2013.07.047

Moore, T., & Christin, N. (2013). Beware the middleman: Empirical analysis of Bitcoin-exchange risk. In A. R. Sadeghi (Ed.), *Financial Cryptography and Data Security: 17th International Conference Revised* (selected papers) (FC 2013) (pp. 25-33). Okinawa, Japan: Springer.

Mršić, L., Surla, G., & Balković, M. (2019). Technology Driven Smart Support System for Tourist Destination Management Organizations. *International Conference on Innovative Computing and Communication (ICICC-2019).* Available at: http://icicc-conf.com/icicc2019.pdf

Nabil, E. (2016). A Modified Flower Pollination Algorithm for Global Optimization. *Expert Systems with Applications, 57*, 192–203. doi:10.1016/j.eswa.2016.03.047

Nadaraya, E. A. (1964). On estimating regression. *Theory of Probability and Its Applications, 9*(1), 141–142. doi:10.1137/1109020

Naik, G. (2014, Sept. 30). Wildlife Numbers Drop by Half Since 1970, Report Says. *The Wall Street Journal.*

Nakamoto, S. (2008). *Bitcoin: A peer-to-peer electronic cash system* (White paper). Bitcoin Project. Retrieved February 15, 2017 from www.bitcoin.org

NASA POWER. (n.d.). *Project Data Sets.* Retrieved from https://power.larc.nasa.gov

NASA. (2019). Retrieved October 7, 2019, from The NASA Surface Meteorology and Solar Energy Data Set: https://eosweb.larc.nasa.gov/sse/

Nasiri, G. R., Zolfaghari, R., & Davoudpour, H. (2014). An integrated supply chain production–distribution planning with stochastic demands. *Computers & Industrial Engineering*, *77*, 35–45. doi:10.1016/j.cie.2014.08.005

Naveen, P., Chandel, A. K., Vedik, B., & Topwal, K. (2016). *Economic dispatch with valve point effect using symbiotic oragnisms search algorithm.* Academic Press.

Niaki, S. T. A., & Hoseinzade, S. (2013). Forecasting S&P 500 index using artificial neural networks and design of experiments. *Journal of Industrial Engineering International*, *9*(1), 1–9. doi:10.1186/2251-712X-9-1

Nielsen, J. (2019). *10 Heuristics for User Interface Design: Article by Jakob Nielsen.* Retrieved 22 November 2019, from https://www.nngroup.com/articles/ten-usability-heuristics/

Nijegorodov, N. (1996). Improved ashrae model to predict hourly and daily solar radiation components in Botswana, Namibia, and Zimbabwe. *WREC,* (9), 1270-1273.

Niu, P. P., Wang, X. Y., Yang, Y. P., & Ming, Y. (2011). A novel color image watermarking scheme in non-sampled contourlet-domain. *Journal of Expert Systems with Applications*, *38*(3), 2081–2098. doi:10.1016/j.eswa.2010.07.147

Nkwe, T. K., & Denko, M. K. (2009). Self-optimizing cooperative caching in autonomic wireless mesh networks. In *IEEE Symposium on Computers and Communications* (pp. 411-416). IEEE. 10.1109/ISCC.2009.5202375

NLTK 3.4.5 documentation. (2019). Retrieved 21 September 2019, from https://www.nltk.org/api/nltk.tokenize.html

Norvig, P. (2019). *How to Write a Spelling Corrector.* Retrieved 1 December 2019, from https://norvig.com/spell-correct

Notton, G., Paoli, C., Vasileva, S., Nivet, M. L., Canaletti, J., & Cristofari, C. (2012). Estimation of hourly global solar irradiation on tilted planes from horizontal one using artificial neural networks. *Energy*, *39*(1), 166–179. doi:10.1016/j.energy.2012.01.038

O'Connor, N., & Madden, M. G. (2006). A neural network approach to predicting stock exchange movements using external factors. *Knowledge-Based Systems*, *19*(5), 371–378. doi:10.1016/j.knosys.2005.11.015

Ober, M., Katzenbeisser, S., & Hamacher, K. (2013). Structure and anonymity of the Bitcoin transaction graph. *Future Internet*, *5*(2), 237–250. doi:10.3390/fi5020237

Obrezkov, V. I. (Ed.). (1981). *Hydropower: textbook for students of higher educational institutions.* Energoizdat.

Odabasi, S. D., & Zaim, A. H. (2010). A survey on wireless mesh networks, routing metrics and protocols. *International Journal of Electronics Mechanical and Mechatronics Engineering*, *2*(1), 92–104.

Olah, C. (2015). *Understanding LSTM Networks -- colah's blog*. Retrieved 22 March 2019, from http://colah.github.io/posts/2015-08-Understanding-LSTMs/

On the development of renewable energy in Russia until 2030. (n.d.). http://www.renen.ru/on-the-development-of-renewable-energy-in-russia-until-2030/

Opritov, V. A., Pyatygin, S. S., & Retivin, V. G. (1991). *Bioelektrogenez u vysshikh rasteniy* [Bioelectrogenesis in higher plants]. Nauka.

Orecchini, F., Santiangeli, A., Zuccari, F., Pieroni, A., & Suppa, T. (2019). Blockchain Technology in Smart City: A New Opportunity for Smart Environment and Smart Mobility. In P. Vasant, I. Zelinka, & G. W. Weber (Eds.), *Intelligent Computing & Optimization. ICO 2018. Advances in Intelligent Systems and Computing* (Vol. 866). Springer. doi:10.1007/978-3-030-00979-3_36

Orgill, J. F., & Hollands, K. G. (1977). Correlation Equation for Hourly Diffuse Radiation on a Horizontal Surface. *Solar Energy, 19*(4), 357–359. doi:10.1016/0038-092X(77)90006-8

Ouammi, A., Zejli, D., Dagdougui, H., & Benchrifa, R. (2012). Artificial neural network analysis of Moroccan solar potential. *Renewable & Sustainable Energy Reviews, 16*(7), 4876–4889. doi:10.1016/j.rser.2012.03.071

Pacheco, X. P. (2018). How Technology Can Transform Wildlife Conservation. In Green Technologies to Improve the Environment on Earth. IntechOpen.

Page, J. K. (1961). The estimation of monthly mean values of daily total short wave radiation on vertical and inclined surfaces from sunshine records for latitudes 40°N-40°S. *Proc. U.N. Conf. on New Sources of Energy, 4*, 378-390.

Page, E. (1966). The imminence of grading essays by computer. *Phi Delta Kappan, 47*(5), 238–243. https://www.jstor.org/stable/20371545?seq=1#page_scan_tab_contents

Palattella, M. R., Dohler, M., Grieco, A., Rizzo, G., Torsner, J., Engel, T., & Ladid, L. (2016). Internet of Things in the 5G era: Enablers, architecture, and business models. *IEEE Journal on Selected Areas in Communications, 34*(3), 510–527. doi:10.1109/JSAC.2016.2525418

Paliwal, M., & Kumar, U. A. (2009). Neural networks and statistical techniques: A review of applications. *Expert Systems with Applications, 36*(1), 2–17. doi:10.1016/j.eswa.2007.10.005

Pan, G., & Xu, Y. (2016). *Chaotic Glowworm Swarm Optimization Algorithm Based on Gauss Mutation*. Academic Press.

Panchenko, V., & Kharchenko, V. (2019). Development and Research of PVT Modules in Computer-Aided Design and Finite Element Analysis Systems. Handbook of Research on Energy-Saving Technologies for Environmentally-Friendly Agricultural Development, 314-342. Doi:10.4018/978-1-5225-9420-8.ch013

Panchenko, V., Chirskiy, S., & Kharchenko, V. V. (2019). Application of the Software System of Finite Element Analysis for the Simulation and Design Optimization of Solar Photovoltaic Thermal Modules. Handbook of Research on Smart Computing for Renewable Energy and Agro-Engineering, 106 – 131. Doi:10.4018/978-1-7998-1216-6.ch005

Panchenko, V. (2019). Photovoltaic solar modules for autonomous heat and power supply. *IOP Conference Series: Earth and Environmental Science*, 317, 9. 10.1088/1755-1315/317/1/012002

Panchenko, V. A. (2018). Solar Roof Panels for Electric and Thermal Generation. *Applied Solar Energy, 54*(5), 350–353. doi:10.3103/S0003701X18050146

Panchenko, V., Izmailov, A., Kharchenko, V., & Lobachevskiy, Y. (2020). Photovoltaic Solar Modules of Different Types and Designs for Energy Supply. *International Journal of Energy Optimization and Engineering, 9*(2), 74–94. doi:10.4018/IJEOE.2020040106

Panchenko, V., Kharchenko, V., & Vasant, P. (2019). Modeling of Solar Photovoltaic Thermal Modules. *Intelligent Computing & Optimization. Advances in Intelligent Systems and Computing, 866*, 108–116. doi:10.1007/978-3-030-00979-3_11

Pankaj, P., Geetan, S., MacDonald, R., Shukla, P., Sharma, A., & Menasria, S., … Judd, T. (2018). Application of data science and machine learning for well completion optimization. *Proceedings of the Annual Offshore Technology Conference, 6*, 4186–4201. 10.4043/28632-MS

Paoletti, S., Casini, M., Giannitrapani, A., Facchini, A., Garulli, A., & Vicino, A. (2011). Load forecasting for active distribution networks. In *Proceeding 2nd IEEE PES International Conference and Exhibition on Innovative Smart Grid Technologies* (pp. 1-6). Manchester, UK: IEEE.

Parishwad, G., Bhardwaj, R. K., & Nema, V. K. (1997). Estimation of hourly solar radiation for India. *Renewable Energy, 12*(12), 303–313. doi:10.1016/S0960-1481(97)00039-6

Paruchuri, V. (2019). *Vik's Blog - Writings on machine learning, data science, and other cool stuff.* Retrieved 10 November 2019, from https://www.vikparuchuri.com/

Patel, J., Shah, S., Thakkar, P., & Kotecha, K. (2015). Predicting stock market index using fusion of machine learning techniques. *Expert Systems with Applications, 42*(4), 2162–2172. doi:10.1016/j.eswa.2014.10.031

Pawan Kotiyal. (2019). *Reasons Why We Need to Save Wildlife.* Retrieved from https://www.tourmyindia.com/blog/reasons-to-save-wildlife/

Pedro, F. (2016). *Understanding Bitcoin, Cryptography, Engineering and Economics.* TJ International Ltd.

Perez, R., Stewart, R., Seals, R., & Guertin, T. (1988). *The Development and Verification of the Perez Diffuse Radiation Model.* Sandia National Labs. doi:10.2172/7024029

Personal, M., Archive, R., & Fleten, S. (2010). *Evaluation of static hedging strategies for hydropower producers in the Nordic market.* Academic Press.

Peša, A. R., & Brajković, A. (2016). Testing the 'black swan effect' on Croatian stock market between 2000 and 2013. *EMAJ: Emerging Markets Journal, 6*(1), 1–16. doi:10.5195/EMAJ.2016.92

Petajajarvi, J., Pettissalo, M., Mikhaylov, K., Roivainen, A., & Hänninen, T. (2015). On the coverage of LPWANs: Range evaluation and channel attenuation model for LoRa technology. In *2015 14th International Conference on ITS Telecommunications (ITST)*. IEEE.

Petäjäjärvi, J., Mikhaylov, K., Yasmin, R., Hämäläinen, M., & Iinatti, J. (2017). Evaluation of LoRa LPWAN technology for indoor remote health and wellbeing monitoring. *International Journal of Wireless Information Networks, 24*(2), 153–165. doi:10.100710776-017-0341-8

Peters, M., Neumann, M., Iyyer, M., Gardner, M., Clark, C., Zettlemoyer, L., and Lee, K. (2018). Deep contextualized word representations. *NAACL 2018.*

Pflaum, I., & Hateley, E. (2014). A bit of a problem: National and extraterritorial regulation of virtual currency in the age of financial disintermediation. *Georgetown Journal of International Law, 45*, 1169–1215.

Phandi, P., Chai, K., & Ng, H. (2015). Flexible Domain Adaptation for Automated Essay Scoring Using Correlated Linear Regression. *Proceedings of the 2015 Conference on Empirical Methods in Natural Language Processing*, 431-439. Retrieved 3 March 2019, from https://www.aclweb.org/anthology/D15-1049

Pocheevsky V. (2016) Ovoshchi na elektrichestve [Vegetables on electricity]. *Priusadebnoye khozyaystvo, 5*, 32–35.

Pollock, J., Stoecker-Sylvia, Z., Veedu, V., Panchal, N., & Elshahawi, H. (2018). Machine learning for improved directional drilling. *Proceedings of the Annual Offshore Technology Conference, 4*, 2496–2504. 10.4043/28633-ms

Popel, O. C., Frid, S. E., & Kiseleva, S. V. (2010). Climate data for renewable energy in Russia. Academic Press.

Popel, O. S., Frid, S. E., Kolomiyets, Yu. G., Kiseleva, S. V., & Terekhova, E. N. (2010). Atlas resursov solnechnoj energii na territorii Rossii, spravochnoe izdanie [Atlas of solar energy resources in Russia, reference publication]. Moskva: izd. OIVT RAN.

Popel, O. C., Frid, S. E., & Kolomiets, Y. G. (2010). *Atlas of solar resources on the territory of Russia*. MIPT.

Porpiglia, F., Amparore, D., Autorino, R., Checcucci, E., Cooperberg, M. R., Ficarra, V., & Novara, G. (2020). Traditional and Virtual Congress Meetings During the COVID-19 Pandemic and the Post-COVID-19 Era: Is it Time to Change the Paradigm? *European Urology*

Powell-Morse, A. (2016). *Waterfall Model: What Is It and When Should You Use It?* Retrieved 30 March 2019, from https://airbrake.io/blog/sdlc/waterfall-model

Power Ledger. (2019). *Technical white paper*. Retrieved from https://www.powerledger.io/wp-content/uploads/2019/11/power-ledger-whitepaper.pdf

Raa, B., Dullaert, W., & Aghezzaf, E. H. (2013). A matheuristic for aggregate production–distribution planning with mould sharing. *International Journal of Production Economics*, *145*(1), 29–37. doi:10.1016/j.ijpe.2013.01.006

Raftopoulou, M., Jorguseski, L., & Litjens, R. (2019, June). Design and Assessment of Low-Latency Random Access Procedures in 5G Networks. In *2019 European Conference on Networks and Communications (EuCNC)* (pp. 406-411). IEEE.

Raghavan, M., Churchill, A., & Tian, J. (2014). *The Effects of Portfolio Capital Flows and Domestic Credit on the Australian Economy* (Unpublished manuscript). University of Tasmania, Launceston, Australia.

Rahouti, M., Xiong, K., & Ghani, N. (2018). Bitcoin concepts, threats, and machine-learning security solutions. *IEEE Access: Practical Innovations, Open Solutions*, *3*(6), 67189–67205. doi:10.1109/ACCESS.2018.2874539

Ramani, P. & Murarka, P. (2013). *Stock market prediction using artificial neural network.* Doi:10.21090/ijaerd.c25

Raschka, S., & Mirjalili, V. (2017). *Python machine learning*. Packt Publishing Ltd.

Rather, A. M., Sastry, V. N., & Agarwal, A. (2017). Stock market prediction and Portfolio selection models: A survey. *OPSEARCH*, *54*(3), 558–579. doi:10.100712597-016-0289-y

Rathore, K. S., Hodges, T. K., & Robinson, K. R. (1988). A refined technique to apply electrical currents to callus-cultures. *Plant Physiology*, *88*(3), 515–517. doi:10.1104/pp.88.3.515 PMID:16666337

Rauch, H. E., Tung, F., & Striebel, C. T. (1965). Maximum likelihood estimates of linear dynamic systems. *AIAA Journal*, *3*(8), 1445–1450. doi:10.2514/3.3166

Raza, U., Kulkarni, P., & Sooriyabandara, M. (2017). Low power wide area networks: An overview. *IEEE Communications Surveys and Tutorials*, *19*(2), 855–873. doi:10.1109/COMST.2017.2652320

Reddy, B. R., Zhang, J., & Ellis, M. (2016). Cement Strength retrogression issues in offshore deep water applications - Do we know enough for safe cementing? *Proceedings of the Annual Offshore Technology Conference*, *2*, 1612–1632. 10.4043/27012-MS

Rehman H.., Yafi E., Nazir M., Mustafa K. (2019). Security Assurance Against Cybercrime Ransomware. In *Intelligent Computing & Optimization. ICO 2018. Advances in Intelligent Systems and Computing*, (vol. 866). Springer.

Reid, F., & Harrigan, M. (2011). An analysis of anonymity in the Bitcoin system. In Y. Altshuler, Y. Elovici, A. B. Cremers, N. Aharony, & A. Pentland (Eds.), Security and privacy in social networks (pp. 197-273). New York: Springer. doi:10.1109/PASSAT/SocialCom.2011.79

Reliable prognosis. (n.d.). *Weather for 243 countries of the world.* Retrieved from http://www.rp5.ru

Ren, Z., Cong, K., Aerts, T., de Jonge, B., Morais, A., & Erkin, Z. (2018). A scale-out blockchain for value transfer with spontaneous sharding. In *Proceedings of Crypto Valley Conference on Blockchain Technology* (vol. 3, pp. 1-10). 10.1109/CVCBT.2018.00006

Revelle, C. S., Eiselt, H. A., & Daskin, M. S. (2008). A bibliography for some fundamental problem categories in discrete location science. *European Journal of Operational Research, 184*(3), 817–848. doi:10.1016/j.ejor.2006.12.044

RF Patent No. 2143775. (1999). Method and Apparatus for Electric Power Transmission. *Inventor's Bulletin, 36.*

RF Patent No. 2172546. (2001). Method and Apparatus for Electric Power Transmission. *Inventor's Bulletin, 23.*

RIA Novosti. (2019, December 30). *Putin signed a law on the development of green micro-generation.* Retrieved from https://ria.ru/20191230/1563033599.html

Rietveld, M. (1978). *A New Method For Estimating The Regression Coefficients.* Agrtcultural Meteorology.

Ritchie, J. R. B., & Crouch, G. I. (2003). *The competitive destination: a sustainable tourism perspective* (J. R. B. Ritchie & G. I. Crouch, Eds.). CABI. doi:10.1079/9780851996646.0000

Rizwan, M., Jamil, M., Kirmani, S., & Kothari, D. P. (2014). (2014). Fuzzy logic based modeling and estimation of global solar energy using meteorological parameters. *Energy, 70*, 685–691. doi:10.1016/j.energy.2014.04.057

RMI. (2018). *Demand flexibility - the key to enabling a low-cost, low-carbon grid.* Retrieved from https://info.rmi.org/demand_flexibility

Ron, D., & Shamir, A. (2013). Quantitative analysis of the full Bitcoin transaction graph. In A. R. Sadeghi (Ed.), *Proceedings of the 17th International Conference on Financial Cryptography and Data Security, LNCS 7859* (pp. 6-24). Springer. 10.1007/978-3-642-39884-1_2

Rosas, P. (2003). *Dynamic Influences of Wind Power on the Power System* (Unpublished doctoral dissertation). Ørsted Institute and Technical University of Denmark, Copenhagen, Denmark.

Rosenfeld, M. (2011). Analysis of bitcoin pooled mining reward systems. *Computing Research Repository* (Technical report). Retrieved May 22, 2016 from https://arxiv.org/pdf/1112.4980.pdf

Ross, W. (n.d.). *The death of Venice: Corrupt officials, mass tourism and soaring property prices have stifled life in the city.* Available at: https://www.independent.co.uk/news/world/europe/the-death-of-venice-corrupt-officials-mass-tourism-and-soaring-property-prices-have-stifled-life-in-10251434.html

Rudner, L., & Gagne, P. (2019). *An Overview of Three Approaches to Scoring Written Essays by Computer.* Retrieved 3 March 2019, from https://pareonline.net/htm/v7n26.htm

Rudner, L., & Liang, T. (2002). Automated Essay Scoring Using Bayes' Theorem. *The Journal of Technology, Learning, andAssessment, 1*(2). Retrieved March 3, 2019, from https://ejournals.bc.edu/ojs/index.php/jtla/article/download/1668/1512

Russia's Weather. (n.d.). *Weather archive.* Retrieved from http://www.meteo.infospace.ru

Saha, B., & Sharma, S. (2012). Steganographic Techniques of Data Hiding using Digital Images. *Defence Science Journal, 62*(1), 11–18. doi:10.14429/dsj.62.1436

Sahinkaya, E., Tuncman, S., Koc, I., Guner, A. R., Ciftci, S., Aygun, A., & Sengul, S. (2019). Performance of a pilot-scale reverse osmosis process for water recovery from biologically-treated textile wastewater. *Journal of Environmental Management, 249*(August), 109382. doi:10.1016/j.jenvman.2019.109382

Salzberg, S. (1985, August). Heuristics for Inductive Learning. IJCAI, 603-609.

Samarin, G.N. (2013). *Sposob elektrostimulyatsii zhiznedeyatel'nosti rasteniy* [Method electrostimulation of plant life]. Patent RF, 2555449.

Sang-Hong, L., & Joon, S. L. (2011). Forecasting KOSPI based on a neural network with weighted fuzzy membership functions. *Expert Systems with Applications*, *38*(4), 4259–4263. doi:10.1016/j.eswa.2010.09.093

Sarbon, S. (2013). *The unresolved patents of Nikola Tesla. The inventions that Tesla did not protect.* Nikola Tesla Museum.

Schot, J., Kanger, L., & Verbong, G. (2016). The roles of users in shaping transitions to new energy systems. *Nature Energy*, *1*(16054).

Schuh, F. J. (1964). Computer makes surge-pressure calculations useful. *Oil & Gas Journal*, *31*, 96.

Schumacker, R. E., & Lomax, R. G. (2004). A beginner's guide to structural equation modeling. Psychology Press.

Schumaker, R. P., & Chen, H. (2009). Textual analysis of stock market prediction using breaking financial news: The AZFin text system. *ACM Transactions on Information Systems*, *27*(2), 12. doi:10.1145/1462198.1462204

Schwarz, G. (1978). Estimating the dimension of a model. *Annals of Statistics*, *6*(2), 461–464. doi:10.1214/aos/1176344136

Şen, Z. (1998). Fuzzy algorithm for estimation of solar irradiation from sunshine duration. *Solar Energy*, *63*(1), 39–49. doi:10.1016/S0038-092X(98)00043-7

Sgora, A., Vergados, D. D., & Chatzimisios, P. (2009). IEEE 802.11s wireless mesh networks: Challenges and Perspectives. In *International Conference on Mobile Lightweight Wireless Systems* (pp. 263-271). Springer. 10.1007/978-3-642-03819-8_25

Shadravan, A., Tarrahi, M., & Amani, M. (2017). Intelligent tool to design drilling, spacer, cement slurry, and fracturing fluids by use of machine-learning algorithms. *SPE Drilling & Completion*, *32*(2), 131–140. doi:10.2118/175238-PA

Shamir, A. (1979). How to share a secret. *Communications of the ACM*, *22*(11), 612–613. doi:10.1145/359168.359176

Shanmugasundaram, R., Pavithra, S., Sangeetha, V., Tamilselvan, S., & Ahmed, A. T. (2017). Iot based animal tracking and monitoring system in zoo. *South Asian Journal of Engineering and Technology*, *3*(2), 162–168.

Shi, J., Liu, Z., Tang, L., & Xiong, J. (2017). Multi-objective optimization for a closed-loop network design problem using an improved genetic algorithm. *Applied Mathematical Modelling*, *45*, 14–30. doi:10.1016/j.apm.2016.11.004

Sichitiu, M. L. (2005). Wireless mesh networks: opportunities and challenges. In *Proceedings of World Wireless Congress* (*vol. 2*, pp. 21). Academic Press.

Siemens. (2018). *Design of distributed energy systems*. Erlangen: Siemens AG.

Simonovic, S. P. (1992). Reservoir systems analysis: Closing gap between theory and practice. *Journal of Water Resources Planning and Management*, *118*(3), 262–280. doi:10.1061/(ASCE)0733-9496(1992)118:3(262)

Sims, C. A. (1986). Are forecasting models usable for policy analysis? Quarterly Review, 2-16.

Sims, C. A. (1980). Macroeconomics and reality. *Econometrica*, *48*(1), 1–48. doi:10.2307/1912017

Sims, C. A., Stock, J. H., & Watson, M. W. (1990). Inference in linear time series models with some unit roots. *Econometrica*, *58*(1), 113–144. doi:10.2307/2938337

Singh, N., & Singh, S. B. (2017). *Hybrid Algorithm of Particle Swarm Optimization and Grey Wolf Optimizer for Improving Convergence Performance*. Academic Press.

Smirnov, A.A., & Kholmansky, A.S. (2017). Zavisimost' fotosinteza pigmentov i produktivnosti tomata ot spektral'nogo sostava obluchatelya [Dependence of pigments photosynthesis and productivity of tomato on spectral composition of irradiator]. *Nauchnaya zhizn, 10*, 14-19.

Smirnov, A., Kholmanskiy, A., & Ukhanova, V. (2018). Optimization of lighting spectrum of greenhouse vegetables by using light emitting diodes. *International Journal of Research in Pharmacy and Biosciences.*, *5*(4), 11–17.

Snaselova, P., & Zboril, F. (2015). Genetic Algorithm using Theory of Chaos. *Procedia Computer Science*, *51*, 316–325. doi:10.1016/j.procs.2015.05.248

Snøtun, H. (2018). Using machine learning to create data. *Offshore Technology Conference Asia 2018. OTCA, 2018.* Advance online publication. doi:10.4043/28587-ms

Söder, L., & Ackermann, T. (2005). Wind Power in Power Systems: An Introduction. In T. Ackermann (Ed.), *Wind Power in Power Systems* (pp. 25–51). John Wiley & Sons, Ltd.

Sokolov A.V. (2018) Analiz i istoriya razvitiya elektricheskikh stimulyatsiy rasteniy [Analysis and history of the development of electrical stimulation of plants]. *Innovatsii v sel'skom khozyaystve, 3*(28), 87–93.

Soleimani, H., & Kannan, G. (2015). A hybrid particle swarm optimization and genetic algorithm for closed-loop supply chain network design in large-scale networks. *Applied Mathematical Modelling*, *39*(14), 3990–4012. doi:10.1016/j.apm.2014.12.016

Soler, A. (1990). Monthly specific Rietveld's correlations. *Solar & Wind Technology*, *7*(2-3), 305–308. doi:10.1016/0741-983X(90)90100-G

Sompolinsky, Y., & Zohar, A. (2015). Secure high-rate transaction processing in Bitcoin. *Proceedings of Financial Cryptography and Data Security: 19th International Conference*. doi: 10.1007/978-3-662-47854-7_32

Song, W., Li, Z., Ding, Y., Liu, F., You, H., Qi, P., Wang, F., Li, Y., & Jin, C. (2018). Performance of a novel hybrid membrane bioreactor for treating saline wastewater from mariculture: Assessment of pollutants removal and membrane filtration performance. *Chemical Engineering Journal*, *331*, 695–703. doi:10.1016/j.cej.2017.09.032

Soroush, H., Arney, D., & Goldman, J. (2016). Toward a safe and secure medical Internet of Things. *IIC Journal of Innovation*, *2*(1), 4–18.

Soualah-Alila, F. (2016). DataTourism: Designing an Architecture to Process Tourism Data. In *Information and Communication Technologies in Tourism 2016* (pp. 751–763). Springer International Publishing., doi:10.1007/978-3-319-28231-2_54

Sousa, T., Soares, T., Pinson, P., Moret, F., Baroche, T., & Sorin, E. (2019). Peer-to-peer and community-based markets: A comprehensive review. *Renewable & Sustainable Energy Reviews*, *104*, 367–378. doi:10.1016/j.rser.2019.01.036

Spencer, J. W. (1971). Fourier series representation of the position of the sun. *Applied Optics*, (10), 2569–2571.

Spethoven, A. (2019). Theory and reality of cryptocurrency governance. *Journal of Economic Issues*, *53*(2), 385–393. doi:10.1080/00213624.2019.1594518

Srianan, T., & Sangsawang, O. (2019). *Path-Relinking for Fire Station Location* (Vol. 866). Advances in Intelligent Systems and Computing.

Statista Research Department. (2020). Internet of Things Units Installed Base by Category 2014-2020. *Statista*. https://www.statista.com/statistics/370350/internet-of-things-installed-base-by-category/

Steven, M. D., & Unsworth, M. H. (1979, July). The diffuse solar irradiance of slopes under cloudless skies. *Quarterly Journal of the Royal Meteorological Society*, *105*(445), 593–602. doi:10.1002/qj.49710544507

Strebkov, D. S. (2002). Possibility of Transmitting Electric Power without Metal Wires. *Proceedings of the Russian Academy of Agricultural Sciences*, *1*, 47–50.

Strebkov, D. S., Nekrasov, A. I., & Avramenko, S. V. (2002). Investigation of Electric Power Transmission through Non-Metal Conducting Channels. *Proceedings of the 3-rd International Scientific-and-Practical Conference "Ecology and Agricultural Techniques"*, *1*, 45–52.

Strebkov, D. S., Nekrasov, A. I., Trubnikov, V. Z., & Nekrasov, A. A. (2018). Single-wire resonant electric power systems for renewable based electric grid. In Handbook of Research on Renewable Energy and Electric Resources for Sustainable Rural Development. doi:10.4018/978-1-5225-3867-7.ch019

Strebkov, D. S. (2014). Perspectives of using technologies of Nikola Tesla in up-to-datas power engineering. *Light & Engineering*, *2*, 4–14.

Strebkov, D. S., Avramenko, S. V., & Nekrasov, A. I. (2002, July–August). Investigation of Electric Energy Transmission Process in non-Metallic Conducting Channels. *New Energy Technologies*, *4*(7), 24–27.

Strebkov, D. S., & Nekrasov, A. I. (2005). Development of Resonant Methods for Electric Power Transmission in Russia. *Proceedings of the 8-th Symposium "Electric Engineering – 2010"*.

Strebkov, D. S., Nekrasov, A. I., & Nekrasov, A. A. (2019). *Renewable Energy and Power Supply Challenges for Rural Regions*. Resonant Power System for Electric Transport.

Subulan, K., Taşan, A. S., & Baykasoğlu, A. (2015). A fuzzy goal programming model to strategic planning problem of a lead/acid battery closed-loop supply chain. *Journal of Manufacturing Systems*, *37*, 243–264. doi:10.1016/j.jmsy.2014.09.001

Suganthi, L., Iniyan, S., & Samuel, A. A. (2015). Applications of fuzzy logic in renewable energy systems – A review. *Renewable & Sustainable Energy Reviews*, *48*, 585–607. doi:10.1016/j.rser.2015.04.037

Šulek, P. (2012). A hydro power system operation using Genetic Algorithms and mixed-integer nonlinear programming. *Slovak Journal of Civil Engineering*, *20*(1), 1–9. doi:10.2478/v10189-012-0001-z

Supit, I., & Van Kappel, R. R. (1998). A simple method to estimate global radiation. *Solar Energy*, *63*(3), 147–160. doi:10.1016/S0038-092X(98)00068-1

Sureshkumar, A., & Ravindran, R. S. (2016). Swarm and fuzzy based cooperative caching framework to optimize energy consumption over multimedia wireless sensor networks. *Wireless Personal Communications*, *90*(2), 961–984. doi:10.100711277-016-3274-0

Suresh, S., & Lal, S. (2016). An efficient cuckoo search algorithm based multilevel thresholding for segmentation of satellite images using different objective functions. *Expert Systems with Applications*, *58*, 184–209. doi:10.1016/j.eswa.2016.03.032

Surfer. (2019). Retrieved October 15, 2019, from Surfer 8 Software: http://www.goldensoftware.com

Surie, D., Laguionie, O., & Pederson, T. (2008). *Wireless sensor networking of everyday objects in a smart home environment. In Intelligent Sensors, Sensor Networks and Information Processing (ISSNIP)2008*. IEEE.

Su, S., Zhan, D., & Xu, X. (2008). An extended state task network formulation for integrated production-distribution planning in supply chain. *International Journal of Advanced Manufacturing Technology, 37*(11-12), 1232–1249. doi:10.100700170-007-1063-3

Swarbrooke, J. (1999). *Sustainable Tourism Management*. CABI.

Swartman, R. K., & Ogunlade, O. (1967). Solar radiation estimates from common parameters. *Solar Energy, 11*(3-4), 170–172. doi:10.1016/0038-092X(67)90026-6

Swati, G. P., & Thakur, I. S. (2019). Biodegradation of pyrene by Pseudomonas sp. ISTPY2 isolated from landfill soil: Process optimisation using Box-Behnken design model. *Bioresource Technology Reports, 8*(September), 100329. doi:10.1016/j.biteb.2019.100329

Taghipour, K., & Ng, H. (2016). A Neural Approach to Automated Essay Scoring. *Proceedings of the 2016 Conference on Empirical Methods in Natural Language Processing*, 1882-1890. Retrieved 29 January 2019, from https://www.aclweb.org/anthology/D16-1193

Taghipour, K. (2017). *Robust Trait-Specific Essay Scoring Using Neural Networks and Density Estimators. The.* National University of Singapore.

Talbi, E. (2016). *A Taxonomy of Hybrid Metaheuristics A Taxonomy of Hybrid Metaheuristics*. Academic Press.

Tawhid, M. A., & Ali, A. F. (2017). *A Hybrid Grey Wolf Optimizer And Genetic Algorithm For Minimizing Potential Energy Function*. Memetic Computing.

TensorFlow. (2019). *RESTful API | TensorFlow Serving | TensorFlow*. Retrieved 30 January 2019, from https://www.tensorflow.org/serving/api_rest

Tesla, N. (1999c). World System of wireless transmission of energy telegraph and telephone Age. Oct.16, 1927. In Nikola Tesla, Articles (vol. 1). Academic Press.

Tesla, N. (1956). *Lectures. Patents. Articles*. N. Tesla Museum.

Tesla, N. (1978). *Colorado Springs Notes 1899 – 1900*. Nolit.

Tesla, N. (1999a). *My Inventions. Electrical Experimenter, February, March, June and October. 1919. In Nikola Tesla. Selected works*. Academic Press.

Tesla, N. (1999b). The Problem of Increasing Human Energy. The Century Illustrated Monthly Magazine. June, 1900. In *Nikola Tesla. Selected works*. Academic Press.

Tesla, N. (2008). *From Colorado to Long island: Research Notes – Colorado Springs 1899-1900 – New York. 1900-1901*. Nikola Tesla Museum.

Thabit, R., & Khoo, B. E. (2014, February). Robust reversible watermarking scheme using Slantlet transform matrix. *Journal of Systems and Software, 88*, 74–86. doi:10.1016/j.jss.2013.09.033

Thamaraiselvan, C., Lerman, S., Weinfeld-Cohen, K., & Dosoretz, C. G. (2018). Characterization of a support-free carbon nanotube-microporous membrane for water and wastewater filtration. *Separation and Purification Technology, 202*, 1–8. doi:10.1016/j.seppur.2018.03.038

Thanh, P. N., Bostel, N., & Péton, O. (2008). A dynamic model for facility location in the design of complex supply chains. *International Journal of Production Economics, 113*(2), 678–693. doi:10.1016/j.ijpe.2007.10.017

Thomas, J. J., Karagoz, P., Ahamed, B. B., & Vasant, P. (2019). *Deep Learning Techniques and Optimization Strategies in Big Data Analytics*. IGI Global.

Tian, Y. Q., Davies-Colley, R. J., Gong, P., & Thorrold, B. W. (2001). Estimating solar radiation on slopes of arbitrary aspect. *Agricultural and Forest Meteorology*, *109*(1), 67–74. doi:10.1016/S0168-1923(01)00245-3

Ting, C. K., Liao, X. L., Huang, Y. H., & Liaw, R. T. (2017). Multi-vehicle selective pickup and delivery using metaheuristic algorithms. *Information Sciences*, *406*, 146–169. doi:10.1016/j.ins.2017.04.001

Ting, D. S. W., Carin, L., Dzau, V., & Wong, T. Y. (2020). Digital technology and COVID-19. *Nature Medicine*, 1–3. PMID:32284618

Ting, Y. W., & Chang, Y. K. (2007, July). A novel cooperative caching scheme for wireless ad hoc networks: Group caching. In *International Conference on Networking, Architecture, and Storage* (pp. 62-68). IEEE. 10.1109/NAS.2007.4

Tong, X., Cui, Y., Wang, Y. H., Bai, Y., Yu, T., Zhao, X. H., Ikuno, N., et al. (2020). Fouling properties of reverse osmosis membranes along the feed channel in an industrial-scale system for wastewater reclamation. *Science of the Total Environment*, *713*, 136673. doi:10.1016/j.scitotenv.2020.136673

Tospornsampan, J., & Kita, I. (2005). Optimization of a multiple reservoir system operation using a combination of genetic algorithm and discrete differential dynamic programming : a case study in Mae Klong system. Academic Press.

Trabea, A., & Shaltout, M. (2000). Correlation of global solar radiation with meteorological parameters over Egypt. *Renewable Energy*, *21*(2), 297–308. doi:10.1016/S0960-1481(99)00127-5

Tschorsch, F., & Scheuermann, B. (2016). Bitcoin and beyond: A technical survey on decentralized digital currencies. *IEEE Communications Surveys and Tutorials*, *18*(3), 2084–2123. doi:10.1109/COMST.2016.2535718

Tsoukalas, I., & Makropoulos, C. (2015). Multiobjective optimisation on a budget: Exploring surrogate modelling for robust multi-reservoir rules generation under hydrological uncertainty. *Environmental Modelling & Software*, *69*, 396–413. doi:10.1016/j.envsoft.2014.09.023

Twomey, P. (2013). Halting a shift in the paradigm: The need for Bitcoin regulation. *Trinity College Law Review*, *16*, 67–90.

Tyagunov, M. (2017). Distributed energy systems is the future of the world's power industry. In *Proceedings of the 2nd International Conference on the Applications of Information Technology in Developing Renewable Energy Processes & Systems (IT-DREPS)* (pp. 1 - 4). Amman, Jordan: IEEE.

UCL. (2019). *Could Colombia lead the peer-to-peer energy revolution?* Retrieved from https://www.ucl.ac.uk/bartlett/sustainable/sites/bartlett/files/policybriefingp2pcolombia.pdf

UNDP. (2015). *Sustainable Development Goals (SDGs)*. Retrieved from https://www.undp.org/content/undp/en/home/sustainable-development-goals.html

UNEP-WTO. (2005). Making tourism more sustainable. A guide for policy makers. *Environment*. doi: 92-807-2507-6

UNFCCC. (2015). *The Paris Agreement*. Retrieved from https://unfccc.int/process-and-meetings/the-paris-agreement/the-paris-agreement

Unrau, S., & Torrione, P. (2017). Adaptive Real-Time Machine Learning-Based Alarm System for Influx and Loss Detection. *SPE Annual Technical Conference and Exhibition*, 11. 10.2118/187155-MS

UNWTO. (2007). *A Practical Guide to Tourism Destination Management*. Available at: https://www.e-unwto.org/doi/book/10.18111/9789284412433

UNWTO. (2018). *'Overtourism'? Understanding and Managing Urban Tourism Growth beyond Perceptions. World Tourism Organization*. UNWTO. doi:10.18111/9789284419999

UNWTO. (n.d.). *Definition of sustainable tourism*. Available at: http://sdt.unwto.org/content/about-us-5

Vallés, M., Reneses, J., Frías, P., & Mateo, C. (2016). Economic benefits of integrating Active Demand in distribution network planning: A Spanish case study. *Electric Power Systems Research, 136*, 331–340. doi:10.1016/j.epsr.2016.03.017

Vallverdu-Queralt, A., Oms-Oliu, G., Odriozola-Serrano, I., Lamuela-Raventos, R. M., Martm-Belloso, O., & Elez-Martinez, P. (2013). Metabolite profiling of phenolic and carotenoid contents in tomatoes after moderate-intensity pulsed electric field treatments. *Food Chemistry, 136*(1), 199–205. doi:10.1016/j.foodchem.2012.07.108 PMID:23017413

Varthanan, P. A., Murugan, N., & Kumar, G. M. (2012). A simulation based heuristic discrete particle swarm algorithm for generating integrated production–distribution plan. *Applied Soft Computing, 12*(9), 3034–3050. doi:10.1016/j.asoc.2012.05.001

Vasant & Voropai. (2016). *Sustaining Power Resources through Energy Optimization and Engineering*. Academic Press.

Vasant, P., & Zelinka, I. G.-W. W. (2019). *Intelligent Computing & Optimization: Advances in Intelligent Systems and Computing*. Academic Press.

Vasant, Weber, & Dieu. (2016). *Handbook of Research on Modern Optimization Algorithms and Applications in Engineering and Economics*. Academic Press.

Vasant, P. (2019). *Intelligent Computing & Optimization* (Vol. 866). Springer. doi:10.1007/978-3-030-00979-3

Vasant, P., Kose, U., & Watada, J. (2017). Metaheuristic techniques in enhancing the efficiency and performance of thermo-electric cooling devices. *Energies, 10*(11), 1703. doi:10.3390/en10111703

Vasant, P., Marmolejo, J. A., Litvinchev, I., & Aguilar, R. R. (2019). Nature-inspired meta-heuristics approaches for charging plug-in hybrid electric vehicle. *Wireless Networks*. Advance online publication. doi:10.100711276-019-01993-w

Vasant, P., Zelinka, I., & Weber, G. W. (2020). *Intelligent Computing and Optimization*. Springer International Publishing. doi:10.1007/978-3-030-33585-4

P. Vasant, I. Zelinka, & G. W. Weber (Eds.). (2019). Intelligent Computing & Optimization. In *Proceedings of the 1st International Conference on Intelligent Computing and Optimization 2018 (ICO 2018)*. New York, NY: Springer International Publishing.

Vasek, M., Thornton, M., & Moore, T. (2014). Empirical analysis of denial-of-service attacks in the bitcoin ecosystem. In R. Böhme, M. Brenner, T. Moore, & M. Smith (Eds.), Lecture Notes in Computer Science: Vol. 8438. *Financial Cryptography and Data Security. FC 2014* (pp. 57–71). Springer. doi:10.1007/978-3-662-44774-1_5

Velkin, V. I. (2015). *Metodologiya rascheta kompleksnyh sistem VIE dlya ispol'zovaniya na avtonomnyh ob"ektah* [Methodology for calculating complex renewable energy systems for use in autonomous facilities]. URFU.

Vellas, F., & Bécherel, L. (1999). *The International Marketing of Travel and Tourism: A Strategic Approach*. Macmillan Education UK. doi:10.1007/978-1-349-27486-4

Venkatanarasimhulu, C., & Satyaprasad, K. (2011). A new SVD based hybrid color image watermarking for copyright protection using contourlet transform. *International Journal of Computers and Applications, 20*(8), 18–27. doi:10.5120/2455-3314

Vezzoli, C. (2018). Distributed/decentralized renewable energy systems. In *Designing Sustainable Energy for All. Green Energy and Technology* (pp. 23–39). Springer. doi:10.1007/978-3-319-70223-0_2

Vicini, S., & Sanna, A. (2012). How to Co-Create Internet of Things-enabled Services for Smarter Cities. *The First International Conference on Smart Systems, Devices and Technologies.*

Vissarionov, V. I., Deryugina, G. V., Kuznetsova, V. A., & Malinin, N. K. (2008). Solnechnaya energetika [Solar power]. Moskva: Izdatel'skij dom MEI.

Vissarionov, V. I., Deryugina, G. V., Kuznetsova, V. A., & Malinin, N. K. (2008). *Solar energy.*

VOX. (2019). *After rising for 100 years, electricity demand is flat. Utilities are freaking out.* Retrieved from https://www.vox.com/energy-and-environment/2018/2/27/17052488/electricity-demand-utilities

Walia, A. (2017). The Vanishing Gradient Problem. *Medium.* Retrieved 29 January 2019, from https://medium.com/@anishsingh20/the-vanishing-gradient-problem-48ae7f501257

Wang, G. (2015). *Elephant Herding Optimization Elephant Herding Optimization.* Academic Press.

Wang, G., Deb, S., & Cui, Z. (2015). Monarch butterfly optimization. *Neural Computing and Applications.*

Wang, R., Hajiaghajani, F., & Biswas, S. (2017, October). Incentive based cooperative content caching in Social Wireless Networks. In *IEEE 28th Annual International Symposium on Personal, Indoor, and Mobile Radio Communications (PIMRC)* (pp. 1-7). IEEE. 10.1109/PIMRC.2017.8292169

Wanga, Z., Wanga, F., & Sub, S. (2011). Solar irradiance short-term prediction model based on BP neural network. *Energy Procedia, 12*, 12. doi:10.1016/j.egypro.2011.10.065

Wang, H., Wu, A., & Tan, X. (2006). A new secure authentication scheme based threshold ecdsa for wireless sensor network. *Security Management.* Advance online publication. doi:10.1007/978-3-662-45472-5_11

Wang, K. J., Makond, B., & Liu, S. Y. (2011). Location and allocation decisions in a two-echelon supply chain with stochastic demand–A genetic-algorithm based solution. *Expert Systems with Applications, 38*(5), 6125–6131. doi:10.1016/j.eswa.2010.11.008

Wang, Y., Li, J., Huang, L., Jing, Y., Georgakopoulos, A., & Demestichas, P. (2014). 5G mobile: Spectrum broadening to higher-frequency bands to support high data rates. *IEEE Vehicular Technology Magazine, 9*(3), 39–46. doi:10.1109/MVT.2014.2333694

Wang, Y.-P. E., Lin, X., Adhikary, A., Grövlen, A., Sui, Y., Blankenship, Y., Bergman, J., & Razaghi, H. S. (2017). A primer on 3GPP narrowband Internet of Things. *IEEE Communications Magazine, 55*(3), 117–123. doi:10.1109/MCOM.2017.1600510CM

Wang, Z., Yu, X., Mu, Y., & Jia, H. (2020). A distributed Peer-to-Peer energy transaction method for diversified prosumers in Urban Community Microgrid System. *Applied Energy, 260*, 1–13. doi:10.1016/j.apenergy.2019.114327

Ward, R. G. (1996). The influence of electric currents on the growth of tomato plants. *Acta Physiologiae Plantarum, 18*, 121–127.

Washburn, D., Sindhu, U., Balaouras, S., Dines, R. A., Hayes, N. M., & Nelson, L. E. (2010). *Helping CIOs Understand "Smart City" Initiatives: Defining the Smart City, Its Drivers, and the Role of the CIO.* Forrester Research, Inc.

Wasserman, P. D., & D., P. (1993). Advanced methods in neural computing. In *Advanced methods in neural computing* (pp. 1–255). Retrieved from https://dl.acm.org/citation.cfm?id=562821

Waterfall Model, S. D. L. C. (2019). Retrieved 30 March 2019, from https://www.tutorialspoint.com/sdlc/sdlc_waterfall_model.htm

Weber, R. H. (2010). Internet of Things–New security and privacy challenges. *Computer Law & Security Review*, *26*(1), 23–30. doi:10.1016/j.clsr.2009.11.008

Werthner, H. (2003). Intelligent Systems in Travel and Tourism. *IJCAI 2003: 18th International Joint Conference on Artificial Intelligence*.

Whitley, D. (2001). An overview of evolutionary algorithms: Practical issues and common pitfalls. *Information and Software Technology*, *43*(14), 817–831. doi:10.1016/S0950-5849(01)00188-4

Whitmore, A., Agarwal, A., & Xu, L. D. (2015). The Internet of Things—A survey of topics and trends. *Information Systems Frontiers*, *17*(2), 261–274. doi:10.100710796-014-9489-2

WHO. (2020). *Myth Busters*. https://www.who.int/emergencies/diseases/novel-coronavirus-2019/advice-for-public/myth-busters

Wolfers, J., & Zitzewitz, E. (2006). Five open questions about prediction markets (No. w12060). National Bureau of Economic Research.

WRDC. (2019). Retrieved October 5, 2019, from World Radiation Data Centre: http://wrdc.mgo.rssi.ru/wwwrootnew/wrdc_ru_new.htm

Wu, W., & Huang, Y. (2010, December). Hierarchical cooperative data caching for wireless mesh networks. In *IEEE/IFIP 8th International Conference on Embedded and Ubiquitous Computing (EUC)* (pp. 289-296). IEEE. 10.1109/EUC.2010.48

Wu, D., Liu, B., Yang, Q., & Wang, R. (2020). Social-aware cooperative caching mechanism in mobile social networks. *Journal of Network and Computer Applications*, *149*(102457), 1–12. doi:10.1016/j.jnca.2019.102457

Wu, Y., Luo, A., & Xu, D. (2019). Forensic analysis of Bitcoin transactions. In *Proceedings of IEEE International Conference on Intelligence and Security Informatics (ISI)* (pp. 167-169). Shenzhen, China: IEEE.

Xiang, Z., & Fesenmaier, D. R. (2017). *Big Data Analytics*. Tourism Design and Smart Tourism., doi:10.1007/978-3-319-44263-1_17

Xiang, Z., Wang, D., O'Leary, J. T., & Fesenmaier, D. R. (2015). 'Adapting to the Internet: Trends in Travelers' Use of the Web for Trip Planning'. *Journal of Travel Research*, *54*(4), 511–527. Advance online publication. doi:10.1177/0047287514522883

Yang, X. (2010). *Firefly algorithm, stochastic test functions and design optimisation*. Academic Press.

Yang, X., & Press, L. (2010). Nature-Inspired Metaheuristic Algorithms (2nd ed.). Academic Press.

Yang, C. H., Huang, C. T., & Wang, S. J. (2011). Reversible steganographybased on side match and hit pattern for VQ compressed images. *Information Sciences*, *181*(11), 2218–2230. doi:10.1016/j.ins.2011.01.015

Yang, K., Zheng, J., Yang, M., Zhou, R., & Liu, G. (2013). Adaptive Genetic Algorithm for Daily Optimal Operation of Cascade Reservoirs and its Improvement Strategy. *Water Resources Management*, *27*(12), 4209–4235. doi:10.100711269-013-0403-3

Yang, S. S., Ong, S. K., & Nee, A. Y. C. (2016). A decision support tool for product design for remanufacturing. *Procedia CIRP*, *40*, 144–149. doi:10.1016/j.procir.2016.01.085

Yang, X. S. (2014). *Nature-inspired optimization algorithms*. Elsevier.

Yan, Y., Qian, Y., Sharif, H., & Tipper, D. (2013). A survey on smart grid communication infrastructures: Motivations, requirements and challenges. *IEEE Communications Surveys and Tutorials*, *15*(1), 5–20. doi:10.1109/SURV.2012.021312.00034

Yao, J., & Tan, C. L. (2000). A case study on using neural networks to perform technical forecasting of forex. *Neurocomputing, 34*(1-4), 78–98. doi:10.1016/S0925-2312(00)00300-3

Yao, J., Tan, C. L., & Poh, H.-L. (1999). Neural networks for technical analysis: A study on KLCI. *International Journal of Theoretical and Applied Finance*, 2(02), 221–242. doi:10.1142/S0219024999000145

Yeh, W. W. (1985). Reservoir Management and Operations Models: A State-of-the-Art Review. *Water Resources Research, 21*(12), 1797–1818. doi:10.1029/WR021i012p01797

Yoo, J. H. (2009). Maximization of hydropower generation through the application of a linear programming model. *Journal of Hydrology (Amsterdam), 376*(1–2), 182–187. doi:10.1016/j.jhydrol.2009.07.026

Youssefzadeh, S. B. (2014). *Future of Payment Platforms* (Thesis). Department of Electrical Engineering, California Polytechnic State University, San Luis Obispo, CA.

Yu, Y., Chambon, S., Liu, Q., & Belaskie, J. P. (2018). Recorded well data enriches the testing of automation systems by using a deep neural network approach. *Society of Petroleum Engineers - IADC/SPE Drilling Conference and Exhibition*. doi:10.2118/189591-ms

Yudaev, I. V., & Nosova, T. A. (2015) Vliyaniye predposevnoy elektrostimulyatsii na semyana risa s nizkimi posevnymi kachestvami [The effect of pre-sowing electrical stimulation on rice seeds with low sowing qualities]. In *Metody i tekhnicheskiye sredstva povysheniya effektivnosti ispol'zovaniya elektrooborudovaniya v promyshlennosti i sel'skom khozyaystve 80-ya* [Methods and technical means to improve the efficiency of use of electrical equipment in industry and agriculture 80th scientific-practical conference] (pp. 464–468). Academic Press.

Yu, L., Wang, S., & Lai, K. L. (2009). A neural-network-based nonlinear metamodeling approach to financial time series forecasting. *Applied Soft Computing*, 9(2), 563–574. doi:10.1016/j.asoc.2008.08.001

Zai Ir Mint. (2013). *Research of energy characteristics of regional wind energy in the Republic of the Union of Myanmar* (Unpublished candidate of engineering sciences dissertation). Moscow Power Engineering Institute (National Research University), Moscow, Russia.

Zamarripa, M., Silvente, J., & Espuña, A. (2012). Supply chain planning under uncertainty using genetic algorithms. *Computer-Aided Chemical Engineering, 30*, 457–461. doi:10.1016/B978-0-444-59519-5.50092-7

Zarifis, A., Cheng, X., Dimitrious, S., & Efthymious, L. (2015). Trust in digital currency enabled transactions model. *Proceedings of the 9th Mediterranean Conference on Information Systems (MCISE)*. Retrieved March 10, 2017 from http://www.icsd.aegean.gr/group/data.php?usrnm=eloukis&data=1b

Zarifis, A., Efthymiou, L., Cheng, X., & Demetriou, S. (2014). Consumer trust in digital currency enabled transactions. In W. Abramowicz & A. Kokkinaki (Eds.), *Business Information Systems Workshops (BIS 2014). Lecture Notes in Business Information Processing* (Vol. 183, pp. 241–254). Springer International Publishing. doi:10.1007/978-3-319-11460-6_21

Zelinka, I. (2019). *Intelligent Computing and Optimization* (Vol. 2019). Academic Press.

Zelinka, I., Tomaszek, L., Vasant, P., Dao, T. T., & Hoang, D. V. (2018). A novel approach on evolutionary dynamics analysis – A progress report. *Journal of Computational Science, 25*, 437–445. doi:10.1016/j.jocs.2017.08.010

Zeng, J., & Qiao, W. (2013). Short-term solar power prediction using a support vector machine. *Renewable Energy, 52*, 118–127. doi:10.1016/j.renene.2012.10.009

Zeng, X. T., Ping, L. D., & Pan, X. Z. (2010). A lossless robust data hiding scheme. *Journal of Pattern Recognition, 43*(4), 1656–1667. doi:10.1016/j.patcog.2009.09.016

Zhai, Y., Wang, Y., Zhang, M., Gittell, J. H., Jiang, S., Chen, B., ... Wang, X. (2020). From Isolation to Coordination: How Can Telemedicine Help Combat the COVID-19 Outbreak? *medRxiv*.

Zhalechian, M., Tavakkoli-Moghaddam, R., Zahiri, B., & Mohammadi, M. (2016). Sustainable design of a closed-loop location-routing-inventory supply chain network under mixed uncertainty. *Transportation Research Part E: Logistics and Transportation Review, 89*, 182-214.

Zhang, K., Zhang, Z., Wang, H., Wang, X., Zhang, X., & Xie, Y. F. (2020). Synergistic effects of combining ozonation, ceramic membrane filtration and biologically active carbon filtration for wastewater reclamation. *Journal of Hazardous Materials, 382*(August), 121091. doi:10.1016/j.jhazmat.2019.121091

Zhang, G., Patuwo, B. E., & Hu, M. Y. (1998). Forecasting with artificial neural networks: The state of the art. *International Journal of Forecasting, 14*, 35–62. doi:10.1016/S0169-2070(97)00044-7

Zhang, H., Liu, N., Chu, X., Long, K., Aghvami, A. H., & Leung, V. C. (2017). Network slicing based 5G and future mobile networks: Mobility, resource management, and challenges. *IEEE Communications Magazine, 55*(8), 138–145. doi:10.1109/MCOM.2017.1600940

Zhao, J., Shen, Y., Chen, W., Zhang, Z., & Johnston, S. (2017). Machine Learning–Based Trigger Detection of Drilling Events Based on Drilling Data. *SPE Eastern Regional Meeting*. 10.2118/187512-MS

Zhao, J., Zhang, P., & Cao, G. (2008). On cooperative caching in wireless P2P networks. In *The 28th International Conference on Distributed Computing Systems* (pp. 731-739). IEEE.

Zhao, S., Zhang, Y., Xiong, X., Botelho, A., & Heffernan, N. (2017). A Memory-Augmented Neural Model for Automated Grading. *Association for Computing Machinery*. Retrieved 2 March 2019, from https://www.researchgate.net/publication/316940225_A_Memory-Augmented_Neural_Model_for_Automated_Grading

Zhao, J., Zhang, P., Cao, G., & Das, C. R. (2010). Cooperative caching in wireless p2p networks: Design, implementation, and evaluation. *IEEE Transactions on Parallel and Distributed Systems, 21*(2), 229–241. doi:10.1109/TPDS.2009.50

Zhao, Y., Li, P., Li, R., & Li, X. (2019). Characterization and mitigation of the fouling of flat-sheet ceramic membranes for direct filtration of the coagulated domestic wastewater. *Journal of Hazardous Materials*, 121557. PMID:31735469

Zhong, X., & Enke, D. (2017). Forecasting daily stock market return using dimensionality reduction. *Expert Systems with Applications, 67*, 126–139. doi:10.1016/j.eswa.2016.09.027

Zhou, Y., Zhao, H., Bai, H., Zhang, L., & Tang, H. (2012). Papermaking Effluent Treatment: A New Cellulose Nanocrystalline/Polysulfone Composite Membrane. *Procedia Environmental Sciences, 16*, 145–151. doi:10.1016/j.proenv.2012.10.020

Zohar, A. (2015). Bitcoin: Under the hood. *Communications of the ACM, 58*(9), 104–113. doi:10.1145/2701411

About the Contributors

J. Joshua Thomas received his PhD. degree from University Sains Malaysia (USM), School of Computer Sciences, in 2015. He worked as research assistant at the Artificial Intelligence Lab in University Sains Malaysia. His research interests include scheduling algorithms, machine learning algorithms, data analytics, deep learning, and visual analytics. Dr. J. Joshua has authored several publications in leading international conferences and journals. Some of his research work were published in conferences including, IEEE, ICONIP, IVIC, IV, COMPSE, ICO. He has funded external, internal grants, short term research projects and industry collaborative projects. He has been invited as plenary speaker at IAIM2019, delivered Workshop's at International conferences. He is an associative editor for the journal of Energy Optimization and Engineering (IJEOE), and invited as guest editor for JVLC-Elsevier, IJDSA-Springer, IJCC-IGI-Global, IJIRR-Inderscience.

Valeriy Kharchenko currently holds the position of Chief Scientific Officer of the Federal Scientific Agroengineering Center Key qualifications: researcher and consultant in a wide range of energy saving fields, including energy policy at national and international level, energy RTD strategy, RUE and RES. He is the official reviewer of a number of international journals, including Sola Energy, Deputy Editor-in-Chief of European Journal of Renewable Energy ", member of the editorial boards of several international journals, honorary professor of the Institute of Hydropower and Renewable Energy of the National Research University" MEI. Under his edition a number of books were published by IGI Global (USA).

Pandian Vasant is a senior lecturer at Department of Fundamental and Applied Sciences, Faculty of Science and Information Technology, Universiti Teknologi PETRONAS in Malaysia. He holds PhD (UNEM, Costa Rica) in Computational Intelligence, MSc (UMS, Malaysia, Engineering Mathematics) and BSc (2nd Class Upper-Hons in Mathematics, UM, Malaysia) in Mathematics. He has co-authored research papers and articles in national journals, international journals, conference proceedings, conference paper presentation, and special issues lead guest editor, lead guest editor for book chapters, edited books, and keynote lecture (257 publications indexed in Web of Science). In the year 2009, Dr. Pandian Vasant was awarded top reviewer for the journal Applied Soft Computing (Elsevier), awarded outstanding reviewer in the year 2015 for ASOC (Elsevier) journal and Top reviewer for Sentinels of Science: Computer Science (Oct. 2015 - Sept. 2016). He has 26 years of working experience at the various universities from 1989-2017. Currently he is Editor-in-Chief of IJEOE, Member of American Mathematical Society (AMS), MERLIN Research Group (TDTU), and NAVY (TUO, Czech Republic) Research Groups.

* * *

Shahbaz Afzal was born on 10th August 1993. He obtained his Bachelor of Technology in Information Technology and Telecommunication Engineering from BGSB University Jammu in 2014. He received his Master of Technology in Information Technology from B S Abdur Rahman University, Chennai in 2016. He is currently pursuing his Ph.D on "Load Balancing in Infrastructure Cloud Computing" at B S Abdur Rahman Crescent Institute of Science and Technology, Chennai. His area of research include Load Balancing in Cloud Computing, Application oriented Cloud Computing, Distributed Computing, Mathematical Computing, Information and Communication Technology, and Machine Learning. He has contributed numerous research articles in various national and international journals and conferences.

Amany Alshawi is the founding director of the King Abdulaziz City for Science and Technology (KACST) Women Program, an initiative to increase the role of KACST women in all domains. In 2014, she was among the leading researchers who established the National Center for Cybersecurity Technology in KACST. She joined KACST in 2011, leading the cryptography research group. She serves as an advisory board member in multiple associations, among them: ArabWIC Saudi chapter and Princess Nora University distinguished Saudi women prize. In 2016 she founded IEEEWiE Riyadh affinity group and is still chairing the committee. Dr. Alshawi holds a PhD in information technology from George Mason University and an MS in computer information systems from the University of Miami. Prior to joining KACST, she held multiple positions in Prince Sultan University - College for Women, among them cooperative training program coordinator, and research center director.

Ohood Althobaiti received her master's degree in computer science with excellent GPA (2012) from College of Computer and Information Sciences, King Saud University, Saudi Arabia. Her research interests lie in the general field of computer networks and cybersecurity with special focuses on security in the Internet of Things (IoT). She published papers. She was awarded reward scientific publishing in ISI journal, first-class honour award from Taif University in 2008 as well as a scholarship for PhD degree. Also, she is a lecturer, Computer Science Department, College of Computers and Information Technology, Taif University, Saudi Arabia. Currently, she is PhD candidate at Centre for Telecommunications Research, King's College London, London, United Kingdom.

Baba Babaev studied at the faculty of heat and power engineering of the Moscow State Engineering and Construction Institute, which graduated in 1982. After graduation worked at the Kola NPP. From 1990 to 1993, working in the Scientific Research Energy Institute G. M. Krzhizhanovskogo, Dagestan branch, increased scientific qualifications to a senior researcher. In 1996 defended the dissertation of the candidate of chemical sciences at the Kuban State Technological University. And in 2016 defended the dissertation of the doctor of technical sciences in "United Institute of High Temperatures". Specialist in the field of energy, development of energy systems based on renewable sources and energy storage, actively introduces new information technologies in various fields of energy in the educational process. More than 168 scientific works have been published, 20 patents and 8 author's certificates for fundamentally new developments in the field of energy, some of which are implemented in practice. At a high scientific and methodological level reads the special courses "Energy accumulators", "Physical principles of renewable energy sources", "Energy facilities of non-traditional renewable energy sources", holds open lectures, participates in the formation of the subject of course and diploma works, training of young specialists. Since 1996 has been working at the Federal State Budget Educational Educational

Institution of Higher Education "Dagestan State University", Professor of the Department "Renewable Energy Sources".

Mislav Balkovic (PhD, Assistant Professor), dean of Algebra University College, is active researcher in the area of National Qualifications Framework and Knowledge management where he participated in number of national projects including development of legislation for CROQF and implementation of various policy and regulatory documents in higher, VET and adult education. He is currently working on research in the area of Validation of non-formal and informal learning and development of modern educational approaches (methodics and didactics). He was representative of Croatian employers (2010 – 2012) in Advisory Committee for Vocational Education and Training, DG Education and Culture, EU Commission, Bruxelles. Since 2006 he has been intensively involved in preparation of different types of development projects in the area of education. Till this end he prepared more than 50 international and national project's applications and managed some 30 different projects. Mislav currently acts as the president of Sector skills council for Electrotechnic and Computing and president of Croatian Employers Association - Association of Employers in Education, during (2010 - 2016) period he was active member of National Council for Science and Higher education and member of Governmet's committee for implementation of education, science and technology Strategy.

Tarun Kanti Bandyopadhyay received his PhD, MTech, BTech in Chemical Engineering in 2011, 1997, and 1994, respectively. He also received BSc Hons. in Chemistry from Calcutta University in 1991. He joined National Institute of Technology Agartala in 2010 as an Assistant Professor in Department of Chemical Engineering. He holds many administrative posts including HOD in the Department of Bio Engineering. He has guided one doctoral student, five students are pursuing PhD under his guidance, 14 Master's thesis, and 20 BTech projects. He has authored one book, three book chapters, and 30 international journal papers. He has reviewed two doctoral thesis and an NPTEL site. His research area includes CFD, fluid dynamics, membrane integrated separation and filtration process, biofuel, enzyme extraction, etc. He has 22 years of academic, research and industrial experience.

Anirban Banik is pursuing his PhD in Civil Engineering Department from National Institute of Technology Agartala. He received his BE in Civil Engineering from G.H. Raisoni College of Engineering Nagpur in 2014 and Master's in Civil Engineering from National Institute of Technology Agartala in 2016.

B. Bazeer Ahamed received a Bachelor of Technology in Vel Tech Engineering College, Affiliated to Anna University, Chennai, India and Master of Computer Science Engineering in Anna University of Technology, Tiruchirapalli, India. Ph.D From Sathyabama Institute of Technology and Science, Chennai India. He has published more than 20 peer reviewed international journals and participated in several high profile conferences. At present he is working as Lecturer, Department of Information Technology, Oman Prof. Bazeer research is mainly focused on Data Mining &Information retrieval; additionally his research includes Networks, Data bases, Big Data. He is a Member of IEEE, ISTE, IAENG, and CSTA. He chaired the several sessions at National and International Conference.

Y. Bevish Jinila is Associate Professor, School of Computing, Sathyabama Institute of Science and Technology, Chennai, India.

Sushant Kumar Biswal received his PhD from Indian Institute of Technology Kanpur in 2012. He joined National Institute of Technology Agartala 2012 as an Assistant Professor in Civil Engineering Department. He is supervising PhD thesis of three students from 2016. He has twelve journal papers, twenty-four conference proceeding. His research interest includes Fluvial hydraulics and sediment transport. He has reviewed three international journal papers. He has 16 years of academic, research and industrial experience.

Sukanta Chakraborty has completed his MCA from IGNOU, India. He completed his M Tech in Computer Science & Engineering from Tripura University. Currently, he is pursuing PhD in Department of Computer Science & Engineering, Tripura University.

Martin Chanza holds an MSc. in Statistics & Actuarial Science from the University of Witwatersrand and is a Lecturer in the Department of Statistics and Operations Research, North West University, Mahikeng, South Africa. He recently submitted his PhD for examination and his area of specialisation includes Time Series Analysis, Econometric Modelling, Data mining, Forecasting Models, Machine Learning and Financial Risk Management.

Amir Ahmad Dar currently works at the Department of Mathematics, B. S. Abdur Rahman Crescent Institute of Science and Technology, Chennai. Amir does research in Econometrics, Business Economics and Financial Derivatives. Their current project is 'Design of experiment'.

Olusegun Sunday Ewemooje, PhD, is a lecturer at the Department of Statistics, Federal University of Technology, Akure, Nigeria. He is currently a postdoctoral research fellow at the Population and Health Research Entity, Faculty of Humanities, North-West University (Mafikeng Campus), South Africa. His research interest includes; Sample Survey Designs and Applications, Population Studies, Demography and Environmental Statistics. He has published many articles both in national and international peer-review journals. School of Sciences Akure, Ondo 340252, NG.

Sergey Kachan was born in the city of Volzhsky, Russia. 08/09/1980 from 1998-2003 Moscow State Forest University (Mytishchi branch of MSTU named after N. E. Bauman). In the specialty "Machinery and equipment of the forest complex" the qualification of an Engineer was awarded. from 1999-2002 Moscow State Forest University (Mytishchi branch of MSTU named after N.E. Bauman). Military department, reserve officer, military rank Lieutenant, profile name command, assigned by order of the MORF No. 630, 07.23.2003 VUS No. 261001. from 2002-2005 Moscow State Forest University (Mytishchi branch of MSTU named after N. E. Bauman). In the specialty "Economics and management at the enterprises of mechanical engineering and instrument-making" qualification the Economist-manager is assigned. from 2003-2006 Moscow State Forest University (Mytishchi branch of MSTU named after N. E. Bauman). Postgraduate study, specialty 05.21.05 Wood science, technology and equipment for wood processing. from 2007-2012 All-Russian Research Institute of Computer Engineering and Informatization (VNIIPVTI) as a senior engineer, 2008 transferred to the position of head of the sector. From 2014 to the present, Junior Researcher, Federal Scientific Agroengineering Center VIM. Participant in many student scientific and practical conferences.

G. Kavitha was born on 31st August 1978. She obtained her Bachelor of Engineering in Electronics and Communication Engineering from Crescent Engineering College, Madras University, Chennai in 2000. She obtained her Master's Degree in Computer Science and Engineering from Crescent Engineering College, Madras University in 2002 and Doctor of Philosophy in Faculty of Information and communication, Anna University in 2013. She joined as a lecturer in the Department of Information Technology at B.S.A Crescent Engineering College, Chennai in 2002. She is presently working as Associate Professor at B.S. Abdur Rahman Crescent Institute of Science and Technology. She has to her credit 24 publications, 14 research papers in referred International Journals with high Impact Factor and 10 in International Conferences. Her research interests focus on Workflows on the Grid and Cloud Computing, Wireless Networks, Artificial Intelligence, IoT and Machine Learning.

Shwetank Krishna is a PhD scholar in Petroleum Engineering Department of Universiti Teknologi PETRONAS. He has received his bachelor's degree from University of Petroleum and Energy Studies (UPES) and master's degree from Andhra University, India in 2013 and 2015, respectively. His research interest includes drilling hydraulics, wellbore stability and predictive modeling of non-Newtonian fluid flow.

Sumit Kumar is currently pursuing his Ph.D. from the Indian Institute of Management Indore in Information Systems. He is a master and bachelor in computer science. He has research interests in Machine Learning, Text Mining, Natural Language Processing, Stock Market Analysis, and Evolutionary Algorithms.

Abhishek Majumder received his B.E. degree in Computer Science & Engineering from National Institute of Technology, Agartala and M.Tech. degree in Information Technology from Tezpur University, Assam in 2006 and 2008, respectively. He has completed PhD from the Department of Computer Science & Engineering, Assam University, Assam in 2018. He is currently working as an Assistant Professor in the Department of Computer Science & Engineering, Tripura University (A Central University), Suryamaninagar, India. His areas of interest are Wireless Network and Cloud Computing.

Baatai Maksatov is a postgraduate student of the hydropower and renewable energy sources department of Moscow Power Engineering Institute (National Research University).

Leo Mršić, PhD, is the experienced top level Corporate Executive, Entrepreneur, Assistant Professor, College Professor, Certified Court Expert Witness with deep expertize in appliance of technology in different areas of business and government, strong background in different areas (law, finance, technology, math) and fields (retail, project management) with focus on analytic methods and data management/data usage with ability to extend its approach from operational through strategic level, strong organizational skillset (teams/companies up to 1.000+ employees) during 20+ years of management experience, Head of AUC@Lab and Chair for Data Management and Analysis at University College Algebra Zagreb, active in edu community with wide range of different experiences (Editor, Lecturer, Reviewer, Editorial Board Member, Advisor, Book Author) able to provide deep and usable approach in many areas especially related to technology, math, structured decision making and educational methodologies.

Elias Munapo is Professor of Operations Research at the Department of Statistics and Operations Research, North-West University, Mafikeng Campus, South Africa. His areas of specialization include Applied Statistics & Operations Research.

Alisher Narynbaev is a graduate student of the Hydropower and renewable energy department of National Research University "Moscow Power Engineering Institute".

Swapna Nath has completed her M Tech in Computer Science & Engineering from the Department of Computer Science & Engineering, Tripura University in 2014.

Alexey Nekrasov was born October 30, 1945, in 1973 he graduated from the Faculty of Agriculture Electrification GSKHI Kazakh engineer-electrician agricultural production. In 2007 he defended his doctoral thesis on the specialty 05.20.02 - Electrotechnology and electrical equipment in agriculture. Currently conducts research on the electrification of agriculture, on the effective use and operation of the electrical equipment of agricultural enterprises. Directly involved in the development of new methods of resonant transfer and application of electric energy, as well as performing other basic and essential applied science and technology programs. Nekrasov AI It has 275 scientific works, including 19 books, monographs, guidelines and brochures, 50 patents and copyright certificates for inventions. Completed development introduced in agricultural production and are recommended for use by scientists, engineering companies and energy specialists, are used in the educational process.

Anton Nekrasov was born 04 November 1981 in Moscow. In 2004 graduated from the Moscow state Agroengineering University n. a. V. P. Goryachkin, for being awarded the qualification of engineer-electrician, specialty – electrification and automation of agriculture. He is the author of 65 published scientific works, including 3 guidelines and 9 patents, of which 40 papers in journals approved by VAK. Working as a researcher. In 2015, he defended his thesis.

Vladimir Panchenko, from 2003 to 2009, studied at "Bauman Moscow State Technical University" at the faculty "Power Engineering " with qualification of engineer. From 2009 to 2012 studied at the graduate school of the ''All-Russian Scientific Research Institute of Agriculture Electrification'', specializing in "Renewable Energy Installations". Passed an internship program on the International Energy Program in the Solar Energy Center (SEC), the Certificate under the Indian Technical and Economic Cooperation (ITEC), the Program of the Ministry of External Affairs (MEA), the Government of India - India, Delhi, November-December 2011. Participated in the International Conference-Exhibition Intersolar Europe 2012, Germany, Munich, June 11 - 15, 2012 as part of the delegation of the Ministry of Energy of the Russian Federation. Ph.D. thesis "Development and research of solar photovoltaic thermal module with a paraboloid-type concentrator" was protected on July 2, 2013. From 2013 to 2017 head of the laboratory ''Solar photovoltaic modules with an experimental - technological site''. The Council of Young Scientists of the VIESH was elected the Chairman of the Council of Young Scientists and was appointed a member of the Academic Council of the VIESH in 2014 - 2017. In 2014 was appointed as a contact person and a member of the International Solar Energy Society of the Russian section of the community. Since September 2015 associate professor at the ''Russian University of Transport''. Since 2017, a senior researcher at the ''Solar Energy Laboratory'' of "Federal Scientific Agroengineering Center VIM". Passed the advanced training in 2017 in ''Russian University of Transport'' the program

"Using e-learning and distance educational technologies in the implementation of educational programs on distance learning". In 2016 awarded the grant ''Young lecturer of Russian University of Transport''. In 2018 awarded the Scholarship of the President of the Russian Federation for 2018 - 2020 for young scientists and graduate students engaged in advanced research and development in priority areas of the modernization of the Russian economy. Since 2019 member of the Council of Young Scientists of the Russian University of Transport MIIT. Has 3 medals of exhibitions, 3 diplomas of the winner of competitions, participated in more than 90 conferences and exhibitions of various levels, has more than 150 publications, including patents , monographs and educational publications. Area of interest includes three-dimensional modeling, use of solar energy and electrical transport.

Daniel Adrian Perez-Moscote is an Electronics Engineer from The National University of Colombia with a master's degree in Sustainable Energy Systems from The University of Edinburgh. With training in renewable energy and energy efficiency in Mexico, Germany, China and Russia. Special interest in sustainable development issues, particularly the promotion of renewable energy use and universal access to modern energy services. Currently pursuing a Ph.D. degree from the National Research University "Moscow Power Engineering Institute" (MPEI).

Yuri Proshkin, Ph.D., is a Senior Researcher FSAC VIM. He has authored more 30 articlesin the newspapers. His scientific works are devoted to the study ofthe effects of electrophysical effects on agricultural and biological objects.

Roman Pugachev, born in 1976, graduated from MPEI in 1999 with a degree in "Hydroelectric Power". For 20 years of pedagogical work he conducted lectures, practical and laboratory classes on the following courses: "Hydroelectric power plant design»; "Hydropower plants"; "Unconventional energy"; "Mathematical problems of hydropower"; "Hydraulic structures"; "Environmental aspects of renewable energy sources". In recent years, he has been intensively engaged in the problems of using renewable energy sources, takes part in a number of research projects. Pugachev R. V. published about 100 printed works, including several textbooks.

Mogari Rapoo holds MSc. in Statistics and he is a Lecturer in the Department of Statistics and Operations Research, North West University, Mafikeng, South Africa. He is going to enroll for PhD next year and his area of specialization are Machine learning, Econometrics and Time series.

Syahrir Ridha is an Associate Professor at Universiti Teknologi PETRONAS (UTP). Currently he holds position as Director for Institute of Hydrocarbon Recovery which consist of 6 (six) research centers including Center of Seismic Imaging, Center of Carbonate, Center for Enhanced Oil Recovery, Basin Studies, Wells Research, and Shale Gas. He has been with the organization for the last 6 years. His research interest includes drilling technology, well cementing, drilling fluids and wellbore integrity. He has secured several national and international research grants as a project leader. He has delivered several short course training at both National and International levels. Ridha has authored and co-authored more than 35 technical papers. He holds a PhD degree in Petroleum Engineering from UTP. Ridha is a member of Society of Petroleum Engineer and Energy Institute.

Mlabika Sarkar has completed her M Tech in Computer Science & Engineering from the Department of Computer Science & Engineering, Tripura University in 2014.

Alexander Smirnov, CSc. (Engineering), is a Senior Researcher FSAC VIM (2009-present). He finished the postgraduate school of All-Russian Research Institute of Electrification of Agriculture in 2009. He has authored 43 articles. His scientific works are devoted to the study of the effects of electrophysical effects (microwave, ozone, optical radiation) on agricultural objects.

Alexander Sokolov, PhD (candidate of technical sciences), is a Leading Researcher, Laboratory of Lighting Engineering Federal Scientific Agroengineering Center VIM.

Dmitry Strebkov is Supervisor VIESH. Academician of the Academy of Sciences of the Russian Academy of Agricultural Sciences, Ph.D. Professor. Born March 11, 1937 in Vinnitsa. He graduated in 1959 the Faculty of electrification of the Moscow Institute of mechanization and electrification of agriculture. DS Strebkov world's first theorized and created a controlled light photovoltaic generator voltage of 30 kV for electrical processes. DS Strebkov - a prominent scientist in the field of electrification and automation of agriculture, renewable energy, new power transmission methods and new fuels, energy conservation. DS Strebkov - a prominent scientist of world renown in the field of renewable energy sources and new energy technologies. He has published more than 1,200 scientific papers, including 407 patents for the invention of the USSR and the Russian Federation and 60 foreign patents (USA, Germany, France, Great Britain, Japan, and others).

R. Subhashini is Dean, School of Computing, Sathyabama Institute of Science and Technology, Chennai, India.

Gorazd Surla is MICE Sales Specialist at Valamar Riviera Master in Business Economy (mag. oec.) at University of Dubrovnik Master of Business Administration (Executive MBA) e-Leadership at Algebra University.

Mikhail Tyagunov is Doctor of Technical Sciences, Professor of Hydropower and Renewable Energy Sources Department of Moscow Power Engineering Institute (MPEI).

Alexey Vaskov has 10 years of experience in the use of renewable energy sources and experience in practical work in the project area (preparation of working documentation, conducting a feasibility study). Research area: mode control of power complexes based on renewable energy. Publications: over 30, 1 copyright certificate.

Index

F

G

H

I

L

M

N

O

P

R

S

T

U

V

W

www.ingramcontent.com/pod-product-compliance
Lightning Source LLC
LaVergne TN
LVHW062104300926
849161LV00010B/2371
* 9 7 8 1 7 9 9 8 3 6 4 5 2 *